1st International Workshop on Materials Science and Mechanical Engineering (IWMSME 2017)

IOP Conference Series: Materials Science and Engineering
Volume 281

Kunming, China
27 – 29 October 2017

Editors:

Manuel F.M. Costa
Hamid Yaghoubi
Maria Giovanna Buonomenna

ISBN: 978-1-5108-5496-3
ISSN: 1757-8981

Printed from e-media with permission by:

Curran Associates, Inc.
57 Morehouse Lane
Red Hook, NY 12571

Some format issues inherent in the e-media version may also appear in this print version.

Printed by Curran Associates, Inc. (2018)

For permission requests, please contact the Institute of Physics
at the address below.

Institute of Physics
Dirac House, Temple Back
Bristol BS1 6BE UK

Phone: 44 1 17 929 7481
Fax: 44 1 17 920 0979

techtracking@iop.org

Additional copies of this publication are available from:

Curran Associates, Inc.
57 Morehouse Lane
Red Hook, NY 12571 USA
Phone: 845-758-0400
Fax: 845-758-2633
Email: curran@proceedings.com
Web: www.proceedings.com

Table of contents

Volume 281

1st International Workshop on Materials Science and Mechanical Engineering

27–29 October 2017, Kunming, China

Accepted papers received: 4 December 2017
Published online: 21 December 2017

Preface

1:Materials Science

1.1 Optical/Electronic/Magnetic Materials

1.2 Materials Physics and Chemistry

1.3 Materials Physics and Chemistry

1.4 Micro / Nano Materials / composites

1.7 Materials machining

1.8 Others

2.2 Thermal Engineering Theory and Applications

2.3 Virtual Manufacturing, and Simulation

2.4 Engineering Optimization

2.5 Theory and Application of Friction and Wear

IOP Conf. Series: Materials Science and Engineering **281** (2017) 011001 doi:10.1088/1757-899X/281/1/011001

IWMSME2017 Preface

The 2017 International Workshop on Materials Science and Mechanical Engineering (IWMSME2017) was held during October 27th -29th, 2017 in Kunming, Yunnan, China. It is an excellent international academic forum for sharing knowledge and results in theory, methodology and applications of materials science and mechanical engineering. The workshop looks for significant contributions to all major fields of materials science and mechanical engineering. The aim of the workshop is to provide a platform to the global researchers and practitioners from both academia as well as industry to meet and share cutting-edge development in the fields.

For this workshop, we have got support of editors, keynote speakers, TPC members and authors from more than 30 countries. The international Program Committee worked very hard to have all papers peer-peer reviewed before the review deadline. The three keynote speakers and other participants have made quite creditable presentations at the workshop. We received 208 paper submissions. After hundrands of TPC members' extremely strict peer-reviewing, 64 papers are accepted for publication by IOP Conference Series: Materials Science and Engineering (Online ISSN: 1757-899X Print ISSN: 1757-8981).

We would like to express our sincere gratitude to all the members of Committee, organizers and authors for their enthusiasm, time, and expertise. Thanks for all of their contribution to the success of IWMSME2017. We are very grateful to the general chair Prof. Huajun Fan from Prairie View A&M University, USA for his brilliant organization work.

Our deep thanks also go to the many volunteers and staff members for the long hours and hard work they have generously given to IWMSME2017.

IWMSME2017 organizer
October 30, 2017

IOP Conf. Series: Materials Science and Engineering **281** (2017) 011002 doi:10.1088/1757-899X/281/1/011002

IWMSME2017 Committee

General Chair

- Prof. Hua-Jun Fan, Prairie View A&M University, USA

Editorial Board

- Dr. Manuel F. M. Costa, University of Minho, Portugal
- Dr. Hamid Yaghoubi, Iran Maglev Technology (IMT), Iran
- Prof. Maria Giovanna Buonomenna, Consiglio Nazionale dei Chimici (CNC), Italy

International Technical Program Committee

Technical Program Committee Chair

- Prof. Zhenyou Li, China NERIN Engineering Co. Ltd, China

- Prof. Jiaying Xin, College of Food Engineering, Harbin University of Commerce, China
- Dr. Yanhuai Ding, Xiangtan Univeristy, China
- Dr. M.MADHIARASAN, Anna University Regional Campus, India
- Dr. Bappa Acherjee, Department of Production Engineering, Birla Institute of Technology, India
- Dr. Liliana-Laura Badita, National Institute of Research and Development in Mechatronics and Measurement Technique, Romania
- Dr. Sudibyo, Indonesian Institute of Sciences, Indonesia
- Dr. Hamid Yaghoubi, Iran Maglev Technology (IMT), Iran
- Prof. Huajun Fan, Prairie View A&M University, USA
- Prof. Alexander N. Herega, Odessa National Academy of Food Technologies, Ukraine
- Dr. Manuel F. M. Costa, University of Minho, Portugal
- Dr. Mohammad Mojammel Al Hakim, East West University, Bangladesh
- Prof. Ana Maria Josceanu, University "POLITEHNICA" Bucharest, Romania
- Dr. Ali Askari, Karaj Islamic Azad Univesity, Iran
- Dr. Krzysztof Kuś, University of Warmia and Mazury in Olsztyn, Poland
- Prof. Osman Adiguzel, Firat University, Turkey
- Dr. Yonggang Tong, Academy of Armored Forces Engineering, China
- Dr. Ching-Tai Ng, The University of Adelaide, Australia
- Dr. Ganggang Sha, Hohai University, China
- Dr. Mohd Yuhazri Bin Yaakob, Universiti Teknikal Malaysia Melaka, Malaysia
- Prof. Sudarshan Rao K, Shri Madhwa Vadiraja Institute of Technology & Management, India
- Dr. Rafał Grzejda, West Pomeranian University of Technology, Poland
- Dr. J.Janet, Principal, Sri Krishna College of Engineering and Technology, India
- Dr. Muhammad Naufal Bin Mansor, Universiti Malaysia Perlis (UniMAP), Malaysia
- Prof. Zhenyou Li, China NERIN Engineering Co. Ltd, China

IOP Conf. Series: Materials Science and Engineering **281** (2017) 011002 doi:10.1088/1757-899X/281/1/011002

- Dr. Vineet Jain, Amity University Gurgaon, India
- Dr. P. Murali krishna , M. S. Ramaiah Institute of Technology, India
- Dr. Capt Kannan Perumal, Universiti Teknologi Malaysia (UTM), Singapore
- Dr. Claudia Pagano, Institute of Industrial Technologies and Automation, Italy
- Dr. Lihui Shi, Centerfield Media Inc, USA
- Dr. Issa Ahmed Abed, Southern Technical University, Iraq
- Dr. Viruntachar Kruefu, Maejo University, Thailand
- Dr. Siva Prasad Kondapalli, ANITS, Visakhapatnam, India
- Master. Fan Menglan , PLA University of Science and Technology, China
- Dr. José Manoel Antelo Gomez, Petrobras, Brazil
- Prof. Arnulfo Luévanos Rojas, Autonomous University of Coahuila, México
- Dr. Ştefan Ţălu, The Technical University of Cluj-Napoca, Romania
- Dr. Jan Ziobro, Technical Institute of the Jan Grodek State Vocational Academy (PWSZ) in Sanok, Poland
- Dr. Krupal Prabhakar Pawar, Sharadchandra Pawar College of Engineering, India
- Dr. Wei Lu, Jiamusi University, China
- Prof. Bin Chen, Xi'an Jiaotong University, China
- Dr. Maciej Dyzia, Silesian University of Technology, Poland
- Prof. Zhenyu Wang, North China Electric Power University, China
- Dr. Imran Memon, Zhejiang University, China
- Dr. Guohua Xie, Wuhan University, China
- Prof. Maria Giovanna Buonomenna, Consiglio Nazionale dei Chimici (CNC), Italy
- Dr. Ghezai T. Musie, Universitu of Texas at San Antonio, USA
- Dr. Manish Bhardwaj, SRM University, India
- Dr. Doan Quang Tri, National Center for Hydro-Meteorological Forecasting (NCHMF), Vietnam
- Dr. Marco Giorgetti, University of Bologna, Italy
- Prof. Amir Azam Khan, Universiti Malaysia Sarawak (UNIMAS), Malaysia
- Dr. Zhaoguang Peng, Society of Reliability Engineers, China
- Dr. Bifeng Cui, Beijing University of Technology, China
- Dr. Yongjuan Jing, Beijing Institute of Aeronautical Manufacturing, China
- Dr. Qiaoli Lin, Lanzhou University of Technology, China
- Prof. Jiankang Huang, Lanzhou University of Technology, China
- Dr. Shuangjie Zhang, Hebei University of Scinece and Technology, China
- Dr. Xiangchen Meng, Harbin Institute of Technology, China
- Dr.Yichun Liu,Kunming University of Science and Technology, China
- Prof. Sawsan Abdul majid, University of Ottawa, Canada
- Prof. Ying Wang , Hangzhou Dianzi University, China
- Dr. Jianguo Zhang, Chinese Academy of Science, China
- Dr. Artur Jedrusyna, Wroclaw University of Science and Technology, Poland
- Prof. Yuechun Fu, Guangxi University, China
- Dr. Yu Liu, Dalian Jiaotong University, China
- Prof. Xianglin Zhang, State key laboratory of Materials processing and die/mould technology, China
- Dr. Dongping Zhao, Xi'an Aeronautical University, China
- Dr. Jee Kian Siong, Multimedia University, Malaysia

IOP Conf. Series: Materials Science and Engineering **281** (2017) 011002 doi:10.1088/1757-899X/281/1/011002

IOP Conf. Series: Materials Science and Engineering **281** (2017) 011002 doi:10.1088/1757-899X/281/1/011002

- Dr. Mahdi Moosazadeh, Western Sydney University, Australia
- Dr. Jianwei Nai, Nanyang Technological University, Singapore
- Dr. AISSA Abderrahmane, University of Mascara, Algeria
- Mr. Prajoy Podder, Ranada Prasad Shaha University, Bangladesh
- Prof. Elhem GHORBEL, University of Cergy Pontoise, France
- Prof. Francesco Marotti de Sciarra, University of Naples Federico II - Dept. of Strucutres, Italy
- Dr. Sigeomi TAKAI, Graduate School of Energy Science, Kyoto University, Japan
- Dr. João Carlos R. Plácido da Silva, University of Sacred Heart, Brazil
- Prof. Bruno Bisceglia, University of Salerno, Italy
- Dr. Ricardo Branco, University of Coimbra, Portugal
- Dr.Manish Bhardwaj, SRM University, India
- Dr. Dipen Kumar Rajak, Sandip Institute of Technology & Research Centre (SITRC), India
- Dr.Haider Zaman, University Of Engineering and Technology, Pakistan
- Dr. Feng Cai, Princeton University, USA
- Prof.SAIDI Hemza, University of chlef UHBC, France
- Prof. SAIDI Hemza, University of chlef UHBC, France
- Dr. Ahmed Eid Moussa SHAHIN, Mansoura University, Egypt
- Dr. Haider Zaman, University Of Engineering and Technology, Pakistan
- Dr. Zhang Ziqiang, Beijing University of Technology, China
- Dr. Grzegorz Ilewicz, University of Rzeszów, Poland
- Prof. Changzhong Wu, University of Jinan, China
- Dr. Michele Calì, University of Catania, Italy
- Dr. Zonggao Mu, Harbin Institute of Technology, China
- Prof. Zi Bin, Hefei University of Technology, China
- Dr. Yanchun Li, Shandong University of Science and Technology, China
- Prof. Yumin Zhang, Harbin Institute of Technology, China
- Dr. Mohammad Bagheridaneshvar, Daneshvar Innovation Group, Iran
- Dr. Jie Yuan, China Aerodynamics Research and Development Center, China
- Dr. Hongwei Cui, Shandong University of Technology, China
- Prof. Sedef Kent, Istanbul Technical University, EEF Maslak, Turkey
- Dr. Hany Hassanin, Kingston University, UK
- Prof. Yong Kang, Wuhan University, China
- Dr. Youbao Tang, National Institutes of Health, USA
- Dr. Changxin Chen, Shanghai Jiao Tong University, China
- Dr. Song Xu, Fuzhou University, China
- Prof. Dayse Iara dos Santos, São Paulo State University, Brasil
- Dr. Maria V. Nikolic, Institute for Multidisciplinary Research, UK
- Dr. Lifang Chen, East China University of Science and Technology, China
- Dr. Ginés Doménech Asensi, Universidad Politécnica de Cartagena, Spain
- Dr.Qingsong Wang, University of Science and Technology of China, China
- Dr. Dipak Rana, University of Ottawa, canada
- Dr. Yuanyuan Zhou, Chinese Academy of Sciences, China
- Prof. Jian Zhang, Wuhan University of Technology, China
- Prof. Gang Chen, Tianjin University, China

IOP Conf. Series: Materials Science and Engineering **281** (2017) 011002 doi:10.1088/1757-899X/281/1/011002

- Dr. SUN Yewang, Beijing Institute of Technology, China
- Prof. Wang Xiaodong, Shougang Jingtang United Iron and Steel Co., Ltd., China
- Dr. Takeshi IWAMOTO, Hiroshima University, Japan
- Prof. Roumen Petrov, Ghent University, Belgium
- Dr. Zijian Qiao, Xi'an Jiaotong University, China
- Dr. Jian-Hua Yang, China University of Mining and Technology, China
- Dr. Zhen Chang, Henan University of Science and Technology, China
- Dr. Stefano Pagliassotto, Tsubaki Nakashima Co. LTD, Italy
- Prof. Dr Qiang Chen, Beijing Institute of Graphic Communication, China
- Prof. Shengli Wu, Xi'an Jiaotong University, China
- Prof. Yubo Jiao, Jilin University, China
- Prof. Mihaela POPESCU, University of Craiova, Romania
- Prof. Jianlei Cui , Xi'an Jiaotong University, China
- Prof. Yi-Ting Mai, Department of Sport Management, National Taiwan University of Sport, Taiwan
- Prof. Francisco Falcone, Universidad Publica de Navarra, Spanin
- Prof. Hu Shi, Xi'an Jiaotong University, China
- Prof. George Papaioannou, University of Athens, Greece
- Dr. Abu Hena Muhammad Nazir Hossain, University of Massachusetts Lowell, USA
- Prof. Manuela PANOIU, University Politechnica Timisoara, Romania
- Prof. Mei Liu, Shandong Normal University, China
- Dr. Ifat Jahangir, Intel Corporation, UK
- Dr. Javier Darío Fernández Ledesma, Universidad Pontificia Bolivariana, Colombia
- Dr. Lei Ma, Guilin University of Electronic Technology, China
- Dr. Guodong Cui, Southwest Jiaotong University, China
- Prof. Xinyu Tan, China Three Gorges University, China
- Prof. Ding Yuan, Beihang University, China
- Prof. Yan Zhang, Capital Normal University, China
- Prof. Junkuo Gao, Zhejiang Sci-Tech University, China
- Prof. WEI YU, Shanghai Polytechnic University, China
- Dr. Hua-yong LIAO, Changzhou University, China
- Dr. Christoph Pospiech, Lenovo Global Technology Germany GmbH, Germany
- Prof. Dayse Iara dos Santos, São Paulo State University, Brazil
- Dr. Marton Benke, University of Miskolc, Hungary
- Prof. Jianlei Cui, Xi'an Jiaotong University, China
- Dr. Guodong Cui, Southwest Jiaotong University, China
- Dr. Christoph Pospiech, Lenovo Global Technology Germany GmbH, Germany
- Dr. Yaolin Zhang, FBInnovations,Canada
- Dr. Huiyu Zhou, Queen's University Belfast, UK
- Dr. Yaqiong LI, University of Science and Technology Beijing, China
- Dr. Bulent Aktas, Harran University, Turkey
- Prof. Chao-Lung Hwang,, Civil and Construction Engineering, Taiwan Tech, Taiwan
- Dr. Sergei Ghyngazov, National Research Tomsk Polytechnic University, Russia
- Prof. Jan Awrejcewicz, The Lodz University of Technology, Poland
- Prof. XiangDong Liu, Zhejiang Sci-Tech University, China

IOP Conf. Series: Materials Science and Engineering **281** (2017) 011002 doi:10.1088/1757-899X/281/1/011002

IOP Conf. Series: Materials Science and Engineering **281** (2017) 011003 doi:10.1088/1757-899X/281/1/011003

Peer review statement

All papers published in this volume of *IOP Conference Series: Materials Science and Engineering* have been peer reviewed through processes administered by the proceedings Editors. Reviews were conducted by expert referees to the professional and scientific standards expected of a proceedings journal published by IOP Publishing.

IOP Conf. Series: Materials Science and Engineering **281** (2017) 012001 doi:10.1088/1757-899X/281/1/012001

Band Gap Optimization Design of Photonic Crystals Material

Y Yu[*1], B Yu[2] and X Gao[1]

[1] School of Science, Beijing University of Posts and Telecommunications, Beijing 100876, People's Republic of China

[2] School of Management Science and Engineering, Dongbei University of Finance and Econimics, Dalian 116025, People's Republic of China

E-Mail: [*]yang.yu@bupt.edu.cn

Abstract. The photonic crystal has a fundamental characteristic - photonic band gap, which can prevent light to spread in the crystals. This paper studies the width variation of band gaps of two-dimension square lattice photonic crystals by changing the geometrical shape of the unit cells' inner medium column. Using the finite element method, we conduct numerical experiments on MATLAB 2012a and COMSOL 3.5. By shortening the radius in vertical axis and rotating the medium column, we design a new unit cell, with a 0.3*3.85e-7 vertical radius and a 15 degree deviation to the horizontal axis. The new cell has a gap 1.51 percent wider than the circle medium structure in TE gap and creates a 0.0124 wide TM gap. Besides, the experiment shows the first TM gap is partially overlapped by the second TE gap in gap pictures. This is helpful to format the absolute photonic band gaps and provides favorable theoretical basis for designing photonic communication material.

1. Introduction

In 1987, when E.Yablonovitch and S.John studied to control the spontaneous radiation and used periodic structures to control the spread of radiation, they first put forward the concept of photonic crystals. The photonic crystal is a novel optical material made of two or more than two different kinds of materials arranged periodically. When the electromagnetic wave propagates in photonic crystals, the wave obeys the refraction, reflection and transmission laws, and the periodical Bragg scattering of electron, which is similar to electronic bands, modulates the wave to form photonic energy bands. With proper lattice constants and dielectric function ration, the photonic crystals can produce gaps between energy bands. Electromagnetic waves of certain frequencies are completely prevented in band gaps; see Figure 1 [1]. The photonic band gap is an important property of photonic crystals material. The research and use of photonic crystals physical material and components are significant to develop the information and technology industry, and also provide a possible way to acquire the large-scale integrated optical path [2]. For these reasons, there are a lot of scholars dedicating to study the property of photonic crystals today; they design and explore various structures to acquire low frequency band gaps, such as the design of unit cells in direct problem [3-7], the search for structures with optimal gaps in inverse problem [8-11] and considering to optimize gaps within a given frequency range [12].

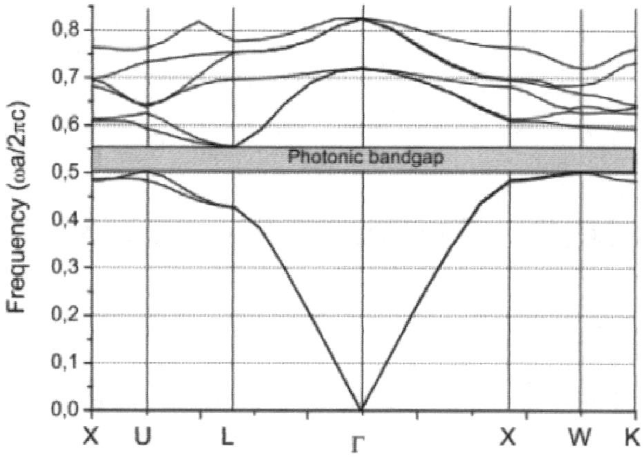

Figure 1. Schematic diagram of photonic band gaps

The two-dimension photonic crystal is a material that the dielectric function changes periodically in two dimensions and be constant in the third dimension; see Figure 2. Different lattice arrangements produce different lattice types. In this paper, we study band gaps of the unit cells with square lattices, while the lattices' inner shape can be circle, square, hexagonal and any other shapes. If you want to acknowledge more concepts and theories about the two-dimension photonic crystal, you can see lecture [13].

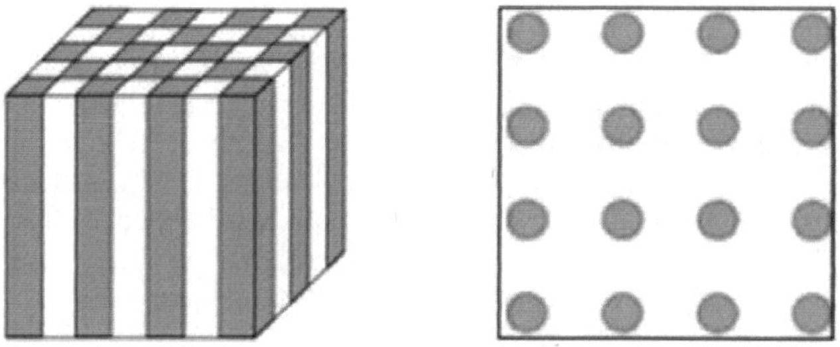

Figure 2. The two-dimension photonic crystal

The finite element method is a numerical method based on calculus of variation, approximation theory and partial differential equations; it is widely used in mathematics and physics. The main idea of this method is to make the continuous physical field discrete by finite elements, approximate the physical field by using basic functions in each sub-region and finally composite the total stiffness matrix. The method has good properties in numerical convergence and stability; it has been widely used in finding the gaps of photonic crystals [14].

The main purpose of this paper is to design the shape of unit cells, having low and wide band gaps, in two-dimension square lattice photonic crystals. Numerical experiment results are presented to show qualitative guidance rules for designing unit cells. In the two-dimension polarization modes, we give the related theoretical derivation of the band gap problem and numerical examples using the finite element method. Using the unit cell with a circle inner medium column as an initial structure, we test the effects relating to radius in vertical axis and angle parameters. As is shown by the numerical results conducting on MATLAB 2012a and COMSOL 3.5 software, we can come to the conclusion that it makes for producing low frequency and wide band gaps destroying the symmetry of medium columns and the whole unit cells.

2. The band gap problem and finite element method

When the waves propagate in photonic crystals, they satisfy Maxwell equations. The interactions between waves and medium are given by Material equations. The effects we consider in photonic

IOP Conf. Series: Materials Science and Engineering **281** (2017) 012001 doi:10.1088/1757-899X/281/1/012001

crystals are mainly electrostatic effects. That means the electromagnetic waves propagate in the uniform, passive, non-magnetic and periodic dielectric. In this case, the model satisfies the simplified Maxwell equations and Material equations as follows:

$$\begin{cases} \nabla\times\mathrm{E}=-\frac{1}{c}\cdot\frac{\partial \mathrm{B}}{\partial t} \\ \nabla\times\mathrm{H}=\frac{1}{c}\cdot\frac{\partial \mathrm{D}}{\partial t} \\ \nabla\cdot\mathrm{D}=0 \\ \nabla\cdot\mathrm{B}=0 \end{cases} \tag{1}$$

$$\begin{cases} \mathrm{B}(\mathrm{r},t)=\mathrm{H}(\mathrm{r},t) \\ \mathrm{D}(\mathrm{r},t)=\varepsilon(\mathrm{r})\mathrm{E}(\mathrm{r},t) \end{cases} \tag{2}$$

where E is electric field intensity, H is magnetic field intensity, D is electric displacement vector and B is magnetic induction intensity, and the four field quantity are all vectors relaying on coordinates and time. The $\varepsilon(\mathrm{r})$ is dielectric function of the medium, t is time, c is light velocity in vacuum, ∇ is nabla operator and r represents the three-dimension location (x, y, z).

Combining with the Material equations (2), the Maxwell equations (1), with the electric field coupled with the magnetic field, can be transferred into wave equations with the separated electric field and magnetic field component. In addition, as the wave equations include the time component, it's difficult to solve by numerical methods. Meanwhile, electromagnetic waves can be decomposed into simple harmonic waves. When E and H are simple harmonic waves, the wave equations can be simplified into Helmholtz equations with no time terms:

$$\begin{gathered} \nabla\times\nabla\times\mathrm{E}(\mathrm{r})=\left(\frac{\omega}{c}\right)^2\cdot\varepsilon(\mathrm{r})\cdot\mathrm{E}(\mathrm{r}) \\ \nabla\times\left(\frac{1}{\varepsilon(\mathrm{r})}\cdot\nabla\times\mathrm{H}(\mathrm{r})\right)=\left(\frac{\omega}{c}\right)^2\cdot\mathrm{H}(\mathrm{r}) \end{gathered} \tag{3}$$

In particular, considering the two cases - TE mode and TM mode in two-dimension, the problem of photonic crystals band gaps can be reduced. Here, the TM mode refers to that the electric field is polarized along z-axis, $\mathrm{E}(\mathrm{r})=(0,0,E(\mathrm{r}_{\parallel}))$, and the $E(\mathrm{r}_{\parallel})$ is a function with variables x, y; the TE mode refers to that the magnetic field is polarized along z-axis, $\mathrm{H}(\mathrm{r})=(0,0,H(\mathrm{r}_{\parallel}))$, and the $H(\mathrm{r}_{\parallel})$ is a function with variables x, y. The $\mathrm{r}_{\parallel}$ here represents the three-dimension location (x, y). So, the Helmholtz equations (3) can be reduced to:

$$\begin{gathered} \mathrm{TE}: -\nabla\cdot\left(\frac{1}{\varepsilon(\mathrm{r}_{\parallel})}\nabla H(\mathrm{r}_{\parallel})\right)=\left(\frac{\omega}{c}\right)^2\cdot H(\mathrm{r}_{\parallel}) \\ \mathrm{TM}: -\nabla\cdot\left(\nabla E(\mathrm{r}_{\parallel})\right)=\left(\frac{\omega}{c}\right)^2\varepsilon(\mathrm{r}_{\parallel})\cdot E(\mathrm{r}_{\parallel}) \end{gathered} \tag{4}$$

Now, the band-gap problem satisfies Neumann boundary conditions.

Furthermore, as the two-dimension photonic crystals are periodic, the two-dimension boundary value problem model (4) can be turned into characteristic equations (5) in the unit cell Ω, basing on the Bloch theory.

$$\mathrm{TE}: -\left(\nabla + i\mathrm{k}_{\parallel}\right)\cdot\left(\frac{1}{\varepsilon(\mathrm{r}_{\parallel})}\left(\nabla + i\mathrm{k}_{\parallel}\right)H_{\mathrm{k}_{\parallel}}(\mathrm{r}_{\parallel})\right) = \left(\frac{\omega}{c}\right)^2 H_{\mathrm{k}_{\parallel}}(\mathrm{r}_{\parallel})$$

$$\mathrm{TM}: -\left(\nabla + i\mathrm{k}_{\parallel}\right)\cdot\left(\left(\nabla + i\mathrm{k}_{\parallel}\right)E_{\mathrm{k}_{\parallel}}(\mathrm{r}_{\parallel})\right) = \left(\frac{\omega}{c}\right)^2 \varepsilon(\mathrm{r}_{\parallel})E_{\mathrm{k}_{\parallel}}(\mathrm{r}_{\parallel}) \tag{5}$$

where $\mathrm{k}_{\parallel} = (a,b)$ are wave vectors in two-dimension place space and chosen along the irreducible Brillouin zone boundary. The following Figure 3 gives the corresponding conversion processes for the first and the irreducible Brillouin zone in two-dimension.

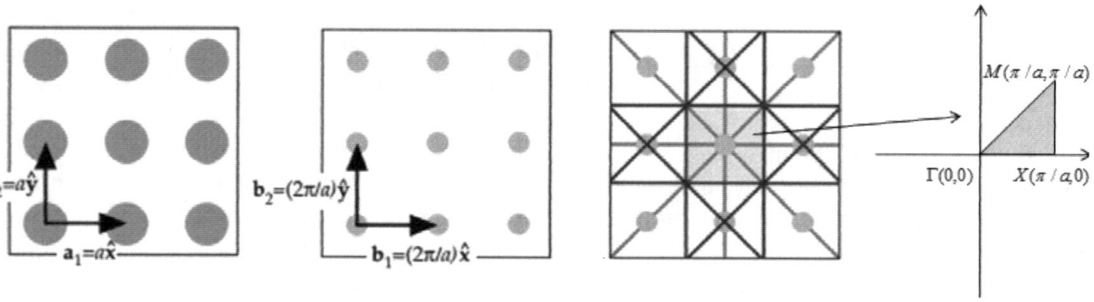

Figure 3. The irreducible Brillouin zone in two-dimension

Now, the electromagnetic field variables satisfy the periodic Floquet boundary conditions on $\Pi = \partial\Omega$. Choosing base vectors of the unit cell $\sigma_1 = (a,0), \sigma_2 = (0,a)$ parallel to the x-axis and y-axis respectively, we can get the following Floquet boundary conditions.

$$\begin{cases} H_{\mathrm{k}_{\parallel}}(\mathrm{r}_{\parallel}+\sigma_i) = H_{\mathrm{k}_{\parallel}}(\mathrm{r}_{\parallel}) \\ \dfrac{1}{\varepsilon(\mathrm{r}_{\parallel}+\sigma_i)}\partial_n H_{\mathrm{k}_{\parallel}}(\mathrm{r}_{\parallel}+\sigma_i) = \dfrac{1}{\varepsilon(\mathrm{r}_{\parallel})}\partial_n H_{\mathrm{k}_{\parallel}}(\mathrm{r}_{\parallel}) \end{cases} \tag{6}$$

and

$$\begin{cases} E_{\mathrm{k}_{\parallel}}(\mathrm{r}_{\parallel}+\sigma_i) = E_{\mathrm{k}_{\parallel}}(\mathrm{r}_{\parallel}) \\ \partial_n E_{\mathrm{k}_{\parallel}}(\mathrm{r}_{\parallel}+\sigma_i) = \partial_n E_{\mathrm{k}_{\parallel}}(\mathrm{r}_{\parallel}) \end{cases} \tag{7}$$

where n is outward normal direction on the boundary of the unit cell.

Then, to derivate the process of calculus of variation related to the finite element method, we write the TE mode and TM mode equation as the following uniform form - the eigenvalue form.

$$\left(\nabla + i\mathrm{k}_{\parallel}\right)\cdot\left[\delta\left(\nabla + i\mathrm{k}_{\parallel}\right)u\right] + \lambda\vartheta u = 0, u \in \Omega$$

$$\begin{cases} u_{\mathrm{k}_{\parallel}}(\mathrm{r}_{\parallel}+\sigma_i) = u_{\mathrm{k}_{\parallel}}(\mathrm{r}_{\parallel}) \\ \varsigma\partial_n u_{\mathrm{k}_{\parallel}}(\mathrm{r}_{\parallel}+\sigma_i) = \delta\partial_n u_{\mathrm{k}_{\parallel}}(\mathrm{r}_{\parallel}) \end{cases} \tag{8}$$

where

$$\delta = \begin{cases} \dfrac{1}{\varepsilon(\mathrm{r}_{\parallel})} & \mathrm{TE} \\ 1 & \mathrm{TM} \end{cases}, \quad \vartheta = \begin{cases} 1 & \mathrm{TE} \\ \varepsilon(\mathrm{r}_{\parallel}) & \mathrm{TM} \end{cases}, \quad \varsigma = \begin{cases} \dfrac{1}{\varepsilon(\mathrm{r}_{\parallel}+\sigma_i)} & \mathrm{TE} \\ 1 & \mathrm{TM} \end{cases}.$$

$\lambda = \left(\frac{\omega}{c}\right)^2$ is the eigenvalue to solve, u is the field quantity to solve and here it is eigenvector.

IOP Conf. Series: Materials Science and Engineering **281** (2017) 012001 doi:10.1088/1757-899X/281/1/012001

Define the periodic Sobolev space

$$H^1_{k_{\|}}(\Omega) = \{f(r_{\|}) \in H^1(\Omega) | f(r_{\|} + \sigma_i) = f(r_{\|}) \quad on \quad \Pi = \partial\Omega, \ i = 1,2\}. \tag{9}$$

The related norm definitions of Sobolev space can be found in lecture [15]. The eigenvalue equations (8) can be transferred into the following form. To $\forall k_{\|} \in B$, find $(\lambda, u) \in C \times H^1_{k_{\|}}(\Omega)$ satisfy

$$\int_{\Omega} \delta(\nabla + ik_{\|})u \cdot \overline{(\nabla + ik_{\|})v} d\Omega = \lambda \int_{\Omega} \vartheta u \cdot \bar{v} d\Omega \tag{10}$$

Here, the $\overline{(\nabla + ik_{\|})v}, \bar{v}$ is the conjugate of $(\nabla + ik_{\|})v, v$.

Then, the finite element method is conducted step by step. First, the unit cell is subdivided by triangular and the element stiffness matrix is developed by choosing suitable basis functions in each triangular cell. Second, sythesize the global stiffness matrix and consider the Floquent boundary conditions. Finally, the generalized eigenvalue equation $AU = \lambda BU$ can be derived in the two-dimension photonic crystals problem. There are several ways to solve the generalized eigenvalue equations; readers who are interested in the problems can consult lecture [16]. Especially, when the matrix A and B are positive definite, we can get the reverse of B or decompose B by Cholesky algorithm.

3. The numerical experiment

In this chapter, we use the finite element method to solve the band-gap problem of two-dimension square lattice photonic crystals numerically. The numerical experiment was initialized by testing the structure with circle inner medium column, and then studied on changing the radius in vertical axis and rotating the medium column to acquire qualitative guiding theories in the design of unit cells with low frequency and wide gaps. We conducted the experiment on MATLAB 2012a and COMSOL 3.5 software. In detail, we used the TE mode and TM mode 2D eigenfrequency analysis modules in COMSOL 3.5 to build unit cells, and then we imported the models into MATLAB 2012a to plot the band gaps figures.

The related parameters used in COMSOL 3.5 are dielectric functions $\varepsilon_1 = 1, \varepsilon_2 = 8.898$ for two-dimension square lattice, the length of the square unit cell a=3.85e-7, the initial search frequency 2.5e14 and the initial circle medium column radius r=0.378*a. First, we change the length of the radius in vertical axis. Then, we rotate the horizontal axis; see Figure 4. In this paper, the seven lowest frequency bands are mainly considered and the gaps between them, namely along the three side of irreducible Brillouin zone boundary.

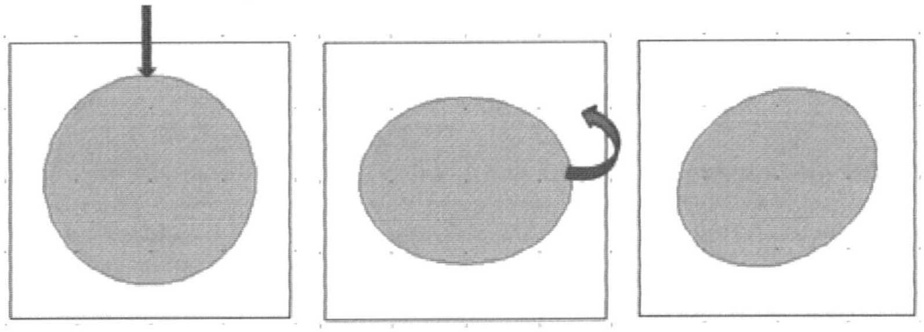

Figure 4. The change in medium column of COMSOL models

Firstly, when the length of the radius in vertical axis changes among the set {0.4*a, 0.378*a, 0.3*a, 0.2*a, 0.1*a}, the corresponding largest TE mode and TM mode band gaps of each unit cell are shown in table 1. The value of band gaps here we referred is the normalized frequency. That is to say, the value will not be affected by the size of the crystals material.

Table 1. The relation between radius in vertical axis and largest band gaps

	0.1*a	0.2*a	0.3*a	0.378*a	0.4*a
TE mode	0.0442	0.0374	0.0595	0.0456	0.0390
TM mode	0	1.0e-03*0.4189	0.0107	0	1.0e-03*0.3153

Table 1 shows that when the vertical axis radius equal 0.3*a, the TE mode gap is 1.39 percent bigger than the corresponding circle structure and a 0.0107 width TM mode gap is created. So, we can draw the conclusion that changing the vertical axis radius is changing the symmetry in some degree which contributes to produce wider gaps.

Then, the changing of angle parameter was considered by rotating the horizontal axis of the structure with a 0.3*a vertical axis radius which have relative largest gaps in table 1. The following table 2 shows the relation between the angles and largest gaps.

Table 2. The relation between angle parameter and largest band gaps

	0°	15°	30°	45°
TE mode	0.0595	0.0607	0.0588	0.0557
TM mode	0.0107	0.0124	0.0230	0.0275

Table 2 shows that when the horizontal axis is rotated 15°, a TE mode gap which is 0.12 percent wider than the non-rotating structure and a TM mode gap which is 0.17 percent wider than the non-rotating structure can be obtained. The subsequent changes in angle make no improvement in TE mode gap, but they help to widen the TM mode gaps. In particular, the TM mode gap in the structure rotated 45° is 1.68 percent wider than the non-rotating structure. So, we can come to the conclusion that rotating the elliptical medium column destroys the symmetry of the unit cell further and produces band gaps wider.

The whole changing process, from the circle medium column to the structure with a 0.3*a vertical axis radius and a 15° horizontal rotating angle, makes the TE mode gap increase 1.51 percent and creates a 0.0124 width TM mode gap. In the following Figure 5, we list the TE mode and TM mode energy band diagrams of the structure with a 0.3*a vertical axis radius and a 15° horizontal rotating angle.

From the gaps' location in Figure 5, we can see that the second gap of TE mode has partly overlapped with the first gap of TM mode. This phenomenon can develop an absolute band gap which provide favourable theoretical basis for designing photonic communication material.

4. Conclusions and future work

We design a novel structure by destroying the symmetry of the circle medium column and the unit cell to acquire low frequency and wide band gaps. The new unit cells with a 0.3*a vertical axis radius and a 15° horizontal rotating angle has both wide TE mode gaps and TM mode gaps.

Many subjects still need to do in this area. They all relate to the optimal of band gaps of two-dimension photonic crystals, including research on more complex butterfly lattices and tessellate lattices structure, exploration to band gaps character of photonic quasi-crystal structure and the reverse problem.

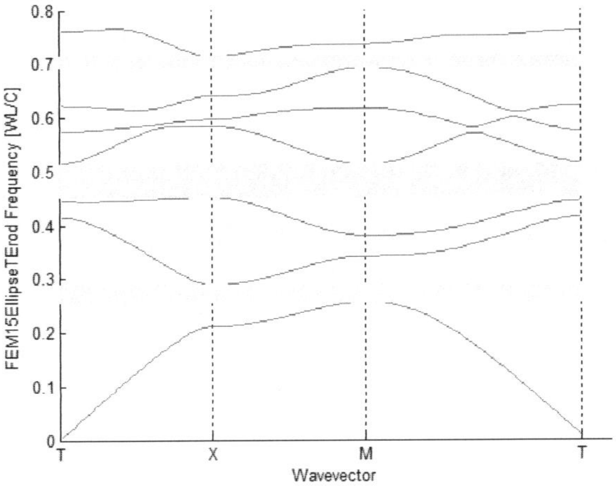

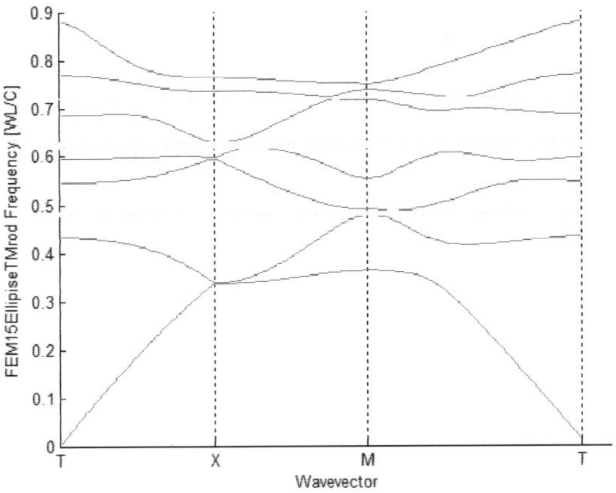

Figure 5. The TE mode and TM mode energy bands of the structure with a 0.3*a vertical axis radius and a 15° horizontal rotating angle

References

[1] Ma X Y 2010 *Photonic Crystal Principle and application* (Beijing: Science Press)
[2] Ye W M 2010 *Photonic Crystal Introduction* (Beijing: Science Press)
[3] Wang X H, Gu B Y, Li Z Y and Yang G Z 1999 *Phys.Rev.B.* **60** 11417
[4] Liu D, Gao Y H, Gao D S and Han X Y 2012 *Opt.Commun.* **285** 1988
[5] Li Z Y, Gu B Y and Yang G Z 1998 *Phys.Rev.Lett.* **81** 2574
[6] Zhu N, Wang J, Cheng C and Yan X 2013 *Optik* **124** 309–312
[7] Liu J J, Fan Z G, Kuang M, He G M, Guan C Y and Yuan L B 2013 *Opt. Commun.* **288** 52
[8] Cox S and Dobson D 1999 *SIAM J. Appl. Math.* **59** 2108
[9] Chiu Y K, Stanley O and Eli Y 2005 *Appl. Phys. B.* **81** 235
[10] Han M, Ngoc N, Robert F, Parrilo P and Jaime P. 2010 *J.Comput.Phys.* **10** 3706
[11] Elizabeth H, Andras S, Djidjel Ki, Marc M, Ken T and Cox S 2012 *Photonics and Nanostructures - Fundamentals and Applications* **10** 25.
[12] Cox S and Dobson D 2000 *J. Comput. Phys.* **158** 214
[13] Igor A S and Igor V G 2009 *Photonic Crystal* (Heidelberg: Springer)
[14] Jin J M 1998 *Finite Element Method of Electromagnetic Field* (Xi'an: Xi'an electronic science and technology university press)
[15] Li R H and Liu B 2009 *Numerical Methods for Differential Equations* (Beijing: Advanced Education Press)
[16] Klaus J B 2016 *Finite Element Method: Theory, Format and Solution*, ed Xuan J P (Beijing: Advanced Education Press)

IOP Conf. Series: Materials Science and Engineering **281** (2017) 012002 doi:10.1088/1757-899X/281/1/012002

MEMS-based Optic Fiber Fabry-Perot Sensor for Underwater Acoustic Measurement with A Wavelength-switched System

J Xia [1, 2], F Y Wang [1], H Luo [1], Y M Hu [1] and S D Xiong [1*]

[1] Academy of Ocean Science and Engineering, National University of Defense Technology, Changsha 410073, P. R. China;
[2] College of Optics Science and Engineering, National University of Defense Technology, Changsha 410073, P. R. China

*Corresponding Email: nudtxsd@163.com

Abstract. In this paper, a MEMS-based extrinsic Farby-Perot Interferometric (EFPI) acoustic pressure acoustic sensor is presented. The diaphragm structure is used as the second reflected surface, and the sensitive surface to acoustic pressure. A wavelength-switched phase demodulation system for EFPI sensors is used for acoustic signal recovery. The modified phase demodulation system has been demonstrated to recover the signal to a stable intensity fluctuation level of ±0.5 dB at the test frequency of 2000 Hz. In the test depth of 50cm, the sensor has a resonant frequency of 3.7 kHz, a flat frequency range of 10-800Hz, and a corresponding acoustic pressure sensitivity of -159 dB re. 1/μPa.

1. Introduction

The miniaturization of the traditional long-arm fiber interferometers is in the state of bottleneck at present, whereas new generation of the micro electro-mechanical system (MEMS) technique brings a light path to the sensor miniaturization. MEMS-based miniaturized fiber optic acoustic sensor reduces the size of sensing element greatly, and its precise processing control fabrication and mass production capacity are far beyond traditional fiber acoustic sensor. At present, the structure types of MEMS-based miniaturized fiber optic acoustic sensors are classified as below. (1) Capillary structure[1-3]: in order to enhance the sensitivity of capillary structure fiber acoustic sensor, the feasible way is to increase the radius of the capillary tube and reduce its wall thickness. However, the sensor with a thin capillary wall has a weak strength, which is easy to damage and limited to use. (2) Diaphragm structure: MEMS-based diaphragm structure can be classified into two types: fiber-end structure[4-5] and fiber-tip structure[6-8]. The most obvious difference between these two types is the size of sensor. Fiber-tip structure is characterized in compact structure and small bulk, and the sensor size is close to the diameter of the fiber. The key point of fiber-tip structure is located at fabricating the sensor diaphragm in a small scale. The common method to solve that problem is to utilize MEMS fabrication technology or coating technology, and the corresponding diaphragm thickness can reach at ~μm and ~10nm respectively.

Compared with the conventional acoustic sensors, fiber optic acoustic sensors have been illustrated and researched for decades for their characteristics, such as immunity to electromagnetic interference, electrically passive, resistance to corrosion, remote sensing operation, light weight and small size.

IOP Conf. Series: Materials Science and Engineering **281** (2017) 012002 doi:10.1088/1757-899X/281/1/012002

Extrinsic Fabry-Perot interferometric (EFPI) fiber optic sensor has attracted the interests of researchers in the acoustic signal detection due to its high sensitivity and wide frequency response. Various kinds of diaphragm-based EFPI fiber optic sensors, such as the structure based on a silica diaphragm spliced with a silica capillary, a glass/silicon diaphragm bonded to a silicon base, and a dipped polymer membrane on a hollow fiber tip have been developed, and some of these sensors are applied in the detection of marine hydro-acoustic waves. But most types of these sensors are not suitable for acoustic detection due to the mismatch of the acoustic impedance with the sensors diaphragm. As the effective diameter of the diaphragm, limited by the fiber itself, is small, the diaphragm must be thin if it is to have considerable sensitivity, which presents a great challenge in sensor fabrication. And it is difficult to adjust and control the F-P cavity length during fabrication. In addition, some EFPI fiber optic sensors based on MEMS diaphragm [9] may encounter high stress in the deep underwater because of the mismatches in coefficients of thermal expansion among the sensing elements with different materials, which may even lead to a failure.

In this paper, we focus on the structure design and production, packaging technology of MEMS-based optical fiber EFPI acoustic sensor. Based on the deficiency of EFPI fiber optic acoustic sensors and the demodulation methods, an optimized wavelength-switched phase demodulation system can be achieved completely in one optical path to track the underwater acoustic signal. MEMS-based optical fiber EFPI acoustic sensor has the characteristics of low noise, high sensitivity and reliability, large dynamic range, and low cost, which can be applied in the miniaturized optical fiber MEMS underwater acoustic detection system.

2. Sensor fabrication and apparatus

2.1. MEMS-based EFPI acoustic pressure sensor

In the schematic diagram of EFPI acoustic pressure acoustic sensor, the diaphragm structure is used as the second reflected surface, and the sensitive surface to acoustic pressure. The sensor is constituted with fiber end, clamp, diaphragm, and sleeve, as shown in Fig. 1. Considering the gas permeability, pressure resistance and assembly process specification, the center of sensitive diaphragm is composed of transition part and connection part. The innermost/centermost part is sensing diaphragm. There are two small holes in the part of diaphragm transition area for the achievement of acoustic pressure, which has the characteristics of temperature compensation and buffering hydrostatic pressure.

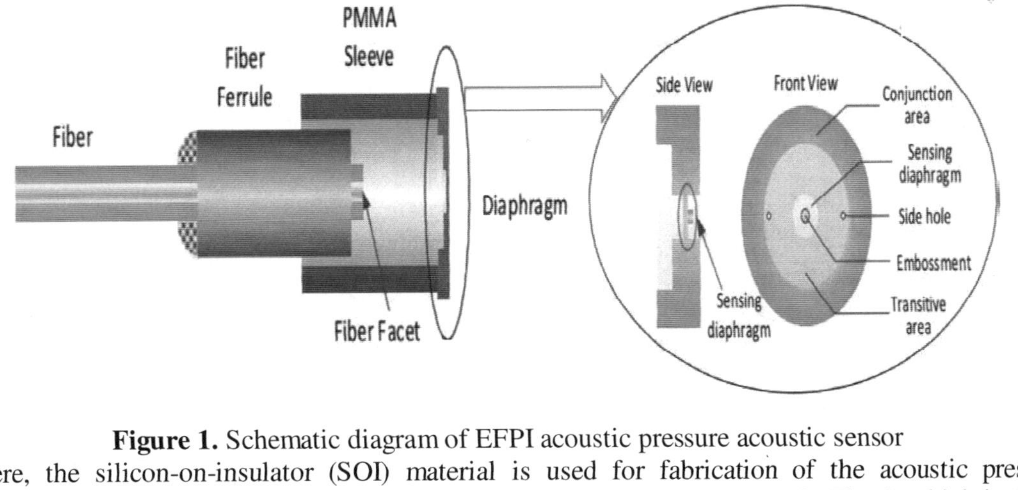

Figure 1. Schematic diagram of EFPI acoustic pressure acoustic sensor

Here, the silicon-on-insulator (SOI) material is used for fabrication of the acoustic pressure sensitive structure. The technological process is detailed as below, as shown in Fig. 2, which includes four lithographic steps and one wet corrosion step. Step 1: Etching the convex diaphragm structure. A circular arrangement is etched on the top silicon using the lithography and reactive ion etching (RIE) etching techniques. In order to obtain the convex diaphragm structure, etching time manipulation is used to reach the pre-set diaphragm thickness. Step 2: Etching holes at the transition zone。 Due to the large depth-to-width ratio of holes, the etching of the silicon and the depositing of the sidewall protective materials are carried out alternatively using the deep reactive ion etching (DRIE) method. At this point, the etching operation on the top silicon has been completed. Step 3: Etching on the bottom silicon at the transition zone using the lithography and RIE etching techniques. Step 4: Etching the frontal area of the diaphragm to SiO_2 layer. This processing technology is carried out on a thicker bottom silicon, and it is not necessary to control the processing time accurately. Step 5: Removing the exposed SiO_2 layer. Step 6: Clean the diaphragm structure.

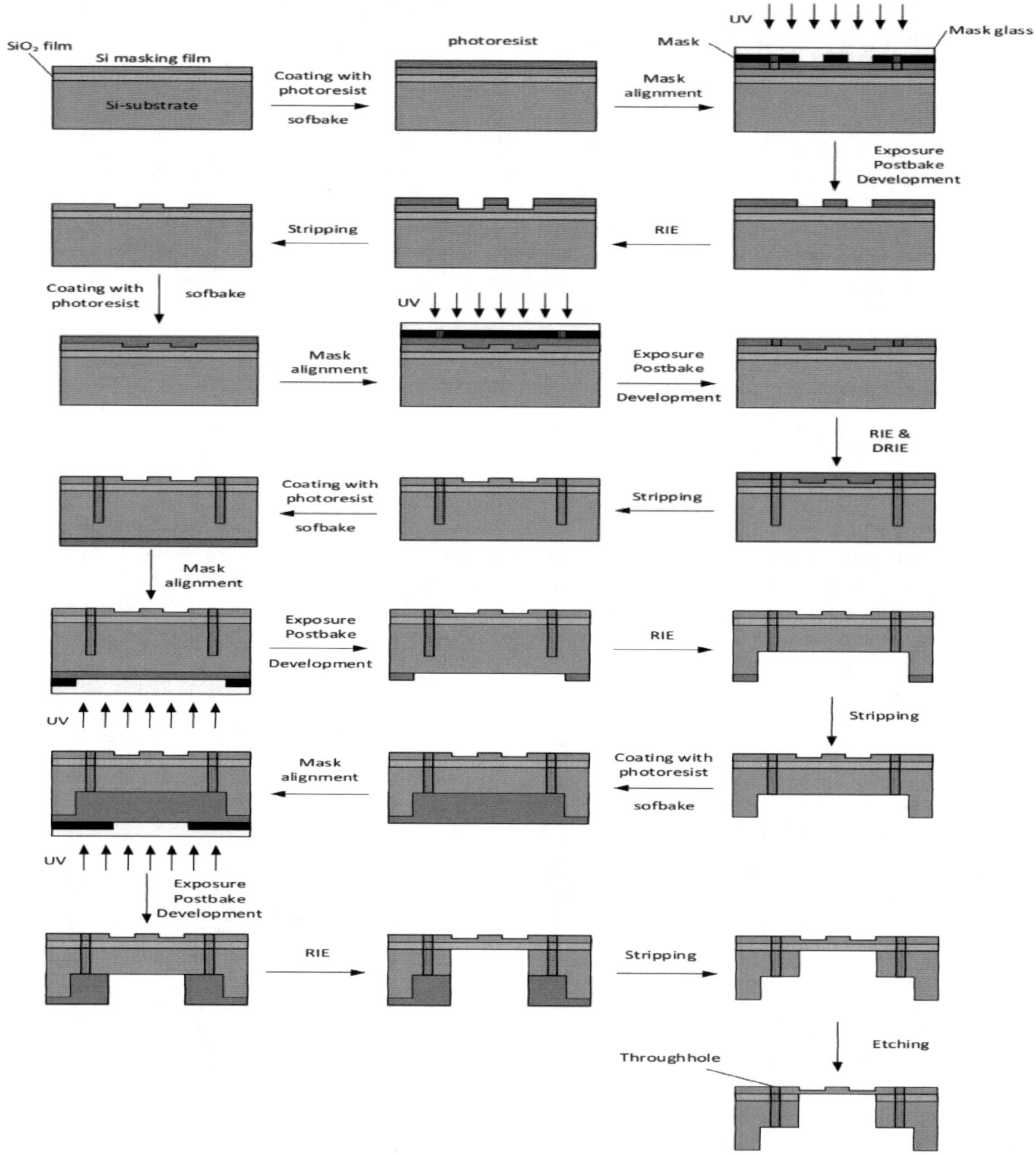

Figire 2. Flow chart of acoustic pressure structure fabrication

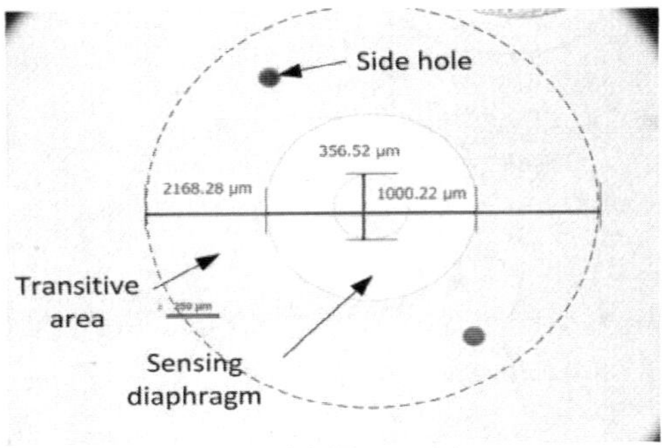

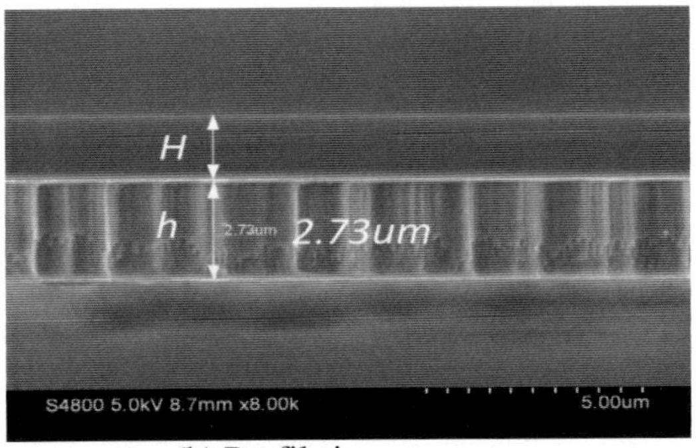

(a) Frontal image (b) Profile image

Figure 3. The optical and SEM micrographs of acoustic pressure structure

The local microscopic images of diaphragm structure is observed with the optical microscope and scanning electron microscope (SEM), as shown in Fig. 3.

In Fig.3(a), it can be seen that the diameter of diaphragm sensitive area is about 1 mm and the diameter of the convex platform is about 0.36 mm. The thickness of diaphragm and convex platform are measured about 2.73 mm and 1.87 mm in In Fig.3(b) respectively. It can be calculated that the acoustics pressure sensitivity of diaphragm is 2.19 nm/Pa and the corresponding deflection angle is 1.22° at 50m depth of water, and its underwater operation resonant frequency can reach at 4.1 kHz.

2.2. Configuration

The proposed configuration is presented in Fig. 4. The optical output from the broad-band source (BBS, with a maximum power of 42.5 mW) injects into a single-mode fiber optical circulator (OC). As explained, the back reflected optical signal from PMFBG has two orthogonal reflection peaks with different central wavelengths and transmits into an electro-optic modulator. Two natural wavelengths of PMFBG are chosen to present the spectral separation, which are set into a manual polarization controller (PC) first for wavelength switching. For the back-reflected bandwidths of PMFBG are narrow, the optical intensity via the OC attenuates to 14.75 μW. To acquire sufficient optical intensity, a polarization-maintaining Erbium-doped fiber amplifier (PM-EDFA) is added into the successive system. The signal generator applies continuous square waves (with a high-level V1 and a low level V2) to EOM. Aim to achieve the separation of two natural wavelengths at different electrical level status, a polarizer (P) is inserted after EOM and only one linearly polarized beam in certain special direction can pass through the polarizer. Two separated optical signals with different Bragg wavelengths are guided into the EFPI sensor probe via another optical circulator (OC). Then, their back reflected signals from the EFPI sensor are detected using the photoelectric detector (PD2) and recorded with another data acquisition device. To verify the dynamic measurement characteristic of the system, all tests are implemented based on the underwater acoustic signal measurement. To detect the underwater acoustic signal, the EFPI sensor probe is hanged in a water tank. The position under the EFPI sensor probe locates an acoustic emission (AE) device. The continuous sine wave in Fig.1 is sent to drive the AE device to modulate the EFPI sensor, and the given frequency of AE device acts as the compared reference with the frequency of detected signal. The polarization control unit is added to realize the operation point of EOM immune to external disturbance.

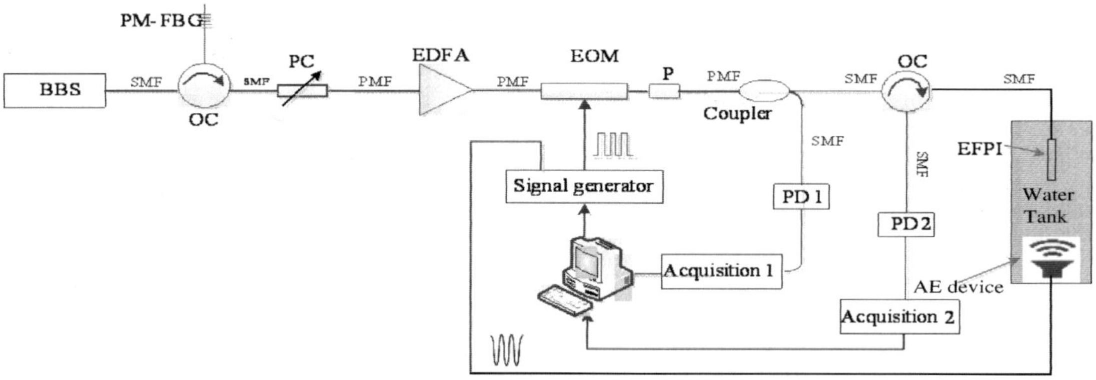

Figure 4. Schematic diagram of the wavelength-switched phase demodulation system for EFPI sensor with polarization control configuration.

3. Experiments and results

To verify the features of the optimized wavelength-switched phase demodulation system, the output of the wavelength-switched structure at different voltage level status are obtained in Fig.6. The continuous DC voltage levels are set as V1=1.2V and V2=-5.4V respectively. It is concluded that the wavelength-switched structure can achieve the separation of two orthogonal reflection peaks with different central wavelengths, which can be taken as two signal lights for EFPI sensor phase recovery. The demodulated phase signal is shown in Fig. 5(a), and the frequency-domain of the phase signal is shown in Fig. 5(b). It can be found that the frequency of the demodulated phase variation is consistent with the 2000 Hz sinusoidal driving signal, and the intensity of the 2000Hz phase signal reaches a stable state and the signal to noise ratio (SNR) level of ~37 dB. In Fig. 5(b), the first harmonic slightly emerges out of the noise at its double frequency, which is due to the self-resonance of the EFPI sensor and the crosstalk between two output channels. In addition, the power spectral density variation of the target signal is measured to be a fluctuation level of ±0.5 dB at the test frequency of 2000 Hz.

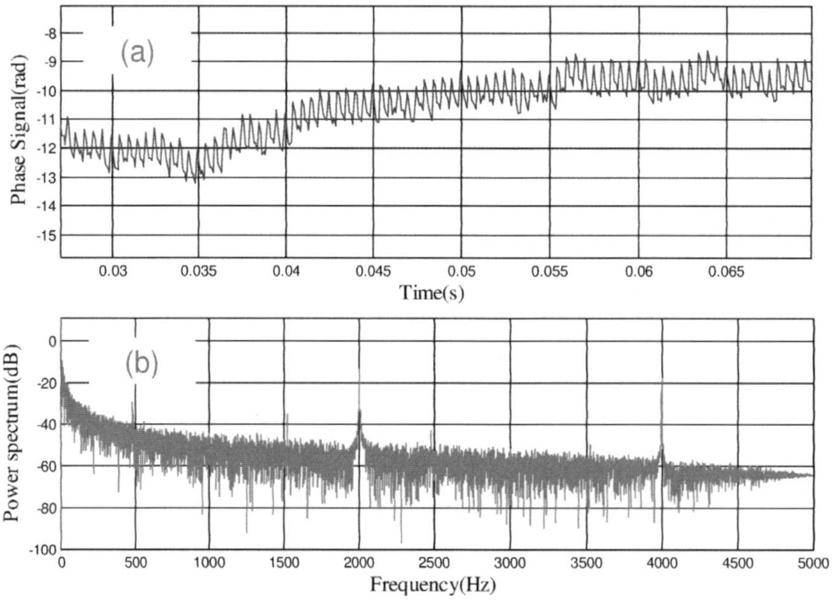

Figure 5. The demodulated phase signal (a) Demodulated phase time-domain signal; (b) Demodulated phase frequency-domain signal

Adjusting the location depth of the acoustic pressure sensor, the acoustic pressure sensitivity of the sensor at different depths are measured and analyzed, as shown in Fig. 6. In the test depth of 50cm, the sensor has a resonant frequency of 3.7 kHz, a flat frequency range of 10-800Hz, and a corresponding acoustic pressure sensitivity of -159 dB re. 1/μPa. When the location depth increases from 50cm to 12.5cm, the acoustic pressure sensitivity at 12.5cm decreases about 10dB, and the resonant frequency decreases from 4000Hz to 2600Hz, which illustrates that diaphragm has been warped due to the hydrostatic pressure when the depth changes. When the location depth of sensor returns back to 12.5cm, however, the resonant frequency of sensitivity curve changes due to the water weeping in the FP cavity. As the added mass effect of sensor increases, the resonant frequency of sensor reduces.

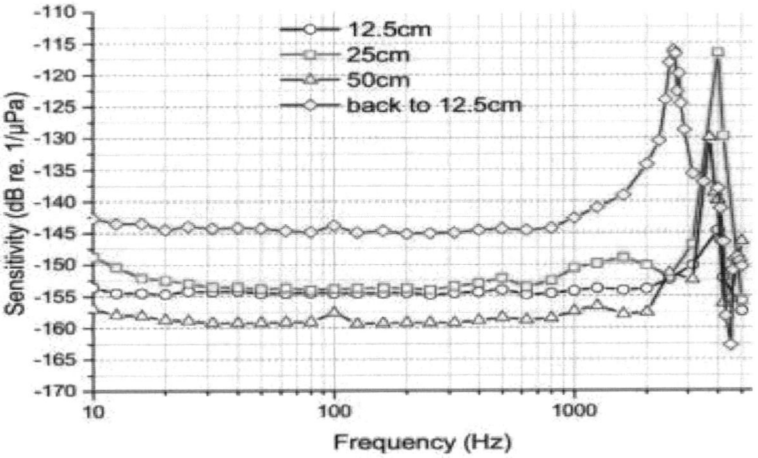

Figure 6. The acoustic pressure sensitivity of the sensor at different depths

4. Conclusions

In this paper, a stability-enhanced wavelength-switched phase demodulation system based on the polarization control unit is presented and demonstrated. A diaphragm-based EFPI sensor is presented for the underwater acoustic measurement. The experiment result shows that the system with polarization control can enhance the stability of power spectral density variation with a fluctuation level of ±0.5 dB. In addition, the demodulation system can operate in a wide frequency range under different water depths. Preliminary experimental results have shown that the wavelength-switched with polarization control could find potential applications in marine acoustic, medical science measurements, etc.

Acknowledgments

This work was supported by the National Natural Science Foundation of China under grant numbers of 11574397 and 11304388; National High Technology Research and Development Program of China (863 Program) (2013AA09A412-1).

References

[1] Albassyiouni M, Yu M and Balachandran B 2002 *Proc. SPIE* B **4693** 396-406.
[2] Lee W, Hall N, Zhou Z 2004 *IEEE Journal of Selected Topics in Quantum Electronics* B **10** 643-651.
[3] Chin K K, Sun Y, Feng G 2007 *Applied Optics* B **46** 7614–7619.
[4] Yu M and Balachandran B 2003 *Journal of Intelligent Material Systems and Structures* B **14** 409-414.
[5] Li M, Wang M, Li H 2006 *Optics Express* B **14** 1497-1504.
[6] Hill G C, Melamud R and Declercq F E 2007 *Sensors and Actuators A-physical* B **138** 52-62.
[7] Xu J, Wang X and Cooper K L 2005 *Optics Letters* B **30** 3269-3271.

IOP Conf. Series: Materials Science and Engineering **281** (2017) 012002 doi:10.1088/1757-899X/281/1/012002

[8] Wang X, Xu J and Zhu Y 2006 *Optics Letters* B **31** 885–887.
[9] Abeysinghe D C, Dasgupta S and Boyd J T 2001 *Photonics Technology Letters IEEE* B **13** 993-995.

IOP Conf. Series: Materials Science and Engineering **281** (2017) 012003 doi:10.1088/1757-899X/281/1/012003

Measurement of the Microwave Refractive Index of Materials Based on Parallel Plate Waveguides

F Zhao[1*], J Pei[1], J S Kan[1] and Q Zhao[2]

[1] Metrology and Testing Centre, China Electronics Standardization Institute, Beijing 100176, China
[2] State Key Laboratory of Tribology, Department of Mechanical Engineering, Tsinghua University, Beijing 100084, China

*Corresponding E-mail: zhaofei@cesi.cn

Abstract. An electrical field scanning apparatus based on a parallel plate waveguide method is constructed, which collects the amplitude and phase matrices as a function of the relative position. On the basis of such data, a method for calculating the refractive index of the measured wedge samples is proposed in this paper. The measurement and calculation results of different PTFE samples reveal that the refractive index measured by the apparatus is substantially consistent with the refractive index inferred with the permittivity of the sample. The proposed refractive index calculation method proposed in this paper is a competitive method for the characterization of the refractive index of materials with positive refractive index. Since the apparatus and method can be used to measure and calculate arbitrary direction of the microwave propagation, it is believed that both of them can be applied to the negative refractive index materials, such as metamaterials or "left-handed" materials.

1. Introduction

Metamaterials are one of the most popular research fields in recent decades. In theory, metamaterials have many peculiar properties, such as backward wave properties, negative refraction properties, resonant wave absorption properties, etc., which is a breakthrough in traditional electromagnetic phenomena. New high-performance electronic components or devices can be designed according to such peculiar properties. For example, a perfect copy of an object exists on the opposite side of the planar slab made of metamaterials with negative refractive index under idealized conditions [1]. In 2006, D. R. Smith *et al.* proposed that the electromagnetic cloaking structure could realize the electromagnetic wave diffraction propagation and achieve the stealth effect on the metal target of the microwave band [2]. In many of the electromagnetic properties of metamaterials, the refractive index is one of the important quantifiable characteristic parameters.

In 2001, R.A. Shelby and D.R. Smith *et al.* [3] developed a device for measuring the negative refractive index for left-handed materials. The prismatic sample was placed between two circular aluminum plates (30 cm in diameter) with a plate spacing of 1.2 cm. The rotating detector was used for measuring the microwave beam refracted by the prismatic samples. In this way, the transmitted power spectrum of the measured sample was measured as a function of angle from the interface normal, and then the refractive index of the measured sample was obtained. However, the device was not widely recognized or used after then.

In 2006, D. R. Smith *et al.* [2, 3] proposed a parallel plate waveguide apparatus at X-band frequencies thus constraining the electromagnetic fields to two dimensions. This apparatus can be used for characterizing the microwave anomalous propagation characteristics of metamaterials.

In this paper, we modified a similar apparatus based on the parallel plate waveguide method, and proposed a method for calculating the refractive index of the measured samples on the basis of measured data.

2. Measurement technique and calculation method

2.1. Measurement theory

The apparatus that we build and utilize is similar in concept to previous literatures [2, 4-7]. In the parallel plate waveguide chamber, the microwave scattering is reduced to two dimensions by confining electromagnetic waves between two conducting planes, so long as only lowest transverse electromagnetic (TEM) mode is excited. For this mode, the electric and magnetic fields do not vary along the axis between the plates (z-axis). Therefore, a map of the fields at any plane within the parallel plate waveguide chamber should provide an equivalent characterization of the microwave scattering. This equivalence is convenient, because it allows us to detect the microwave scattering just above the measured samples rather than inside, where it would be difficult to insert a probe antenna that would not also alter the field pattern.

2.2. Measurement system

The core of the apparatus that we build is the parallel plate waveguide chamber (as shown in Figure 1), consisting of an upper and a lower aluminium plate made from 10 mm tooled aluminium sheet. The mapping area within the chamber is set by a ring of microwave absorber (Dalian Tianyu microwave products) cut into a saw-toothed pattern and fixed to the lower plate, serving to reduce reflections from the waveguide edges back into the mapping region.

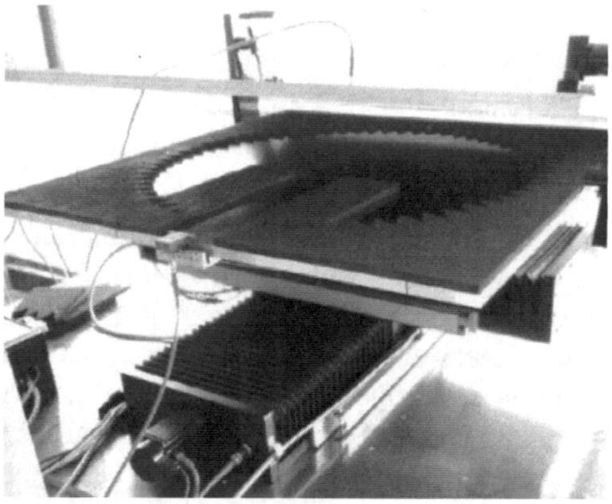

Figure 1. Photograph of the parallel plate waveguide chamber in the open position.

A planar beam can be excited by introducing microwaves into the chamber by an X-band waveguide adapter fixed to the lower plate. The adapter was connected to port 1 of the Vector Network Analyzer (VNA) (Agilent, model number E8363B). Since the measured sample is also place on the lower plate, the samples and the sources do not move relative to each other during the scan. The bottom plate was fixed to a customized 2D automatic displacement platform. The electromagnetic fields in the chamber are detected by a coaxial antenna mounted onto the upper plate. The coaxial antenna was connected to port 2 of the VNA. The VNA provides the source microwave signal and phase sensitive detection of the return signal. A customized LabVIEW program coordinates the motion of the platform and scan with the data acquisition of the VNA. The transmitted signal (S21) data, along with the position and frequency information, are stored as complex values in matrices.

2.3. Calculation Method of the Refractive Index

To calculate the refractive index of the measured sample, the field of the test area as shown in Fig. 2 was scanned and then the matrices were stored as shown below.

$$\begin{bmatrix} f & x_1 & x_2 & \cdots & x_{n-1} & x_n \\ y_1 & a_{1\,1} & a_{1\,2} & \cdots & a_{1\,n-1} & a_{1\,n} \\ y_2 & a_{2\,1} & a_{2\,2} & \cdots & a_{2\,n-1} & a_{2\,n} \\ \vdots & \vdots & \vdots & \ddots & \vdots & \vdots \\ y_{n-1} & a_{n-1\,1} & a_{n-1\,2} & \cdots & a_{n-1\,n-1} & a_{n-1\,n} \\ y_n & a_{n\,1} & a_{n\,2} & \cdots & a_{n\,n-1} & a_{n\,n} \end{bmatrix} \text{ and } \begin{bmatrix} f & x_1 & x_2 & \cdots & x_{n-1} & x_n \\ y_1 & \varphi_{1\,1} & \varphi_{1\,2} & \cdots & \varphi_{1\,n-1} & \varphi_{1\,n} \\ y_2 & \varphi_{2\,1} & \varphi_{2\,2} & \cdots & \varphi_{2\,n-1} & \varphi_{2\,n} \\ \vdots & \vdots & \vdots & \ddots & \vdots & \vdots \\ y_{n-1} & \varphi_{n-1\,1} & \varphi_{n-1\,2} & \cdots & \varphi_{n-1\,n-1} & \varphi_{n-1\,n} \\ y_n & \varphi_{n\,1} & \varphi_{n\,2} & \cdots & \varphi_{n\,n-1} & \varphi_{n\,n} \end{bmatrix}$$

where f is the test frequency; x_i and y_j is the x coordinates of the ith row and the y coordinates of the jth column of the test point, respectively. a_{ij} and φ_{ij} (i = 1, 2,... n-1, n; j = 1, 2,... n-1, n) are the amplitude and phase of S21 data acquired by the VNA, corresponding to the test point in the ith row and the jth column.

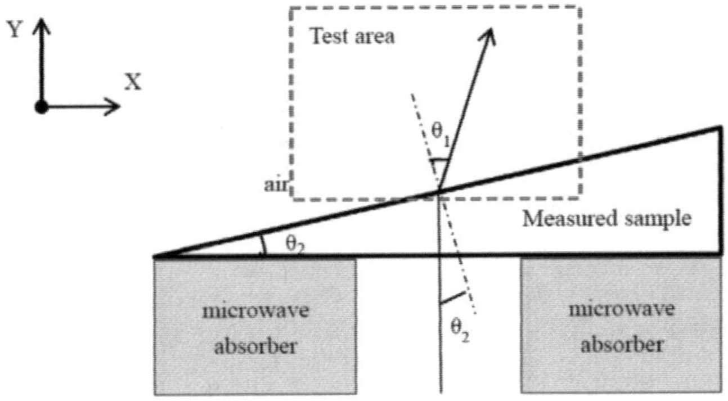

Figure 2. Schematic diagram of microwave propagation direction.

The direction of the ejection wave of a certain frequency, f, as shown in Figure 2 is determined by the following methods:

- iterate through the data in the amplitude matrix and find the maximum and its corresponding coordinates, namely (a_{I}, x_{I}, y_{I});
- in the phase matrix, identify the corresponding phase value at the same coordinates, namely (φ_{I}, x_{I}, y_{I});
- in the phase matrix, find out all the phase points of the values within a certain range, such as ±0.02 rad, And then record the corresponding coordinates of such phase points, which formed a series of approximate circular arcs that are centred at the point of (x_{I}, y_{I});
- according to the coordinates of the above selected points, identify the corresponding amplitude value at the same coordinates in the amplitude matrix, and then iterate through the data in each "circular arcs" and find the maximum of each "circular arcs" and its corresponding coordinates, namely (a_{II}, x_{II}, y_{II}), (a_{III}, x_{III}, y_{III}), etc.
- Based on the above identified coordinate points, such as (x_{I}, y_{I}), (x_{II}, y_{II}), (x_{III}, y_{III}), etc., the linear fitting is performed using the least square method. If the slope of the proposed line is k, the relationship between the slope k and the refractive angle θ_1 and the incident angle θ_2 is as shown below.

$$\theta_1 = \operatorname{arccot}(k) + \theta_2 \tag{1}$$

The refractive index, n, can then be calculated according to the Snell's law.

IOP Conf. Series: Materials Science and Engineering **281** (2017) 012003 doi:10.1088/1757-899X/281/1/012003

$$n = \sin\theta_1/\sin\theta_2 \tag{2}$$

3. Results and discussion

Two samples of PTFE samples with a wedge of 15° and 30° sharp angles have been measured using the apparatus that we build. Each sample has a thickness of about 10 mm. The permittivity at the frequency of about 12 GHz were measured by a split-cavity resonator method [8] and shown around $\varepsilon_{\mathrm{PTFE}}$ = 2.31±0.06. When the permittivity, ε, and permeability, μ, of a material are simultaneously positive, the refractive index is given by $n = \sqrt{\varepsilon\mu/\varepsilon_0\mu_0}$ (ε_0 and μ_0 are the free-space permittivity and permeability, respectively) [9]. Therefore, the refractive index of the present PTFE samples would be predicted for n_{PTFE} = 1.52±0.02.

The normalized amplitude and phase map at the frequency of 12 GHz of the 15° wedge sample are shown in Figure 3 and Figure 4, respectively. Similar results were also obtained for another 30° wedge sample, but due to the length of the article, they were not given herein.

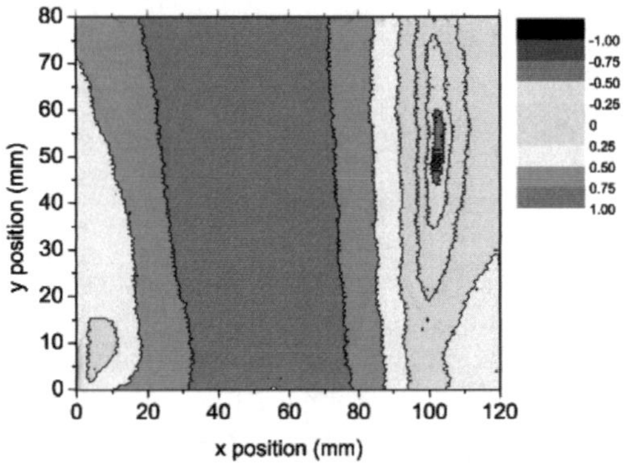

Figure 3. The normalized amplitude map at the frequency of 12.0GHz of the 15° wedge sample.

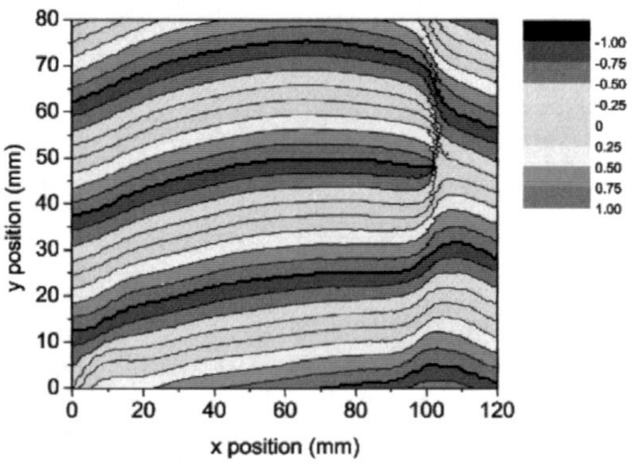

Figure 4. The normalized phase map at the frequency of 12.0GHz of the 15° wedge sample.

The refractive index of the measured wedge sample was calculated according to the above-mentioned method. The slopes of the linear fittings, *k*, were shown in Figure 5 and Figure 6, respectively. And the refractive indexes, along with the process parameters in the calculation were listed in Table 1. It is clear that the refractive index measured by the apparatus is substantially consistent with the refractive index inferred with the permittivity of the sample. Obviously, this apparatus and the proposed method can be applied to measure and calculate arbitrary direction of the microwave propagation in the parallel plate waveguide chamber. Therefore, it is believed that this apparatus and the corresponding calculation method of the refractive index can be applied to the negative refractive index materials, such as metamaterials or "left-handed" materials.

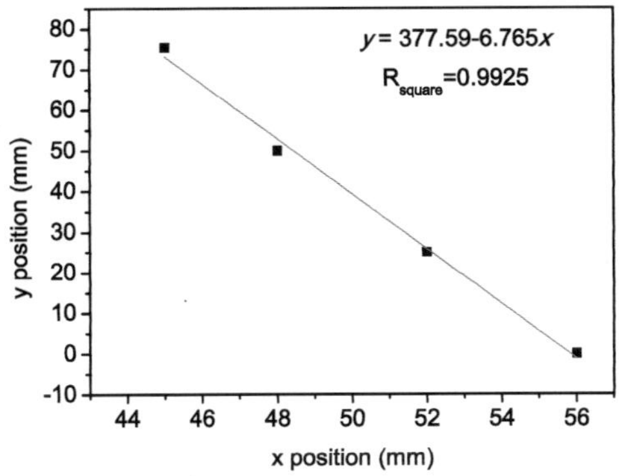

Figure 5. The least square linear fitting result of the 15° wedge sample.

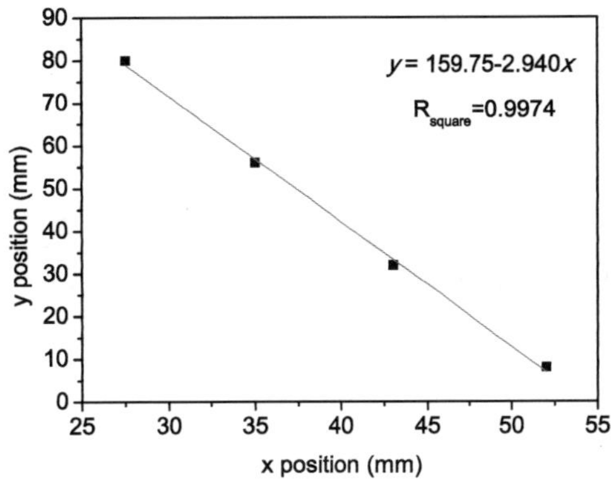

Figure 6. The least square linear fitting result of the 30° wedge sample.

Table 1. The process parameters of the refractive index calculated on the basis of the amplitude and phase matrices at the frequency of 12.0GHz of the wedge samples measured using the apparatus that we build.

Sample	incident angle θ_2	slope k	refractive angle θ_1	Refractive index n
15° wedge	15°	-6.765	8.409°	1.535
30° wedge	30°	-2.940	18.785°	1.504

4. Conclusions

We constructed a parallel-plate-waveguide based measurement setup to scan the amplitude and phase of the parallel-plate field. Measurement results of PTFE positive refractive index materials show that the refractive index calculation method proposed in this paper is a competitive method for the characterization of metamaterials with negative refractive index.

References

[1] Smith D R, Pendry J B and Wiltshire M C K 2004 *Science* **305** 788.
[2] Schurig D, Mock J J, Justice B J, Cummer S A, Pendry J B, Starr A F and Smith D R 2006 *Science* **314** 977.
[3] Shelby R A, Smith D R and Schultz S 2001 *Science* **292** 77.
[4] Justice B J, Mock J J, Guo L H, Degiron A, Schurig D and Smith D R 2006 *Optics Express* **14** 8694.
[5] Smith D R, Schurig D, Mock J J, Kolinko P and Rye P 2004 *Appl. Phys. Lett.* **84** 2244.
[6] Parimi P V, Lu W T, Vodo P, Sokoloff J, Derov J S and Sridhar S 2004 *Phys. Rev. Lett.* **92** 127401.
[7] Vodo P, Parimi P V, Lu W T and Sridhar S 2005 *Appl. Phys. Lett.* **86** 201108.
[8] Janezic M D and Baker-Jarvis J 1999 *IEEE Trans. Microw. Theory Tech.* **47** 2014.
[9] Pendry J B 2000 *Phys. Rev. Lett.* **85** 3966.

Synthesis and Characterization of Polyurethanes Based on Hydroxylated Tung Oil (HTO) and PAPI

W Q Yang[1,2], J X Zhang[2], Z Y Ren[2*], L Jiang[2], Y Fu[2] and S Q He[1]

[1] School of Materials Science and Engineering, Zhengzhou University, Zhengzhou, China

[2]High & New Technology Research Center of Henan Academy of Sciences, Zhengzhou, China

*zyren23@163.com

Abstract. Two novel polyurethane films were successfully synthesized. One was from hydroxylated tung oil (HTO) and polymethylene polyphenyl isocyanate (PAPI), which was designated as HTO-PAPI; and the other was from HTO end-capped a traditional polyurethane prepolymer and PAPI, which was designated as PP-HTO-PAPI. FTIR and NMR results show that the prepared HTO-PAPI and PP-HTO-PAPI are the expected structures. Mechanical testing result shows that the Young's modulus and elongation at break for HTO-PAPI and PP-HTO-PAPI are 282.13 MPa, 1.8 MPa and 14.15%, 218.23%, respectively, indicating that the two polyurethane films have very distinctive properties. And the properties can be controlled by incorporating different ratio of HTO and prepolymer. The present work provides a good foundation for the synthesis of novel polyurethane adhesives that will be hopefully used in different applications.

Keywords: polyurethane; hydroxylated tung oil (HTO); mechanical properties; characterization; PAPI.

1. Introduction

Polyurethane is a typical segmented copolymer generally composed of soft and hard segments. The soft segment has a long chain polyol (usually polyether or polyester polyol) with molecular weight of 1000-3000g/mol while the hard segment consists of an isocyanate and a small molecular chain extender, usually a diol or diamine. The soft segment offers elasticity while the hard segment acts as the physical crosslinking point and plays the key role in determining mechanical properties[1].

Hydroxyl-modified tung oil or hydroxylated tung oil (HTO), which is prepared from a natural plant oil, namely tung oil, is a type of new polyol[2-3]. According to its structure and molecular weight (Figure 1), it is neither a typical polyol as soft segment, nor the typical chain extender as one part of hard segment. The same conjugated triene structure as tung oil makes HTO have post-polymerization function [4] while the long pendent chain possibly has some plasticization. Therefore HTO is a type of special new polyol that have shown its special function in polyurethane dispersion in our previous work [5]. Among the different isocyanates, polymethylene polyphenyl isocyanate (PAPI) has functionality of over 2, which can be used for preparing crosslinked materials [6-7]. However, there are few reports on PAPI related polyurethane, possibly because the reaction involved is hard to control.

IOP Conf. Series: Materials Science and Engineering **281** (2017) 012004 doi:10.1088/1757-899X/281/1/012004

In the present work, we briefly report the synthesis and structure/mechanical characterization of two novel polyurethanes HTO-PAPI and PP-HTO-PAPI. It is of important significance to incorporate both HTO and PAPI into polyurethane chain not only because HTO is from a renewable source, but also the function of HTO is special and worth to be deeply studied.

2. Experimental

Synthesis of HTO-PAPI. The HTO dried in a vacuum oven was added to a four-necked flask with nitrogen protection, and PAPI was added at a certain ratio. An appropriate amount of solvent dimethyl carbonate (DMC) was added. A drop of phosphoric acid was added when the temperature is about 40°C. Then temperature was raised to 60~70 °C and was maintained for one and half hours.

Synthesis PP-HTO-PAPI. The dry HTO was added to a four-necked flask, and the PP (prepared from polytetrahydrofuran (PTMG) and diphenylmethane diisocyanate (MDI) was added at a ratio of –NCO: -OH = 1:2. After this step of reaction was over, PAPI was added at a ratio of -NCO:-OH = 1:1.

NCO content was determined to trace the extent of polymerization during above syntheses. When the reactions were over at certain condition, the product was poured into polytetrafluoroethylene (PTFE) disc to dry at room temperature.

Figure 1 shows the pathway for the preparation of two polyurethanes.

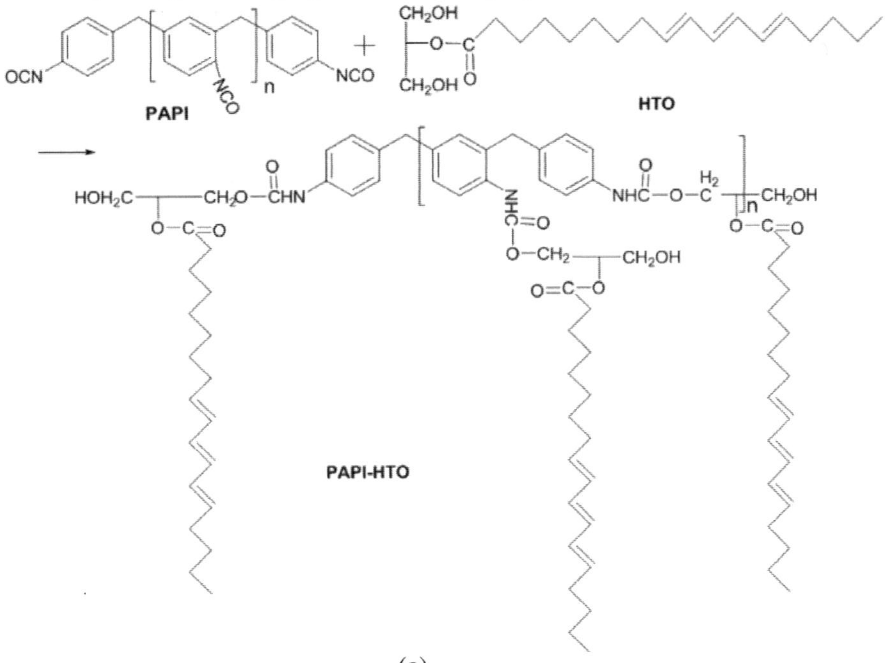

(a)

IOP Conf. Series: Materials Science and Engineering **281** (2017) 012004 doi:10.1088/1757-899X/281/1/012004

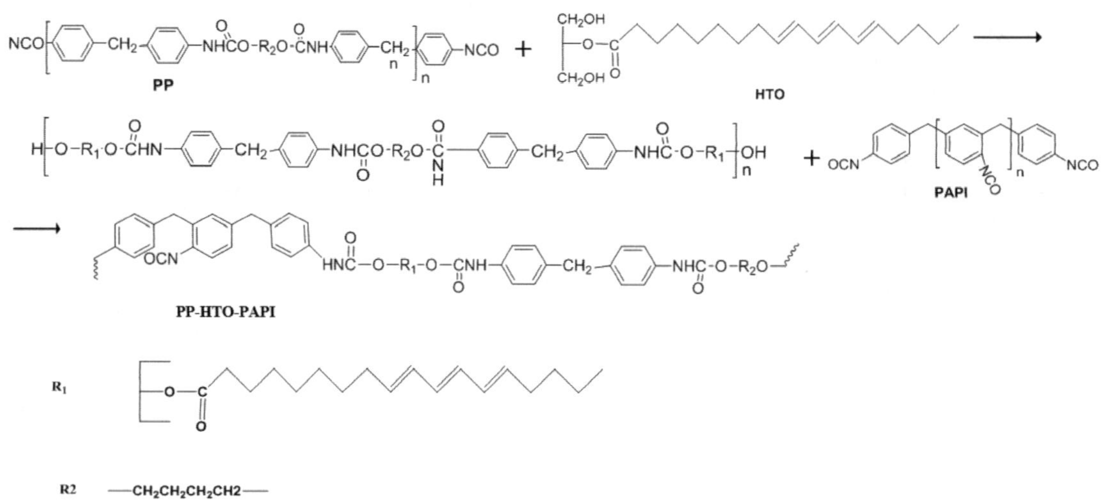

(b)

Figure. 1. Synthesis routes of two polyurethanes **a:** PAPI-HTO; **b:**PP-HTO-PAPI

3. Results and discussion

3.1. Structural characterization

Figure 2 is the comparative FTIR spectra of two polyurethanes based on different compositions. It can be observed that the characteristic bands of urethane in both PP-HTO-PAPI and PAPI-HTO are very obvious, including the νNH band at 3326cm^{-1}, the urethane νC=O band at 1727 cm^{-1} as well as the amide II and amide III bands [8-9] at 1534cm^{-1} and 1219cm^{-1}, respectively. It is also seen that there is small band at around 993cm^{-1} [4], showing the triene structure remains in two polyurethanes.

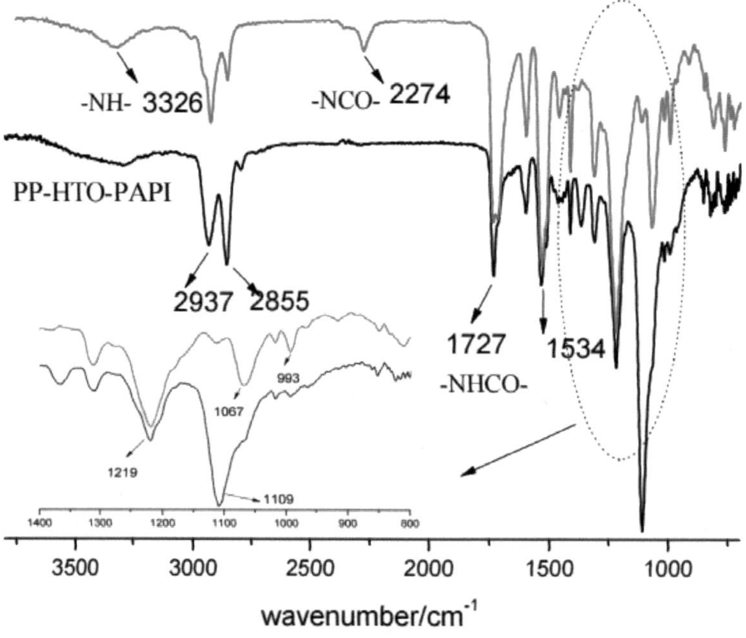

Figure. 2. FTIR spectra of PP-HTO-PAPI and PAPI-HTO

Moreover, it can be noted that the band intensity at 2855cm^{-1} and at 1109cm^{-1} are significantly stronger in PP-HTO-PAPI than that in PAPI-HTO. This result shows that PP prepared by MDI and PTMG has been introduced into PP-HTO-PAPI, as PTMG contains more CH_2 group and ether oxygen bond.

In addition to FTIR, ^{1}HNMR was also used to characterize the two structures. Among the chemical shifts, the peak standing for carbamate hydrogen at δ 6.8 ppm as well as the group peaks representing conjugated double bonds at δ 5.38-6.35ppm from HTO [4, 10] all show that the HTO structure bearing triene bonds have been introduced to two polyurethane chains.

Above results show that the expected polyurethane structures have been successfully synthesized in PP-HTO-PAPI and PAPI-HTO films.

3.2. Mechanical properties

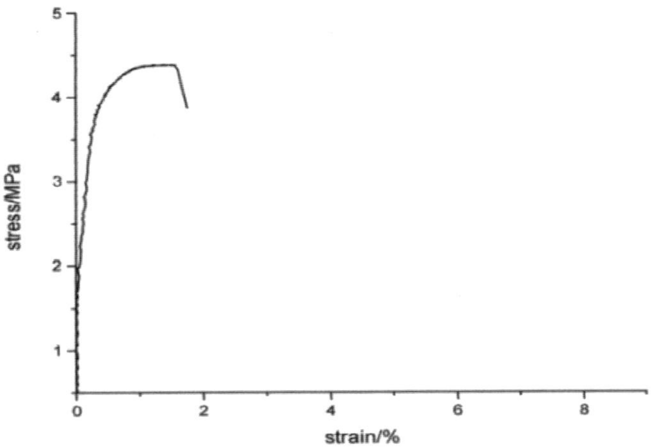

Figure. 3. Stress-strain curve of PAPI-HTO film

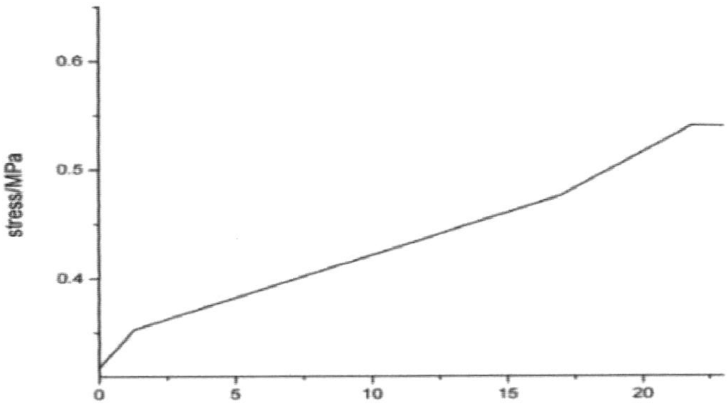

Figure. 4. Stress-strain curve of PP-HTO-PAPI film

Table 1 .Mechanical Properties of two types of polyurethane films

Sample	Tensile strength (MPa)	Young's modulus (MPa)	Elongation at break (%)
PAPI-HTO	41.37	282.13	14.15
PP-HTO-PAPI	4.56	1.80	218.23

Figure 3 and Figure 4 are the stress-strain curves of both PAPI-HTO and PP-HTO-PAPI films; Table 1 presents the Young's modulus and elongation at break. It is seen that the two polyurethane films have very distinctive properties. The Young's modulus and elongation at break of HTO-PAPI are 282.13 MPa and 14.15%, while those of PP-HTO-PAPI are 1.80 MPa and 218.23%, respectively. These results indicate that the HTO-PAPI film is a hard and strong material that breaks before the material cracks whereas the PP-HTO-PAPI film is soft and tough material. The results also show that the structure of soft and hard segments in polyurethane has obvious effect on their properties. For PAPI-HTO, the post-polymerization [11-12] of the three conjugated double bonds in HTO is expected to make additional contribution on the hard and strong properties in addition to PAPI's polyfunctional cross-linking[13-14], while for PP-HTO-PAPI, the introduction of soft segment reduces the crosslinking density and increases the elasticity. Therefore, polyurethane adhesives with different properties based on above compositions can be hopefully used for different applications. More research work is under way.

4. Conclusions

In summary, two types of polyurethane PAPI-HTO and PP-HTO-PAPI were successfully synthesized, and FTIR and ^{1}H-NMR were used to characterize their structures. The mechanical properties can be adjusted from hard and strong to soft and tough via selection of compositions, whereas the multi-functional NCO groups in PAPI and the post-polymerization properties of the conjugated triene bonds in HTO would give the film good properties. The results can provide a good theoretical basis for the preparation of novel polyurethane adhesive.

Acknowledgements

This study has been supported by the Henan Science and Technology Project Foundation (the basic and frontier research project) (162300410023).

References

[1] Petrović Z S, Ferguson J. Polyurethane elastomers, *Prog. Polym. Sci.*, *Rubber Technology*, 1991, **16**(5), 695–836.

[2] Xiao H X, U.S. Pat., 7,462,679, 2008.

[3] Mosiewicki M A, Casado U, Marcovich N E, et al. Polyurethanes from tung oil: Polymer characterization and composites, *Polym. Eng. Sci.,* 2009, **49**(4), 685–692.

[4] Ren Z Y, Liu L, Wang H F, Fu Y, Jiang L, Ren B X, Novel Amphoteric Polyurethane Dispersion with Postpolymerization Crosslinking Function Derived from Hydroxylated Tung Oil: Synthesis and Properties, *RSC Advances, (Communication)* 2015, **5**(35), 27717-27721.

[5] Ren Z Y, Zhang C L, Jiang L, Wang H F, Zhao W, Fu Y and Zhang Y B. Facile Synthesis and Newly Found Biological Activity of Aminolized Tung Oil Components with Conjugated Double Bonds, *RSC Advances (Communication)*, 2013, **3** (39), 17712-17716.

[6] Jr C U P, Xu X and Wang L, et al. Characterizing semi-interpenetrating polymer networks composed of poly(vinyl chloride) and 5-15% of oligomeric MDI isocyanate cross-linked networks, *Polymer*, 2000, **41**(14) , 5405-5413.

[7] Lubguban A A, Tu Y C and Lozada Z R, et al. Functionalization via glycerol transesterification of polymerized soybean oil. *J. Appl. Polym. Sci*, 2009, **112**(1), 19-27.

[8] Q Ge, H Wang, Y She, et al. Synthesis, characterization, and properties of acrylates-modified tung-oil waterborne insulation varnish. *J. Appl. Polym. Sci,* 2015, **132**(10), pp 27-47.

[9] Zhang C X, Ren Z Y, Qian H Y, Ma D Z, Amide II and Amide III Bands in Polyurethane Model Soft and Hard Segments, *Polym. Bull.*, 2008, **60**, 97-101.

[10] Zhao W, Wang H F, Ren Z Y, Jiang L, Fu Y and Wang X L, Structural Confirmation of A New Hydroxylated Tung Oil Compound by NMR, *Magn. Reson. Chem.* 2015, **53**(2), 160-162.

IOP Conf. Series: Materials Science and Engineering **281** (2017) 012004 doi:10.1088/1757-899X/281/1/012004

[11] Huang Y, Pang L and Wang H, et al. Synthesis and properties of UV-curable tung oil based resins via modification of Diels-Alder reaction, nonisocyanate polyurethane and acrylates, *Prog. Org. Coat.,* 2013, **76**(4), 654-661.

[12] Thanamongkollit N, Miller K R and Soucek M D. Synthesis of UV-curable tung oil and UV-curable tung oil based alkyd, *Prog. Org. Coat.,* 2012, **73**(4), 425-434.

[13] Kessler M R. Bio-based Polyurethanes Prepared from Different Vegetable Oils, *ACS Appl. Mater.Inter.,* 2014, **7**, 1226-1233.

[14] Lu Y S, Larock R C. Soybean oil-based, aqueous cationic polyurethane dispersions: Synthesis and properties, *Prog. Org. Coat.,* 2010, **69**(1), 31-37.

IOP Conf. Series: Materials Science and Engineering **281** (2017) 012005 doi:10.1088/1757-899X/281/1/012005

Research on the Bond Anchorage Properties of Alkali-Activated Slag Cementitious Material

J Zhu[*1], W Z Zheng[2], Y F Leng[1], C Z Qin[1] and Z Z Xu[1]

1.College of Civil Engineering and Architecture, Harbin University of Science and Technology, Harbin 150080

2 Key Lab of Structures Dynamic Behavior and Control of the Ministry Education, Harbin Institute of Technology, Harbin 150090

Corresponding author, J Zhu, Email: [*]zhujing02@126.com

Abstract. By bond-anchorage property tests at 20°C ~500°C, the distribution of shear stress between carbon fiber sheets and concrete at all levels of loading and anchorage lengths were measured, which means the bond lengths during CFRP sheets are pulled off at the same time when the concrete is torn and stripped were gotten. The failure modes were obtained. In addition, the failure loads were measured, and the calculated formulas of anchorage lengths were identified by fitting at high temperature. It can be seen that the anchorage lengths of carbon fiber sheets increase with increasing temperature at 20°C ~100°C, the anchorage lengths of carbon fiber sheets decrease with increasing temperature at 100°C ~500°C. Tests prove that AASCM has favorable high-temperature resistant and bond anchorage properties.

1. Introduction

With the rapid development of China's steel industry, slag has become a huge industrial by-product. It is known that China produces about 240 million tonnes of slag per year[1-3], which accounts for 50 per cent of global production. Alkali-activated slag cementitious material (AASCM) is based on industrial by-products - slag as raw material, with sodium silicate, NaOH as the main alkaline activators, with or without a certain cold heat history of aluminate natural minerals containing silicon or industrial waste residue (e.g metakaolin, phosphorus slag, steel slag, coal gangue, fly ash, red mud, etc) and alkaline activators (such as alkali metal hydroxide, alkali metal carbonate, alkali metal sulfate, etc). Thus, AASCM is obtained by chemical reactions[4-6]. It is known that the compressive strength of AASCM is not less than 50 MPa. On the other hand, organic epoxy matrices are commercially available and have been extensively used to the FRP reinforcing technique. Nevertheless, the major disadvantages of organic matrices are their lack of heat and fire resistance, low glass transition temperature Tg (60°C to 82°C), susceptibility to UV radiation, and impossibility of application on humid surfaces. On the contrary, the compressive strength of AASCM after 600 °C is not reduced, which means AASCM has excellent high-temperature resistant[7-10], Although Taljsten studies and defines anchor lengths of steel and CFRP plates bonded to concrete with epoxy[11], the bond anchorage properties of the inorganic adhesives still have been published very few, and the guide specifications and standards on the bond anchorage properties of AASCM are rarely seen at home and abroad. Moreover, the lack of detailed and reliable technical indicators greatly hinders its popularization and application. Therefore, this essay studies on the bond anchorage properties of

AASCM at room temperature, at high temperature, and after high temperature. The failure modes and failure loads were obtained. In addition, this essay can be used as the main basis for determining the parameters of various engineering designs.

2. Materials and formulation

2.1. Raw materials and mix proportion

S95 slag (Quality grade is S95, which means specific surface area of slag is $400m^2/kg$) is proposed by mass fraction of 36.9% SiO_2, mass fraction of 15.66% Al_2O_3, mass fraction of 37.57% CaO, mass fraction of 9.3% MgO, mass fraction of 0.36% Fe_2O_3. The activity indexes of slag (Which means the compressive strength ratio of the test samples and comparison samples of the same age (%)) are mainly: the quality coefficient 1.69, the alkaline coefficient 0.97, and the activity coefficient 0.42.The main parameters of potassium silicate solution are shown in Table 1. The mass fraction of sodium hydroxide is not less than 96.0%.

Table 1. Parameters of potassium silicate solution

Baume degree	Density	Modulus	mass fraction, %	
			K_2O	SiO_2
46.3	1.465	2.76	15.98	28.15

Test uses one relatively optimal matching ratio W42 of AASCM, and main parameters are shown in Table 2.

Table 2. The proportion parameters used for test

Matching ratio	Slag	Modulus of potassium silicate	Potassium silicate amount, %	Sodium hydroxide,%	Water amount %
W42	1	1.0	12	5.35	42

2.2. CFRP sheets and concrete

Two commercially available types of CFRP sheets (UT70-20 and UT70-30, Toray, Japan) and C30 concrete with a water/ binder ratio 0.40 were used for bond tests. The tensile strengths of CFRP sheets were determined according to JG/T 167-2004 on the basis of the nominal thicknesses. The nominal thicknesses were also used to all other calculations of the present study. The CFRP sheet (UT70-30) has tensile strength of 4125 MPa, an ultimate tensile strain of 1.71%, and a modulus of elasticity of 244 GPa. The nominal thickness of UT70-30 is 0.167 mm and fiber weight is 300 g/m^2. On the other hand, the CFRP sheet (UT70-20) has tensile strength of 4114 MPa, nominal thickness of 0.111 mm and fiber weight of 200 g/m^2.

3. The bond anchorage properties of AASCM

3.1. The bond anchorage properties of AASCM at room temperature

In order to achieve CFRP sheets and concrete working together, the CFRP-adhesive-concrete interfaces bonding strengths should be satisfactory; For the CFRP sheets giving full play to their strengthening effects, the interface of adhesive should have sufficient bond anchorage strength to transfer shear stress. If the anchorage length of CFRP sheets is insufficient, there is comparatively large stress concentration at the CFRP end, which will lead to debonding of the CFRP strip from the concrete, so CFRP sheets are difficult to give full play to their strengthening effects; By contrast, if the anchorage length of CFRP sheets is too long, the unnecessary materials will be wasted. Therefore, identifying appropriate anchorage lengths are very important to practical engineering.

The two types of anchorage failure modes summarized from twenty specimens were:

- Anchorage failure (AF). Sixteen specimens were failed by peeling of the concrete cover adjacent to the CFRP sheets. The failure process started with visible concrete cracks near the loading end of the concrete prism along the longitudinal axis of the CFRP sheets. As the load increased, visible cracking in the concrete initiated interfacial debonding of the CFRP strip from the loading end, and the majority of debonding occurred at the middle part of the test specimen along its longitudinal axis; A thin layer of concrete was peeled off and attached to the CFRP strip accompanied by a loud popping sound. When the anchorage length was insufficient, the anchorage failure would happen as shown in Figure 1(a).
- Tensile failure of CFRP sheets (TF). Four specimens were failed by CFRP rupture. When the applied load P increased to about 80% of the ultimate load Pu, a crisp sound was heard due to some extent of debonding in concrete adjacent to the adhesive-concrete surface, which indicated that the CFRP sheets were pulled uniformly; Then a sharp crackling sound might be heard due to CFRP rupture. The fracture surface of the CFRP sheets generally presented zigzag shapes, and the location of the fracture surface was in the non-anchorage zone of CFRP sheets. When the anchorage length was sufficient, the tensile failure of the CFRP sheets would happen as shown in Figure 1(b).

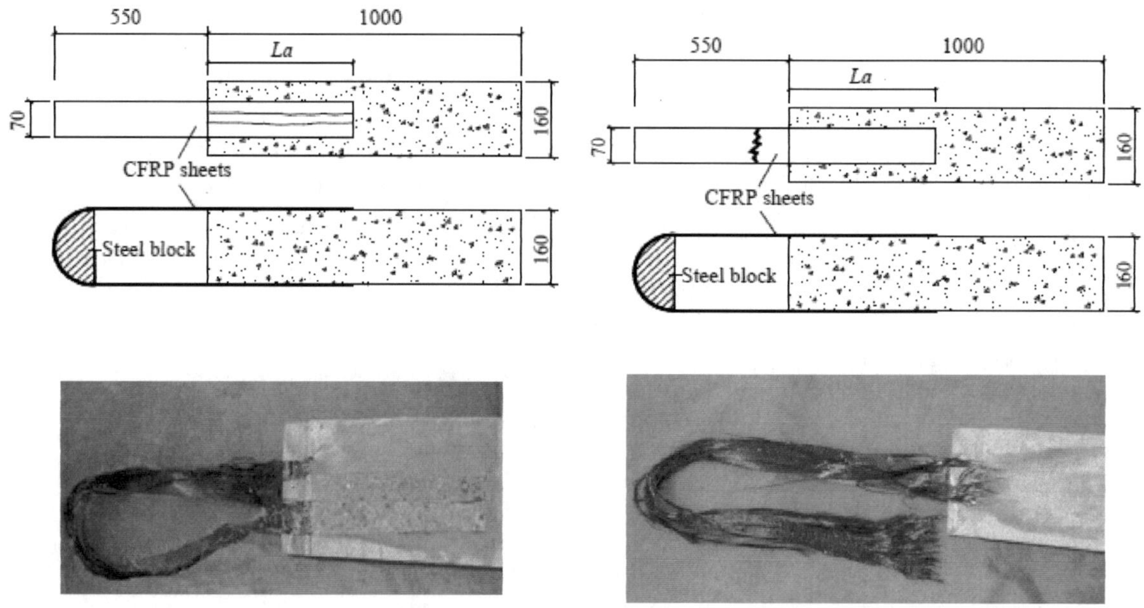

(a) Anchorage failure (b) Tensile failure of CFRP sheets

Figure 1 Schematic diagram of two types of anchorage failure modes

Therefore, the anchorage length of CFRP sheets can be identified more clearly using the test results as shown in Table 3. Which explains precisely the concept of anchorage length: the stress is transferred along the length of CFRP sheets, and the strain is not zero along the length of the CFRP sheets. The typical strain distribution and development for specimens T-10 and T-20 along the CFRP strip are as shown in Figure 2, where P is the applied load; Pu is the ultimate load; s is the slip at the loaded end.

IOP Conf. Series: Materials Science and Engineering **281** (2017) 012005 doi:10.1088/1757-899X/281/1/012005

The anchorage length La increases almost linearly with the increase of the applied load P, but the average shear strength τ decreases with the increase of the anchorage length La (as shown in Table 3). By observing the strain development along the length of the CFRP sheets (Figure 2), it is clear that the CFRP sheets do not attain their fracture strain. Consequently, a difference in the length of the high stress area reveals a difference in the anchorage length value: the higher the length of the high stress area, the smaller the basic anchorage length.

Table 3 Test results of the anchorage lengths of CFRP sheets

Type	No.	L_a(mm)	P(kN)	Anchorage failure modes	R (%)	$\overline{P}$(kN)	τ (MPa)
UT70-20 CFRP sheets	T-1	120	16.9	AF	95%	17.20	1.02
	T-2		17.5		80%		
	T-3	140	18.4	AF	75%	18.25	0.93
	T-4		18.1		80%		
	T-5	160	18.8	AF	65%	19.01	0.85
	T-6		19.2		85%		
	T-7	180	19.7	AF	45%	19.55	0.78
	T-8		19.4		30%		
	T-9	200	19.8	TF	—	19.95	0.71
	T-10		20.1		—		
UT70-30 CFRP sheets	T-11	220	20.8	AF	80%	21.05	0.68
	T-12		21.3		75%		
	T-13	240	21.9	AF	65%	22.20	0.66
	T-14		22.5		70%		
	T-15	260	23.4	AF	45%	23.65	0.65
	T-16		23.9		30%		
	T-17	280	24.2	AF	30%	24.00	0.61
	T-18		23.8		35%		
	T-19	300	24.0	TF	—	24.35	0.58
	T-20		24.7		—		

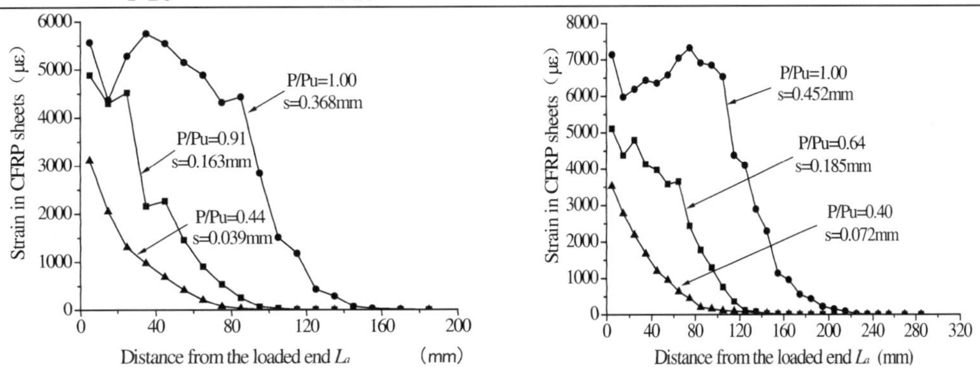

(a) Typical strain distribution for T-10 （b）Typical strain distribution for T-20

Figure 2 Strain development along the length of CFRP sheets

3.2. The bond anchorage properties of AASCM at high temperature

In order to identify the constant temperature time, thermocouples were embedded in AASCM specimen surface center to 4 mm, and the heating curve of furnace and measuring point temperature are shown in Figure 3 and Figure 4. Lastly, the constant temperature time is identified as 30min.

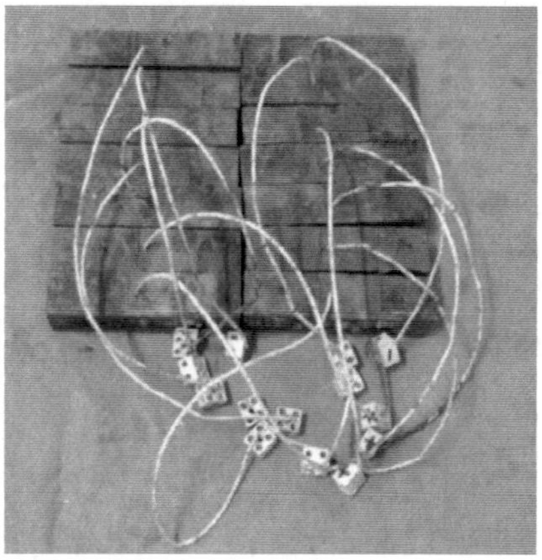

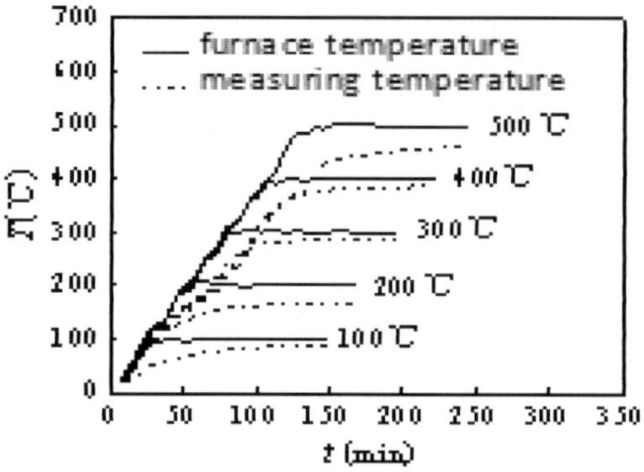

Figure 3 Specimens with thermocouples **Figure 4** Furnace and measuring point temperature

The heating rate was chosen 4 °C /min and target temperatures were 100°C, 200°C, 300°C, 400°C, 500°C. The test phenomenon is shown as Figure 5.

(a) Anchorage failure (b) Tensile failure of CFRP sheets

Figure 5 Schematic diagram of two types of anchorage failure modes at high temperature

When target temperatures are 100°C, 200°C, 300°C, 400°C, 500°C, anchorage lengths are measured with 387.5 mm, 337.5 mm, 312.5 mm, 300 mm and 300 mm, respectively. The anchorage length of the measured values can be relatively high temperature CFRP anchor length ($L_{a.T}/L_a$) with temperature change law of unified use Type (1) express, fitting curve and test curve are as shown in Figure 6.

$$\frac{L_{a.T}}{L_a}=\begin{cases}0.91+4.80\left(\dfrac{T}{1000}\right) & 20°C\le T\le 100°C,\ R^2=0.999\\[2ex] 1.73-4.53\left(\dfrac{T}{1000}\right)+11.65\left(\dfrac{T}{1000}\right)^2-11.17\left(\dfrac{T}{1000}\right)^3 & 100°C<T\le 500°C,\ R^2=0.997\end{cases} \tag{1}$$

In the formula,

$L_{a.T}$——the anchor length of CFRP at temperature T (mm);

L_a——the anchor length of CFRP at room temperature (20 °C) (mm);

T——experience temperature (°C);

R^2——the correlation coefficient of the fitting accuracy

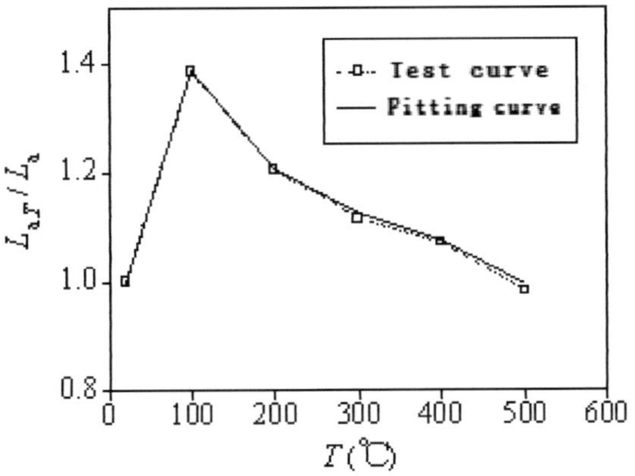

Figure 6 Relative anchorage length at high temperature

As shown in Figure 6, it is seen that the anchorage lengths of CFRP sheets increase with increasing temperature at 20°C ~100°C, but the anchorage lengths of CFRP sheets decrease with increasing temperature at 100°C ~500°C. In short, the basic anchorage lengths at high temperature were higher than those at room temperature.

3.3. The bond anchorage properties of AASCM after high temperature

When target temperatures are 100°C, 200°C, 300°C, 400°C, 500°C after high temperature, anchorage lengths are measured with 487.5 mm, 475 mm and 450 mm, 437.5 mm and 400 mm, respectively. Comparative analysis shows that the damage form of double shear specimens after high temperature are the same with room temperature damage form, but AASCM and concrete color becomes shallow, and concrete stripping areas after high temperature are bigger than that at room temperature .Compared with the double shear specimens under high temperature failure pattern, at the same temperature, the stripping areas of the concrete after high temperature are bigger than that at high temperature, From Figure 1 and Figure 5, it can be seen that AASCM provides the anaerobic protection for CFRP after high temperature and at high temperature, and AASCM avoids the CFRP wire from being oxidized. The anchorage length of the measured values after high temperature can be relatively CFRP anchor length at high temperature, which can be expressed by Type (2) express, and fitting curve and test curve are as shown in Figure 7.

$$\frac{L'_{a.T}}{L_a}=\begin{cases}0.82+9.26\left(\dfrac{T}{1000}\right) & 20°\text{C}\le T\le 100°\text{C},\ R^2=0.999\\ 1.83-1.09\left(\dfrac{T}{1000}\right) & 100°\text{C}<T\le 500°\text{C},\ R^2=0.988\end{cases} \tag{2}$$

In the formula,

$L`_{a.T}$——the anchor length of CFRP after temperature T (mm).

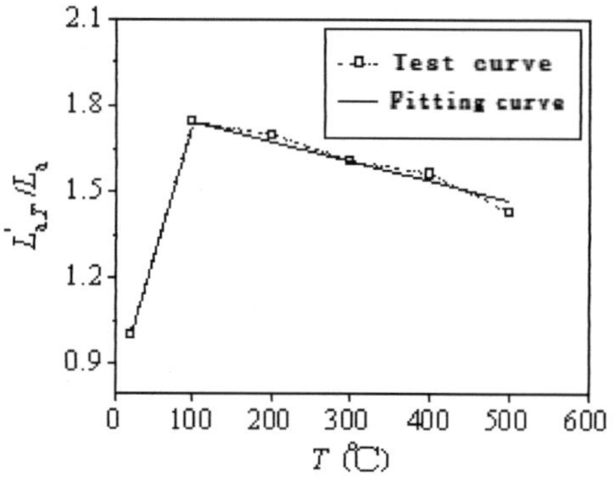

Figure 7 Relative anchorage length after high temperature

As shown in Figure 7, it is seen that the relative anchorage lengths of CFRP sheets grow linearly after 20°C ~100°C, but the relative anchorage lengths of CFRP sheets decline gradually at 100°C ~500°C. In short, the basic anchorage lengths after high temperature were higher than those at room temperature and those at high temperature. Figure 7 shows that the fitting curve fits well with the test curve.

4. Conclusions

By bond-anchorage property tests at 20°C ~500°C, the all levels of loading and anchorage lengths were measured. It can be seen that the anchorage lengths of carbon fiber sheets increase with increasing temperature at 20°C ~100°C, and the anchorage lengths of carbon fiber sheets decrease with increasing temperature at 100°C ~500°C.

- It is identified that the basic anchorage length at room temperature, at high temperature and after high temperature meet expectations. Tests prove that AASCM has favorable high-temperature resistant and excellent bond anchorage properties.
- The two types of anchorage failure modes summarized from twenty specimens were anchorage failure and tensile failure of CFRP sheets. AASCM provides the anaerobic protection for CFRP after high temperature and at high temperature, and AASCM avoids the CFRP wire from being oxidized.
- The failure loads and anchorage lengths were measured at high temperature and after hith temperature, and the calculated formulas of anchorage length were obtained by fitting. The fitting curve fits well with the test curve

Acknowledgements

This work was financially supported by the National Natural Science Foundation of China (NO. 51508140).

References

[1] Bisby L A. *Fire Behaviour of Fibre-reinforced Polymer (FRP) Reinforced or Confined Concrete[D]*. Kingston, Ontario; Dissertation of the Queen's University, 2003; 10-81.

[2] Gamage J C P H, Al-Mahaidi R, Wong M B. Bond Characteristics of CFRP Plated Concrete Members under Elevated Temperature[J]. *Composite Structures*, 2006(75); 199-205.

[3] Bisby L A, Green M F, Kodur V K R. Modeling the Behavior of Fiber Reinforced Polymer-Confined Concrete Columns Exposed to Fire[J]. *Journal of Composites for Construction,* 2005(9); 15-24.

[4] Ri-On Oh, Sang-Sun Cha, Seong-Yong Park, et al. Mechanical properties and water purification characteristics of natural jute fiber-reinforced non-cement alkali-activated porous vegetation blocks[J],*Paddy and Water Environment*, 2014,12(1): 149-156.

[5] Zheng W. Z, Zhu J. The effect of elevated temperature on bond performance of alkali-activated GGBFS paste[J]. *Journal of Wuhan University of Technology (Materials Science Edition)*, 2013,28(4):721-725.

[6] Palomo A, Alonso S, Fernández-Jiménez A, Alkaline activation of fly ashes, a NMR study of the reaction products[J]. *Journal of the Amerian Ceramic Society,* 2004,87(6):1141-1145.

[7] Zheng W. Z, Zhu J. *Application foundation of alkali-activated slag cementitious material in structural engineering[M]*, Harbin Institute of Technology Press, 2015.

[8] Pimraksa K, Chindaprasirt P, Rungchet A,et al. Lightweight Geopolymer Made of Highly Porous Siliceous Materials with Various Na_2O/Al_2O_3 and SiO_2/Al_2O_3 Ratios[J]. *Materials Science and engineering A-Structural materials properties microstructure and processing*, 2011,528(21): 6616-6623.

[9] Zhu J, Zheng W. Z. "Shear performance of alkali-activated slag cementitious material used for FRP-strengthened concrete structures", *Journal of Southeast University.*2012,42(5):962-969.

[10] Won J P, Kang H B, Lee S J, et al. Eco-Friendly Fireproof High-Strength Polymer Cementitious Composites[J]. *Cement and Concrete Composites*, 2012,30(4):406-412.

[11] Taljsten B. Defining anchor lengths of steel and CFRP plates bonded to concrete[J]. *International Journal of Adhesion and Adhesives*, 1997, 3(17); 319-327.

IOP Conf. Series: Materials Science and Engineering **281** (2017) 012006 doi:10.1088/1757-899X/281/1/012006

Low-temperature Selective Catalytic Reduction of NO with NH_3 over CuO_x/CNTs Catalyst

B N Ren

Institute of Oceanology, SanYa University, Sanya 572022, Hainan, China

E-mail: renbingnan@126.com

Abstract. The metal oxide catalyst was prepared by loading CuO_x on carbon nanotubes (CNTs) with impregnation method. The catalyst was characterized by BET, TEM and XPS, and the catalytic activity of the catalyst for selective catalytic reduction (SCR) of NO was investigated. The results showed that the species of active components loaded on the catalyst was given priority to with CuO. The NO conversion was improved with temperature increase under the range of 150 to 300°C. The oxygen content had an outstanding influence on the NO conversion of catalysts at lower concentration range. Once the oxygen content was enhanced over 5%, there was no significant increase. Increasing of mole ratio of NH_3/NO could increase the NO conversion. When mole ratio of NH_3/NO was continued to exceed 1.1, the NO conversion decreased. With the increasing of space velocity, the NO conversion was decreased under the reaction conditions.

1. Introduction

NO_x is one of the most predominant pollutants. It has become important to reduce the emission of nitrogen oxides from both stationary and mobile combustion technologies in order to keep the health of humans and protect environment. Among various NO_x emission control techniques, selective catalytic reduction (SCR) of NO with NH_3 in the presence of excess oxygen has been proven to be the most effective which is widely used. It has been proved that the catalyst being prepared through supported metallic oxides are with good catalytic performance. Various metal oxide based catalysts have been shown to possess some potential for low temperature application, including Fe_2O_3, CuO, Co_2O_3, Mn_2O_3, CeO_2, V_2O_5, etc.. Due to the carbon material with excellent surface properties, the carbon-supported catalysts have the very good performance. Among the reported catalysts, metal oxides supported on carbon materials (activated carbon, activated coke, activated carbon fiber) have shown very high activity. As a novel carbon material, carbon nanotubes (CNTs) have attracted much attention due to their unique electric, mechanical and structural characteristics. It has been widely studied in fields of high strength carbon fiber materials[1], hydrogen storage materials [2-4], new catalytic materials [4-8], etc. With the tunable pore structure and surface properties and the large specific surface area, CNTs have been regarded as the perfect supporting materials for catalysts. Meanwhile, due to one-dimensional tubular structure of CNTs, catalytic materials with specific properties can be prepared in the cavity of tubular. So as a new catalytic material, CNTs have caused attention of many researchers [9-12]. Fe/CNTs, Mn-FeO_x/ CNTs, Fe_2O_3/CeO_2 catalysts have been investigated the catalytic performance for the selective catalytic reduction of NO with NH_3[13-16]. Cu is one of the typically studied transition metals in SCR catalysts. Also, copper oxides are found to be active in medium temperature SCR of NO. The reduction of Cu^{2+} to Cu^{+} is a possible reason for catalyst deactivation at low temperatures [17].

Therefore, it can be an issue deserved intensive study to make use of CNTs to develop low temperature de-NO_x catalyst. This research prepared CuO_x/CNTs catalyst through supported copper oxide on the CNTs after purification. We have investigated copper oxide supported on the CNTs catalysts as low-temperature SCR catalysts.

2. Materials and methods

2.1. Catalyst preparation

The raw CNTs were purified by means of immersion in a 65% of HNO_3 solution and refluxed at boiling point for 4 to 6 h. Then the sample was washed with distilled water till pH 6 to 7, and dried at 110 °C for 12 h. The supported CuO_x/CNTs catalysts were prepared by impregnation method. The CNTs after purification was impregnated in different concentrations of cupric nitrate solution at room temperature for 24 h, then dried at 110 °C for 12 h. And after grinded, the CNTs were calcified at 500 °C nitrogen atmosphere for 5 h.

2.2. Catalyst characterization

The specific surface area and pore volume of the CNTs and CuO_x/CNTs catalysts were analyzed by an TRISTAR II 3020M which can be determined through N_2 absorption method. The pore volume was calculated using the BET method. A JEM-1200EX transmission electron microscope was used for TEM which acceleration voltage is 120 kv and magnification is 0~200000. XPS analysis was carried out by making use of AXIS Ultra-type X-ray photoelectron spectroscopy (England Kratos).

2.3. Catalytic activity tests

The catalytic activity tests were carried out in a fixed bed reactor made of stainless steel in the 150 to 300°C range. The gas composition was 450 ppm NO, 500 ppm NH_3, 5% O_2 balanced by N_2. The total flow rate was 500 ml•min^{-1}. The experimental apparatus of evaluation of the catalytic activity tests included sections of gas configuration, catalytic reaction and flue gas analysis, which was shown in Figure 1.

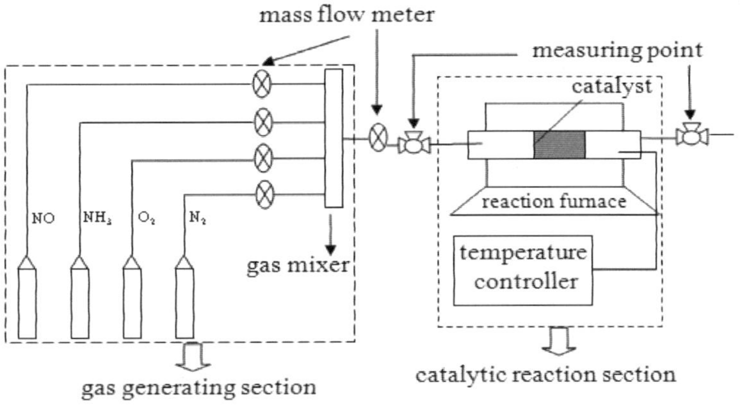

Figure 1. Experimental apparatus for activity tests

NO concentrations in the inlet and outlet gases were measured by flue gas analyzer (KM940, Kane). The NO conversion was calculated from the NO concentration at the inlet and outlet of the reactor. Before starting the activity measurements, the CuO_x/CNTs catalysts were degasified under N_2 flow at 120°C for 1 h. Afterwards, the activity tests were performed under isothermal conditions. Steady-state the NO conversion was observed when the NO concentration at the outlet of the reactor became steady, several hours after the reaction started.

3. Results and discussion

3.1. Catalyst characterization

The surface area, pore volume and pore size (inner diameters) of the CNTs after purification and CuO_x/CNTs catalyst were listed in Table 1. The surface area, pore volume, and mean pore size of catalyst slightly decreased about 34% compared with the data of CNTs after purification, which illustrated that the metallic oxide was loaded on the surface of carrier after impregnation and calcification. Due to the block of pore by metallic oxide, the surface area, pore volume, and mean pore size all decreased. It can be proposed that the active component was already loaded on the surface of CNTs.

Table 1. The surface area and pore structure analyses of CNTs

Samples	S_{BET} / m^2 g^{-1}	S_{meso} / m^2 g^{-1}	Pore volume / cm^3 g^{-1}	Mean pore size / nm
CNTs after purification	145.1517	123.4085	0.776334	23.7064
CuOx /CNTs catalyst	96.2612	84.0927	0.482603	16.3087

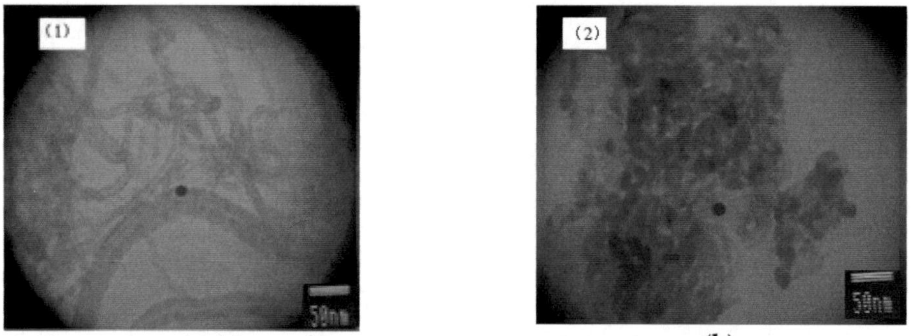

(a) (b)

Figure 2. TEM image of carbon nanotubes: (**a**) after oxidized by nitrate acid; (**b**) 20% CuO_x/CNTs

Figure 2 showed the TEM image of CNTs-after purification and CuO_x/CNTs catalyst, in which the HNO_3 treatment made most of the carbon nanotubes open (Figure 2a), which leads to an increase of the surface areas. For CuO_x/CNTs catalyst, CuO_x particles were located on the wall of the tubes (Figure 2b). The metal oxide particles loaded on the CNTs was evenly on the surface of the carrier. Future researches should be worthy to be carried out to reveal the effect between the catalyst activity and the metal oxide particle dispersion.

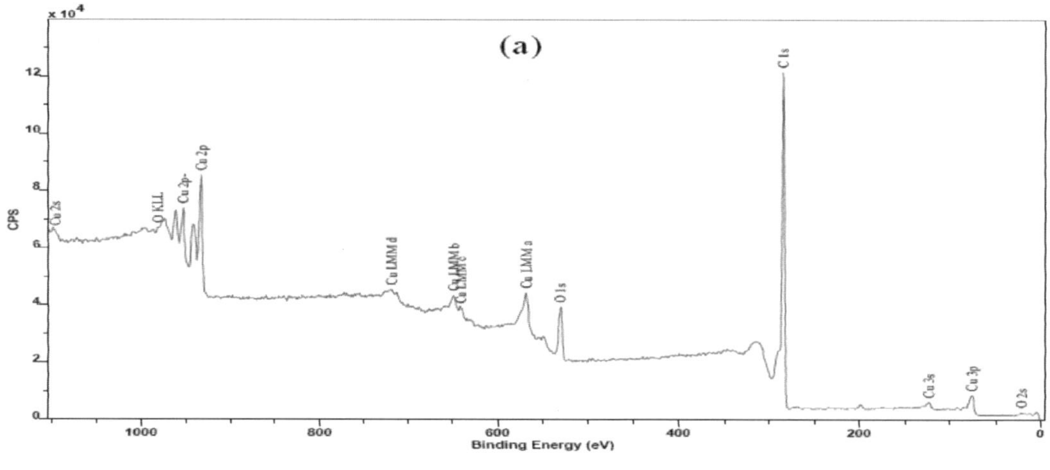

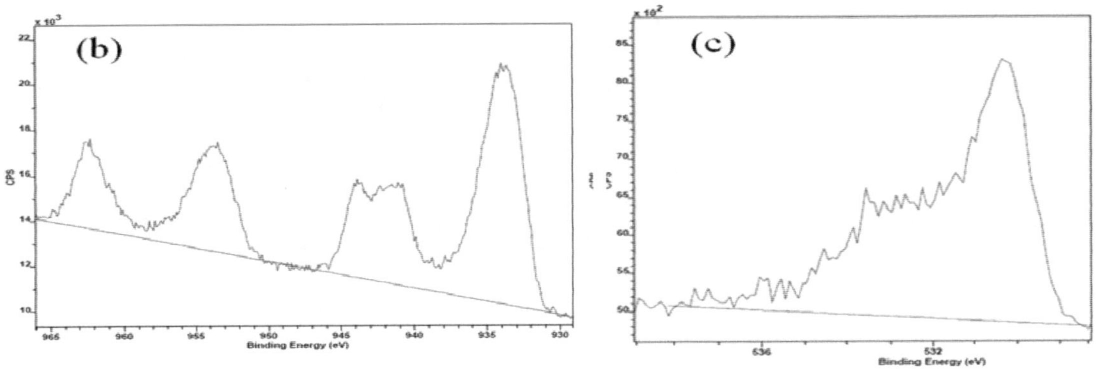

Figure 3. XPS spectra of the CuO_x/CNTs: (a) CuO_x/CNTs, (b) Cu (2p), (c) O(1s)

Figure 3 illustrated the surface element concentration of CuOx/CNTs catalyst by XPS spectra, in which (a), (b), and (c) were separately CuO_x/CNTs catalyst spectra, Cu(2p) spectra and O(1s) spectra. The catalyst XPS spectra showed three obvious peaks, the peak of 934 eV was the absorption peak of cuprum; the binding energy in the middle peak was 530 eV, which was the absorption peak of oxygen; peak with the biggest binding energy was 284 eV, which was the absorption peak of carbon. The surface element concentrations of CuO_x/CNTs catalyst sample measured by XPS were 90.44% of C, 5.9% of O, 3.3% of Cu. It can be seen from Figure 3b that the biggest absorption peak was 934 eV corresponded to CuO. Therefore, the combination form of metallic oxide was given priority to with CuO.

3.2. SCR activity of CuO_x/CNTs catalyst

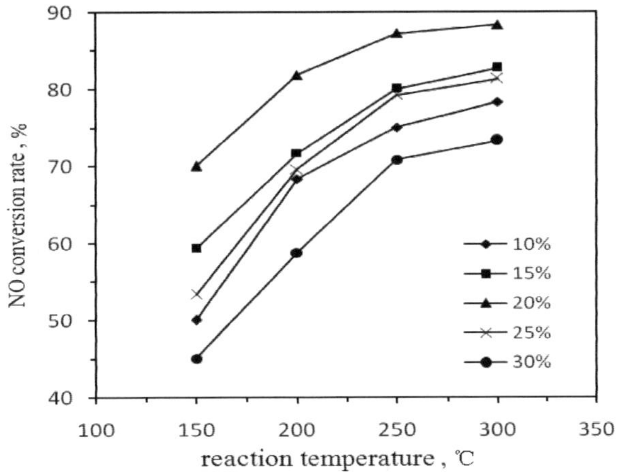

Figure 4. Effect of reaction temperature on the activity of CuO_x/CNTs

Figure 4 presented the NO conversion of CuO_x/CNTs catalyst at different temperature. The result indicated that all catalysts reached to the highest NO conversion rate when in temperature of 300°C. The NO conversion of catalyst prepared with different impregnation concentrations increased rapidly with the rise of temperature under the temperature range of 150 to 300°C. During the process of the SCR, rising temperature would lead to the reaction rate of the NO conversion increased. The consumption of NH_3 was mainly for NO reduction reaction, and NO conversion rate increased rapidly. After that, continue to rise the reaction temperature, NO reaction would be restricted for the occurrence of NH_3 oxidation reaction because rising temperature made the NH_3 oxidation reaction easily occurred.

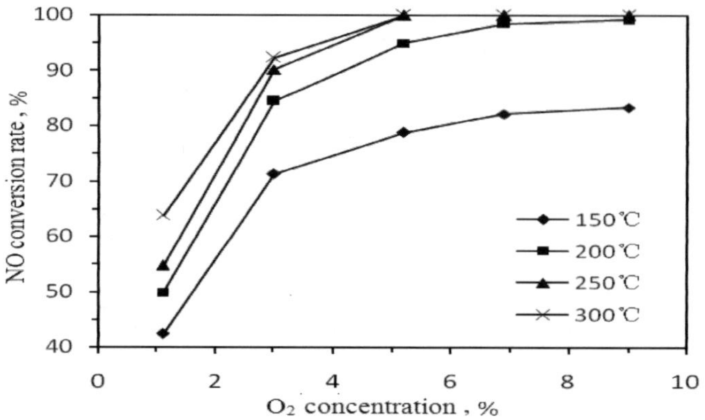

Figure 5. Effect of O_2 concentration on NO removal over CuO_x/CNTs

Figure 5 showed the effect of oxygen content on SCR activity of CuO_x/CNTs catalyst. The result showed that the NO conversion increased with increasing of oxygen content at each reaction temperature. With increasing oxygen content from 1% to 5%, the NO conversion increases rapidly. At 300°C the NO conversion increased rapidly from 39% to 87%. However, no obvious change was observed when the O_2 concentration exceeded 5%. The NO conversion remained stable basically. This indicated that O_2 played a significant promoting role in the SCR reaction. Gas phase oxygen and adsorbed oxygen species achieved a balanced state when the O_2 concentration reached 5%.

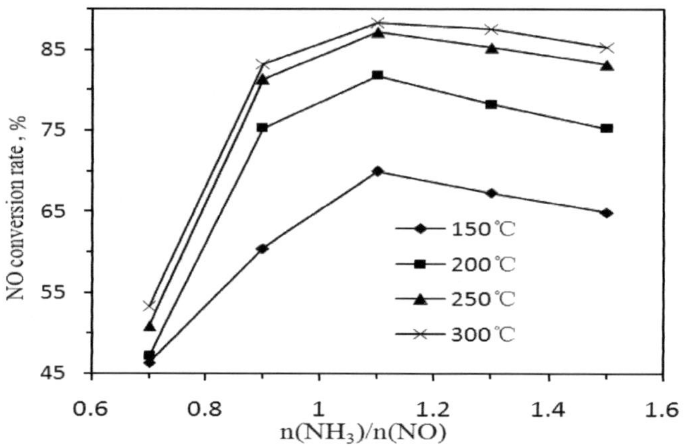

Figure 6. Effect of $n(NH_3)/n(NO)$ on NO removal over CuO_x/CNTs

Figure 6 illustrated the influence of NH_3/NO mole ratio on NO reduction at different temperatures (150 to 300°C). The result showed that the NO conversion increased with increasing of NH_3/NO mole ratio in the range of ratios from 0.7 to 1.1. When the NH_3/NO mole ratio continued to exceed 1.1, the NO conversion decreased. For NH_3/NO mole ratio <1.1, increasing of NH_3/NO mole ratio could increase the unit mass of NH_3 adsorption quantity on the catalyst. So the SCR reaction rate was accelerated. But excessive NH_3 cannot be adsorbed on catalyst. NH_3 was reacted with NO, which could lead to the decease of the NO conversion.

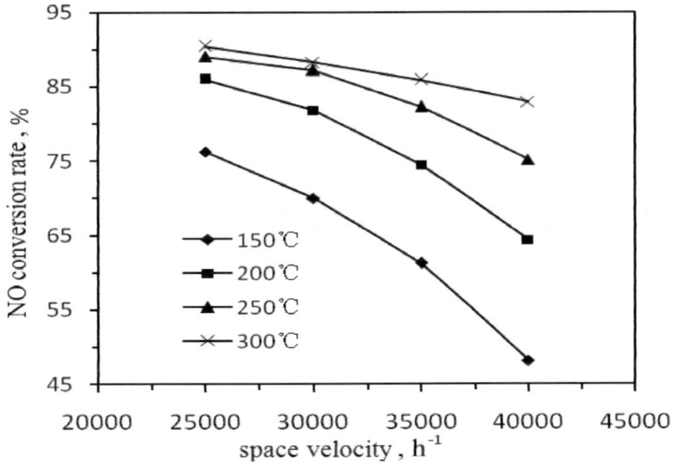

Figure 7. Effect of space velocity on NO removal over CuO_x/CNTs

Figure 7 presented the effect of space velocity on the NO conversion at different temperatures (150 to 300°C). The NO conversion all tended to decrease with the increase of space velocity under all the experimental temperatures. However, when reaction happened in 150°C, the NO conversion decreased rapidly with the increase of space velocity. While the NO conversion remained unchanged and then decreased slowly under the temperature range of 150 to 300°C. Which illustrated that higher reaction temperature can weaken the effect of space velocity on the NO conversion. Furthermore, the NO conversion rose constantly with the rising of temperature under each space velocity. The higher the space velocity was, the shorter time the reactant gas stayed on catalyst. At a high space velocity, the decrease in the NO conversion may result from insufficient contact time between the reactants and the catalyst or from insufficient amounts of active sites to react.

4. Conclusions

The CNTs after purification was impregnated in different concentrations of cupric nitrate solution. The metallic oxide was loaded on the surface of CNTs after impregnation and calcification. Due to the block of pore by metallic oxide, the surface area, pore volume, and mean pore size all decreased. The metal oxide particles loaded on the CNTs was evenly on the surface of the carrier. Future researches should be worthy to be carried out to reveal the effect between the catalyst activity and the metal oxide particle dispersion.

During the SCR reaction process of CuO_x/CNTs catalyst, rising temperature would lead to the reaction rate of the NO conversion increased. Continue to rise the reaction temperature, NH_3 oxidation reaction would happen, the reduction reaction of NO would be restrained, the NO conversion rate started to slow down. O_2 played a significant promoting role in the SCR reaction. The NO conversion increased with increasing of oxygen content at each reaction temperature. The oxygen content caused a prominent affect on catalyst at lower oxygen content. Once the O_2 concentration exceeded 5%, the NO conversion remained stable basically. For NH_3/NO mole ratio<1.1, increasing of NH_3/NO mole ratio could increase the unit mass of NH_3 adsorption quantity on the catalyst. The NO conversion increased. When the NH_3/NO mole ratio continued to exceed 1.1, the NO conversion decreased. The NO conversion all tended to decrease with the increase of space velocity under all the experimental temperatures. Higher reaction temperature can weaken the effect of space velocity on the NO conversion. The higher the space velocity was, the shorter time the reactant gas stayed on catalyst.

IOP Conf. Series: Materials Science and Engineering **281** (2017) 012006 doi:10.1088/1757-899X/281/1/012006

Acknowledgments
This study was funded by the Natural Science Foundation of Hainan Province of China (413128), the Academy-Locality Cooperation Project of Sanya (2012YD50).

References

[1] Poncharal P., Wang Z.L., Ugarte D. and de Heer W.A. 1999 *Science* **283** p1513-1516
[2] Dillon A.C., Jones K.M., Bekkedahl T.A., Kiang C.H., Bethune D.S. and Heben M.J. 1997 *Nature* **386** p377-379
[3] Ye Y., Ahn C., Witham C., Fultz B., Liu J., Rinzler A., Colbert D., Smith K. and Smalley R. 1999 *Appl. Phys. Lett.* **74** p2307-2309
[4] Liu C., Fan Y.Y., Liu M., Cong H.T., Cheng H.M. and Dresselhaus M.S. 1999 *Science* **286** p1127-1129
[5] Planeix J.M., Coustel N., Coq B., Brotons V., Kumbhar P.S., Dutartre R., Geneste P., Bernier P. and Ajayan P.M. 1994 *J. Am. Chem. Soc.* **116** p7935-7936
[6] Coq B., Planeix J.M., Brotons V. 1998 *Appl. Catal. A* **173** p175-183
[7] Zhang Y., Zhang H.B., Lin G.D., Chen P., Yuan Y.Z. and Tsai K.R. 1999 *Appl. Catal. A* **187** p213-224
[8] Liang C., Li Z., Qiu J.S. and Li C. 2002 *J. Catal.* **211** p278-282
[9] Zhang H.Z., Qiu J.S., Liang C., Wang X., Wang Y.P., Feng Z.C. and Li C. 2005 *Catal. Lett.* **101** p211-214
[10] De Jong K.P., Geus J.W. 2001 *Cat. Rev.* **42** p481-510
[11] Serp P., Corrias M., Kalck P. 2003 *Appl. Catal. A* **253** p337-358
[12] Wildgoose G.G., Banks C.E., Compton R.G. 2006 *Small* **2** p182-193
[13] Ma Z., Yang H., Li Q., Zheng J. and Zhang X. 2012 *Appl. Catal. A- Gen.* **427-428** p43-48
[14] Zhang Y., Zheng Y., Wang X. and Lu X. 2015 *Catal. Commun.* **62** p57-61
[15] Cai S., Hu H., Li H., Shi L. and Zhang D. 2016 *Nanoscale* 8 p3588-3598
[16] Fang C., Shi L., Hu H., Zhang J. and Zhang, D. 2015 *RSC Adv.* 5 p11013-11022
[17] Ma Z., Yang H., Liu F. and Zhang, X. 2013 *Appl. Catal. A- Gen.* 467 p450-455

IOP Conf. Series: Materials Science and Engineering **281** (2017) 012007 doi:10.1088/1757-899X/281/1/012007

Effect of Cooling Rate on the Microstructure and Mechanical Properties of C-Mn-Al-Si-Nb Hot-Rolled TRIP Steels

B Fu[1*], M Y Lu[2], W Y Yang[3], L F Li[4] and Z Y Zhao[1]

[1] Beijing Institute of Aeronautical Materials, Beijing 100095, China
[2] Shougang Jingtang Iron & Steel Co., Ltd., Tangshan 063200, China
[3] School of Materials Science and Engineering, University of Science and Technology Beijing, Beijing 100083, China
[4] State Key Laboratory for Advanced Metals and Materials, University of Science and Technology Beijing, Beijing 100083, China

Email: fuboustb@163.com

Abstract. A novel thermomechanical process to manufacture hot-rolled TRIP steels has been proposed based on dynamic transformation of undercooled austenite (DTUA). The cooling rate between DTUA and isothermal bainitic treatment in the novel process is important. In the present study, effect of this cooling rate on the final microstructures and mechanical properties of a C-Mn-Al-Si-Nb TRIP steel was investigated. The results showed that the volume fractions of acicular ferrite and retained austenite were increased with the increment of cooling rate. As a consequence, higher yield strength and larger total elongation were obtained for the investigated steel with higher cooling rate. In addition, a value of 30.24 GPa% for the product of tensile strength and total elongation was acquired when the cooling rate was 25 K/s. This value has met the standard of the "Third Generation"of advanced high strength sheet steels.

1. Introduction

Transformation induced plasticity (TRIP) steels have been a promising candidate to reduce the automotive body weight due to its good combination of strength and ductility. This excellent performance is ascribed to its so-called TRIP effect occurring during deformation. The typical microstructure of TRIP steels consists of ferrite matrix and a dispersion of bainite, martensite and retained austenite. The retained austenite is metastable at room temperature and can transform into martensite induced by straining, leading to the postponement and stabilization of necking [1-3]. A novel thermomechanical process to manufacture hot-rolled TRIP steels has been proposed based on dynamic transformation of undercooled austenite (DTUA) [4-7]. The main feature of this process is that the formation of such mixed microstructure can be well controlled by the applied strain of hot deformation of the undercooled austenite. Moreover, the TRIP steels based on DTUA have a refined and homogeneous microstructure, usually with the average ferrite grain size of 1~2 μm, which is considered beneficial for the improvement of mechanical properties of TRIP steels [4-7].

In this novel process, cooling rate after dynamic transformation is an important process parameter and has an obvious influence on the final microstructure and mechanical properties of TRIP steels. In the present study, a C-Mn-Al-Si-Nb TRIP steel was selected to investigate the effect of cooling rate after dynamic transformation on the microstructure and mechanical properties of the steel.

2. Materials and experimental method

The chemical composition of the steel used in this study is shown in Table 1, where the A_3 temperature calculated by using Thermo-Calc and the A_{r3} temperature obtained by dilatometry under cooling rate of 5 K/s are also displayed, respectively. The casting blanks were first reheated for 2 hours at a temperature of 1473 K, and then hot-forged at 1423 K to 1173K followed by air-cooled to room temperature. The samples for compression were machined as the shape shown in [4]. The thermomechanical process to manufacture hot-rolled TRIP steels based on DTUA is indicated schematically in figure 1. After austenization at 1523 K for 5 min, the samples were cooled at 5 K/s to 1373 K, and deformed at 1 s^{-1} to strain of 30%. Then, the samples with fully recrystallized austenite microstructure were directly cooled at 5 K/s to 1053 K, where austenite was between A_3 and A_{r3} and considered to be undercooled. Subsequently the samples were deformed at 1 s^{-1} to strain of 50% at 1053 K, and then cooled at different rates (35, 25, 15 K/s) to 723 K where a 3 min hold was taken before quenched to room temperature. The whole processes above were performed in a Gleeble1500 hot simulation test machine.

Microstructure examinations were conducted using a LEICA DC 100 optical microscope (OM) with the etchant of LePera reagent [8]. The volume fraction and lattice parameter of retained austenite were determined by the method of X-ray diffraction [9], conducting by a MXP21VAHF diffractometer. Room temperature tensile tests were performed in a Reger 3010 tensile tester at a speed of 1.2mm/min, using the samples machined from the compressed samples with the dimension as shown in [4].

Table1. Chemical composition of the investigated steel (wt.%)

C	Si	Mn	Al	Nb	P	S	N	Fe	A_3/K	A_{r3}/K
0.2	0.5	1.49	1.04	0.038	0.02	< 0.0045	0.034	Bal.	1247	1013

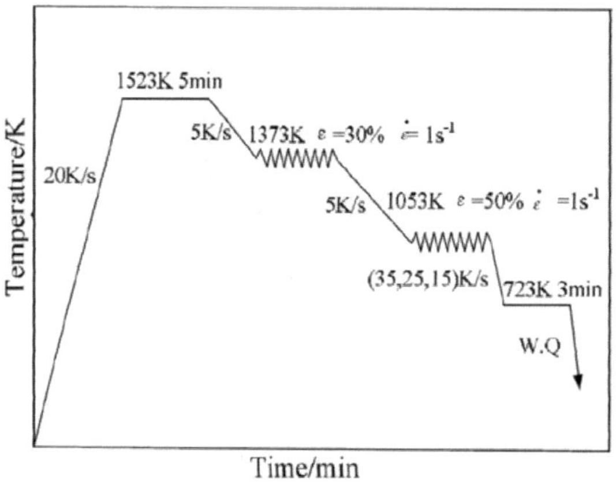

Figure 1. Schematic diagram of thermomechanical process

3. Results and discussions

Variation of microstructure as a function of cooling rate after dynamic transformation is displayed in figure 2. As can be seen, when the cooling rate was high (35 K/s), a great amount of acicular ferrite was observed in the microstructure (figure 2(a)). This kind of ferrite is considered as a bainitic structure with retained austenite and martensite layers between the ferrite laths. The acicular ferrite was also observed by Timokhina et al. in the TRIP steels with Nb addition [10]. As the cooling rate was reduced, the amount of acicular ferrite decreased, while the fraction and grain size of equiaxed ferrite increased due to the sufficient growth time at low cooling rate (as shown in figure 2(b) and (c)). The volume fraction of retained austenite also varied with the cooling rate, indicated in figure 3. Much more retained austenite could be obtained as the increment of cooling rate. This trend may be explained as follows: the austenite located between acicular ferrite was usually considered more stable owing to the geometrical restrictions of acicular ferrite laths [10]. Thus, this relatively stable austenite was more prone to be remained finally at room temperature. Because more acicular ferrite was formed at high cooling rate as discussed above, more retained austenite was expected to be acquired as well at high cooling rate. Moreover, for high cooling rate, the C and Nb atoms may not have enough time to precipitate in the form of carbides during cooling. Therefore, these solute C and Nb atoms could further increase the stability of austenite by solid solution strengthening [11-13].

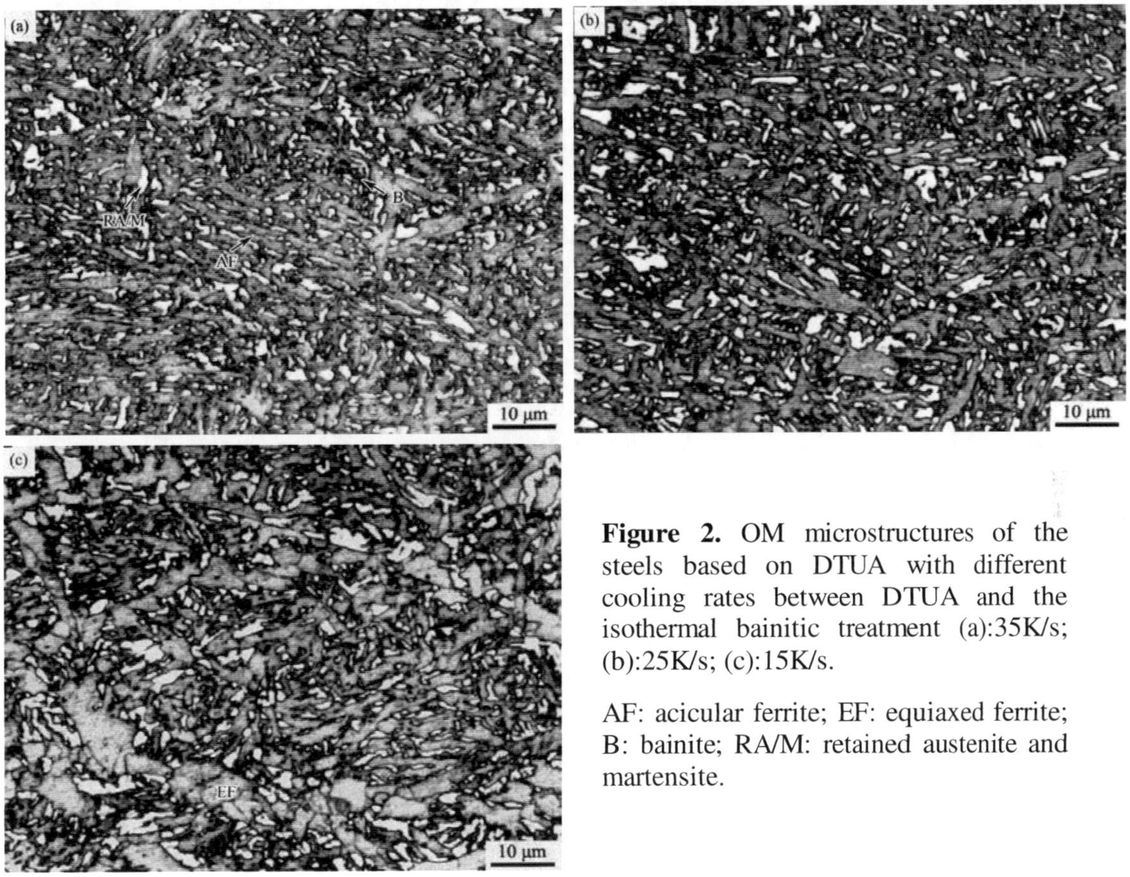

Figure 2. OM microstructures of the steels based on DTUA with different cooling rates between DTUA and the isothermal bainitic treatment (a):35K/s; (b):25K/s; (c):15K/s.

AF: acicular ferrite; EF: equiaxed ferrite; B: bainite; RA/M: retained austenite and martensite.

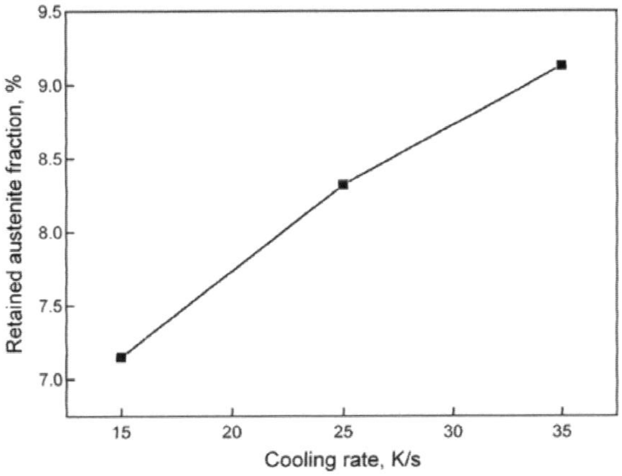

Figure 3. Variation of volume fraction of retained austenite with different cooling rates

The mechanical properties of the investigated steels were displayed in figure 4 with different cooling rates. The total elongation in figure 4(a) showed an increasing tendency with cooling rate, which was similar to the variation of retained austenite (as shown in figure 3). This consistency indicated the importance of TRIP effect on the improvement of ductility for TRIP steels. The yield strength, shown in figure4(b), was also raised with the increase of cooling rate, which may be owing to the increment of acicular ferrite and the refined microstructure at high cooling rate [10]. The tensile strength increased first and then decreased with cooling rate, leading to a maximum strength at cooling rate of 25 K/s. As a result, the product value of tensile strength and total elongation was also exhibited the maximum value at cooling rate of 25 K/s (indicated in figure 4(c)). This value is used to represent the ability of energy absorption of steels during crash and an important parameter to evaluate the comprehensive mechanical properties of TRIP steels. It is worth noting that the maximum value of product for tensile strength and total elongation obtained in the present study is 30.24 GPa%, which has met the standard of the" Third Generation" of advanced high strength sheet steels [14].

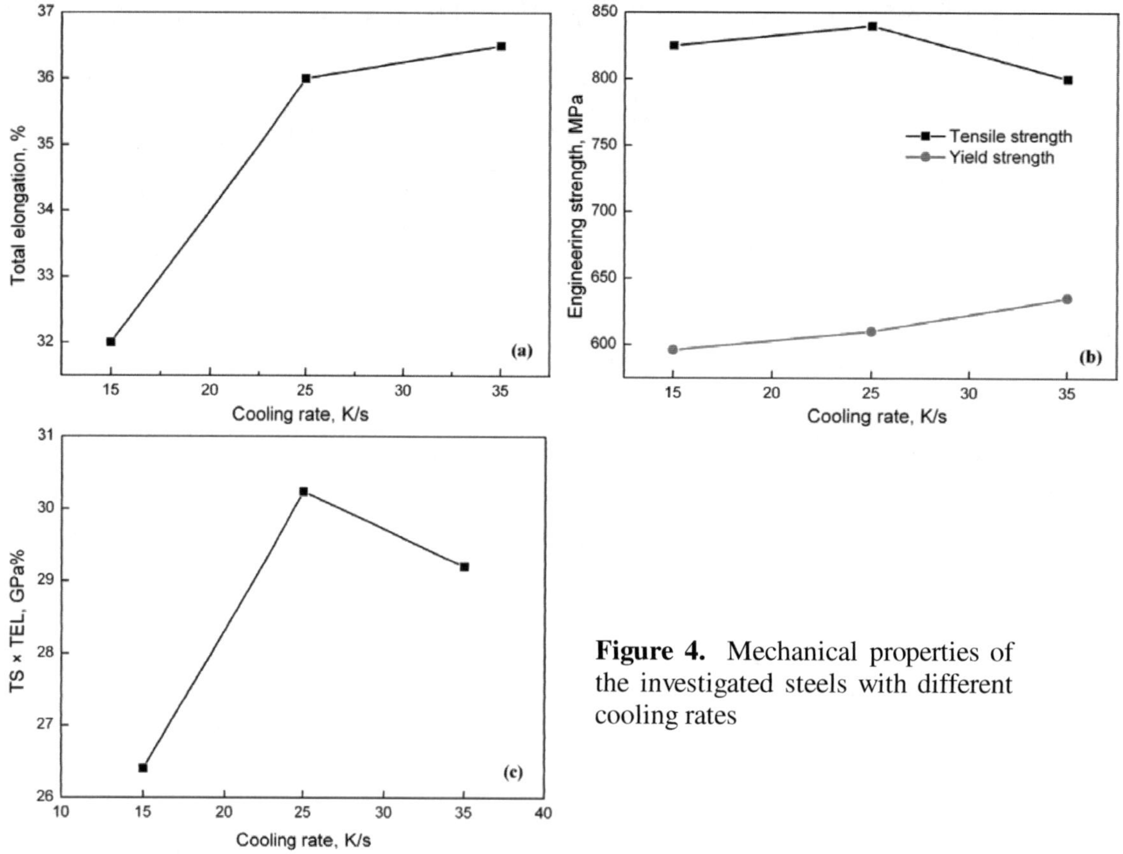

Figure 4. Mechanical properties of the investigated steels with different cooling rates

4. Summary

In the present study, the effect of cooling rate after dynamic transformation for a novel thermomechanical process to manufacture hot-rolled TRIP steels was investigated using a C-Mn-Al-Si-Nb steel. A great amount of acicular ferrite and fine grain size of microstructure were formed at high cooling rate. These microstructure features led to higher yield strength and larger volume fraction of retained austenite, as well as more excellent total elongation, at high cooling rate. As the decrease of cooling rate, the equiaxed ferrite with relatively larger grain size would be the dominant matrix phase due to enough growth time during cooling. In addition, when the cooling rate was 25 K/s, a maximum value of 30.24 GPa% for the product of tensile strength and total elongation was acquired. This value has met the standard of the " Third Generation"of advanced high strength sheet steels and indicates the superiority of this novel thermomechanical process.

IOP Conf. Series: Materials Science and Engineering **281** (2017) 012007 doi:10.1088/1757-899X/281/1/012007

Acknowledgments
Financial support of State Key Laboratory for Advanced Metals and Materials of China is gratefully acknowledged.

References
[1] Zackay V F, Parker E R, Fahr D and Busch R 1967 *Trans. A.S.M.* **60** 252
[2] De Meyer M, Kestens K and De Cooman B C 2001 *Mater. Sci. Tech.* **17** 1353
[3] De Cooman B C 2004 *Curr. Opin. Solid State Mater. Sci.* **8** 285
[4] Sun Z Q, Yang W Y, Qi J J and Hu A M 2002 *Mater. Sci. Eng.* A **334** 201
[5] Yang W Y, Qi J J, Sun Z Q and Yang P 2004 *Acta Metall. Sin.* **40** 135
[6] Yang W Y, Li L F, Yin Y Y, Sun Z Q and Wang X T 2010 *Mater. Sci. Forum* **654-656** 250
[7] Fu B, Yang W Y, Lu M Y, Feng Q, Li L F and Sun Z Q 2012 *Mater. Sci. Eng.* A **536** 265
[8] Lepera F S 1980 *J.Met.* **32** 38
[9] Miller R L 1964 *Trans A.S.M.* **67** 892
[10] Timokhina I B, Hodgson P D and Pereloma E V 2004 *Metall. Mater. Trans.* A **35** 2331
[11] Fu B, Yang W Y, Li L F and Sun Z Q 2013 *Acta Metall. Sin.* **49** 408
[12] Zhang Z R, Zhao Z Y, Li C Z, Jiang Z H and Li H B 2013 *Appl. Mech. Mater.* **395-396** 284
[13] Feng Q, Li L, Yang W and Sun Z 2014 *Mater. Sci. Eng.* A **603** 169
[14] Matlock D K and Speer J G 2009 *Microstruct. Texture Steels* 185

Investigation into Inhibitory Effect of C Incorporation on Diffusion of B in SiGe

J Zhang[1*], J N Wang[1], G B Chen[1], Z H Tang[1], K Z Tan[1], X Chen[1] and W Cui[1]
[1]Science and Technology on Analog Integrated Circuit Laboratory Chongqing, 400060, China

*zhangjing1009@163.com

Abstract. In this paper, the effect of C incorporation on the out-diffusion of B dopant in SiGe during high-temperature annealing and the out-diffusion of B dopant at different annealing temperatures were studied. The experimental results indicated that more than 0.1% C incorporation had a significant inhibitory effect on the out-diffusion of B, but that the inhibition was obviously reduced when the processing temperature was over 1050 ℃. In addition, because the atomic radius of C is smaller than that of Si, Ge, the C incorporation compensates the strain of SiGe, and adjusts the lattice mismatch, thus enhancing the thermal stability of SiGe.

1. Introduction

One of the key issues about the NPN SiGe HBT technology is that after the SiGe base epitaxial layer is grown, the narrow-band distribution of boron in the SiGe layer is maintained during the subsequent thermal processes [1]. In the 0.35um or less process technology, an ultra-thin, highly doped SiGe base width is usually only twenty or thirty nm, B dopant in the SiGe base tends to produce out-diffusion in the subsequent processes such as annealing, and so the boron distribution widens or extends to the adjacent Si region, which results in an additional conduction band barrier, and which significantly reduces the device performance [2]. The out-diffusion of boron may be produced by thermal annealing and by transient diffusion enhancement (TED) induced by external implantation/annealing [2].

To solve this problem, the out-diffusion of boron can be suppressed by doping C atom in SiGe layer so as to prevent device performance from degrading [3].

In B-doped SiGe layer, B impurity out-diffusion is carried out through the interstitial displacement mechanism, partial B impurities are changed into the mobile interstitials B atoms (Bi) by replacement reaction (1), and B interstitial Bi moves outward, resulting in the out-diffusion of B impurities.

$$Bs+I <=> Bi \tag{1}$$

The diffusion of the C atoms in the SiGe layer is similar to that of the B atom diffusion, and is also caused by displacement reaction between Si self-interstitial atoms. In the region with doping concentration of C higher, substitution reaction between substitutional C atom and self-interstitial Si atom is carried out, as shown in formula (2), which forms a movable interstitial C atom (Ci). The Ci atoms produced by the substitution reaction (2) reduces the number of self-interstitial atoms in the Si, thus reducing the chance of interstitial Bi formation by reaction of B atoms with Si self-interstitials.

$$Cs+I <=> Ci \tag{2}$$

As a result, the out-diffusion of C in the region of high C concentration will cause the concentration of undersaturated self-interstitial Si in the C-rich region to decrease, which in turn inhibits the diffusion of boron[4].

2. Experiments and results

The experiment used multilayer structure, as shown in **Figure 1.**which was deposited by RPCVD (reduced pressure chemical vapor deposition). At first, the 10nm Si buffer was deposited on Si substrate, and then the 35nm undoped SiGe was deposited, followed by 10nm p^+ SiGe and 35nm undoped SiGe. P-type doping source used B_2H_6 for in-situ doping, C was doped in all the SiGe layers, the C source was SiH_3CH_3, and finally the 10nm Si cap was deposited. The B dopant profile and its change after annealing (B out-diffusion) were characterized by SIMS.

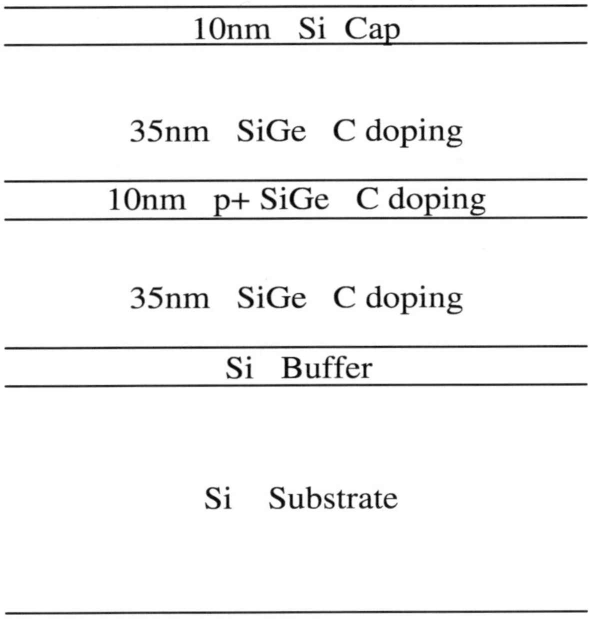

Figure 1. RPCVD-deposited multilayer SiGe sample structure

2.1. Inhibitory effect of different C content on the out-diffusion of B impurity

The first set of experiments was a comparison of the inhibitory effects of different C concentrations on the out-diffusion of B. The content of doped C was 0.1%, 0.2%, and 0.6%, respectively. All the samples were processed under the same annealing conditions at a temperature of 950 °C for 15 minutes. **Figure 2.** shows the B out-diffusion in sample without C doping after annealing, and apparently B out-diffusion was very serious whose full width at half maximum (FWHM) expanded by 2 times. **Figures 3, 4** and **5** indicated the B out-diffusion in the samples with 0.1%, 0.2%, and 0.6% C doped, respectively, after annealing. As can be seen from the figures, the incorporation of 0.1% C has completely inhibited the B out-diffusion, FWHM almost did not expand, while the FWHM of the 0.2% and 06% samples roughly expanded by about 5%.

IOP Conf. Series: Materials Science and Engineering **281** (2017) 012008 doi:10.1088/1757-899X/281/1/012008

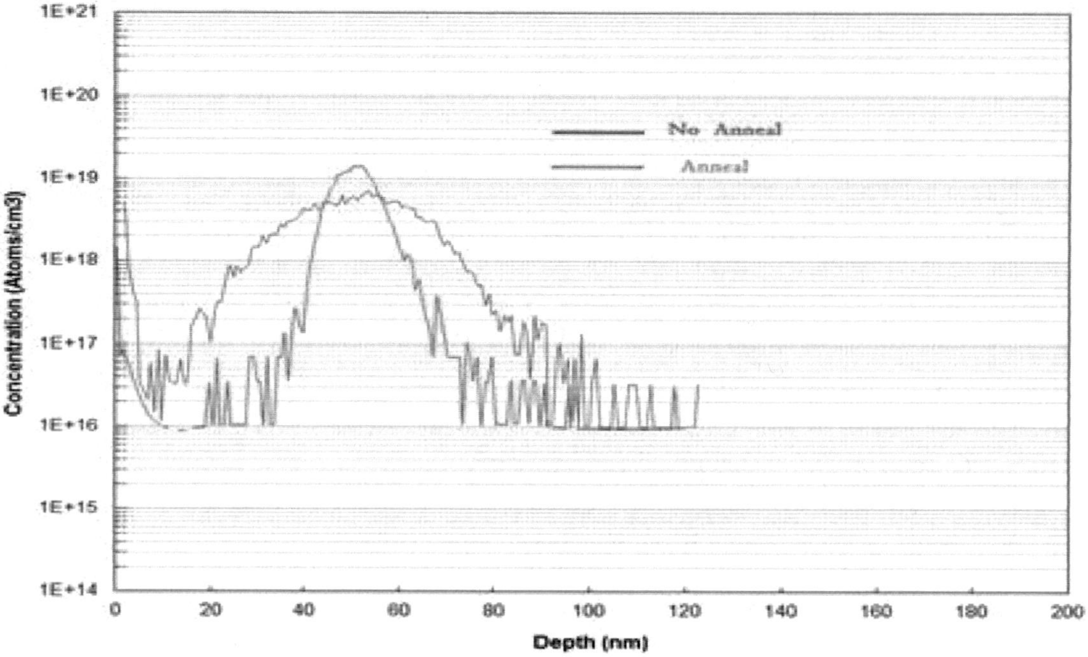

Figure 2. Sample without C doping

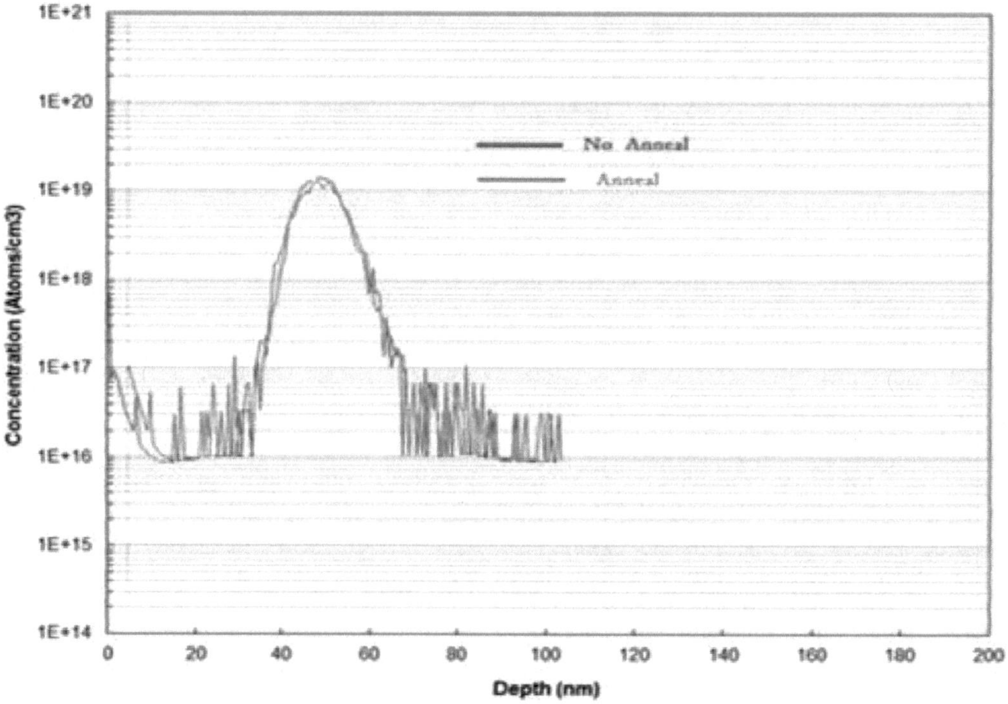

Figure 3. Sample with 0.1% C doped

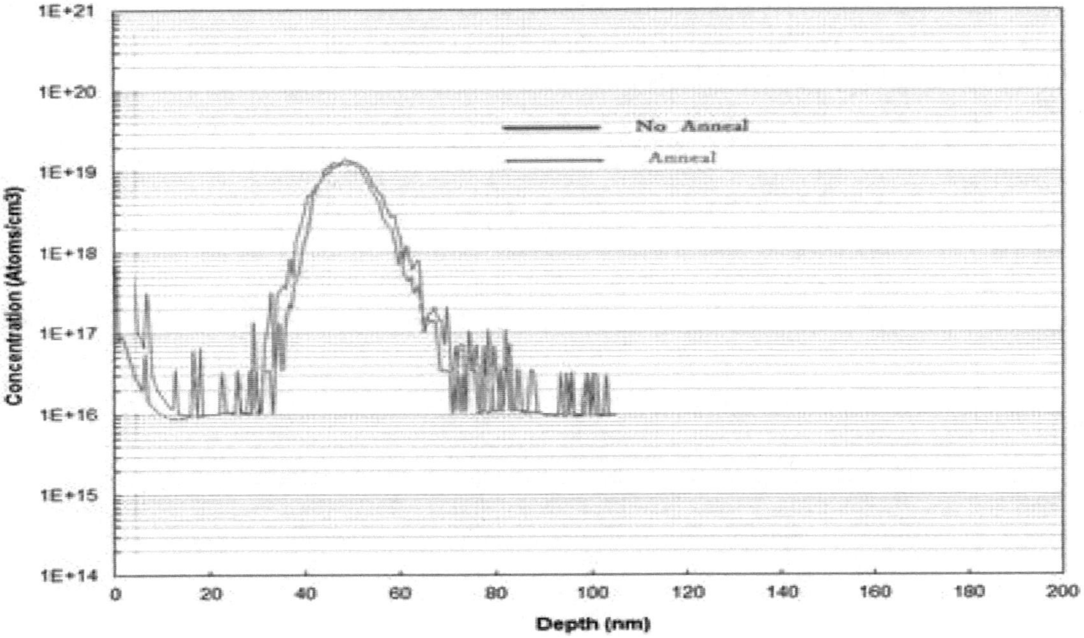

Figure 4. Sample with 0.2% C doped

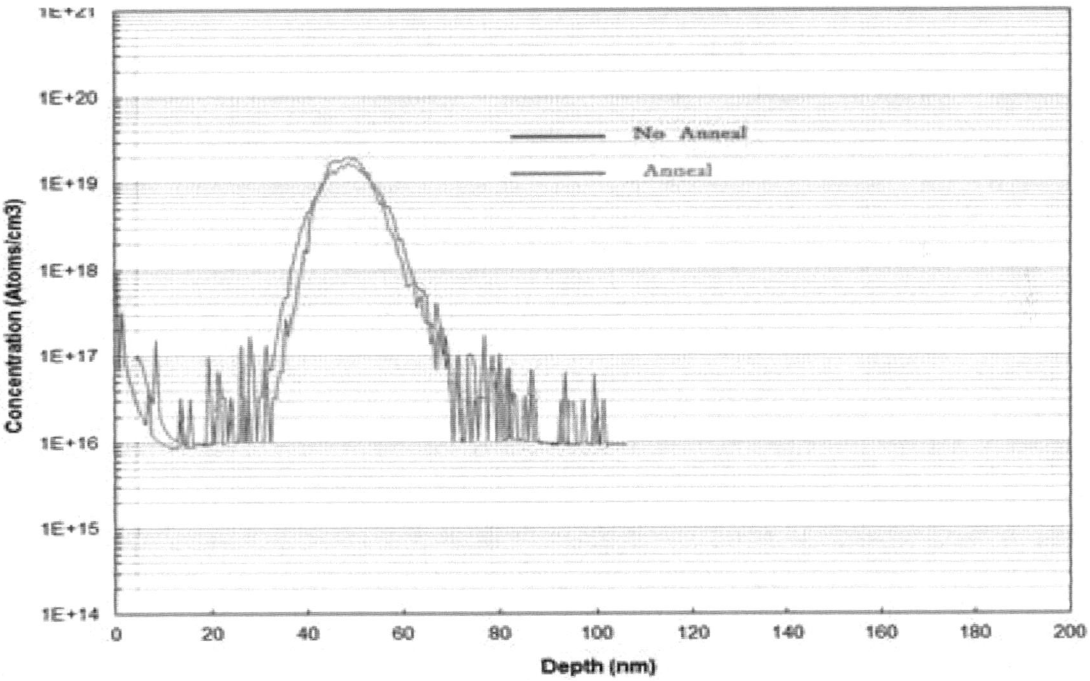

Figure 5. Sample with 0.6% C doped

2.2. B out-diffusion of C doped samples at different annealing temperatures

According to the actual situation of HBT process, 0.1% C-doped samples were annealed at different temperatures, and then the out-diffusion of B was tested. **Figure 6.** showed annealing at 1000°C for 20s, and there was almost no B out-diffusion in the sample. **Figure 7.** showed annealing at 1050°C for

20s, and there was about 10% B out-diffusion due to FWHM expansion. When the annealing temperature is 1100 °C with duration 20sec, the out-diffusion of B is up to nearly 200% as result of FWHM expansion, as shown in **Figure 8.**

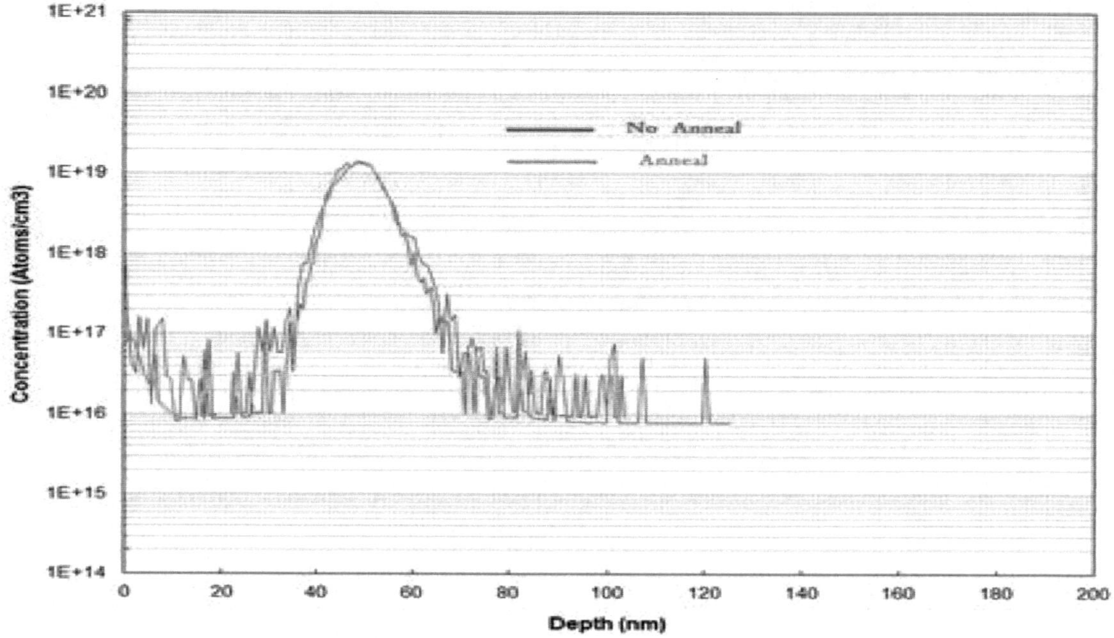

Figure 6. Annealing at 1000°C for 20s

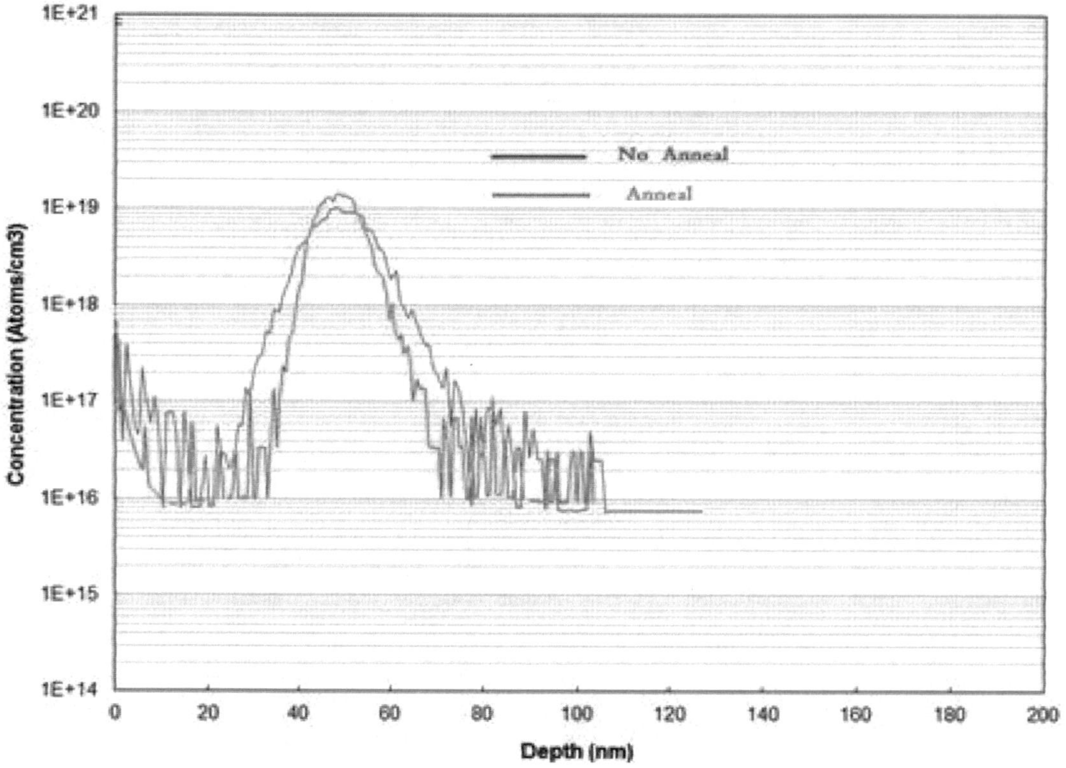

Figure 7. Annealing at 1050°C for 20s

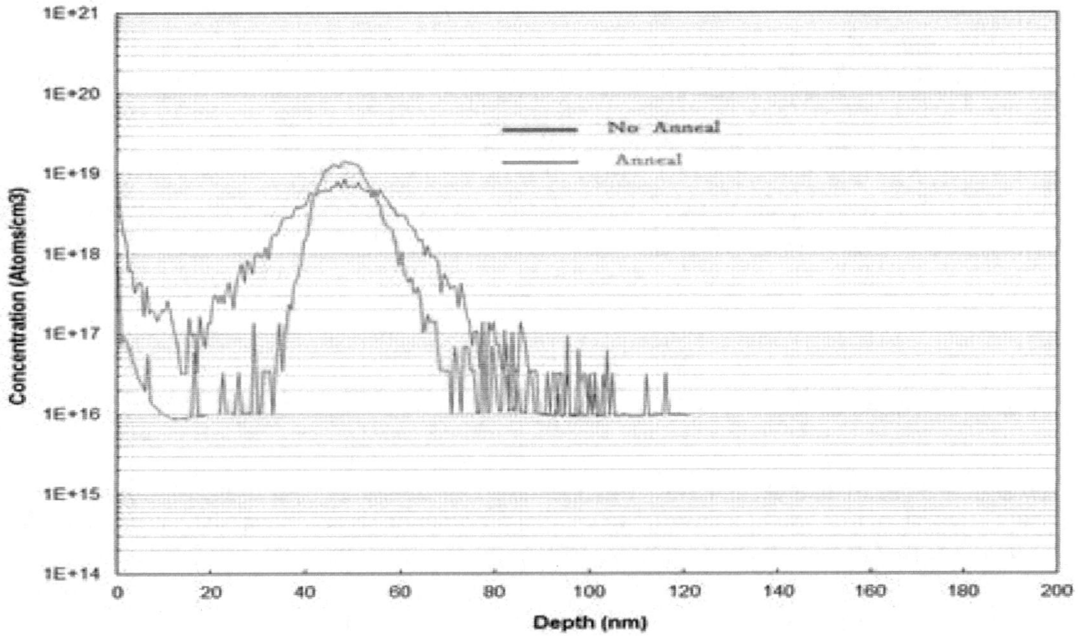

Figure 8. Annealing at 1100°C for 20s

2.3. The strain compensation regulation of C incorporation

X - ray diffraction (XRD) analysis was performed on samples with different C contents. **Figure 9.** showed the XRD rocking curve of SiGe samples with C content of 0, 0.1%, 0.2% and 0.6%. It can be seen from Fig.9 that, as the C atom radius is smaller than Si, Ge atom radius, the incorporation of C compensates the lattice mismatch between Si and Ge, and with increase in C doping, the compensation for the strain of SiGe increases.

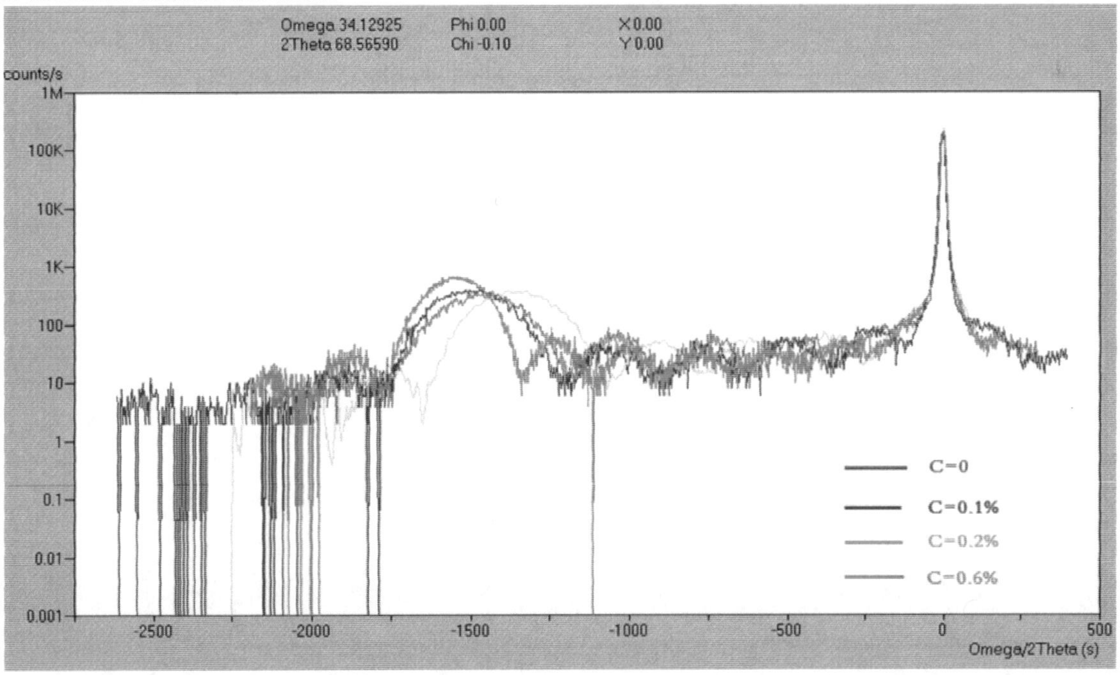

Figure 9. XRD rocking curve of SiGe samples with C content of 0.1%0.2%, 0.6% and 0

IOP Conf. Series: Materials Science and Engineering **281** (2017) 012008 doi:10.1088/1757-899X/281/1/012008

3. Conclusions

The inhibitory effect of different C doping on the out-diffusion of B impurity in SiGe and the out-diffusion of B impurity in SiGe at different temperatures were investigated. The results showed that more than 0.1% C incorporation can effectively and fully inhibit the doped B out-diffusion in SiGe in the annealing process. In the process of SiGe HBT devices, the dopant of the emitter was usually activated by rapid thermal annealing. The experiments demonstrated that there was no serious B out-diffusion effect in C-doped SiGe in rapid annealing at 1050°C for 20s. Therefore, incorporation of C greatly improves the heat budget of the process. Meanwhile, as the C atom radius is smaller than Si, Ge atom radius, the incorporation of C compensates the lattice mismatch of SiGe and enhances the thermal stability of SiGe.

References

[1] Babich F, Orlando M D', Vatta F. A comprehensive study of boron and carbon diffusion models in SiGeC heterojunction bipolar transistors, *Journal of Applied Physics*, **6** (2005) 063530-063530-9

[2] Barbalat B, Schwartzmann T, Chevalier P, Vandelle B, Rubaldo L. The effect of carbon on neutral base recombination in high-speed SiGeC heterojunction bipolar transistors. *Semiconductor Science & Technology*, **1** (2007) S99-S102(4)

[3] Hållstedt J, Parent A, östling M, Radamson H H. Incorporation of boron in SiGe(C) epitaxial layers grown by reduced pressure chemical vapor deposition. *Materials Science in Semiconductor Processing* **8** (2005) 97–101

[4] Maya L, Saïda L, Christian G. Mathematical approach and optimisation of nanometric base thickness for a SiGeC HBT dedicated to radiofrequency applications. *Journal of Computational and Applied Mathematics* **259** (2014) 925-936

IOP Conf. Series: Materials Science and Engineering **281** (2017) 012009 doi:10.1088/1757-899X/281/1/012009

Study on Adsorption Mechanism of Acid-Modified Bentonite

L G Peng, J H Zhu*, J Han, X X Wang, M Zhao, H Q Cheng and X Zhang

School of Material Science and Engineering, Xi'an University of Science and Technology, Xi'an, China 710054

*Corresponding email: 18192393026@163.com

Abstract: The acid-modified bentonite was prepared by sulfuric acid acidification, we compared it in chemical composition, layer structure and adsorption performance before and after the adsorption of Cu2+ by joint analysis of FT-IR, XRD, and ICP. The results indicated that: XRD showed that Cu2+ went into the bentonite layer, the analysis of ICP showed: Cu2+ exchange reaction didn't take place, only coordination reaction occurred, i.e. acid-modified bentonite adsorbs cooper irons mainly in a way of cooperation reaction; the adsorption of acid-modified bentonite conforms to the Langmuir monolayer model and the pseudo-second-order kinetic model.

1. Introduction

It's been a very interesting issue to remove heavy metal irons from polluted waste water in environment field. A number of studies on the removal of Cu^{2+} from waste water or aqueous solutions using different kinds of adsorbents have been reported [1-3].

The removal of heavy metals from contaminated wastewater is a matter of great interest in the field of environmental restoration, especially in this era of environmentally sustainability. Yu Ruilian carried out a study with the acid-modified bentonite prepared by activation modification of natural bentonite with sulfuric acid, it was found that under the suitable condition of removing Cu^{2+}, the removal ratio of Cu^{2+} by acid-modified bentonite can reach greater than 99.9%[4].

However, it lacked the systematic research and effective method of characterization on the ion structure, electronic effects, as well as the framework in the adsorption. In this paper, we studied the adsorption behavior and mechanism, and characterized the relationship between the adsorption process and material structure effect with methods of FT-IR, XRD, and ICP, etc.

2. Experimental

1.1. Preparation of the acid-modified bentonite

A layer of purified bentonite was added at an ark with thickness less than 1 cm, and then was baked in a resistance furnace at 100°C for 3.5h. After that, it was dispersed in a boiling flask-3-neck by adding sulfuric acid solution (ρ=15%) and keeping solid-to-liquid ratio of 1:3, and continued to stir at 95°C for 8h, then centrifuged and rinsed several times with water until pH reached 4-5. Afterwards, the solution was kept at 120°C for 2h for activation. Finally, it was dried at 105°C, ground and sieved to 120 meshes.

1.2. Characterizations of adsorbents
The FT-IR spectra were obtained (KBr) using a Nicolet 5700 Fourier Transform Infrared pectrometer (Thermo Nicolet Ltd, USA) to observe the surface functional groups at ranging from 400 to 4000 cm^{-1}. The XRD patterns of the sample before adsorption and adsorption were obtained with an X-ray diffract meter (X'TRA, ARL Co. Ltd, Switzerland) in the scanning range of 2θ =0°-90°. The elements analysis of the solution was performed by using Inductively Coupled Plasma generating apparatus (ICP).

1.3. Adsorption studies
All batch Cu^{2+} adsorption experiments were carried out in conical flask. First, a series of different concentrations of copper ion solution were prepared. A certain amount of copper ion solution was placed in a 50mL cuvette respectively, diluted to 25mL to measure the absorbance, and the working curve of the solution was prepared. Afterwards, the acid-modified bentonite and copper ion solution were added to the conical flask, and mixed on a shaker at certain speed. At pre-determined time intervals, certain amount of the mixture was taken to the cuvette, diluted to 25mL to measure the absorbance. The adsorption data of copper ions in acid-modified bentonite were calculated using the working curve. Then copper ions were adsorbed on the bentonite adsorption material at room temperature (20°C) with copper ion concentration as a variable, and the adsorption data of copper ions were calculated.

3. Result and discussion

1.4. FT-IR analysis
Figure 1 shows that in the FT-IR spectra of the acid-modified bentonite sample, the position and shape of the band are slightly changed. Stretching vibration originated from AlAlOH, AlFeOH and AlMgOH form broad band at about 3620 cm−1[4]. The presence of three bending vibrations of hydroxyl groups (AlAlOH at 920 cm−1, AlFeOH at 885 cm−1 and AlMgOH at 845 cm−1) indicated that Al3+ was substituted by Fe2+ or Mg2+ in octahedral layer, respectively. The adsorption mechanism of the acid-modified bentonite on the metal ions is as follows: sulfate acid acts on positive ions between bentonite first, then acts on positive ions in octahedron. The dissolution of calcium, magnesium, aluminum, iron and other cations in the interlayers and octahedrons causes the bentonite to become a skeleton with a lot of voids and enhance the electronegativity of acid-modified bentonite.

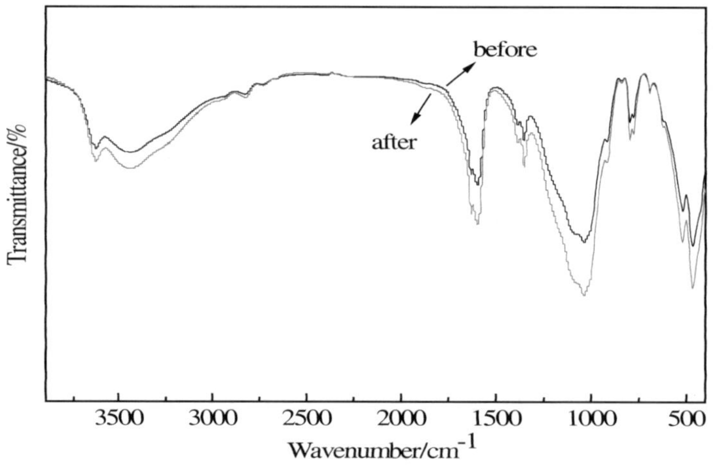

Figure 1. FT-IR spectrogram for acid-modified bentonite before and after adsorption

From Figure 1 we know that acid-modified bentonite adsorption before the wave number of 472cm^{-1}, the absorption wave number after adsorption of 468cm^{-1}. Radius of aluminum ions in alumina octahedral sheet is 0.62 and radius of magnesium ions is 0.74. Nevertheless, radius of cooper ions is 0.96, which proves cooper ions enter alumina octahedral sheet aluminum (magnesium)-oxygen octahedron lattice, then forms coordination complex with magnesium hydroxyl or aluminum hydroxyl and occurs coordination reaction between tiers.

1.5. XRD analysis

Bentonite in the four sections with X-ray diffraction peaks, respectively for: d(001)=(1.2-1.5)nm, 2θ=(5.88-7.36)°; d(003)=(0.4-0.5)nm, 2θ=(17.71-22.20)°; d(005)=(0.24-0.3)nm, 2θ=(29.74-37.43)°; d(060)=(0.149-0.150)nm, 2θ=(61.77-62.24)°[5]. Among them, the intensity of d(001) was the largest. The peak at 26.5°is ascribed to SiO_2.

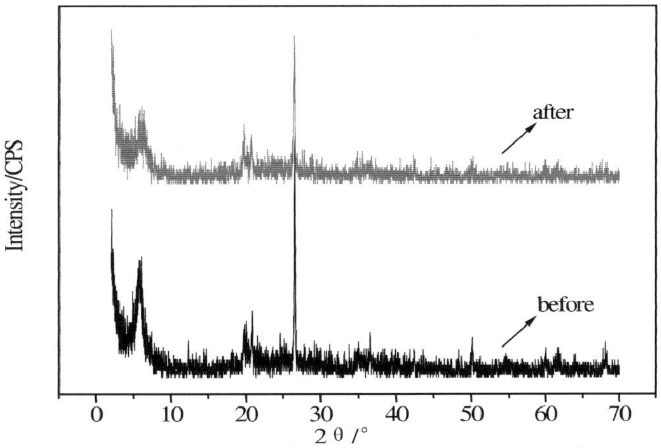

Figure 2. X-ray spectrogram for acid-modified bentonite before and after adsorption

From Figure 2 and Table 1 , after acid-modified bentonite adsorbing cooper ions, diffraction peak of (001) plane and (006) plane mainly have no excursion. But the diffraction peak of the (001) plane is much more sharp than that of the adsorbed surface, the interlayer spacing d(001) of (001) plane before and after adsorption are 1.462nm and 1.452nm separately, and the interlayer spacing d(060) of (060) plane before and after adsorption are 0.150nm. Diffraction peak of (003) plane shifted to the left after adsorption, and d(003) increased from 0.426 nm to 0.450 nm.

Table 1. the interlayer spacing of acid-modified bentonite before and after adsorption

status	interlayer spacing		
	d(001)/nm	d(003)/nm	d(060)/nm
Before adsorption	1.462	0.426	0.150
After adsorption	1.452	0.450	0.150

1.6. ICP analysis

Table 2. The contents of main ions in Cu^{2+} solution

	Cu^{2+}/(mg/L)	Ca^{2+}/(mg/L)	K^{+}/(mg/L)	Mg^{2+}/(mg/L)	Na^{+}/(mg/L)
Before adsorption	10	0	0	0	0
After adsorption	1.97346	2.21889	0.097168	0.079996	0.104356

In the case of constant volume, if copper ions are adsorbed in the form of ion exchange, when acid-modified bentonite adsorbs 1mg/L cooper ions, 1 mg/L of cation (Ca^{2+}, Mg^{2+} etc.) or 2 mg/L of cation (K^+, Na^+ etc.) can be desorbed. If copper is adsorbed in the form of surface coordination or interlayer coordination, the bentonite adsorption material desorbs water while adsorbing copper ions and does not cause an increase in cation in the solution.

As can be seen from Table 2, after adsorption, the concentration of copper ions in the acid-modified bentonite was reduced by 8.027 mg/L, and the total amount of Ca^{2+}, K^+, Mg^{2+}, Na^+ desorbed was 2.500mg/L, which indicated that the adsorption of copper on bentonite mainly relied on complexation, and only a small amount of ion exchange existed. This is consistent with FT-IR and XRD analysis.

1.7. Adsorption mechanism

(1) Kinetic research

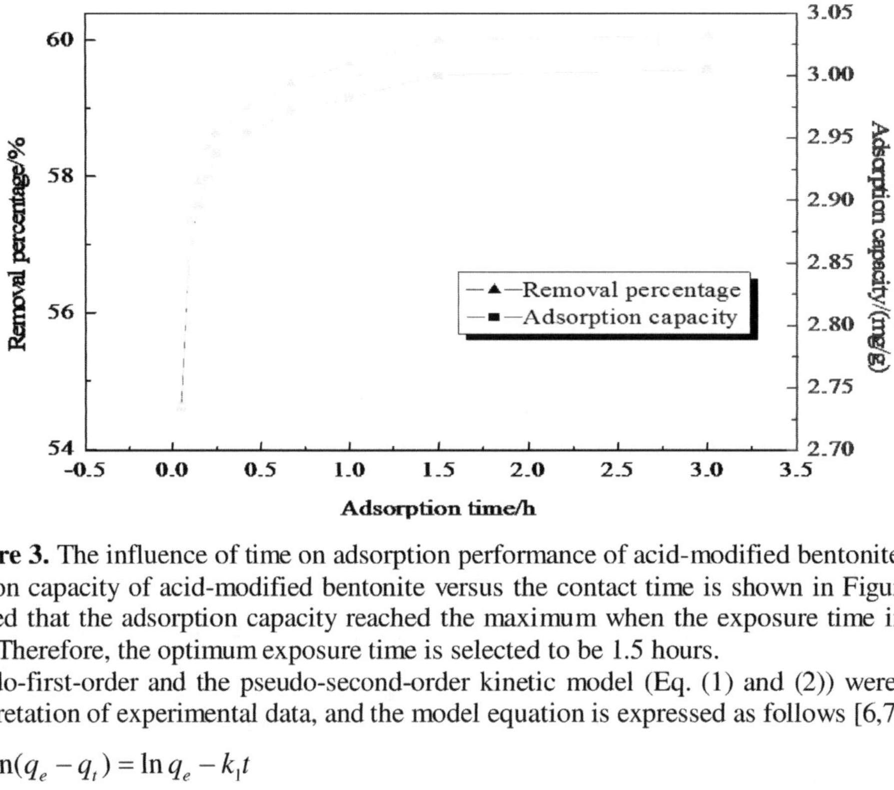

Figure 3. The influence of time on adsorption performance of acid-modified bentonite

The adsorption capacity of acid-modified bentonite versus the contact time is shown in Figure 3 The figure revealed that the adsorption capacity reached the maximum when the exposure time increased to 1.5 hours. Therefore, the optimum exposure time is selected to be 1.5 hours.

The pseudo-first-order and the pseudo-second-order kinetic model (Eq. (1) and (2)) were applied for the interpretation of experimental data, and the model equation is expressed as follows [6,7]:

$$\ln(q_e - q_t) = \ln q_e - k_1 t \tag{1}$$

$$\frac{t}{q_t} = \frac{t}{q_e} + \frac{1}{k_2 q_e^2} \tag{2}$$

Table 3. The fitting result of order kinetic equation

pseudo-first-order kinetic equation	Y=-2.09941-1.23516 X	R^2 0.85923	k_1 1.23516	q_e 0.123
pseudo-second-order kinetic equation	Y=0.00219+0.33215 X	R^2 1	k_2 50.37608	q_e 3.011

Comparing table 3, the process of cooper ions adsorbed by acidic bentonite and pseudo-second-order kinetic reach high fitting, correlation index R^2 is greater than 0.9999 , and fitted saturated adsorption capacity q_e is familiar to the capacity tested in experiment. Therefore, adsorption of acid-modified bentonite on cooper ions conforms to pseudo-second-order kinetic model.

(2) Adsorption isotherm

The Langmuir and Freundlich adsorption isotherm models were applied to describe the RB19-modified clay system. The linear forms of Langmuir (Eq. (3)) and Freundlich (Eq. (4)) adsorption isotherm models can be described with the following equations, respectively [8,9]:

$$\frac{C_e}{q_e} = \frac{C_e}{q_{\max}} + \frac{1}{K_L q_{\max}} \tag{3}$$

$$\lg q_e = \lg K_F + \frac{1}{n}\lg C_e \tag{4}$$

Table 4. The linear fitting result of adsorption isotherm

Langmuir	Y=0.59208+0.13475 X	R^2 0.99701	q_{max} 7.42	K_L 0.2276
Freundlich	Y=0.29817+0.34490 X	R^2 0.84036	K_F 1.3474	n 2.90

Comparing table 4, the correlation index R^2 in linear fitting of Langmuir isothermal adsorption equation is larger than which in Freundlich isothermal adsorption equation, which means that Equilibrium data were fitted well to the Langmuir isothermal model in the studied concentration range of copper ion.

Known from above, adsorption of acidulated bentonite conforms to Langmuir adsorption model means it is adsorption of monomolecular layer. Saturated adsorption capacity of single molecular tier of acid-modified bentonite is 7.42mg/g and limit of adsorption capacity of bentonite is saturated adsorption capacity of monomolecular layer. This illustrates saturated adsorption capacity doesn't increase as the time extend, which indicates that adsorption of acid-modified bentonite to cooper ions is mainly chemical adsorption [10].

4. Conclusions

This study carries out study on the adsorption of copper ions in wastewater by acid-modified bentonite, and the results are as follow:

1) Combining results of chemical composition, interlayer structure and link structure of acid-modified bentonite which be researched by FT-IR, XRD, with ICP indicate that the adsorption of cooper ions by acid-modified bentonite mainly depends on surface and interlayer coordination.
2) Saturated adsorption capacity of acid-modified bentonite is 7.42mg/g.
3) The adsorption of cooper ions by bentonite adsorbent fits the pseudo-second-order kinetics equation and Langmuir adsorbing model, which means the adsorption, is monomolecular layer adsorption and mainly be chemical adsorption.

References

[1] Zhao G X, Zhang H X, Fan Q H, et al. Sorption of Copper(II) onto Super-adsorbent of Bentonite-polyacrylamide Composites[J]. 2010 *J. Hazard. Mater.* **173** pp 661–8.

[2] Daniel A. Glatstein, Franco M. Francisca. Influence of pH and ionic strength on Cd, Cu and Pb removal from water by adsorption in Na-bentonite [J]. 2015 *Appl. Clay . Sci.* **118** pp 61–67.

[3] Xiaoying Jina, Min Zhenga, Binoy Sarkar, et al. Characterization of bentonite modified with humic acid for the removal of Cu (II) and 2,4-dichlorophenol from aqueous solution[J]. 2016 *Appl. Clay. Sci.***134** pp 89-94.
[4] Yu R L , Hu G R.. Experimental Study on Treatment of Wastewater Containing Cu^{2+} by Acid-Modified Bentonite [J]. 2004 *Non-Metallic Mines.* **27** pp 48-49. (In Chinese)
[5] A.V.Kostin, L.V.Mostalygina, O.L.Bukhtoyarov. The mechanism of adsorption of zinc and cadmium ions onto bentonite clay[J]. 2015*Prot. Met. Phys. Chem+.***5** pp 773-8.
[6] Ho Y.S., McKay G.. Kinetic Models for the Sorption of Dye from Aqueous Solution by Wood [J]. 1998*Process. Saf. Environ.* **76** pp 183–191.
[7] Azizian S., Kinetic Models of Sorption: a Theoretical Analysis [J]. 2004*J. Colloid Interf. Sci.* **276** pp 47–52.
[8] Langmuir I., The Adsorption of Gases on Plane Surfaces of Glass, Mica and Platinum [J]. 1918 *J.Am. Chem. Soc.* **40** pp 1361–403.
[9] Freundlich,U H.M.F. Die Adsorption in Lusungen[J]. 1906*J. phys. chem.* **57** pp 385–470.
[10] B.I.Olu-Owolabi, D.B.Popoola, E.I.Unuabonah. Removal of $Cu^{2+}$and $Cd^{2+}$from Aqueous Solution by Bentonite Clay Modified with Binary Mixture of Goethite and Humic Acid[J].2010 *Water. Air. Soil. Poll.***1** pp 459–74.

Electron Localization States in Asymmetric Shape Carbon Nanotubes Caused by Hydrogen Adsorption

L J Pan* and W G Chen

College of physics and electronic engineering, Zhengzhou Normal University, Zhengzhou, China

E-mail: *lijunpan515@163.com

Abstract. In this paper, we presented pseudopotential-based density functional theory studies on energy, structure, energy band structure of hydrogenated single-walled carbon nanotube. The stability of the configuration mainly depends on hydrogen coverage. According to the adsorption energies, the stability deteriorates with the increase of the hydrogen adsorption. The cross section of configurations become various shapes such as "beetle" or "lip" appearance without the balanced effects of hydrogen atoms. We also explored the energy band structures of configurations in three typical adsorption patterns, showing that the disparate trends of energy band gap as the hydrogen atoms concentrate. For $C_{32}H_{24}$, the band gap may reach the large value of 2.79 eV for the adsorption pattern A configuration and reduce to be zero for the adsorption pattern C case, the values of band gap for pattern A configurations decrease, which is opposite of the pattern B configurations as the adsorption hydrogen becomes more disperse. It is deduced that the hydrogen adsorption has significant effect on the electrical properties of the carbon nanotube.

Keywords: carbon nanotube; hydrogen; density functional theory

1. Introduction

Hydrogen has been considered as one of best fuels in the future, which may be free of pollution and substitute for fossil fuels. For practical application with hydrogen, the U.S. Department of Energy (DOE) has estimated up to 6.5 mass% of hydrogen (density of 62.5 kg/m3), however, the best materials such as TiV2 can storage only up to 2.6 mass% at experiment condition [1]. Recently, researches claimed that carbon nanostructure materials in the form of tubes [2-7], graphite [8], and fibers [9-11], may be used for efficient hydrogen storage. Ye et al. used high-purity single-walled carbon nanotubes and obtained 8.0mass% H_2 adsorption at 80 K above 100 bars [2]. Dollin et al. found that single wall carbon nanotube (SWNT) soots could absorb about 5 to 10 weight % of H_2 at 133 K and 300 torr [4]. Browning et al. reported that 6.5mass% hydrogen can be stored in carbon nanofibers under conditions of 120 bars pressure and ambient temperature [9]. Chambers et al. observed that at 120 atmospheric pressure and room temperature, graphite nanofibers with herringbone structure could store 67 weight % of H_2 [11]. Chen et al. reported a H_2 storage system with which H_2 uptake can achieve 20 weight% for Li-doped CNT 653 K or 14 weight% for K-doped CNT at room temperature [5]. But later researches revealed that the high H_2 uptakes in Ref. 9 were due to the impurity water present in the hydrogen feedstream rather than to H_2 [6-7].

Besides physisorption, chemisorption has also been proposed for hydrogen storage in carbon nanotubes. The C-H bond forms when atomic hydrogen adsorbs on SWCNT being shown by studies [12,13]. Recently, theoretical researches [14-20] reported that high coverage up to 7.5 weight% storage can be achieved through chemisorption on the nanotubes. Yang et al. found that the storage capacity of hydrogen depends on the diameters of carbon nanotubes and the hydrogen coverage up to 100% for is stable the SWCNT tube [19]. Nikitin et al. showed approximately 5.1±1.2 weight % hydrogenation of SWCNTs with diameters ranging from 1 to 1.8 nm has been realized, which is a reversible process [20]. The stable structures and mechanically properties of the hydrogenated nanotubes which are necessary to practical applications also are meaningful for exploration. In effect, the stability and band structure of the hydrogenated nanotubes for different adsorption pattern and hydrogen coverage is needed to be investigated. The effect of distribution of hydrogen atoms under the same adsorption pattern and coverage is interesting. This will provide the comprehensive understanding of the effect of hydrogen adsorption on the carbon nanotube.

2. Methods

In this paper, we investigated the structural and electrical properties of hydrogenated single-walled carbon nanotube, using SIESTA (Spanish Initiative for Electronic Simulations with Thousands of Atoms) [21-23] approach within local density approximation (LDA). On a mesh with an energy cutoff of 210 Ry, we adopted the variable cell conjugate gradient algorithm to relax the structures. We set the force convergence criterion on each atom to be 0.005 eV/Å, and the stress tolerance was 0.01 GPa. The intertubular distance is 10 Å, which is large enough to eliminate the interplay for neighboring carbon nanotubes. We also obtained the energy band structures of optimized geometries for different coverage in order to compare the effect of hydrogen adsorption at different levels.

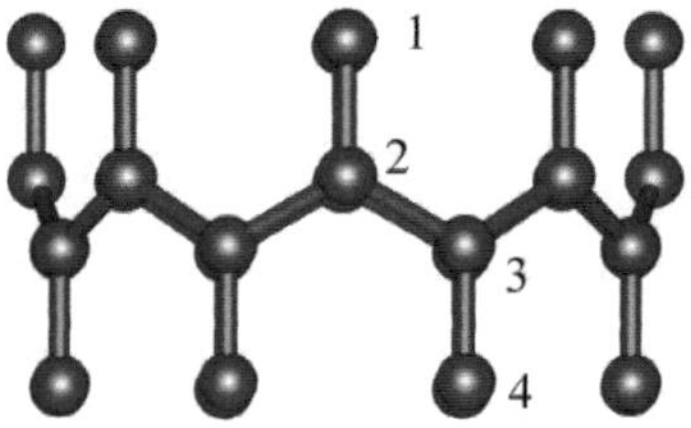

Figure 1. Side view of the SWCNT with the sites of atomic hydrogen labelled.

Previous studies indicated that the hydrogen atoms might prefer to form dimer and chain along the axis of the tube [15, 19]. The sites of atomic hydrogen on the wall of carbon nanotube are showed in Figure 1. The hydrogen atoms trend to form hydrogen dimer [15, 19]. We focused on the specific coverage on nanotubes, considering three typical patterns: pattern A, the vertical chain pattern [24]; pattern B, the dimer pattern [24]; pattern C, the parallel polyacetylene-like chains pattern [25]. The hydrogen atoms adsorb 1-2 sites, 2-3 sites and 1-2-3-4 sites forming pattern A, pattern B and pattern C, respectively. Herein, the formulas expressing the atom number per unit of the hydrogenated SWCNTs are $C_{28}H_8$, $C_{28}H_{16}$, $C_{28}H_{20}$, $C_{28}H_{28}$, $C_{32}H_8$, $C_{32}H_{16}$, $C_{32}H_{24}$, $C_{32}H_{32}$, $C_{36}H_8$, $C_{36}H_{20}$, $C_{36}H_{28}$, $C_{36}H_{36}$ and so on. For $C_{32}H_{16}$, we considered nine hydrogenated configurations in one pattern A, one pattern B and seven pattern C configurations. The adsorption energy per hydrogen atom is defined as below

$$E_{ad} = \left(E_{H-SWCNT} - E_{SWCNT} - n_H E_H\right) / n_H$$

Where E_{SWCNT} is the total energy of carbon nanotube, E_H is the total energy of an isolated H atom, $E_{H-SWCNT}$ is the energy of the hydrogenated nanotube, n_H is the number of hydrogen atom. The low value of E_{ad} indicates the high stability.

3. Results and discussion

3.1. The adsorption energies and structures of the hydrogenated SWCNTs.

For the comparison of stability, adsorption energies per hydrogen atom of hydrogenated carbon nanotube (8,0) at different coverages are plotted in Figure 2, the dotted line shows the global tendency of the adsorption energy. We found that the length of C-H bond is 1.11-1.13 Å, the length of C-C bond ranges from 1.40 to 1.57 Å and the adsorption energies increase linearly as the hydrogen coverage increased, which is in agreement with previous reports [15], indicating the weak stability. The configurations with a small quantity of hydrogen atoms show high stability, such as $C_{32}H_8$, the bottom left in Figure 2, have the lowest value in our calculated structures.

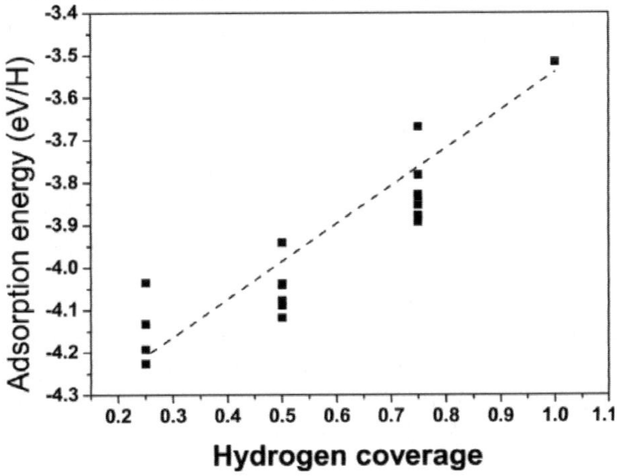

Figure 2. The adsorption energies per hydrogen atom at three adsorption coverages

It is noted that the adsorption energies belonged to three patterns have a variation at the same coverage as shown in Figure 2. The adsorption configurations in pattern C have obvious fine performances which have presented in our studies [26, 27]. In addition, there are two transitions in pattern C configurations [26, 27] and the structures with the hydrogen atoms in disperse manner having lower adsorptions energies. It indicates that hydrogen atoms do not prefer to form clusters on the surface of SWCNT. The diameters of hydrogenated SWCNTs vary with the hydrogen coverage. For the coverage of 100%, the diameter of circular cross section of the adsorption configuration is 7.280 Å which is larger than that of fully relaxed geometry of nanotube (8,0). This is caused by the forming of C-H bond with sp3 hybridization, which is in agreement with report [28], the∠CCH bond angles ranges from 104.1o to 111.3o which are close to the ideal tetrahedral sp3 bond angle of 109.5o. The hydrogen atoms are not symmetrical distributed, thus the cross section of stable configurations becomes collapsed due to the carbon atoms relatively move towards inside.

The shapes of the hydrogenated carbon nanotube are listed in Figure 3, it is seen that the structure deformation are induced by the adsorption. For high hydrogenation, the configurations in pattern A are listed in Figure 3a and Figure 3e, the configurations in pattern C are showed in Figure 3b, Figure 3f and Figure 3h, the atomic hydrogen stretches outward leading to “beetle” shape. For low hydrogenation, as shown in Figure 3c, Figure 3d and Figure 3g, the structures form in the shape of “lip”. It is deduced that the hydrogen adsorption affects obviously the structure of nanotube for all adsorption patterns. The asymmetric distribution of hydrogen atoms affects the structure largely without the balanced effects of hydrogen atoms.

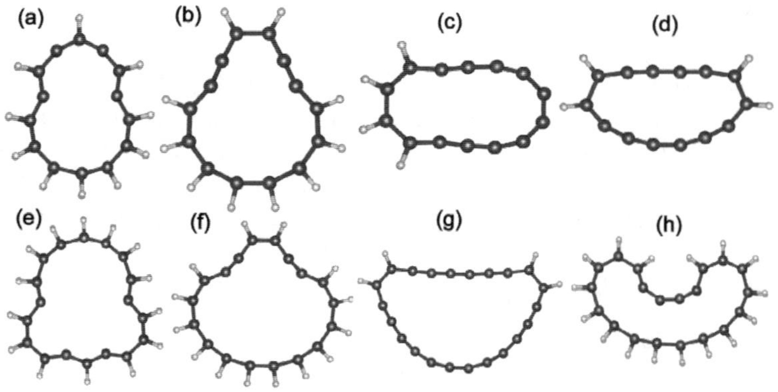

Figure 3. The top views of the hydrogenated nanotubes for (a) $C_{28}H_{20}$ configuration in pattern A, (b) $C_{28}H_{20}$ configuration in pattern C, (c) $C_{28}H_8$ configuration in pattern C with hydrogen atoms together (d) $C_{28}H_8$ configuration in pattern C with disperse hydrogen atoms, (e) $C_{36}H_{28}$ configuration in pattern A, (f) $C_{36}H_{28}$ configuration in pattern C, (g) $C_{40}H_8$ configuration in pattern C, (h) $C_{40}H_{32}$ configuration in pattern C.

3.2. The energy band structures of the hydrogenated SWCNTs.

We have presented the electrical properties of hydrogenated SWCNTs and discussed two transitions [26]. In this paper, the energy band structures of configurations in pattern A, pattern B and pattern C were investigated, the variations related with the disperse extent of atomic hydrogen were also discussed. For example, as shown in Figure 4, the typical three band structures in pattern A with the formula as $C_{36}H_{28}$ were chosen and plotted.

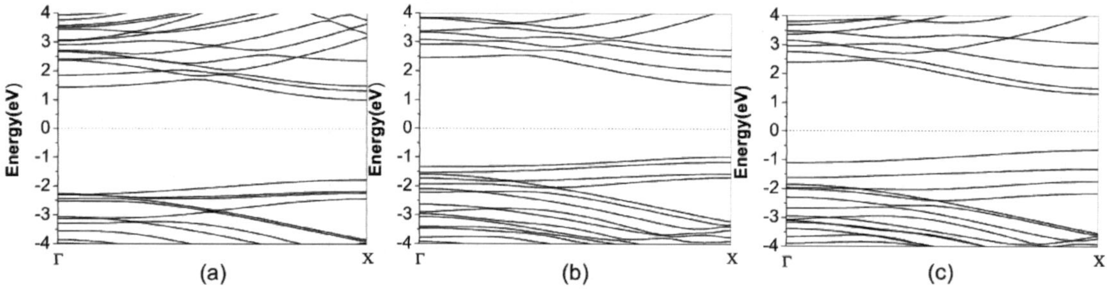

Figure 4. The three band structures of $C_{36}H_{28}$ in pattern A.

Firstly, the adsorbed hydrogen atoms are gathered, the energy bands of configuration are depicted in Figure 4a, which has a band gap of 2.79 eV. The value of gap has reduced to 2.51 eV as the adsorbed hydrogen atoms divide into two parts with a small interval showed in Figure 4b. If the interval is further enlarged, the gap may continue to decrease to 1.97 eV in Figure 4c. In summary, the band gap has the largest value for configuration with hydrogen atoms together, and decreases as the hydrogen atoms are apart. The analogous results are found in all of the hydrogenated carbon nanotubes with the adsorbed manner of pattern A. But this is not applied to the case of hydrogenated carbon nanotubes in pattern B and pattern C.

Figure 5 shows the band gap values of three patterns in the formula of $C_{32}H_{24}$. As shown, there is an opposite tendency for pattern B configurations compared with pattern A ones. It is found that the value of band gap decreases as the hydrogen atoms are together, which is reduced from 2.66 eV for all hydrogen atoms disperse as far as possible to 1.84 eV for the hydrogen atoms close together. For pattern C configurations, the band gap value also decreases as the degree of dispersing increases except two values signed in dotted circle in Figure 5 corresponding to the transitions [26, 27]. It is shown that the band gaps for zigzag nanotubes are sensitive to the adsorption manner and the distribution of the atomic hydrogen. It shows that the value of band gap may be controlled by adjusting the adsorption manner or the distribution of the hydrogen in the certain adsorption manner, the gap even may be zero in pattern C configurations. The similar conclusions are obtained from the other hydrogenated carbon nanotubes with different coverage.

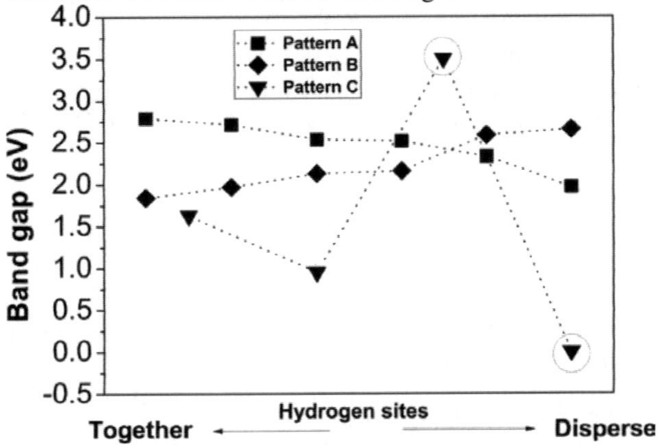

Figure 5. The band gap values of three patterns in the formula of $C_{32}H_{24}$.

In general, the stability of hydrogenated nanotubes deteriorates with the increasing of the hydrogen adsorption, but to some extent, the manner of adsorption may improve it. On the other hand, the electrical properties may be designed by choosing the manner of adsorption and the distribution of hydrogen atoms in forming of structure constriction. Therefore, our calculations showed that the effect of hydrogenation may be comprehensive treated in the aspects of stability and band structure.

4. Conclusions

We researched the stable adsorption configurations by altering the hydrogen coverage, and found that the adsorption energy ascends with the increasing of the hydrogen adsorption, indicating the poor stability. The adsorption pattern and the distribution of the atomic hydrogen may affect the adsorption energy and structural deformation. The cross section of configurations become various shapes such as "beetle" or "lip" appearance without the balanced effects of hydrogen atoms. The electrical properties of hydrogenated nanotubes demonstrate varied trends as the distribution and adsorption pattern have altered. For $C_{32}H_{24}$, the band gap may reach the large value of 2.79 eV in the adsorption pattern A configuration and reduce to be zero in the adsorption pattern C case, the values of band gap for pattern A configurations decrease, which is opposite of the pattern B configurations as the adsorption hydrogen becomes more disperse. It provides the possibility that the band gap may be controlled by adjusting the adsorption pattern and the distribution of the atomic hydrogen.

References

[1] Schlapbach L, Zuttel Z 2007 Hydrogen-storage materials for mobile applications *Nature (London)* 414 353-8
[2] Ye Y, Ahn C C, Witham C and Fultz B 1999 Hydrogen adsorption and cohesive energy of single-walled carbon nanotubes *Appl. Phys. Lett.* 74(16) 2307-2309.
[3] Hirscher M et al. 2001 Hydrogen storage in sonicated carbon materials *Appl. Phys. A* 72 129-132.
[4] Dillon A C, Jones K M, Bekkedahl T A, Kiang C H, Bethune D S and Heben M J 1997 Storage of hydrogen in single-walled carbon nanotubes *Nature (London)* 386 377-379.
[5] Chen P, Wu X, Lin J and Tan K L 1999 High H2 Uptake by Alkali-Doped Carbon Nanotubes under Ambient Pressure and Moderate Temperature *Science* 285 91-93.
[6] Pinkerton F E, Wicke B G, Olk C H, Tibbetts G G, Meisner G P, Meyer M S and Herbst J F 2000 Thermogravimetric Measurement of Hydrogen Absorption in Alkali-Modified Carbon Materials *J. Phys. Chem. B* 104 9460-9467.
[7] Yang R T 2000 Hydrogen storage by alkali-doped carbon nanotubes-revisited *Carbon* 38 623-626.

IOP Conf. Series: Materials Science and Engineering **281** (2017) 012010 doi:10.1088/1757-899X/281/1/012010

[8] Orimo S, Majer G, Fukunaga T, Züttel A, Schlapbach L and Fujii H 1999 Hydrogen in the mechanically prepared nanostructured graphite *Appl. Phys. Lett.* 75 3093-3095.

[9] Browning D J, Gerrard M L, Lakeman J B, Mellor L M, Mortimer R J and Turpin M C 2002 Studies into the storage of Hydrogen in Carbon Nanofibers: Proposal of a Possible Reaction Mechanism Nano *Lett.* 2 201-205.

[10] Ahn C C, Ye Y, Ratnakumar B V, Witham C, Bowman R C and Fultz B 1998 Hydrogen desorption and adsorption measurements on graphite nanofibers *Appl. Phys. Lett.*73 3378-3380.

[11] Chambers A, Park C, Baker R T K and Rodriguez N M 1998 Hydrogen Storage in Graphite Nanofibers *J. Phy. Chem. B.* 102 4253-6

[12] Ruffieux P, Gröning O, Bielmann M, Mauron P, Schlapbach L and Gröning P 2002 Hydrogen adsorption on sp2-bonded carbon: Influence of the local curvature Phys. Rev. B 66 245416.

[13] Nilsson A 2002 Applications of core level spectroscopy to adsorbates *J. Electron Spectr.* 126 3-42.

[14] Park S, Srivastave D and Cho K 2003 Generalized Chemical Reactivity of Curved Surfaces: Carbon Nanotube *Nano Lett.* 3 1273-1277.

[15] Park K A, Seo K and Lee Y H 2005 Adsorption of Atomic Hydrogen on Single-Walled Carbon Nanotubes *J. Phys. Chem. B* 109 8967-8972.

[16] Bauschlicher C W 2000 Hydrogen and fluorine binding to the sidewalls of a (10,0) carbon nanotube *Chem. Phys. Lett.* 322 237-241.

[17] Lee E C, Kim Y S, Jin Y G, Chang K J 2002 First-principles Study of Hydrogen Adsorption on Carbon Surfaces *Phys. Rev. B* 66 072415.

[18] Volpe M, Cleri F 2003 Chemisorption of Atomic Hydrogen in Graphite and Carbon Nanotubes *Surf. Sci.* 544 24-34.

[19] Yang X B and Ni J 2006 Calculations of Hydrogen Coverage on Single-walled Carbon Nanotubes: Dependence on Nanotube Size, Temperature, and Pressure *Phys. Rev. B* 74 195437.

[20] Nikitin A, Ogasawara H, Mann D, Denecke R, Zhang Z, Dai H, Cho K J and Nilsson A 2005 [] Hydrogenation of Single-walled Carbon Nanotubes *Phys. Rev. Lett.* 95 225507.

[21] Ordejón P, Artacho E and Soler J M 1996 Self-consistent order-N density-functional calculations for very large systems *Phys. Rev. B* 53 R10441.

[22] Sánchez-Portal D, Ordejón P, Artacho E and Soler J M 1997 Density-functional Method for very Large Systems with LCAO Basis Sets *Int. J. Quantum Chem.* 65(5), 453-461.

[23] Soler J M, Artacho E, Gale J D, Garcia A, Junquera J, Ordejón P and Sánchez-Portal D 2002 The SIESTA Method for ab initio Order-N Materials Simulation *J. Phys. Condens. Matter.* 14 2745-2781.

[24] Gülseren O, Yildirim T and Ciraci S 2002 Effects of Hydrogen Adsorption on Single-wall Carbon Nanotubes: Metallic Hydrogen Decoration *Phys. Rev. B* 66 121401.

[25] Seifert G, Köhler T and Frauenheim T 2000 Molecular Wires, Solenoids, and Capacitors by Sidewall Functionalization of Carbon Nanotubes *Appl. Phys. Lett.* 77 1313-1315.

[26] Volpe M and Cleri F 2003 Chemisorption of Atomic Hydrogen in Graphite and Carbon Nanotubes Surf. Sci. 544 24-34.

[27] Pan L J, Chen W G, Zhang R Q , Hu X and Jia Y 2010 Influence of high atomic hydrogenation on the electronic structure of zigzag carbon nanotubes: A first-principles study Chin. Phys. Lett. 27 077304.

[28] Pan L J, Chen W G, Sun Q, Hu X, Wang F and Jia Y 2011 Influence of high atomic hydrogenation on the electronic structure of zigzag carbon nanotubes: A first-principles study *physica B* 406 2772-2776.

IOP Conf. Series: Materials Science and Engineering **281** (2017) 012011 doi:10.1088/1757-899X/281/1/012011

Numerical Control Device for Preparation Nano-Carbon Granule Coating Superhydrophobic Template and Its Application

G R Shang* and Y Li

Jilin Teachers Institute of Engineering and Technology, China Changchun 130052

* Corresponding email :shgr888@163.com

Abstract. It is one of the ways for changing surface property by fabricating superhydrophibic coating with the help of template that is made of depositing nano-carbon particles of fuel flame on substrate such as pure copper or aluminium alloy. In the process of making template, it is difficult to keep the deposition layer uniformed. In this work, the problem was solved by manufacturing a set of numerical control equipment. It has been proved by application test that the deposition layer was uniformed by means of this facility. The contact angle is more than 150°. A new way has been developed for making superhydrohibic template.

1. Introduction

In the area of material surface modification, one major goal is to design and prepare self-cleaning surfaces[1-3]. Many surfaces in nature are superhydrophobic, such as the surface of the lotus leaf and butterfly wings[4,5]. Mimicking its surface topography has resulted in a considerable amount of artificial materials[6,7] and developed many applications in industrial and biological fields[8~10]. Micron-scale air "clusters" are trapped in the depressions of the rough surface beneath the water droplets[11~12]. This compound interface results in the increase of micro-contact angle and the decrease of the contact angle hysteresis, which makes the rolling of the water droplets easier and can take the pollutants away from the surface. However, organic liquids such as alcohols or oils can significantly reduce their own surface tension and induce even surface wetting. The droplet will change state from the hole above the concave (Cassie - Baxter state) to fully wet (Wenzel state)[11]. No natural surface can show that it is greater than 150° of the static contact angle and below 10° of the rolling angle at the same time for water and organic liquid. Such surface is superhydrophobic and superoleophobic[13,14].

In recent years, there have been literature reports in the world, it has been reported that using the candle soot as the medium to prepare the nanometer coating on metal or non-metal substrate. However, in the preparation process, the uniformity of the coating is difficult to guarantee. In this paper, a set of numerical control device has been designed to keep the coating uniformity. The speed regulating motor drives the mechanical gearbox and the speed reduction gear for transmission output speed. The preparation of superhydrophobic template with the substrate of pure copper, or aluminum alloy has been made by burning, nano-scale carbon particles produced by the outer flame as a medium through kerosene, soybean oil, the candle, and laid the foundation for the further preparation of super hydrophobic surface.

IOP Conf. Series: Materials Science and Engineering **281** (2017) 012011 doi:10.1088/1757-899X/281/1/012011

2. Basic parameters of the numerical control device

Precise displacements were afforded within horizontal and vertical directions by the numerical control device. The uniform motion in horizontal direction is important as the lower surface of samples which have suitable time to keep a touch with outer flame. The nano-carbon particles will have a deposition layer on the lower surface of samples. The velocity of the movement is between 3mm per minute and 60mm per minute. The maximum displacement is 300mm.The accuracy is 50micrometers.

The motion in the vertical direction is used to adjust the height of flame in order to keep the flame position.

Substrates of samples are from Aluminium alloy and pure copper or other materials. The size of sample is between $(5mm \times 10mm \times 1mm) and (50mm \times 50mm \times 1mm)$ .The mass of the load is between 0.135g and 22.25g according to the size of sample.

3. Principle of the design

The schematic diagram of the numerical control device is shown in Figure 1.

Samples13 evenly arranged is placed on board7 in test. When the programmable controller24 gives a signal, the gear is driven by stepping motor20 and rack16 is driven to make a uniform rectilinear motion. Sample layer board7 is linked with rack16 by mechanical separator system in order to do a uniform rectilinear motion. Rectangular holes are processed in board7 to match samples for keeping a touch between samples lower surface and the certain position of the flame.

The flame will gradually weaken with burning, and it makes a effect on concentration of nano-carbon particles on the lower surface of samples. The temperature of flame in certain position is detected by sensor10 for keeping it in a suitable range. When it is in a designed section, any single will be not sent; when it goes beyond a reasonable range, an order will be sent by PLC8 to stepping motor25. Ballscrew21 will be driven by it and makes a straight line motion in Z direction by means of a set of machine combination which consists of nut22, screw21 and other parts in order to keep the temperature.

Nano-carbon particles from the flame of candles, kerosene and soya-bean oil are deposited on the lower surface of samples13 in this device and the superhydroph0bic template is formed.

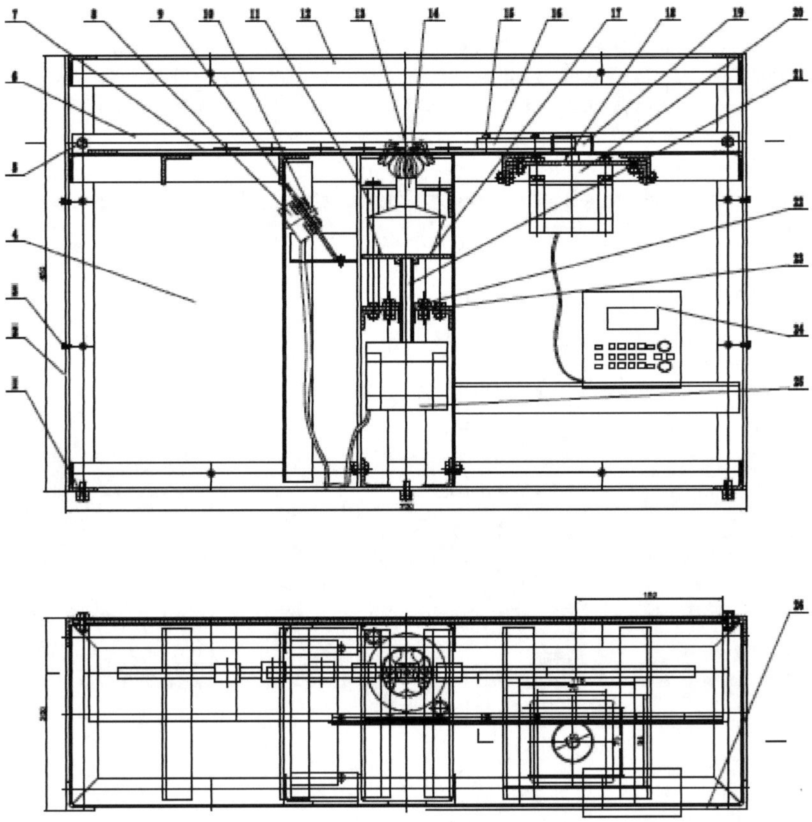

Figure 1. The movement diagram of the mechanism.

1- Horizontal adjustment nail;2- Side guard plate; 2- Side shield;3- Fixed screw; 4- Back shroud;5- Union bolt and nut combination;6- Guide strip;7- Sample board;8-PLC;9- Temperature measuring instrument mounting plate;10- Temperature transmitter;11- Guide cylinder;12- Welded truss;13- Sample;14- fuel source(for example: alcohol burner);15- Binding screw;16- Rack;17- Movable support plate;18- Motor support plate;19-Gear;20- Motor;21- Balls crew;22-Nut;23- Fixed bearing plate; 24-PLC;25-Motor;26- front track fender

4. Materials and methods

4.1. Materials

Aluminium Alloy(5A05),pure copper(TU2),candle, kerosene, soya-bean oil. Kerosene and soybean oil are placed in the alcohol lamp respectively. Lamp wick is cut neatly before burning.

4.2. Characterization

The morphology of the soot particles was characterized by Scanning Electron Microscopy (low voltage LEO 1530 Gemini, Germany, and SU8000,Hitachi, Japan).Samples were washed by high speed centrifuge(TG618,Shanghai,China), and were dried in electrothermal constant-temperature dry box(202-3A,Nanjing,China). Contact angles were tested by Optical Contact Angle Measuring instrument(DSA100,KRUSS,Germany).

4.3. Methods
The superhydrophibic deposition layer with nano-carbon particles was made by Numerical Control Device(Fabricating by Project Group). The ambient temperature stays at the rang 25 degrees Celsius. Outer flame was used when burning. Samples were placed in an airtight plastic box.

4.4. Rules of samples labeling
Each sample is marked with three set of numbers. The first set of numbers represents the fuel flame type. For example: 1- Kerosene flame; 2-Candle flame; 3-Soya-bean flame. The second set of numbers represents the base materials. For instance: 1-Pure copper substrate(TU2); 2- Aluminium Alloy(5A05). The third group represents deposition time(senond).

5. Rusult and discussion
Morphology of template and static contact angle were shown in Figure 2~4.

Figure 2.(a) 1-1-10 Top view

Figure 2. (b) 1-2-10 Top view

Figure 2. (c) 1-2-10 left view

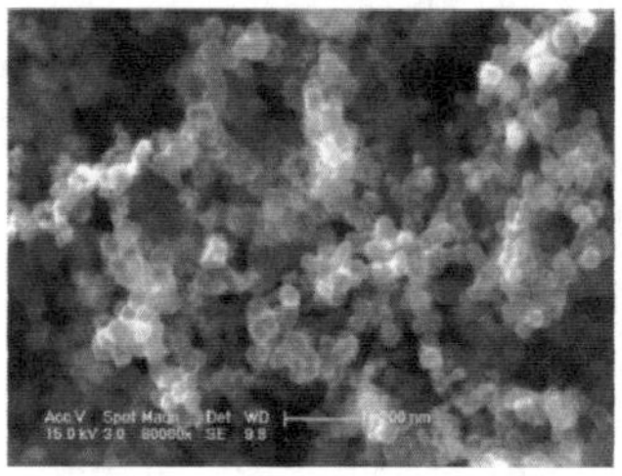

Figure 3. (a) 2-1-30 Top view

Figure 3. (b) 2-2-30 Top view

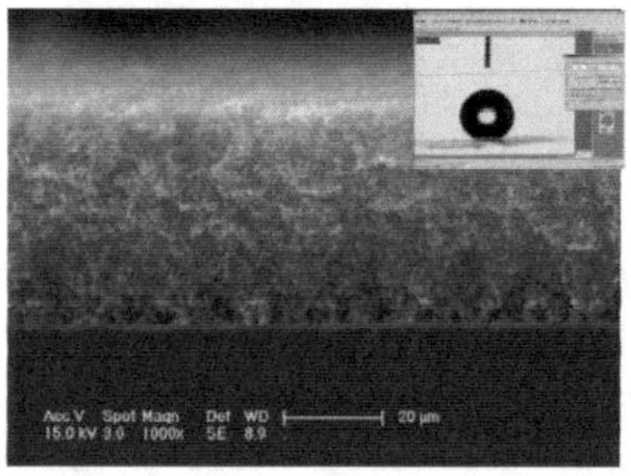

Figure 3.(c) 2-2-30 left view

Figure 4. (a) 3-1-30 Top view

Figure 4. (b) 3-2-30 Top view

Figure 4.(c) 3-2-30 left view

It is shown from morphology of SEM imagine that the diameter of carbon particles on the deposition layer is between 50 nanometer and 70 nanometer. The distribution of particles is random. The carbon particles deposited on the pure copper substrate have higher densities; The porosity of carbon particles deposited on the base of aluminum alloy is higher (Shown in Figure 2). With the deposition time increases, the density augment, the porosity reduced, and the thickness of the sediments increases synchronously (Shown in Figure 3~4).The thickness of the sediment is uniformed and is about 50~70micrometer(Shown in Figure 2~4c). It has been indicated that static contact angle is above 150°(Shown in Figure 2~4c). Because of the surface energy of aluminum alloy is lower than that of pure copper, the contact angle is higher correspondingly.

6. Conclusions and prospect

It is indicated in test that the uniform motion in the horizontal direction of the template afforded by the numerical control devices ensures the uniformity of nano-carbon particles on substrates and the static contact angle more than 150°. The superhydrophobic characteristics are presented. It is a foundation for manufacturing the superhydrophobic surface further.

Acknowledgments

The authors acknowledge the supports by the Science Research Project of Jilin Higher Education (No. 2015387) and Doctor Foundation of Jilin Institute of Teacher and technology(132016001).

References

[1] Barthlott W, Neinhuis C, *Purity of the sacred lotus, or escape from contamination in biological surfaces, Planta.* 202 (1997) 1-8. http://dx.doi.org/10.1007/s004250050096;b) Angew. Chem. 2015, 127, 3448.
[2] Nakajima A, Abe K, Hashimoto K, Watanabe T, *Preparation of hard super-hydrophobic films with visible light transmission, Thin Solid Films.* 376 (2000) 140-143.http://dx.doi.org/10.1016/S0040-6090(00)01417-6.
[3] Nakajima A, Hashimoto K, Watanabe T, Takai K, Yamauchi G and Fujishima A, *Ical Transparent.*
[4] Claudia Z,etc.Bioinspir. *Biomim.* 11 (2016) 056003.
[5] Wang L M and Thomas J. McCarthyAngew. C. *Int. Ed.* 2015, 54, 1 – 6.
[6] Gao L C, McCarthy T J, *Langmuir* 22, 2966 (2006).
[7] Deng X. et al., Adv. Mater. (*Deerfield Beach Fla.)* 23,2962 (2011).
[8] Genzer J, Efimenko K, *Science* 290, 2130 (2000).
[9] Kim S H, Lee S Y, Yang S M, Angew C. *Int. Ed.*49, 2535 (2010).
[10] Yoshimitsu Z, Nakajima A, T. Watanabe, Hashimoto K, *Langmuir* 18, 5818 (2002).
[11] Cassie A B D, Baxter S, *Trans. Faraday Soc.* 40,0546 (1944).
[12] Lafuma A, Quéré D, Nat. Mater. 2, 457 (2003).
[13] Xie Q et al., Adv. Mater. *(Deerfield Beach Fla.)* 16, 302(2004).
[14] Deng X, Lena M, Hans-Jürgen B, *Doris Vollmer Candle Soot as a Template for a Transparent Robust Superamphiphobic Coating,* SCIENCE VOL 335 6 JANUARY 2012,67~69.

IOP Conf. Series: Materials Science and Engineering **281** (2017) 012012 doi:10.1088/1757-899X/281/1/012012

Optimized Preparation of Carboxymethyl Chitosan/Sodium Alginate Hemostatic Microcapsule

Z Hu[*], J Guang, S T Lu, Y Chen, S D Li and Y Cai

Department of Applied Chemistry, Faculty of Chemistry and Environmental Science, Guangdong Ocean University, Zhanjiang 524088, China

*E-mail: (Zhang HU)huzhangqyx@126.com.

Abstract. In order to prepare and optimize a novel and efficient hemostatic material, a hemostatic microcapsule was prepared by piercing granulation technique with carboxymethyl chitosan and sodium alginate as substrate materials. Three factors in the process (mass percentage of carboxymethyl chitosan, concentration of calcium chloride and pH value of cross-linking system) were investigated through single-factor experiment to explore the effect of experimental factors on water absorbance of hemostatic microcapsule. On the basis of single-factor experiment, orthogonal experiment was conducted to optimize the preparation process of hemostatic microcapsule. Optimized process conditions were obtained as following: mass percentage of carboxymethyl chitosan 25%, concentration of calcium chloride 3.0% and cross-linking system pH=9.0. Microcapsule with normal morphology, uniform particle size, good dispersibility and porous structure was prepared under these conditions. And its water absorbance ratio reached 16.17. In vitro procoagulant test showed that the microcapsule had satisfactory procoagulant effect and was in a dose dependent manner.

1. Introduction

In wars, traffic accidents and other accidents, traumatic hemorrhage might cause severe results such as wound infection, shock, coagulation disorder, sepsis, multi-organ failure or even death [1, 2]. Therefore, quick and effective hemostasis has become one of the research focuses in medicine today. With the development of biomedical engineering, more and more attention has been attracted to the research and application of various biological hemostatic materials. There are many kinds of local biological hemostatic materials commonly used in clinic, such as collagen, oxidized cellulose and fibrin glue. Though with advantages, each type of these traditional materials has its disadvantages too. For example, collagen materials, which are made from the skin of pig or cow, have species difference from human and are likely to cause allergy when used for human hemastasis. While oxidized cellulose dressing has coarse surface and poor adhesiveness and is likely to cause wound affection. The clinic application of these traditional hemostatic dressings is greatly limited by their disadvantages [3, 4]. The ideal hemostatic material should have the following combination characteristics: rapid hemostasis, biocompatibility, biodegradability, non-toxic side effects and cheap costs. However, there are no ideal materials to be found today. So, it is of great practical value to develop novel, safe and effective hemastatic materials.

Chitosan is the only basic polysaccharide found in nature at present. It is safe, non-toxic, biocompatible and biodegradable and has broad-spectrum antibacterial function. However, the application fields of chistan are greatly limited by its poor water solubility [5, 6]. Carboxymethyl chitosan (CMCS), a carboxymethylation product of chitosan, is a safe and non-toxic material with satisfactory water solubility, film forming capacity and moisture absorption and retention. It is widely used in agriculture, food, chemical industry and medicine [7, 8].

Sodium alginate (SA) is a sodium salt of polyanionic polysaccharides extracted from natural brown alga and composed of mannuronic acid and guluronic acid units. For its unique physical and chemical properties and good biocompatibility, sodium alginate is widely applied in pharmaceutical preparation, tissue engineering, clinic therapy, cell culture and food processing [9, 10].

The clinical application of composite functional materials has been one of the most attractive research fields at present. This paper optimized the preparation of carboxymethyl chitosan/sodium alginate composite hemostatic microcapsule by piercing granulation technique and provides a reference for industrialized production.

2. Materials and methods

2.1. Materials

Carboxymethyl chitosan (degree of substitution, 63%) was donated by Lianjiang Taixing Marine Biotechnology Co., Ltd. Sodium alginate was purchased from Sinopharm Chemical Reagent Co., Ltd. Unless otherwise specified, the other reagents are analytical reagents.

2.2. Preparation of microcapsule

Hemostatic microcapsule was prepared by piercing granulation technique with carboxymethyl chitosan and sodium alginate as the substrate materials. Carboxymethyl chitosan solution and sodium alginate solution (2%, w/v) were prepared with distilled water, respectively. The two solutions were mixed and stirred well in specified proportions of carboxymethyl chitosan in the substrate materials. Anhydrous $CaCl_2$ powder was dissolved in distilled water to prepare $CaCl_2$ solution with certain mass concentration and its pH was adjusted with dilute acid and alkali solutions. Then $CaCl_2$ solution was placed on a magnetic stirrer and stirred at constant speed. The mixture solution of carboxymethyl chitosan and sodium alginate was slowly injected into $CaCl_2$ solution with a syringe. After continuously stirred for 30 min, the solution was allowed to stand, centrifuged, washed and freeze-dried to obtain microcapsule.

2.3. Water absorbance property of microcapsule

A certain amount of microcapsule was weighed, soaked in distilled water, allowed to swell for 2h, filtered with 150-mesh filters and the surface water was absorbed with filter paper and then weighed. Water absorption ratio was calculated with the following equation:

$$\text{Water absorption ratio} = (W_1 - W_0)/W_0 \quad (1)$$

where W_0 and W_1 are the mass (g) of microcapsule before and after soaking in distilled water, respectively.

2.4. Observation of microcapsule morphology

Microcapsule was fixed on the specimen table with conductive adhesive and treated for conductive coating with vacuum ion sputter. Afterwards, the surface of the microcapsule was observed with a scanning electron microscope (SEM, Hitachi S-4800).

2.5. In vitro procoagulant effect of microcapsule

The in vitro procoagulant effect of microcapsule with different mass was investigated as described below. 80, 60, 40 and 20 mg of microcapsules were weighed and put into 4 test tubes, respectively, and another tube was used as blank control. The five tubes were warmed in water bath at 37 ^{0}C for 1h. One mL of fresh anticoagulant rabbit blood was added into each of the above five tubes, shaken well and continuously warmed in water bath at 37 ^{O}C. The tubes were taken out to check per 30s until coagulation of blood. The coagulation time was recorded.

2.6. Statistical analysis
All the tests were parallel performed for three times and the mean values were given. All statistical analyses were performed using Statistica 7.0 (StatSoft Inc., USA)..

3. Results and discussion

3.1. Single-factor tests

3.1.1. Mass percentage of carboxymethyl chitosan. Carboxymethyl chitosan and sodium alginate mixed solutions were prepared with varied carboxymethyl chitosan percentage in the substrate material (0%, 25%, 50%, 75%, and 100% respectively) and were slowly dropped into 2% CaC12 solution (pH=7.0) to obtain microcapsule. The water absorbance ratio of the microcapsule was measured and showed in Figure 1. As illustrated in Figure 1, water absorbance of microcapsule was significantly influenced by carboxymethyl chitosan mass percentage. The water absorbance ratio of microcapsule increased first and then decreased as the carboxymethyl chitosan mass percentage increased. It peaked at 11.53 when the carboxymethyl chitosan mass percentage increased to 25%. Sodium alginate, containing a great number of –OH, -COOH and -COO^{-}, can directly or indirectly form hydrogen bonding with water and thus has good water absorbance. Microcapsule prepared with sodium alginate alone could absorb water quickly and showed high water absorbance, but was partially cracked later. These showed that microcapsule prepared with sodium alginate alone had poor stability and addition of carboxymethyl chitosan may increase the stability of microcapsule. When the mass fraction of carboxymethyl chitosan reached 50% or even higher, the water absorption ratio decreased. It might be due to the formation of a denser structure by cross-linking with carboxyl and amino groups in carboxymethyl chitosan with calcium chloride.

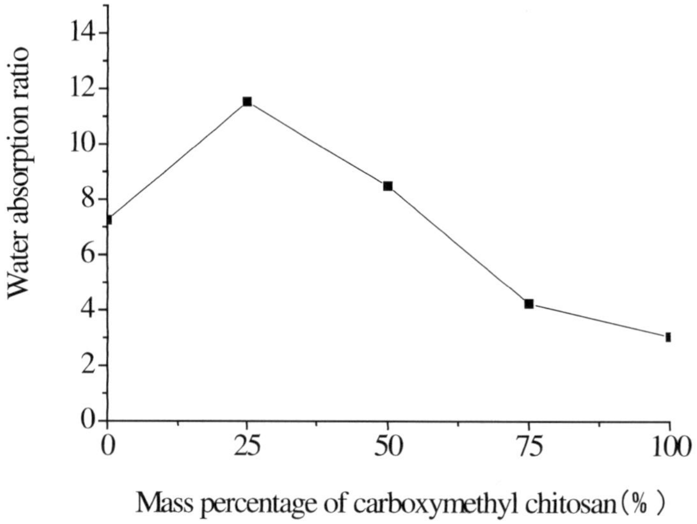

Figure 1. Effect of mass percentage of carboxymethyl chitosan on water absorption of microcapsule

3.1.2. Concentration of $CaCl_2$. Carboxymethyl chitosan/sodium alginate mixed solution was prepared with carboxymethyl chitosan in mass percentage 50% and then added dropwise to $CaCl_2$ solutions

IOP Conf. Series: Materials Science and Engineering **281** (2017) 012012 doi:10.1088/1757-899X/281/1/012012

(pH=7.0) with mass concentration of 1%, 2%, 3.0%, 4.5%, 6.0% and 7.5%, respectively to obtain microcapsules. As shown in Figure 2, the influence of $CaCl_2$ concentration on the water absorption of microcapsules increased first and then decreased, and finally tended to be stable. At a low concentration of $CaCl_2$, the water absorbance was low due to coacervation of a portion of microcapsules caused by association of hydrogen bonds between carboxylate and amino groups in molecule chains of polysaccharide when the speed of complexation of calcium ion and carboxylate ion was slow. As the $CaCl_2$ concentration increased, microcapsules were cross-linked with good dispersibility and loose mesh structure, which provided large and effective free space for water molecules [11]. However, with the concentration of $CaCl_2$ further increasing, the cross-linking density between molecular chains also increased, resulting in inhibiting movement of macromolecules, reducing mesh space, and constricting effective pathway for water molecules to permeate microcapsules [12]. When the concentration of $CaCl_2$ was further increased thereafter, as the quantity of free carboxylate ions that can be contacted was decreased dramatically, there was no significant change in the spatial structure of polymers, and thus the water absorbance ratio showed a flat trend.

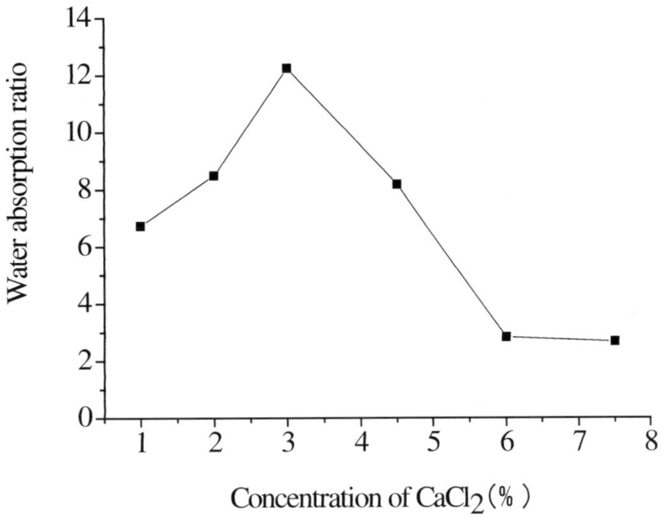

Figure 2. Effect of concentration of $CaCl_2$ on water absorption of microcapsule

3.1.3. Cross-linking system pH. Carboxymethyl chitosan/sodium alginate mixed solution was prepared with carboxymethyl chitosan in mass percentage 50% and then slowly added dropwise to 2% $CaCl_2$ solutions with cross-linking system pH values of 6.0, 7.0, 8.0, 9.0 and 10.0, respectively, to obtain microcapsules. As shown in Figure 3, the change of pH values of the cross-linking system had significant influence on water absorbance of the microcapsules. The water absorbance ratio was low under acid conditions, gradually increased from neutral to basic conditions, peaked at pH=9.0 and decreased as the pH values further increased. The change of water absorbance of microcapsule is closely related with its structure. When the pH values of the cross-linking system changed, due to the different degree of dissociation of sodium alginate and carboxymethyl chitosan, the amino groups and carboxyl groups on carboxymethyl chitosan molecular chains and the carboxyl groups on sodium alginate molecular chains will significantly change and crosslink with $CaCl_2$ to different extent to form microcapsules of different spatial structure. Carboxyl groups in carboxymethyl chitosan and sodium alginate existed completely in the form of $-COO^-$ when pH reached 10.0, and they cross-linked with calcium chloride entirely, resulting in the low water absorption ratio.

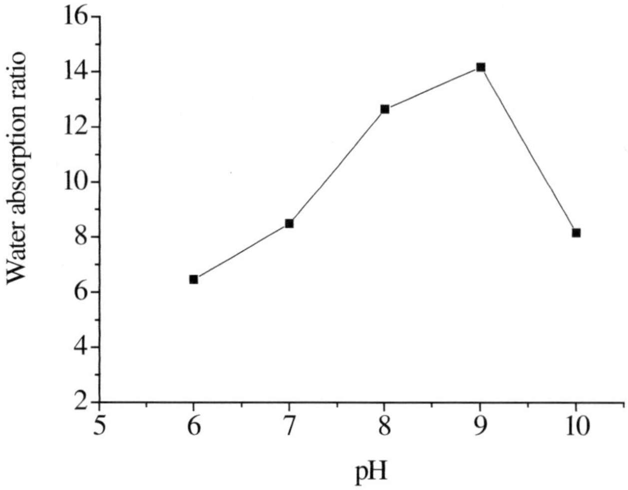

Figure 3. Effect of cross-linking system pH on water absorption of microcapsule

3.2. Preparation optimization of hemostatic microcapsule

On the basis of single-factor experiment, $L_9(3^4)$ orthogonal experiment was designed to optimize the preparation process of hemostatic microcapsule. The water absorbance ratio of hemostatic microcapsule was used as the investigation index for the optimization.

Table 1. L_9 (3^4) orthogonal experimental design and results

Run	Factors			Water absorption ratio
	A: Mass percentage of carboxymethyl chitosan (%)	B: Concentration of $CaCl_2$ (%)	C: Cross-linking system pH	
1	25	2.0	7	11.13
2	25	3.0	8	15.03
3	25	4.5	9	14.11
4	50	2.0	8	11.63
5	50	3.0	9	14.82
6	50	4.5	7	13.19
7	75	2.0	9	12.90
8	75	3.0	7	13.82
9	75	4.5	8	12.34
K1	40.27	35.66	38.14	
K2	39.64	43.67	39.00	
K3	39.06	39.64	41.83	
R	1.21	8.01	3.69	

The design process and results of the orthogonal experiment were shown in table 1. Range analysis revealed that the extent of three factors impact on microcapsule water absorbance descended in the order of B>C>A. The concentration of $CaCl_2$ has the maximum influence, while mass percentage of carboxymethyl chitosan has the minimum influence. Analysis of K values showed that the optimum condition for preparation of microcapsule is $A_1B_2C_3$, i.e., mass percentage of carboxymethyl chitosan 25%, concentration of $CaCl_2$ 3%, and cross-linking system pH=9.0. As verified by parallel experiments, the average water absorbance ratio of microcapsule prepared under these conditions was 16.17.

3.3. Morphology of hemostatic microcapsule

The surface morphology of microcapsule was observed by SEM (Figure 4). As shown in Figure 4(a), the microcapsule prepared has a regular near-spherical shape, uniform and appropriate particle size, good dispersibility and no coacervation. From Figure 4 (b), it is observed that there are numerous small pores of different diameter on the surface of the microcapsule, indicating a porous structure of microcapsule, making for easy hemostasis by water absorbing and swelling.

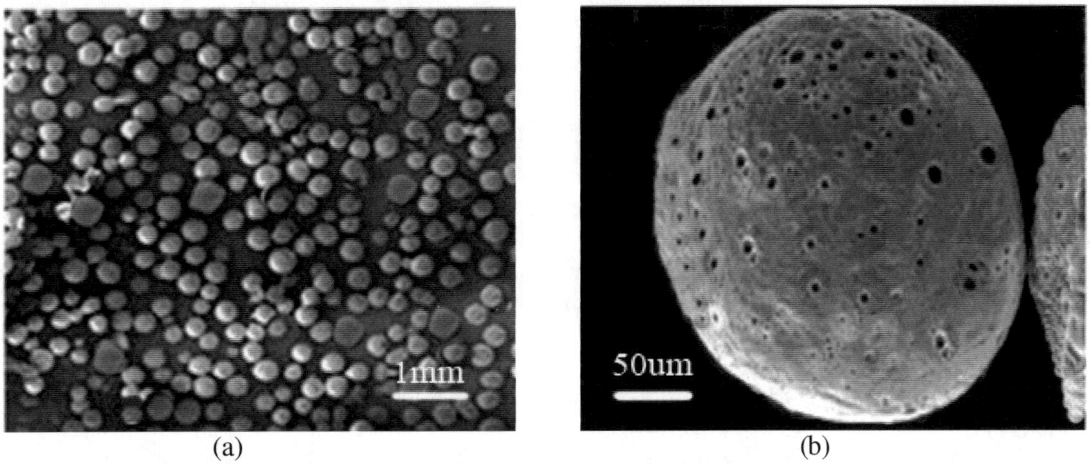

Figure 4. The surface morphology of microcapsule observed by SEM

3.4. Procoagulant property of microcapsule in vitro

Traditional hemostasis research is mostly carried out through clinical practice, but there are some shortcomings such as long time consuming and high cost. Screening of potential hemostatic medicine by in vitro hemostatic experiment is the trend for developing new hemostatic medicines. Using in vitro hemostasis, microcapsule was added into fresh anticoagulant rabbit blood, the coagulation effect of microcapsule was investigated by measuring the blood coagulation time. The results were shown in Table 2. It was observed that the coagulation time of the blank control group exceeded 30min, which means no coagulation. After addition of microcapsule the anticoagulant blood rapidly coagulated, indicating strong in vitro procoagulant effect of microcapsule. Besides, with the dose of microcapsule increasing, the coagulation time shortened, indicating that the procoagulant effect of microcapsule was dose dependent to certain extent.

Table 2. Procoagulant effect of the microcapsule in vitro.

Amount of microcapsule (mg)	0	20	40	60	80
Coagulation time (mins)	>30 (not coagulation)	8.7	6.2	4.8	4.3

4. Conclusions

The composite hemostatic microcapsule was successfully prepared by piercing granulation technique with carboxymethyl chitosan and sodium alginate as substrate materials. The effects of experimental

IOP Conf. Series: Materials Science and Engineering **281** (2017) 012012 doi:10.1088/1757-899X/281/1/012012

factors in the preparation process on water absorbance of hemostatic microcapsule were explored by single-factor experiment. An orthogonal experiment was conducted to optimize the preparation process of hemostatic microcapsule. In vitro procoagulant tests showed that the microcapsule had significant procoagulant effects. Further researches are needed to develop a clinical medical hemostatic material.

Acknowledgments
The authors gratefully acknowledge the financial support by Natural Science Foundation of Guangdong Province of China (2016A030308009), Scientific and Technological Planning Project of Guangdong Province of China (2015A020216019) and Guangdong Ocean University Student's Platform for Innovation and Entrepreneurship Training Program (CXXL2016086).

Author contributions
Zhang Hu designed the experiments and wrote the paper; Jie Guang and Si Tong Lu performed the experiments; Yu Chen and Si Dong Li analyzed the data; Ying Cai contributed reagents, materials and analysis tools.

References
[1] Kauvar D S, Lefering R and Wade C E 2006 Impact of hemorrhage on trauma outcome: an overview of epidemiology, clinical presentations, and therapeutic considerations. *J. Trauma* **60** S3-S11
[2] McManus J G, Eastridge B J, Wade C E and Holcomb J B 2007 Hemorrhage control research on today's battlefield: lessons applied. *J. Trauma* **62** S14.
[3] Spotnitz W D 2014 Fibrin Sealant: The only approved hemostat, sealant, and adhesive — a laboratory and clinical perspective. *ISRN Surgery* **2014** 203943.
[4] Wang X, Yan Y and Zhang R 2006 A comparison of chitosan and collagen sponges as hemostatic dressings. *J. Bioact. Compat. Polym.* **21** 39-54.
[5] Ravi Kumar M N V, Muzzarelli R A A, Muzzarelli C, Sashiwa H and Domb A J 2004 Chitosan chemistry and pharmaceutical perspectives. *Chem. Rev.* **104** 6071-84
[6] Patrulea V, Ostafe V, Borchard G and Jordan O 2015 Chitosan as a starting material for wound healing applications. *Eur. J. Pharm. Biopharm.* **97** 417-26
[7] Fonseca-Santos B and Chorilli M 2017An overview of carboxymethyl derivatives of chitosan: Their use as biomaterials and drug delivery systems. *Mater. Sci. Eng. C* **77** 1349-62
[8] Upadhyaya L, Singh J, Agarwal V and Tewari R P 2013 Biomedical applications of carboxymethyl chitosans. *Carbohydr. Polym.* **91** 452-66
[9] Venkatesan J, Bhatnagar I, Manivasagan P, Kang K H and Kim S K 2015 Alginate composites for bone tissue engineering: a review. *Int. J. Biol. Macromol.* **72** 269-81
[10] Giri T K, Thakur D, Alexander A, Ajazuddin, Badwaik H and Tripathi D K 2012 Alginate based hydrogel as a potential biopolymeric carrier for drug delivery and cell delivery systems: present status and applications. *Curr. Drug Deliv.* **9** 539-55
[11] Rodríguez D E, Romero-García J, Ramírez-Vargas E, Ledezma-Pérez A S and Arías-Marín E 2006 Synthesis and swelling characteristics of semi-interpenetrating polymer network hydrogels composed of poly(acrylamide) and poly(γ-glutamic acid). *Mater. Lett.* **60** 1390-3
[12] Rokhade A P, Patil S A and Aminabhavi T M 2007 Synthesis and characterization of semi-interpenetrating polymer network microspheres of acrylamide grafted dextran and chitosan for controlled release of acyclovir. *Carbohydr. Polym.* **67** 605-13

High-temperature Oxidation of Electrodeposited Ni-Co-SiC Composite Coatings

Y Wang, W C Sun*, J M Xu, F Guo and Z W Jia

College of Materials Science and Engineering, Xi'an University of Science and Technology, Xi'an, Shaanxi 710054, P.R. China

*Corresponding author: sunwanchang@tsinghua.org.cn

Abstract. Ni-Co-SiC composite coatings were prepared on 45 carbon steel by electrodeposition for high temperature anti-oxidation. The high-temperature oxidation was studied by circulating oxidation test. The prepared composite coating shows continuous Ni-Co matrix and uniformly dispersed SiC particles, and there are no cracks and pores and other flaws in the composite coating. The results also reveal that due to the reactive-phase effect of the embedded SiC particles, the oxidation rate of the substrate decreases largely at high temperature. The Ni-Co-SiC composite coating exhibits better high-temperature oxidation compared with Ni-Co coating and 45 carbon steel.

1. Introduction

In various engineering applications, the long term performances under high-temperature, high-pressure and high-speed conditions in aggressive solution or atmosphere are crucial to mechanical products and parts, made of carbon steels. For example, major failure for hot forging process is originated from high temperature wear, and proportion of which can reach 70-80% [1]. Moreover, the majority of engines and cutting tools are also damaged due to the reason as just mentioned. For the purpose of improving oxidation and wear resistance at high-temperature, corrosion resistance of engineering materials, surface modification methods are required, in which preparing high-temperature coating is a good choice. Composite coatings incorporated with fine particles, owning satisfying wear resistance and oxidation resistance, which is composed of matrix metal and uniformly dispersed fine ceramic particles such as ZrO_2 [2], B_4C [3], Si_3N_4 [4], WC [5], Al_2O_3 [2,4,6], CeO_2 [7],and SiC particles [4,8], among which SiC particles is widely used as the reinforcing particles owing to its high hardness and high-temperature stability. In addition, Ni-Co alloys show brilliant thermal, physical and mechanical properties than pure Ni [9]. Therefore, a novel composite coating with excellent physical and chemical properties, achieved by electrodeposition, is led to by the incorporation of different particles within Ni-Co matrix [6-8]. At present, a large amount of research report concentrate on enhancing the microindentation hardness, wear resistance and corrosion resistance of Ni-Co coatings [6-8,10]. Nevertheless, there is little report about their oxidation behavior.

In our study, Ni-Co thin film incorporated with micron SiC particles which can overcome a lot of restrictions under severe work conditions was made by electrodeposition. The oxidation resistance at high temperature of the three kinds of samples was investigated.

IOP Conf. Series: Materials Science and Engineering **281** (2017) 012013 doi:10.1088/1757-899X/281/1/012013

2. Experimental procedures

2.1. *Electrodeposition of Ni-Co-SiC coatings*
A DC power supply was used in the electrodeposition. A 45 carbon steel substrate (12 mm×10 mm×6 mm) was used as the cathode, and two nickel plates were used as the anode. The SiC particles with an average size of about 3.5 μm were pretreated with 37.5% hydrochloric acid for 6 h and cleaned by distilled water. Before the codeposition, the suspension was treated with ultrasonication for 4 h. At the same time, the suspension was stirred by a magnetic stirrer and the temperature of the bath is controlled to be 328 K. For the purpose of obtaining a coating with a thickness of 185 μm, the plating duration was maintained 80 min. Table 1 shows that the process parameters of electrodeposition.

Table 1. Plating bath composition and experimental conditions

Composition of plating solution [g/L]	
$NiSO_4{\cdot}6H_2O$	400
$CoSO_4{\cdot}7H_2O$	25
H_3BO_4	30
$CH_3(CH_2)_{11}OSO_3Na$	0.2
SiC (about 3.5 μm)	80
Experimental conditions	
Temperature	328 [K]
pH	2.5
Current density	5 [A/dm^2]
Magnetic stirring rate	250 [rpm]
Plating duration	80 [min]

2.2. *High temperature oxidation tests*
We performed oxidation test at high temperature in a muffle furnace with an automatic temperature control device (precision ±2 K). 2 hours of oxidation experiments were performed at 573 K and 1173 K, while at 873K for 2h, 4h, 6h, and 8h, respectively. Subsequently, the oxidation weight increment of the specimens before and after high-temperature oxidation was weighed by using an analytical balance (precision 0.01 mg) and compared with that of the 45 carbon steel and Ni-Co film.

3. Results and discussion
As shown in the insets of Figure 1, the oxidation weight increment of three kinds of specimens at different temperature. We observed that when the temperature increased, the oxidation weight gain of three kinds of samples was also increased. We can observed that the oxidation weight increment of the samples is obvious when the temperature was 1173 k. Compared with 45 steel, both coatings show lower weight gain. And the weight gain of the composite coating is the least at different temperature. This suggested that Ni-Co-SiC coatings would result in less oxide. The relative weight gain at 573K for Ni-Co film and Ni-Co-SiC multilayer films relative to 45 carbon steel is 49% and 6%. The relative weight gain at 873K remains the same for Ni-Co coating at 49%, but increases to 35% for Ni-Co-SiC composite coating. The relative oxidation weight gain decreases to 36% and 26% at 1173K for Ni-Co film and Ni-Co-SiC multilayer films respectively. Therefore, the composite coating exhibits a better high-temperature oxidation resistance.

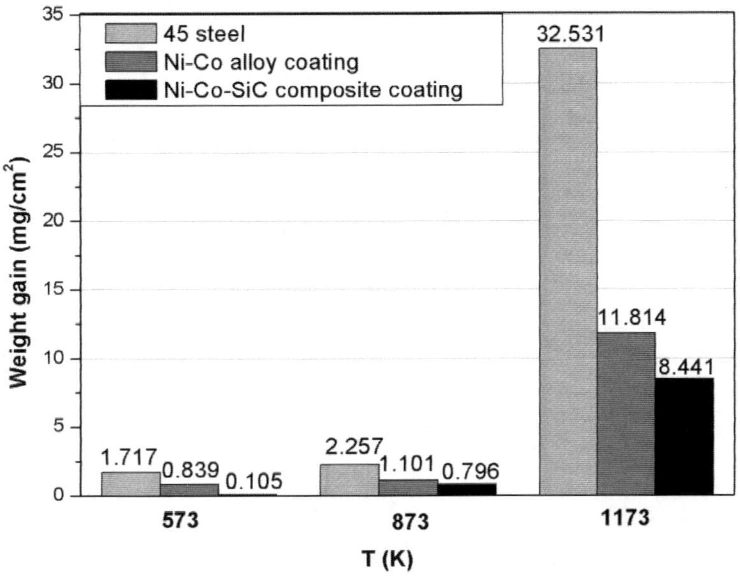

Figure 1. Oxidation weight increment of 45 carbon steel, Ni-Co film and Ni-Co-SiC multilayer films at different temperature

Figure 2 shows the circulation oxidation weight increment curves of three kinds of samples after heating in the air to 873K. As can be seen from the picture, the three samples had a rapid oxidation weight gain at the first two hours. The reason for the rapid weight gain is the formation of oxide film. At this stage, the weight growth depends mainly on the oxidation reaction, rather than the atomic diffusion. In the meantime, we can also see from the image that the mass gain and oxidation rate of the Ni-Co-SiC multilayer films were the smallest. After two hours oxidation, the coatings surface was oxidized. And the oxide layer becomes more and more compact and thicker. So the mass gain of the coated samples decreased obviously. What is more, the compact oxide layer turns to a barrier for oxygen diffusion, and the inner metal alloy was protected by the oxide layer, so that the oxidation rate of multilayer films and composite coating decreases after 2 h. In the meantime, the mass gain of the Ni-Co-SiC multilayer films is less than that of the Ni-Co film. This is due to the formation of a protective oxide layer on the surface of the film, hindering the diffusion of oxygen, and reduced the oxidation rate. Thus, the Ni-Co-SiC multilayer films exhibit a better oxidation resistance at high temperature.

The experimental results show that the oxidation weight increment of the Ni-Co-SiC composite coatings is obviously reduced. The high temperature oxidation resistance of the Ni-Co-SiC multilayer films is improved because of the SiC particles with high thermal stability are dispersed in the Ni-Co matrix. Therefore, it is easy to form an oxide film with fine grain and uniformly compact structure. SiC particles can not decompose at high temperatures, but uniformly dispersed in the Ni-Co matrix. These fine particles dispersed in the oxide film, inhibiting the growth of the film and reducing the grain size, leading to a compact oxide layer. The formation of a dense protective layer can effectively prevent the further reaction of the matrix and oxygen and reduce the oxidation rate. The compactness of the Ni-Co-SiC multilayer films was significantly improved, the porosity was reduced, and the internal diffusion rate of oxygen in the coating was slowed down at the initial stage of oxidation, which directly reduced the oxidation rate, and also reduced the oxidation weight gain.

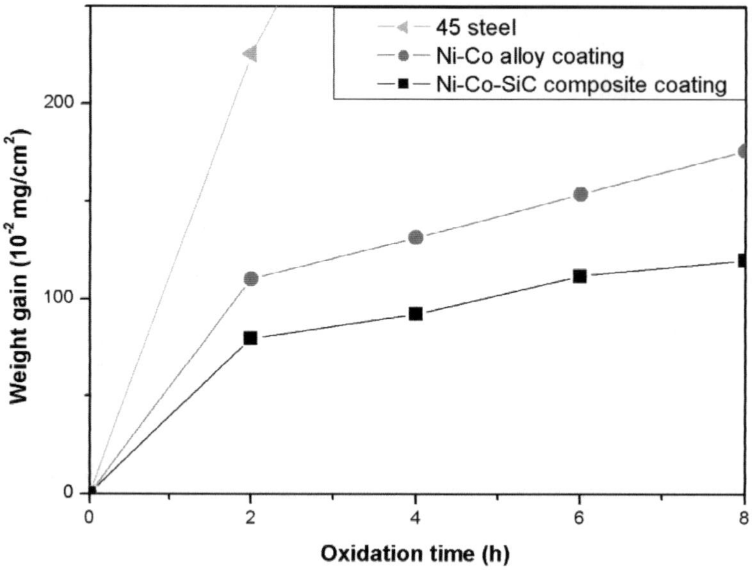

Figure 2. Oxidation weight increment of 45 carbon steel, Ni-Co film and Ni-Co-SiC multilayer films at 873 K with different oxidation time

4. Conclusions

Ni-Co-SiC multilayer films were prepared on 45 carbon steel substrate by electrodeposition. The prepared composite coating shows continuous Ni-Co matrix and uniformly dispersed SiC particles, and there are no cracks and pores and other flaws in the composite coating. Due to the reactive-phase effect of the embedded SiC particles, the oxidation weight increment of Ni-Co-SiC multilayer films is lower than that of Ni-Co film and 45 carbon steel. Therefore the Ni-Co-SiC coatings exbihit an excellent oxidation resrstance at high temperature.

Acknowledgments

This research are financially supported by Shaanxi Industrial Science and Technology Research (2014K08-09), the Scientific Research Program Funded by Yulin city (2012).

References

[1] Jiang X, Liu W J, Dong S Y and Xu B S 2005 *Surf. Coat. Technol* **194** 10-15
[2] Gao J, He Y D and Wang D R 2010 *Mater. Chem. Phys* **123** 731-4
[3] Araghi A and Paydar M H 2014 *Tribology-Materials Surfaces & Interfaces* **8** 146-53
[4] Bai J J, Song L, Wang J Y and Jian X G 2013 *Paint & Coatings Industry* **43** 14-7
[5] Rezaei-Sameti M, Nadali S, Falahatpisheh A and Rakhshi M 2013 *Solid State Communications* **15** 18-21
[6] Wu G, Li N, Zhou D R and Mitsuo K 2004 *Surf. Coat Technol* **176** 157-62
[7] Srivastava M, William Grips V K and Rajam K S 2010 *Appl Surf Sci* **257** 717-22
[8] Shi L, Sun C F, Gao P, Zhou F and Liu W M 2006 *Appl Surf Sci* **252** 3591-9
[9] Srivastava M, William Grips V K and Rajam K S 2008 *J Appl Electrochem* **38** 669-76
[10] Wu G, Li N, Dai C S and Zhou D R 2004 *Mater. Chem. Phys* **83** 307-14

Structure and Dielectric Properties of $(Sr_{0.2}Ca_{0.488}Nd_{0.208})TiO_3$-$Li_3NbO_4$ Ceramic Composites

C C Xia[1*] and G H Chen[2]

School of Materials Science and Engineering, Guangxi Key Laboratory of Information Materials, Guilin University of Electronic Technology, Guilin 541004, P.R. China

[1]2324514656@qq.com; [2]chengh@guet.edu.cn

Abstract. The new ceramic composites of (1-*x*) Li_3NbO_4-*x* $(Sr_{0.2}Ca_{0.488}Nd_{0.208})TiO_3$ were prepared by the conventional solid state reaction method. The sintering behavior, phase composition, microstructure and microwave dielectric properties of the ceramics were investigated specially. The SEM and XRD results show that (1-x) Li_3NbO_4-x $(Sr_{0.2}Ca_{0.488}Nd_{0.208})$ TiO_3 ($0.35 \leq x \leq 0.5$) composites were composed of two phase, i.e. perovskite and Li_3NbO_4. With the increase of x, the ε_r increases from 27.1 to 38.7, $Q\times f$ decreases from 55000 GHz to 16770 GHz, and the τ_f increases from -49 ppm/°C to 226.7 ppm/°C. The optimized dielectric properties with ε_r~31.4, $Q\times f$~16770GHz and τ_f~-8.1ppm/°C could be obtained as x=0.4 sintered at 1100°C for 4h. The as-prepared ceramic is expected to be used in resonators, filters, and other microwave devices.

1. Introduction

With the rapid development of microwave communication technology, the widespread usage of several different wireless systems has required the development of materials and process technologies that can provide the rapid production of low-cost, lightweight, small, multifunctional and highly reliable devices[1]. As the development direction of high performance microwave dielectric materials, perovskite structure materials with high ε_r, high $Q\times f$ value and low τ_f quickly become the research focus. It is well known that an effective way to achieve near-zero τ_f is to mix two compounds, one with a positive and the other with a negative τ_f to form a solid solution or composite.

Recently, $CaTiO_3$ with the perovskite structure and $SrTiO_3$ materials, are widely employed as dielectric resonators because of their higher dielectric constant (ε_r)[2,3]. In theory, $CaTiO_3$-$Li_{0.5}RE_{0.5}TiO_3$ (RE=Sm, Nd), $Ca_{1-x}Re_{2x/3}TiO_3$-$Li_{1/2}Ln_{1/2}TiO_3$(Re=Sm、Nd、La; Ln = Sm、Nd) system showed a higher dielectric constant[4]. Unfortunately, although the resonant frequency temperature coefficient of the material (τ_f) close to zero, but the Qxf reduced rapidly (2000~7500 GHZ), it is difficult to meet the high demand of high-end communication electronic components on the properties of microwave dielectric ceramic dielectric. Therefore, domestic and foreign scholars have studied not only reduce device size, but also has low loss performance of high Q value, high dielectric of microwave dielectric ceramics. These ceramic materials meet the requirements of thermal stability and have high $Q\times f$ value at the same time. In the past few years, F. Liu et al reported that $(Sr_{0.2}Ca_{0.488}Nd_{0.208})$ TiO_3 ceramic showed following dielectric properties: ε_r=130.4, Qxf =9500 GHz, and τ_f =332.5 ppm/°C[5]. D. Zhou et al reported that Li_3NbO_4 exhibited an excellent dielectric properties sintered at 930 °C/2h: ε_r = 15.8, a high Qxf =55,009 GHz and τ_f = -49 ppm/°C [6]. H. F. Zhou et al found that $CaTiO_3$-Li_3NbO_4 ceramic composites sintered at 1020°C/2h possessed outstanding dielectric properties of ε_r=21.9, Qxf=24,900 GHz, and τ_f =5.6 ppm/°C [7]. However, its sintering temperature (≥1400 °C) and τ_f is too high to use practically. The purpose of the present study is to achieve near-zero τ_f along with a high ε_r and high Qxf by incorporating $(Sr_{0.2}Ca_{0.488}Nd_{0.208})$ TiO_3 into Li_3NbO_4, and to investigate the effects of $(Sr_{0.2}Ca_{0.488}Nd_{0.208})TiO_3$ addition into the Li_3NbO_4 on the structure and microwave dielectric properties of the ceramics.

2. Experimental

2.1. Sample preparation

Analytical reagent Sr_2CO_3, $CaCO_3$, TiO_2, Nd_2O_5, Li_2CO_3 and Nb_2O_5 (>99.0%) were used as raw materials. The suppliers of the raw materials are Guangdong Shantou West Long Chemical Company. The Li_3NbO_4 (LN) and $(Sr_{0.2}Ca_{0.488}Nd_{0.208})$ TiO_3 (SCNT) were prepared according to the stoichiometric ratio and ball milled in ethanol medium for 24 h in nylon jars using yttria stabilized zirconia balls. After drying, the LN and SCNT mixtures were calcined at 900 °C and 1150 °C in air for 4 h, respectively. Samples of (1-*x*) LN-*x*SCNT with x=0.35, 0.4, 0.45, 0.5 were prepared from pure calcined LN and SCNT powders by weight ratio. The mixtures were ball-milled in alcohol for 12 h and dried. The slurries were dried, then mixed with 5wt% polyvinyl alcohol (PVA) as binder and granulated. The granulated powders were pressed into disks of 12 mm diameter and 6 mm in thickness. The specimens obtained were then heat-treated at 600°C for 3 h to eliminate PVA, followed by sintering at 1060-1160°C in air for 4 h at a heating and cooling rate of 4°C /min.

2.2. Performance measuring

The bulk density was measured by using the Archimedes method. The crystal structure and phase purity of the powdered samples were analyzed by using X-ray diffraction (XRD, D8-ADVANCE, Bruker, Karlsruhe, Germany) with CuKa radiation. The microstructure observation of the samples was performed by using scanning electron microscopy (FE-SEM; Quanta FEG450, America). The microwave dielectric properties were measured by a Vector Network Analyzer (N5230C, Agilent Technologies, Palo Alto, CA). The temperature coefficient of resonant frequency (τ_f) was measured in the temperature range of 25°C to 75° C using the following equation [8]:

$$\tau_f = \frac{(f_{75} - f_{25}) \times 10^6}{50 \times f_{25}} \quad \text{ppm/°C}$$

where f_{75} and f_{25} are the resonant frequencies at 75°C and 25°C, respectively.

3. Results and discussion

3.1. Phase analysis

Figure 1 shows the XRD patterns of (1-x) LN-xSCNT (0.35≤x≤0.5) composites sintered at optimum sintering temperature for 4h. From the XRD results, it is seen that all samples show two phase coexistence state, i.e. LN and perovskite phase. With the increase of x value, the diffraction peaks of perovskite phase obviously enhance.

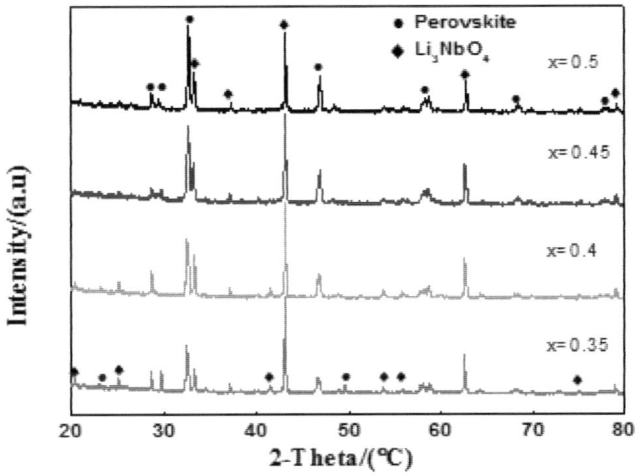

Figure 1. XRD patterns of (1-x) LN-xSCNT specimens

3.2. Microstructure

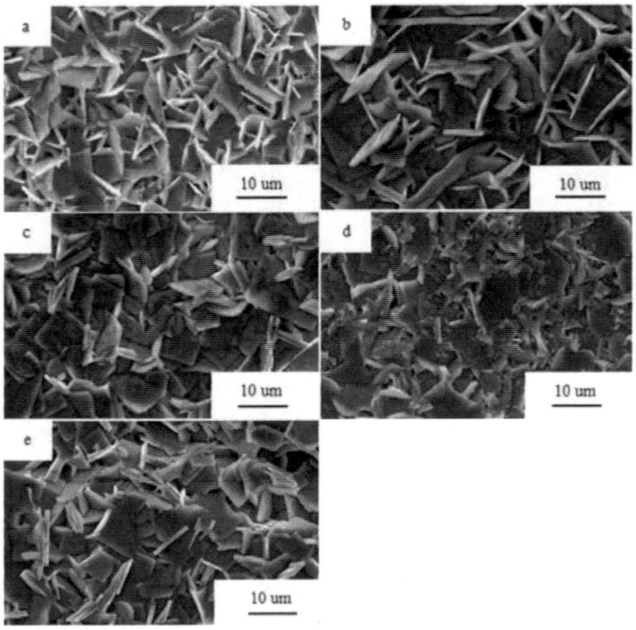

Figure 2. SEM micrographs of 0.6LN-0.4SCNT specimens sintered at different temperatures: (a)1060 °C, (b)1080 °C, (c)1100 °C, (d)1120 °C, (e)1140 °C

Figure 2 shows the SEM micrographs of the 0.6LN-0.4SCNT ceramic sintered at different temperatures. At 1060°C, the ceramic sample has a few of porosity shown in Figure 2a. With the increase of sintering temperature, the ceramic sample gets compact and the grain size is increased. When the sintering temperature reaches 1100°C, the grains grow up quickly, the uniform and densest microstructure is obtained, as shown in Figure 2c. With further increase of the sintering temperature, the grain boundaries become blurred and the porosity gradually is increased, the asymmetrical microstructure is formed, which may be due to excessive sintering. The crystal structure of SCNT and LN have big difference, good stability and easy to form a composite ceramics.

3.3. Sintering characteristics

Figure 3 shows the relative densities of (1-x) N-xSCNT (0.35≤x≤0.5) ceramic composite as a function of sintering temperature. Firstly, the relative densities of samples display a similar variation trend with the increment of temperature and reach a maximum value, then decrease with the further increase of temperature. Raising sintering temperature is beneficial to improving sintering densification of the ceramic materials. The different sample has various optimal sintering temperatures, as shown in Figure 3. At x = 0.4, the optimal sintering temperature is 1100°C and the relative density of the ceramic sample can reach 99.7%. With x increasing, optimal sintering temperature increases, which is due to that the sintering temperature of SCNT (1400 °C) is higher than that of LN (930 °C).

Figure 4 shows the ε_r of (1-x) LN-xSCNT (0.35≤x≤0.5) ceramics sintered at different temperatures. For fixed x value, the ε_r of ceramics first increases and then decreases. This change trend is consistent with that of relative density seen in Figure 3. At their optimum temperatures, the ε_r value is 27.13, 31.38, 34.75, 38.71 for x=0.35, 0.4, 0.45, 0.5 sample, respectively. When x value is increased from 0.35 to 0.5, the dielectric constant is increased from 27.1 to 38.71. We know that the dielectric constant of SCNT (~130.4) is higher than that of Li_3NbO_4. According to compound rule, the higher the content of compound with high ε_r, the bigger the ε_r of the composite.

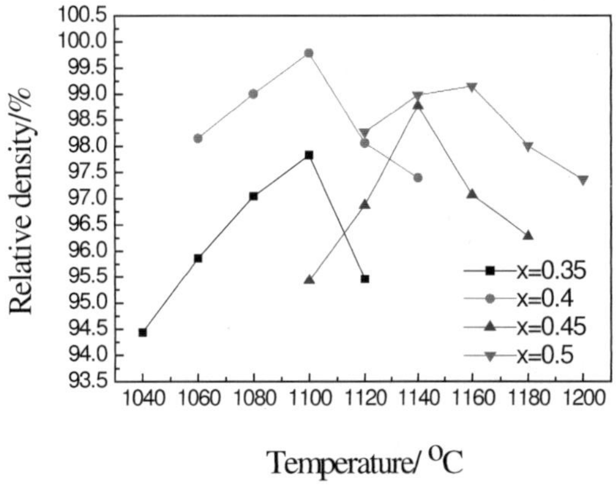

Figure 3. Relative densities of (1-x)LN-xSCNT specimens as a function of sintering

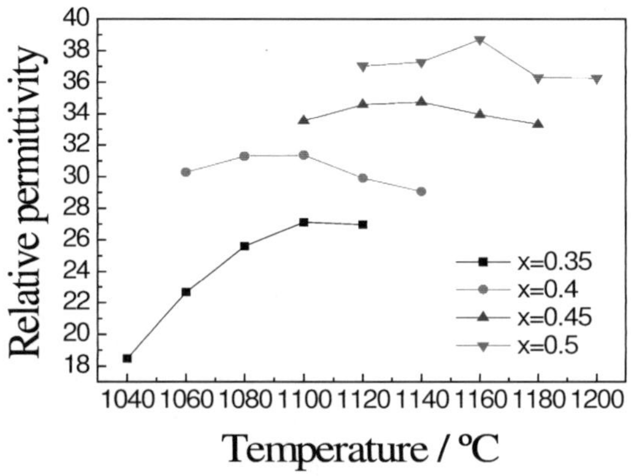

Figure 4. Relative permittivity of (1-x)LN-xSCNT specimens as a function of sintering temperature

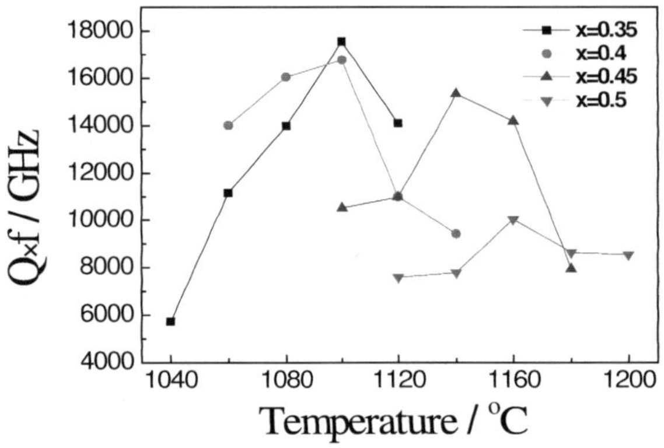

Figure 5. Q×*f* values of (1-x)LN-xSCNT specimens as a function of sintering temperature

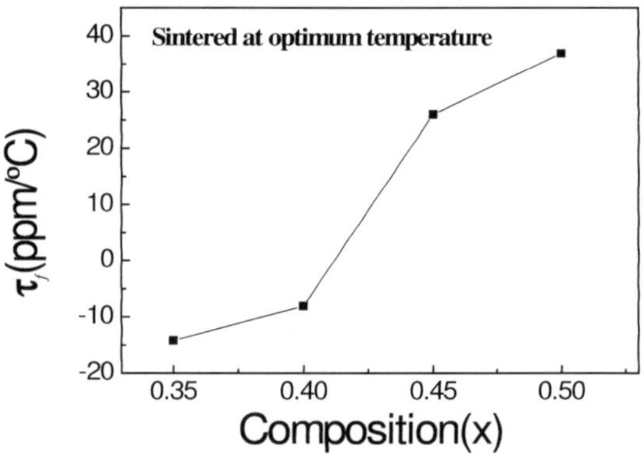

Figure 6. τ_f of the specimens as a function

Figure 5 shows the $Q \times f$ values of (1-x)LN-xSCNT (0.35≤x≤0.5) ceramics as a function of sintering temperature. The change of $Q \times f$ value is associated with the relative density of samples. As the composition keeps unchanged, the $Q \times f$ values show first increase and then decrease with an increase in sintering temperature. For different compositions, the $Q \times f$ value is distinctly decreased with the increase of x. This is owing to the quality factor of SCNT is lower than LN. Therefore, the $Q \times f$ value is related to not only the densification but also the composition.

Figure 6 shows the τ_f of the composites as a function of x value. It is observed that the τ_f increases from negative to positive with increasing x. One knows that the τ_f of LN is negative (–49 ppm/°C) and the τ_f of SCNT is positive (~332.5 ppm/°C). Therefore, the τ_f of the ceramic composites can be easily adjusted within a certain range. When x=0.4, the τ_f value is -8.1 ppm/°C.

4. Conclusions

The crystal structures and microwave dielectric properties of (1-x)LN-xSCNT (0.35≤x≤0.5) ceramics were investigated in this paper. The ceramic composites have been synthesized by the solid state reaction route. The research results show that the dielectric constant, relative density and quality factor of the composites firstly increase and then decrease with the increase of sintering temperature, but the temperature coefficient of resonant frequency (τ_f) change from negative to positive with the increase of x. At 1100 °C, the 0.6 LN-0.4 SCNT ceramic composite has the best microwave dielectric properties: ε_r=31.4, $Q \times f$=16,770 GHz, τ_f=-8.1 ppm/°C. The as-prepared ceramic is expected to be used in resonators, filters, and other microwave devices.

Acknowledgments
This work was supported by the Research funds of The Guangxi Key Laboratory of Information Materials (Nos.131018-Z and 131004-Z).

References

[1] Kell R C, Greenham A C, Olds G C E. Cheminform abstract: high-permittivity temperature-stable ceramic dielectrics with low microwave loss. *Chem Inform*, 4(1973), 352-354.
[2] Liu F , Yuan C L, Liu X Y , Qu J J, Chen G H, Zhou C R, Effects of structural characteristics on microwave dielectric properties of (Sr0.2Ca0.488Nd0.208)Ti1− xGa4x/3O3 *ceramics. Materials Research Bulletin*, 70(2015) 678-683.
[3] Kell R C, Greenham A C , Olds G C E , Am J. *Ceram. Soc.* 56 (1974) 352–354.
[4] Qu Q Q, Wei X, Jing B J, Liu F, Yuan C L. *Microstructures and Microwave Dielectric Properties of (1-x)(Sr0.2Nd0.208Ca0.488)TiO3- xNd(Ti0.5Mg0.5)O3 Ceramics with High Quality Factor. J Inorg Mater*, 11(2015) 1213-1217.
[5] Liu F, Yuan C, Liu X, Yang T, Chen G, Zhou C. Microstructures and dielectric propertiesof (1-x) SrTiO3-xCa0.61Nd0.26TiO3 ceramic system at microwave frequencies. J. Mater. *Sci- Mater. Electron.* 26(2014) 128-133.
[6] Zhou D, Wang H, Pang L X, Yao X, Wu X G . *Microwave Dielectric Characterization of a Li3NbO4 Ceramic and Its Chemical Compatibility with Silver. J. Am. Ceram. Soc.* 91(2008) 4115–4117.
[7] Zhou H, Wang W, Chen X, Miao Y, Liu X, Fang L, He F. *Sintering behavior, phase evolution and microwave dielectric properties of thermally stable (1-x) Li3NbO4-xCaTiO3, composite ceramic. Ceram. Int.* 40(2014) 2103-2107.
[8] Nishikawa T, Wakino K, Tamura H, *Ishikawa Y, in IEEE MTT-S International Microwave Symposium Digest*, 87(1987) 277-280.

IOP Conf. Series: Materials Science and Engineering **281** (2017) 012015 doi:10.1088/1757-899X/281/1/012015

Preparation of ZnO/SnO_2 Composite Nanometer Photocatalyst and Photocatalytic Treatment of Marine Diesel Pollution

J Zhang, X C Yu*, Z W Nie, M C Guo, J H Liu and L P Wang

College of Ocean Technique and Environment department, Dalian Ocean University, China

*Xiaocai Yu, xiaocyu@dlou.edu.cn

Abstract. The ZnO/SnO_2 composite nanophotocatalyst studied in this paper was prepared by a chemical precipitation method, which were characterized by XRD and SEM. The results show that the prepared samples were rutile SnO_2 particles and the average grain size is8.41 nm. In this paper, the factors for the degradation efficiency of marine diesel oil degraded by ZnO/SnO_2 composite nanophotocatalyst are the catalysts' doping ratio, the initial concentration of oil, the pH value of seawater, the dosage of catalyst and the dosage of hydrogen peroxide. The results show that the ZnO/SnO_2 composite nanophotocatalyst can effectively degrade seawater diesel oil under UV light. When the doping ratio of ZnO and SnO_2 is 0.35, the reaction time is 2.5 hours, the pH value of seawater with oil is 7, The concentration of diesel oil is 0.1g/L, the dosage of catalyst is 0.3g/ L and the dosage of hydrogen peroxide is 0.1 g/L, the highest degradation rate is 91.54%.

1. Introduction

With the continuous strengthening of national power, China's resource utilization is in transition gradually. More attention is paid to the ocean with treasure, but a lot of offshore oil extraction and oil transport also exacerbated the risk of marine oil pollution. Marine oil pollution is sudden mostly, and the oil pollution would spread with the ocean currents, waves and other spread[1]. Petroleum hydrocarbons consist of tens of thousands of ingredients, and different fractions will be toxic to creatures. The effects of teratogenic death are caused by petroleum hydrocarbon contaminants in the waters on marine life. Petroleum hydrocarbons also affect the higher organisms through the food chain. Because of the seriousness of this pollution, mankind began to seek the treatment methods of marine oil pollution. At present, the methods of governance are mainly divided into physical, biological and chemical management. It is well known that chemical method is a cost saving and fast processing method. The method used in this paper is photocatalytic degradation method in chemical treatment method. Photocatalytic degradation is an environmentally friendly treatment method that does not undermine the background environment and has the advantage of efficient treatment [2-5].

In this paper, self-made ZnO/SnO_2 composite nano-photocatalystis used for photocatalytic degradation of oil. ZnO is widely used as a nanophotocatalyst and has a remarkable effect[6]. Nano-SnO_2 is usually used to modify other materials, but studies have shown that it also has a good catalytic degradation effect. It is expected to obtain more efficient photocatalyst by using ZnO to modify other materials.

2. Materials and methods

2.1 Materials, apparatus and reagents

The main reagents and equipment are the $SnCl_4 \cdot 5H_2O$, $NH_3 \cdot H_2O/NH_4Cl$, $ZnCl_2$, magnetic stirrer, muffle furnace, UV photocatalytic reaction box, transmission electron microscopy, X-ray diffraction, 752 UV spectrophotometer etc[7].

2.2 Preparation of Composite Photocatalyst

In this experiment, nano composite photocatalyst was prepared by chemical precipitation method with tin tetrachloride pentahydrate and $ZnCl_2$.The different doping ratios of the catalyst were determined by different molar ratios of tin ion and zinc ion in the catalysts. The specific process of preparation is as follows in Figure 1:

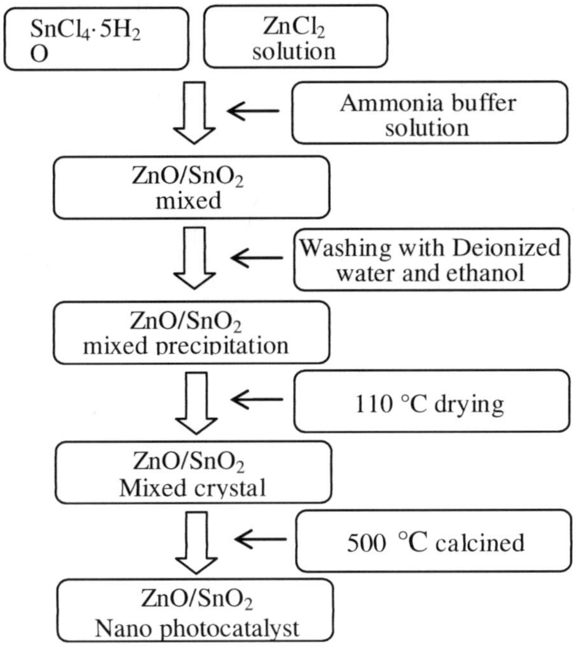

Figure 1.preparation process of photocatalyst

3. Experimental method

Take 50mL of a certain concentration of oil contaminated seawater and measure the pH value. The addition of a certain amount of SnO_2/ZnO composite photocatalyst was carried out under ultraviolet light, after a certain period of time of irradiation to extract. The amount of residual diesel in seawater was determined by UV-Vis spectrophotometry method, and the photocatalytic degradation efficiency was calculated.

4. Results

4.1 Products of characterization by transmission electron microscopy

4.1.1 Analysis of transmission electron microscopy

From the above two figuresfor the doping ratio of 0.2,0.25 composite nano-tin dioxide catalysts, it can be clearly seen that the composite nano-tin dioxide particles are irregular balls. Its size is relatively uniform. The larger particles are gathered by many small particles to form the catalyst particle size

IOP Conf. Series: Materials Science and Engineering **281** (2017) 012015 doi:10.1088/1757-899X/281/1/012015

between 10-30nm. The resulting particle size is consistent with the scherrer formula. Please see Figure 2.

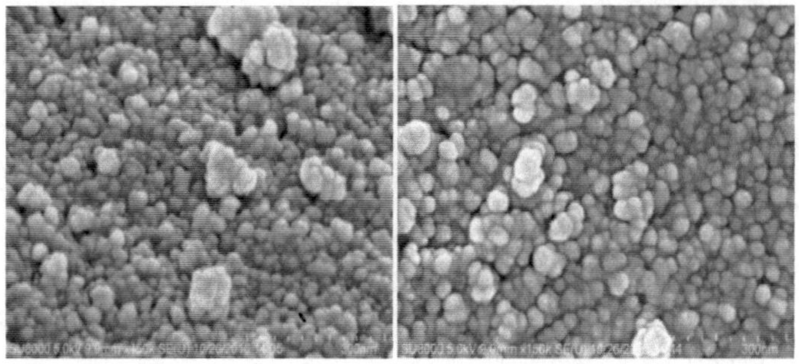

Figure 2. SEM images of composite photocatalyst particles

4.1.2 X-ray diffraction analysis (XRD)

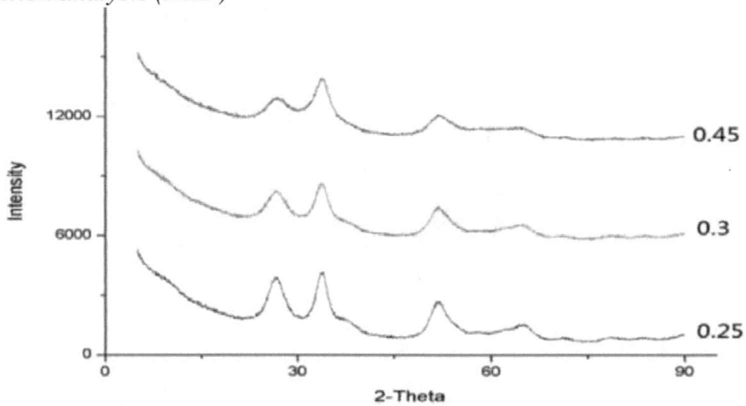

Figure 3. XRD pattern of composite nano-SnO_2

In this paper, by using XRD analysis method for the preparation of compound SnO_2 / ZnO photocatalyst doped ratio 0.25 when the particle size and particle shape is analyzed, as shown in Figure 3.

Doping ratio 0.25, when 2 theta is 26.5569 °, has the obvious peak value, the average crystallite size of nanometer SnO_2 light catalyst Scherrer formula can be used.The lattice constants a, b and c were 8.7148Å, 8.7013Å and 6.2862Å, (211) diffraction peak, by calculation can be concluded that nano light catalyst SnO_2 average crystallite size of 8.41 nm, with the JCPDS card in SnO_2 (01-079-1164) card, determine the preparation of samples for the rutile structure morphology of SnO_2 particles.

4.2 Effect of pH value of seawater on degradation of marine diesel pollutant

10mg of 0.2g/L simulated oil seawater was taken in six 100mL beakers and 50mL of seawater was added. The calcination temperature of nanometer composite photocatalyst was 500 °C, the dosage was 0.2 g/L, the pH values were 6, 7, 8, 9, 10 and 11. 1 mL of 0.3 g/L hydrogen peroxide was added. The absorbance was measured after stirring for 2 hours under ultraviolet light. The effect of different doping ratios on the removal of oil in seawater is calculated by the degradation efficiency, as shown in Figure 4.

It can be seen from Figure 4 that the pH value in the solution is getting higher and higher, and the removal rate of the oil in the solution gradually increases and then decreases. It can be seen that the degradation rate of the composite catalyst to the seawater is best when the pH of the solution is 10, and the removal rate is the fastest when the pH value between 9 and 10. But when the pH value increases again, the removal rate drops rapidly and the removal effect is poor.

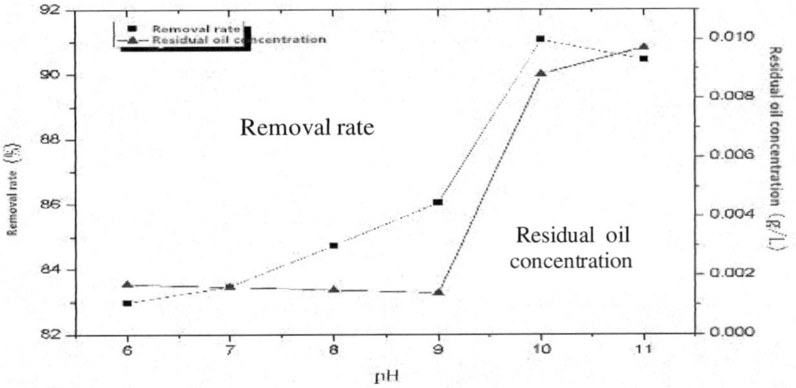

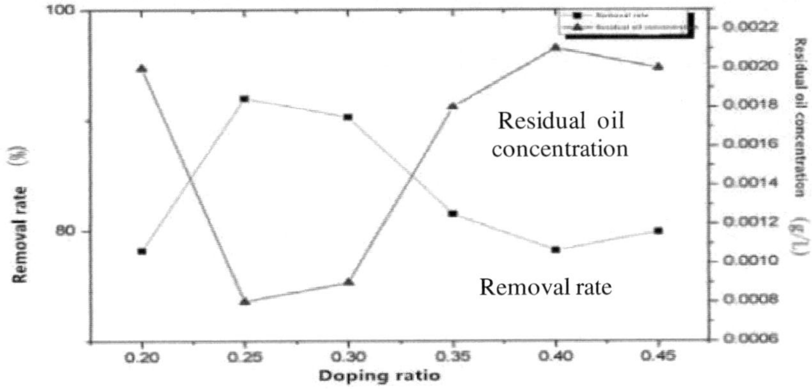

Figure 4. Effects of different pH on the reaction **Figure 5.** Effects of doping ratios on the reaction

4.3 Effect of doping ratio of photocatalyst on degradation of marine diesel pollutant

10mg of 0.2g/L simulated oil seawater was taken in six 100mL beakers and 50mL of seawater was added. The calcination temperature of nanometer composite photocatalyst was 500 °C, the doping ratio of Zn and Sn was 0.2, 0.25, 0.3, 0.35, 0.4 or 0.45. 1 mL of 0.3 g/L hydrogen peroxide added. The absorbance was measured after stirring for 2 hours under ultraviolet light. The degradation efficiency was calculated to obtain the effect of different doping ratios on the removal of oil from seawater, as shown in Figure 5.

As can be seen from Figure 5, when the doping ratio is 0.25, the treatment efficiency is better than 0.2. But when the doping ratio increases again, the removal rate begins to decrease. The reason for the decrease in the removal rate may be: When the doping ratio is gradually increasing, the separation efficiency of the dopant carriers on the composite catalyst decreases and the recombination rate decreases, reducing the activity of the catalyst and decreasing the removal rate [9]. When the doping ratio is 0.45, it can be seen that the removal rate is increased in a small range. In order to study and verify, the validation test are carried out with doping ratio of 0.5, 0.55 and 0.6. It is seen from the data that the removal rate of oil is relatively constant and tends to be gentle.

4.4 Effect of photocatalytic reaction time on degradation of marine diesel pollutant

10mg of 0.2g/L simulated oil seawater was taken in six 100mL beakers and 50mL of seawater was added. The calcination temperature of nanometer composite photocatalystwas 500 °C, the dosage was 0.2 g/L and doping ratio was 0.25. Reaction time was 0.5, 1, 1.5, 2, 2.5 or 3 h. 1 mL of 0.3 g/L

IOP Conf. Series: Materials Science and Engineering **281** (2017) 012015 doi:10.1088/1757-899X/281/1/012015

hydrogen peroxide was added. After extraction, the absorbance was measured and the degradation efficiency was calculated to obtain the effect of different doping ratios on the removal of oil in seawater, as shown in Figure 6.

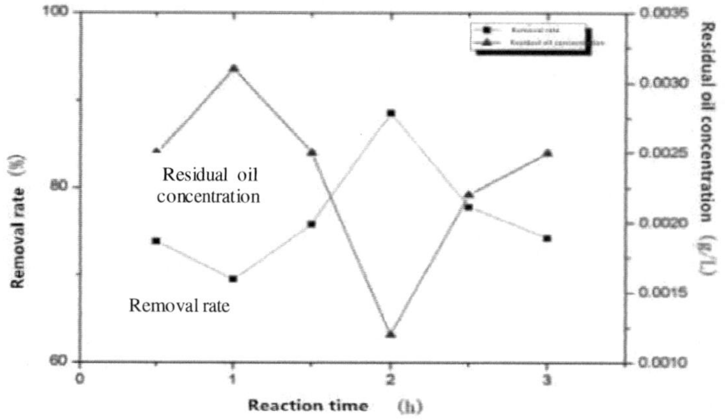

Figure 6. Effects of illumination time on the reaction

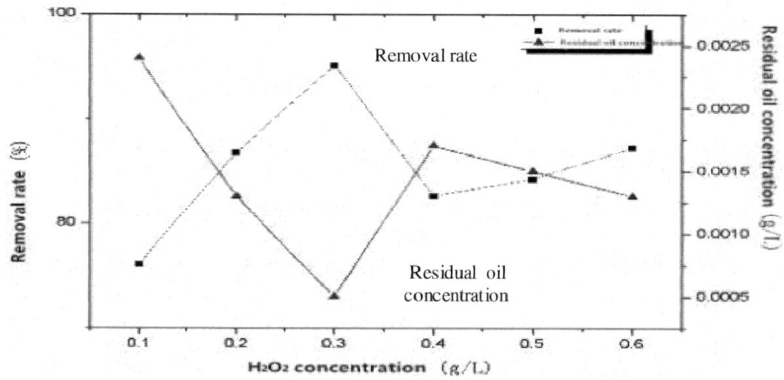

Figure 7. Effects of the hydrogen peroxide concentration on reaction

It can be seen from Figure 6 that the amount of photocatalytic oxidation will increase with the increase of ultraviolet light time in the initial stage of the reaction, but the removal rate is high at 0.5h. But when the UV light time is higher than 2 hours, the seawater oil removal rate began to decline slowly, may be due to the mixed oil and catalyst that prevent the catalyst for catalytic reaction.

4.5 Effect of Hydrogen Peroxide Concentration on Degradation of Diesel pollutant

10mg of 0.2g/L simulated oil seawater was taken in six 100mL beakers and 50mL of seawater was added. The calcination temperature of nanometer composite photocatalyst was 500°C, the dosage was 0.2 g/L and doping ratio was 0.25. 0.1, 0.2, 0.3, 0.4, 0.5 and 0.6 g/L hydrogen peroxide 1 mL was add and stirred for 2 hours under ultraviolet light. The degradation efficiency was calculated to obtain the effect of different concentration of hydrogen peroxide on the removal of oil from seawater, as shown in Figure 7.

It can be seen from Fig. 7 that the removal rate is the highest when the concentration of H_2O_2 is 0.3g/L. The possible reason is that H_2O_2 is the electron acceptor, which competes with the catalyst for electrons, making the tin dioxide surface electron-hole pairs difficult to compound and generating strong oxidizing ions. Thus, the photocatalytic activity was enhanced.

4.6 Effect of Composite Photocatalyst Dosage on Degradation of Marine Diesel pollutant

10mg of 0.2g/L simulated oil seawater was taken in six 100mL beakers and 50mL of seawater was added. The calcination temperature of nanometer composite photocatalyst was 500°C, the dosage was

0.1, 0.2, 0.3, 0.4, 0.5 or 0.6 g/L and doping ratio was 0.25. 0.3 g/L hydrogen peroxide 1 mL was added and stirred for 2 hours under ultraviolet light. The degradation efficiency was calculated to obtain the effect of different dosage on the removal of oil from seawater, as shown in Figure 8.

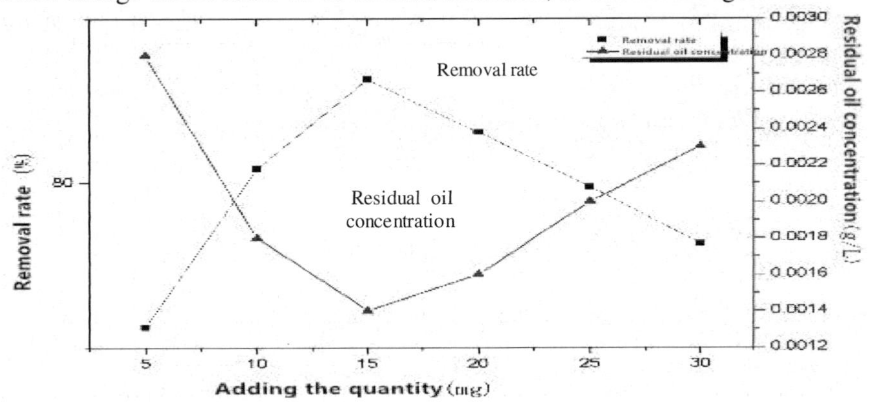

Figure 8. Effects of dosage on the reaction

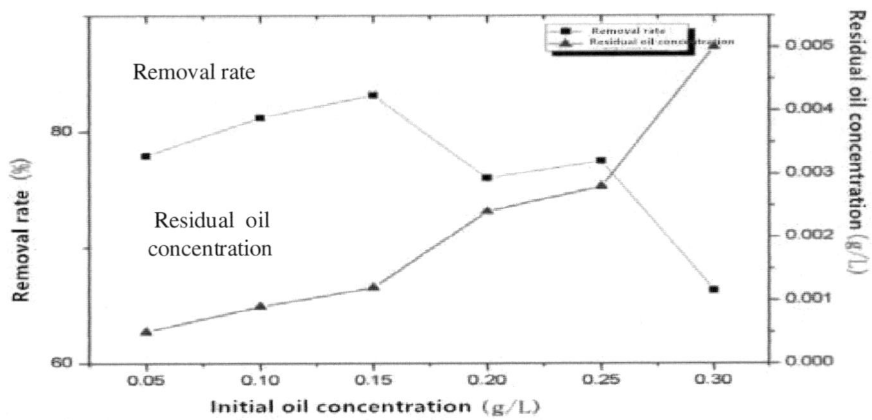

Figure 9. Effects of the initial concentration of diesel on the reaction

It can be seen from Figure 8 that at the beginning of the reaction, the removal rate of oil would increase with the increase of the dosage of nanometer tin dioxide composite photocatalyst. When the dosage of composite photocatalyst is more than 0.3g/L, the removal rate began to declineslowly. The reason for the decrease in the removal rate may be that when the concentration of catalyst in seawater increases to a certain extent, the addition of the catalyst will result in enhanced scattering of light by the tin dioxide, which may result in a decrease in the utilization of solar energy and a reduction in electrons.

4.7 Effect of initial oil concentration on degradation of marine diesel pollutant

10mg of 0.1, 0.2, 0.3, 0.4, 0.5 and 0.6 g/L simulated oil seawater was taken in six 100mL beakers and 50mL of seawater was added. The calcination temperature of nanometer composite photocatalyst was 500 °C, the dosage was 0.2 g/L and doping ratio was 0.25. 0.3 g/L hydrogen peroxide 1 mL was added and stirred for 2 hours under ultraviolet light. The degradation efficiency was calculateedto obtain the effect of different dosage on the removal of oil from seawater, as shown in Figure 9.

It can be seen from Figure 9: at the beginning of the reaction, the oil removal rate would increase with the increase of the concentration of diesel oil. But when the diesel concentration exceeds 0.15 g/L, the seawater oil removal rate began to decline slowly. The reason for the decrease in the removal rate may be that the oil concentration is too high and the degradation rate is reduced.

IOP Conf. Series: Materials Science and Engineering **281** (2017) 012015 doi:10.1088/1757-899X/281/1/012015

5. Orthogonal experiment

In order to determine the optimal reaction conditions of nano-SnO_2 composite photocatalyst for oil degradation in seawater, the factors such as the doping ratio and dosage of catalyst, the initial concentration of diesel, the pH value of solution, the time of ultraviolet light and the concentration of H_2O_2 solution were selected. Sixfactorsfive levels of orthogonal verification experimentswas designed. And the effect of these six factors on the catalytic oxidation rate of nano-SnO_2 composite photocatalyst was analyzed according to the orthogonal test table.

Table.1 Orthogonal experiment table

No.	Reaction time (h)	Concentration of dieselg/L	Doping ratio	Dosage (g/L)	Concentration of H_2O_2(g/L)	pH value	Removal rate
1	0.5	0.05	0.2	0.1	0.1	6	86.05%
2	0.5	0.1	0.25	0.4	0.5	7	85.71%
3	0.5	0.15	0.3	0.5	0.4	8	75.25%
4	0.5	0.2	0.35	0.3	0.2	10	78.06%
5	0.5	0.25	0.4	0.2	0.3	9	73.76%
6	1	0.05	0.25	0.2	0.2	7	80.00%
7	1	0.1	0.3	0.3	0.4	10	78.14%
8	1	0.15	0.2	0.1	0.5	9	76.83%
9	1	0.2	0.4	0.4	0.3	6	70.12%
10	1	0.25	0.35	0.5	0.1	8	77.17%
11	1.5	0.05	0.3	0.3	0.3	8	83.08%
12	1.5	0.1	0.35	0.1	0.5	9	90.89%
13	1.5	0.15	0.25	0.2	0.2	7	75.68%
14	1.5	0.2	0.4	0.5	0.1	6	78.06%
15	1.5	0.25	0.2	0.4	0.4	10	72.69%
16	2	0.05	0.35	0.4	0.4	9	89.97%
17	2	0.1	0.3	0.2	0.5	6	77.93%
18	2	0.15	0.25	0.5	0.3	7	84.65%
19	2	0.2	0.4	0.1	0.2	8	84.44%
20	2	0.25	0.2	0.3	0.1	10	78.64%
21	2.5	0.05	0.4	0.5	0.5	10	74.74%
22	2.5	0.1	0.35	0.3	0.1	7	91.54%
23	2.5	0.15	0.2	0.4	0.2	6	79.11%
24	2.5	0.2	0.3	0.2	0.4	8	79.22%
25	2.5	0.25	0.25	0.1	0.3	9	77.68%
k1	398.83	413.84	393.32	415.89	411.47	391.28	
k2	382.27	424.22	403.71	386.58	397.29	417.58	
k3	400.40	391.52	393.63	409.48	389.29	399.17	
k4	415.64	389.91	427.64	397.60	395.27	409.12	
k5	402.29	379.93	381.12	389.88	406.10	382.27	
R	33.37	44.29	46.52	29.31	22.18	35.30	

One of the best reaction conditions for the determination of oil pollution in seawater by orthogonal test is that the doping ratio of nanometer tin dioxide composite photocatalyst is 0.35. The dosage of catalyst is 0.3g/L, the initial concentration of diesel is 0.1g/L, The concentration of hydrogen peroxide solution is 0.1g/L, the pH value of the solution is 7, the time of ultraviolet light is 2.5 hoursand the oil removal rate in seawater is 91.54%.

In addition, the orthogonal test table can also be obtained that the effect ofcatalyst doping ratio. The catalyst dosage, the initial concentration of diesel, the concentration of hydrogen peroxide, the pH value of the solution and the UV irradiation time on the degradation rate are as follows: catalyst doping ratio> initial concentration of diesel> solution pH> UV light time> catalyst dosage> hydrogen peroxide concentration.

In order to verify whether the removal rate was stable under the best experimental conditions in orthogonal experiments, three sets of parallel experiments were carried out as follows:

Table.2 Experimental verification table

No.	Concentration of diesel (g/L)	Removal rate
1	0.102	92.98%
2	0.102	92.34%
3	0.098	92.69%

From the validation test data, it can be obtained that the removal efficiency is relatively stable and the effect is good, the average removal effect can reach 92.67%.

6. Conclusions

(1) ZnO/SnO_2 composite nanometer photocatalyst prepared by chemical precipitation method was characterized by XRD and SEM. The results show that the composite nano-tin dioxide is an irregular sphere with uniform size.

(2) ZnO/SnO_2 composite nanometer photocatalyst can effectively photocatalytic oxidize diesel pollutant in seawater under ultraviolet light.

(3) When the doping ratio of ZnO/SnO_2 composite nanometer photocatalyst is 0.35, the dosage of catalyst is 0.3 g/L, the initial concentration of diesel is 0.1 g/L, the concentration of hydrogen peroxide solution is 0.1 g/L, the pH value of solution is 7, UV light time was 2.5 hours, the seawater oil removal rate can reach 91.54%.

(4) The influence degree of each factor on the catalytic degradation rate of nano-tin dioxide composite photocatalyst is: catalyst doping ratio> initial concentration of diesel oil> pH value> UV irradiation time> catalyst dosage> hydrogen peroxide concentration..

Acknowledgments

This work has been supported by a grant from marine research and special funds public service sectors of the State Oceanic Administration People's Republic of China (201305002), Liaoning Province large equipment and equipment sharing service platform capacity building funds, Department of Science and Technology of Liaoning(2016LD0105), Science Foundation of Department of Ocean and Fisheries of Liaoning Province (201733), and Liaoning Science and Technology Public Welfare Fund(20170002).

References

[1] Gu E H. *Numerical Simulation of Oil Spill Behavior and Destination at Sea* [D]. Zhejiang University, 2013.

[2] Zhang X F, Yin L. Study on Photocatalytic Performance of Catalysts TiO_2-SnO_2 [J]. *Journal of Chifeng University,* 2011, 03:9-11.

[3] Yu X C, X L Shang, Ji Q Y etc. Photocatalytic Activity of Nano Ce/SnO_2 in Aquaculture Wastewater Treatment [J]. *Marine Environmental Science,* 2016(4): 501-506.

[4] Liu Y Q, Yu X C, Wu Y Y etc. Application of UV-Visible Upconversion/TiO_2 Composite Photocatalyst in Offshore Oil Pollution Treatment [J]. *Journal of Dalian Ocean University,* 2014,29(4):420-424.

[5] Look D C. Recent advances in ZnO materials and devices [J]. *Materials Science & Engineering B,* 2001, 80(1–3):383-387.

[6] Jin X J, Yu X Ci, Wu Y Y etc. Preparation of Nano SnO_2 Photocatalyst and Its Application in Aquaculture Wastewater Treatment [J]. *Journal of Oceanography,* 2015, 33(2):64-69.

[7] Zhang J, Liu H Y, Ji Q Y, et al. Preparation of nano SnO_2 photocatalyst and Study on its photocatalytic degradation of marine diesel oil [J]. *Journal of Dalian Ocean University,* 2016, 31 (4): 438-443.

IOP Conf. Series: Materials Science and Engineering **281** (2017) 012016 doi:10.1088/1757-899X/281/1/012016

Transversely Isotropic Hyperelastic Constitutive Model of Short Fiber Reinforced EPDM Based on Tensor Function

Q L Feng[1,2], C Li[1*] and Y F Liao[1]
[1]Taiyuan University of Science and Technology, Taiyuan 030024, China
[2]Shanxi Architectural College,Taiyuan 030006, China

E-mail: tydz_lc@126.com

Abstract. Short fiber reinforced EPDM is a new kind of composite material used in solid rocket motor winding and coating. It has relatively large deformation under the small stress condition, and the physical non-linear characteristic is obvious. Due to the addition of fiber in the specific direction of the rubber, the macroscopic mechanical properties are expressed as transversely isotropic properties. In order to describe the mechanical behavior under the impact and vibration, the transversely isotropic hyperelastic constitutive model based on tensor function is proposed. The symmetry of the transversely isotropic incompressible material limits the stress tensor '$\boldsymbol{K}$' to be characterized as a function of 5 tensor invariants and 4 scalar invariants. The third power constitutive equations of the model give 12 independent elastic constants of the transversely isotropic nonlinear elastic material. The experimental results show that the non-zero elastic constants are different in the fiber direction and at the different strain rate. Number and value of adiabatic layer and related products R & D has a reference value.

1. Introduction

Ethylene Propylene Diene Monomer (EPDM) is a kind of polymer rubber which has good heat insulation performance. The thermal stability of EPDM can be attributed to its saturated main chain structure. EPDM exhibits outstanding resistance to oxidation, ozonization, weathering effects and excellent low temperature properties. EPDM also has the lowest density among elastomer (0.85 g/cm3) [1,2]. In order to adapt to the complex working environment, the mechanical properties of materials is improved by adding reinforcing agent, organic and inorganic filler[3,4,5]. The short fibre added to the isotropic rubber matrix can improve the tensile performance. Multiple theory model is used to describe the mechanical behavior of rubber materials, such as the Neo-Hookean model[6], the Mooney-Rivlin model[7], the Yeoh model[8], the Ogden model[9],etc. Due to the addition of fiber in the specific direction of the rubber, the macroscopic mechanical properties are expressed as transversely isotropic properties[10]. This paper presents a new constitutive model based on tensor function transverse isotropic elastic material.

2. Experiment and constitutive theory

The internal structure of short fiber reinforced EPDM material is shown in Fig.1. It defines parallel to the fiber direction of 0º, perpendicular to the fiber direction of 90º. The thickness of the thin film coating layer is 0.5mm. The strip specimen's size is 80mm×10mm. Its gage length is 40mm. The tensile testing was performed along the direction 0º and 90º with different tensile strain rates for 0.125/min, 0.5/min, 2.5/min under the condition of 293k and 50% humidity[11].

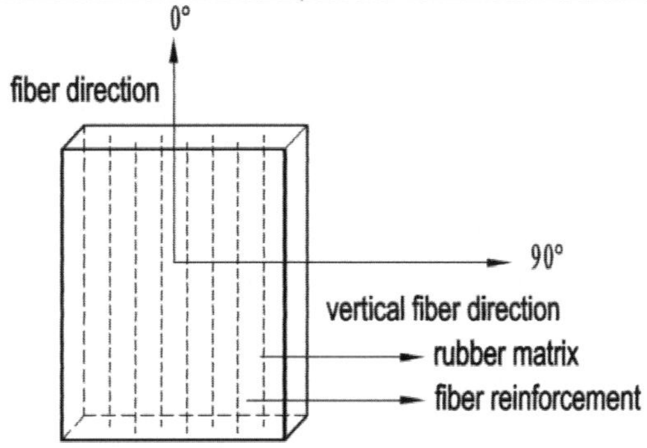

Fig.1 Schematic of short fiber reinforced EPDM.

Tensor function in describing nonlinear material mechanics behavior model is very effective. The general nonlinear constitutive theory research and the theory of tensor functions in the application of continuum mechanics began in Rivlin[12] and Reiner's[13] work, and the nonlinear tensor theory has carried on the detailed research by Mr Chen Li [14]. The symmetry of the transversely isotropic incompressible material limits the stress tensor '*K*' to be characterized as the function of 5 tensor invariants and 4 scalar invariants in equation 1.

$$\boldsymbol{K} = \boldsymbol{K}(\boldsymbol{1}, \boldsymbol{E}, \boldsymbol{E}^2, \boldsymbol{a} \otimes \boldsymbol{a}, \boldsymbol{a} \otimes \boldsymbol{a} \cdot \boldsymbol{E} + \boldsymbol{E} \cdot \boldsymbol{a} \otimes \boldsymbol{a}, \mathrm{tr}\boldsymbol{E}, \mathrm{tr}\boldsymbol{E}^2, \boldsymbol{a} \cdot \boldsymbol{E} \cdot \boldsymbol{a}, \boldsymbol{a} \cdot \boldsymbol{E}^2 \cdot \boldsymbol{a}) \tag{1}$$

'*E*' is strain tensor. '*a*' is meaning of transverse isotropic materials on the surface of the isotropic normal vector, and it is an one-order tensor invariant. '$\boldsymbol{a} \otimes \boldsymbol{a}$' is meaning of a two-order tensor invariant. The nonlinear transverse isotropic incompressible elastic material constitutive in equation 1 can be described as follows:

$$\boldsymbol{K} = \varphi_1 \boldsymbol{1} + 2\varphi_2 \boldsymbol{E} + 3\varphi_3 \boldsymbol{E}^2 + \varphi_4 \boldsymbol{aa} + \varphi_5 (\boldsymbol{aa} \cdot \boldsymbol{E} + \boldsymbol{E} \cdot \boldsymbol{aa}) \tag{2}$$

$\{\boldsymbol{e}_i\}$ is the natural base vector strain tensor of Cartesian coordinate system.

$$\begin{aligned} &\boldsymbol{e}_{\mathrm{a}} := \boldsymbol{e}_1 \otimes \boldsymbol{e}_1, \quad \boldsymbol{e}_{\mathrm{b}} := \boldsymbol{e}_2 \otimes \boldsymbol{e}_2, \quad \boldsymbol{e}_{\mathrm{c}} := \boldsymbol{e}_3 \otimes \boldsymbol{e}_3 \\ &\boldsymbol{e}_{\mathrm{d}} := \boldsymbol{e}_1 \otimes \boldsymbol{e}_2 + \boldsymbol{e}_2 \otimes \boldsymbol{e}_1, \quad \boldsymbol{e}_{\mathrm{e}} := \boldsymbol{e}_2 \otimes \boldsymbol{e}_3 + \boldsymbol{e}_3 \otimes \boldsymbol{e}_2, \quad \boldsymbol{e}_{\mathrm{f}} := \boldsymbol{e}_3 \otimes \boldsymbol{e}_1 + \boldsymbol{e}_1 \otimes \boldsymbol{e}_3 \end{aligned} \tag{3}$$

$\{\boldsymbol{e}_\alpha\}(\alpha = a, \cdots, f)$ make up the two-order symmetric tensor space Sym orthogonal basis, and $\boldsymbol{a} = \boldsymbol{e}_3$.

$$\begin{cases} \boldsymbol{1} = \boldsymbol{e}_a + \boldsymbol{e}_b + \boldsymbol{e}_c \\ \boldsymbol{E} = E_1\boldsymbol{e}_a + E_2\boldsymbol{e}_b + E_3\boldsymbol{e}_c + E_4\boldsymbol{e}_d + E_5\boldsymbol{e}_e + E_6\boldsymbol{e}_f \\ \boldsymbol{E}^2 = (E_1^2 + E_4^2 + E_6^2)\boldsymbol{e}_a + (E_2^2 + E_4^2 + E_5^2)\boldsymbol{e}_b + (E_3^2 + E_5^2 + E_6^2)\boldsymbol{e}_c \\ \quad +(E_1E_4 + E_2E_4 + E_5E_6)\boldsymbol{e}_d + (E_2E_5 + E_3E_5 + E_4E_6)\boldsymbol{e}_e \\ \quad +(E_1E_6 + E_3E_6 + E_4E_5)\boldsymbol{e}_f \\ \boldsymbol{a} \otimes \boldsymbol{a} = \boldsymbol{e}_3 \otimes \boldsymbol{e}_3 = \boldsymbol{e}_c \\ \boldsymbol{a} \otimes \boldsymbol{a} \cdot \boldsymbol{E} + \boldsymbol{E} \cdot \boldsymbol{a} \otimes \boldsymbol{a} = E_{3k}(\boldsymbol{e}_3 \otimes \boldsymbol{e}_k + \boldsymbol{e}_k \otimes \boldsymbol{e}_3) = 2E_3\boldsymbol{e}_c + E_5\boldsymbol{e}_e + E_6\boldsymbol{e}_f \end{cases} \tag{4}$$

$$\begin{cases} \bar{I}_1 = tr\boldsymbol{E} = E_1 + E_2 + E_3 \\ \bar{I}_2 = tr\boldsymbol{E}^2 = E_1^{\ 2} + E_2^{\ 2} + E_3^{\ 2} + 2(E_4^{\ 2} + E_5^{\ 2} + E_6^{\ 2}) \\ J_1 = \boldsymbol{a}\cdot\boldsymbol{E}\cdot\boldsymbol{a} = E_3 \\ J_2 = \boldsymbol{a}\cdot\boldsymbol{E}^2\cdot\boldsymbol{a} = E_{3k}E_{k3} = E_3^{\ 2} + E_5^{\ 2} + E_6^{\ 2} \end{cases} \tag{5}$$

Making the equation 4,5 into equation 2, get 1 to 3 power of constitutive equation 6,7,8.

$$\begin{aligned} \boldsymbol{K}_{(1)} &= \varphi_1\boldsymbol{I} + 2\varphi_2\boldsymbol{E} + \varphi_4\boldsymbol{a}\otimes\boldsymbol{a} + \varphi_5(\boldsymbol{a}\otimes\boldsymbol{a}\cdot\boldsymbol{E} + \boldsymbol{E}\cdot\boldsymbol{a}\otimes\boldsymbol{a}) \\ &= (k_1\bar{I}_1 + k_2J_1)\boldsymbol{I} + 2k_3\boldsymbol{E} + (k_2\bar{I}_1 + k_4J_1)\boldsymbol{a}\otimes\boldsymbol{a} + k_5(\boldsymbol{a}\otimes\boldsymbol{a}\cdot\boldsymbol{E} + \boldsymbol{E}\cdot\boldsymbol{a}\otimes\boldsymbol{a}) \end{aligned} \tag{6}$$

$$\begin{aligned} \boldsymbol{K}_{(2)} &= \varphi_1\boldsymbol{I} + 2\varphi_2\boldsymbol{E} + 3\varphi_3\boldsymbol{E}^2 + \varphi_4\boldsymbol{a}\otimes\boldsymbol{a} + \varphi_5(\boldsymbol{a}\otimes\boldsymbol{a}\cdot\boldsymbol{E} + \boldsymbol{E}\cdot\boldsymbol{a}\otimes\boldsymbol{a}) \\ &= (k_1\bar{I}_1^{\ 2} + k_2\bar{I}_2 + 2k_3\bar{I}_1J_1 + k_4J_1^{\ 2} + k_5J_2)\boldsymbol{I} + 2(k_2\bar{I}_1 + k_6J_1)\boldsymbol{E} + 3\mathrm{k}_7\boldsymbol{E}^2 \\ &\quad + (k_3\bar{I}_1^{\ 2} + k_6\bar{I}_2 + 2k_4\bar{I}_1J_1 + k_8J_1^{\ 2} + k_9J_2)\boldsymbol{a}\otimes\boldsymbol{a} + (k_5\bar{I}_1 + k_9J_1)(\boldsymbol{a}\otimes\boldsymbol{a}\cdot\boldsymbol{E} + \boldsymbol{E}\cdot\boldsymbol{a}\otimes\boldsymbol{a}) \end{aligned} \tag{7}$$

$$\begin{aligned} \boldsymbol{K}_{(3)} &= \varphi_1\boldsymbol{I} + 2\varphi_2\boldsymbol{E} + 3\varphi_3\boldsymbol{E}^2 + \varphi_4\boldsymbol{a}\otimes\boldsymbol{a} + \varphi_5(\boldsymbol{a}\otimes\boldsymbol{a}\cdot\boldsymbol{E} + \boldsymbol{E}\cdot\boldsymbol{a}\otimes\boldsymbol{a}) \\ &= (k_1\bar{I}_1^{\ 3} + 2k_2\bar{I}_1\bar{I}_2 + k_3\bar{I}_1 + 3k_4\bar{I}_1^{\ 2}J_1 + k_5\bar{I}_1J_1^{\ 2} + 2k_6\bar{I}_1J_2 + k_7\bar{I}_2J_1 + k_8J_1^{\ 3} \\ &\quad + k_9J_1J_2 + k_{17}\bar{I}_1^{\ 2} + k_{18}\bar{I}_2)\boldsymbol{I} + 2(k_2\bar{I}_1^{\ 2} + k_{10}\bar{I}_2 + k_7\bar{I}_1J_1 + k_{11}J_1^{\ 2} + k_{12}J_2)\boldsymbol{E} \\ &\quad + 3(\mathrm{k}_3\bar{I}_1 + k_{13}J_1)\boldsymbol{E}^2 + (k_4\bar{I}_1^{\ 3} + k_7\bar{I}_1\bar{I}_2 + k_{13}\bar{I}_1 + k_5\bar{I}_1^{\ 2}J_1 + 3k_8\bar{I}_1J_1^{\ 2} + k_9\bar{I}_1J_2 \\ &\quad + 2k_{11}\bar{I}_2J_1 + k_{14}J_1^{\ 3} + 2k_{15}J_1J_2 + k_{19}\bar{I}_1^{\ 2} + k_{20}\bar{I}_2)\boldsymbol{a}\otimes\boldsymbol{a} + (k_6\bar{I}_1^{\ 2} + k_{12}\bar{I}_2 + k_9\bar{I}_1J_1 \\ &\quad + k_{15}J_1^{\ 2} + k_{16}J_2)(\boldsymbol{a}\otimes\boldsymbol{a}\cdot\boldsymbol{E} + \boldsymbol{E}\cdot\boldsymbol{a}\otimes\boldsymbol{a}) \end{aligned} \tag{8}$$

$$\begin{aligned} \boldsymbol{K} &= \boldsymbol{K}_{(1)} + \boldsymbol{K}_{(2)} + \boldsymbol{K}_{(3)} \\ &= (k_1\bar{I}_1^{\ 3} + 2k_2\bar{I}_1\bar{I}_2 + k_3\bar{I}_1 + 3k_4\bar{I}_1^{\ 2}J_1 + k_5\bar{I}_1J_1^{\ 2} + 2k_6\bar{I}_1J_2 + k_7\bar{I}_2J_1 + k_8J_1^{\ 3} \\ &\quad + k_9J_1J_2 + k_{17}\bar{I}_1^{\ 2} + + k_{18}\bar{I}_1J_1 + k_{19}\bar{I}_2 + k_{20}J_1^{\ 2} + k_{21}J_1 + k_{22}J_2)\boldsymbol{I} + 2(k_2\bar{I}_1^{\ 2} + k_{10}\bar{I}_2 \\ &\quad + k_7\bar{I}_1J_1 + k_{11}J_1^{\ 2} + k_{12}J_2 + k_{23} + k_{24}\bar{I}_1 + k_{25}J_1)\boldsymbol{E} + 3(k_3\bar{I}_1 + k_{13}J_1 + k_{26})\boldsymbol{E}^2 \\ &\quad + (k_4\bar{I}_1^{\ 3} + k_7\bar{I}_1\bar{I}_2 + k_5\bar{I}_1^{\ 2}J_1 + 3k_8\bar{I}_1J_1^{\ 2} + k_9\bar{I}_1J_2 + 2k_{11}\bar{I}_2J_1 + k_{13}\bar{I}_1 + k_{14}J_1^{\ 3} \\ &\quad + 2k_{15}J_1J_2 + k_{27}\bar{I}_1^{\ 2} + k_{28}\bar{I}_2 + k_{29}\bar{I}_1J_1 + k_{30}J_1^{\ 2} + k_{31}J_1 + k_{32}J_2)\boldsymbol{a}\otimes\boldsymbol{a} + (k_6\bar{I}_1^{\ 2} \\ &\quad + k_{12}\bar{I}_2 + k_9\bar{I}_1J_1 + k_{15}J_1^{\ 2} + k_{16}J_2 + k_{33} + k_{34}\bar{I}_1 + k_{35}J_1)(\boldsymbol{a}\otimes\boldsymbol{a}\cdot\boldsymbol{E} + \boldsymbol{E}\cdot\boldsymbol{a}\otimes\boldsymbol{a}) \end{aligned} \tag{9}$$

To simplify the equation 9 to uniaxial tensile state specific equation are obtained.

IOP Conf. Series: Materials Science and Engineering **281** (2017) 012016 doi:10.1088/1757-899X/281/1/012016

$$K_{ii} = (k_1 \bar{I}_1 + k_2 \bar{I}_2 + k_3 \bar{I}_1^{\,2} + k_4 \bar{I}_1 J_1 + k_5 \bar{I}_1^{\,3} + k_6 \bar{I}_1 \bar{I}_2) + 2(k_7 + k_8 \bar{I}_2 + k_9 J_2 + k_{10} \bar{I}_1^{\,2})\varepsilon_{ii} + 3(k_{11} + k_{12} \bar{I}_1)\varepsilon_{ii}^2 \quad (10)$$

3. Experimental analysis and discussion

The engineering stress and strain curve are obtained through 0° and 90° direction of uniaxial tensile test, and the elastic constants of the nonlinear elastic constitutive equation is obtained by REGRESS function in MATLAB software [15]. The elastic constants 'k_1' –'k_{12}' are shown in Tab.1 and Tab.2 . Non-zero elastic constants are 'k_7', 'k_8', 'k_{11}'. Among them 'k_7' is a low order elastic constant in constitutive equation, and it occupies an important position in the material mechanics performance of elastic deformation. Its value decreases with the increasing of tensile strain rate. 'k_8' and 'k_{11}' are high order elastic constants. Their values are affected by the nonlinear mechanical properties of the material, could be positive or negative. It can draw stress and strain curve through bringing these strain data and the constants 'k_1' –'k_{12}' into equation 10. Better fitting effect as can be seen from Fig.2. The equation 10 for describing transverse isotropic materials under uniaxial tensile constitutive relation is very effective.

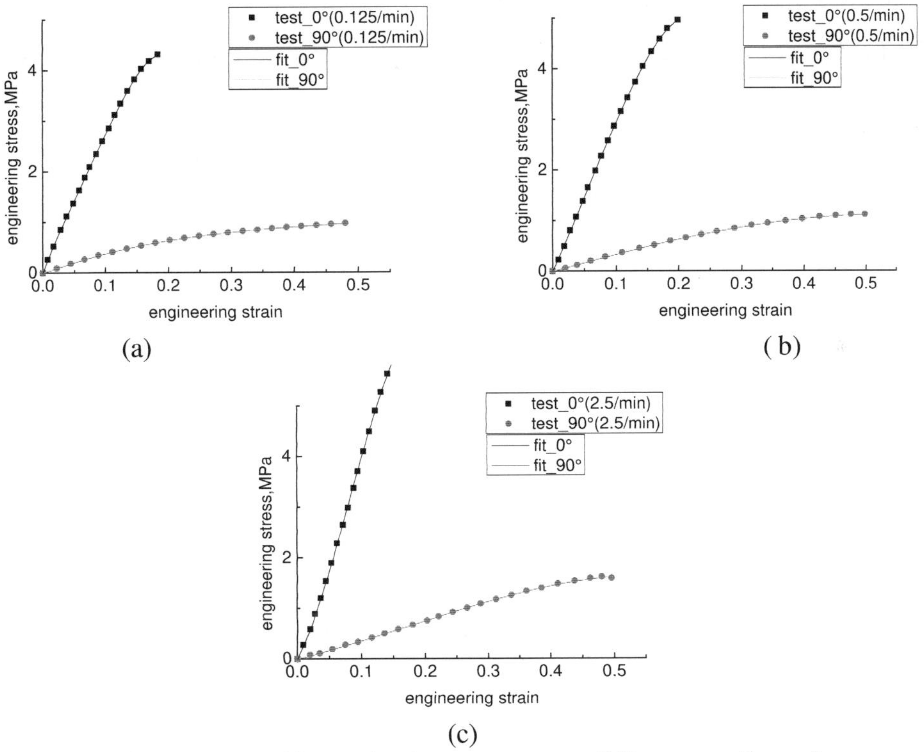

(c)

Fig 2. Fitting results of hyperelastic parameters at different tensile strain rates.

Table 1. The non-zero elastic constants of short fiber reinforced EPDM 0° at different tensile strain rates.

	0.125/min (MPa)	0.5/min (MPa)	2.5/min (MPa)
k_7	14.2339	13.74827	12.25978
k_8	-69.9581	-117.13	-366.889
k_{11}	3.884324	18.17015	85.62068

Table 2. The non-zero elastic constants of short fiber reinforced EPDM 90° at different tensile rates.

	0.125/min (MPa)	0.5/min (MPa)	2.5/min (MPa)
k_7	2.269142	1.766896	1.604102
k_8	1.924606	-0.82582	-3.06809
k_{11}	-2.63598	-0.4654	1.496165

4. Conclusions

The following formula shows the short fiber reinforced EPDM uniaxial tensile constitutive equation.

$$K_{ii} = (k_1 \bar{I}_1 + k_2 \bar{I}_2 + k_3 \bar{I}_1^2 + k_4 \bar{I}_1 J_1 + k_5 \bar{I}_1^3 + k_6 \bar{I}_1 \bar{I}_2) + 2(k_7 + k_8 \bar{I}_2 + k_9 J_2 + k_{10} \bar{I}_1^2)\varepsilon_{ii} + 3(k_{11} + k_{12} \bar{I}_1)\varepsilon_{ii}^2$$

It only has 3 elastic constants in direction of specimen 0°and 90° uniaxial tensile in 0.125 / min to 2.5 / min tensile strain rate of short fiber reinforced EPDM.

The low order term elastic constants of 'k_7' value decreases with the increasing of tensile strain rate.

Acknowledgments

The authors thanks for the support of the National Natural Science Foundation of China(No.11372207)
Responsible author; Chen Li, Professor of Taiyuan University of Science and Technology, China. tydz_lc@126.com.

References

[1] Jiang J, Xu J S, Zhang Z S, et al. Rate-dependent compressive behavior of EPDM insulation: Experimental and constitutive analysis[J]. *Mechanics of Materials,* 2016, 96:30-38.

[2] Bhuvaneswari C M, Sureshkumar M S, Kakade S D, et al. Ethylene-propylene Diene Rubber as a Futuristic Elastomer for Insulation of Solid Rocket Motors[J]. *Defence Science Journal*, 2006, 56(3):309-320.

[3] Wang Z, Lu Y, Liu J, et al. Preparation of nano - zinc oxide/EPDM composites with both good thermal conductivity and mechanical properties[J]. *Journal of Applied Polymer Science*, 2011, 119(2):1144-1155.

[4] Biagiotti J, López-Manchado M A, Arroyo M, et al. Ternary composites based on PP-EPDM blends reinforced with flax fibers. Part II: Mechanical properties/morphology relationship[J]. *Polymer Engineering & Science*, 2003, 43(5):1031–1043.

[5] Das A, Ghosh A K, Pal S, et al. The role of thiophosphoryl disulfide on the co - cure of CR - EPDM blends[J]. *Polymers for Advanced Technologies*, 2004, 15(4):551-559.

[6] Mohd Yusop S H, Ab Patar M N A, Abdul Majeed A P P, et al. A Parametric Investigation on the Neo-Hookean Material Constant[J]. *Advanced Materials Research*, 2014, 915-916:853-857.

[7] Kim B, Lee S B, Lee J, et al. A comparison among Neo-Hookean model, Mooney-Rivlin model, and Ogden model for chloroprene rubber[J]. *International Journal of Precision Engineering & Manufacturing*, 2012, 13(5):759-764.

[8] Reuter T, Ponomarev I. Biomechanical parameter determination of scaffold-free cartilage constructs (SFCCs) with the hyperelastic material models Yeoh, Ogden and Demiray[J]. *Current Directions in Biomedical Engineering*, 2015, 1(1):442-445.

[9] Yaagoubi M E, Juhre D, Meier J, et al. Prediction of Energy Release Rate in crack Opening Mode (Mode I) for Filled and Unfilled Elastomers Using the Ogden Model[J]. *Engineering Fracture Mechanics*, 2017, 182.

[10] Tran A, Pham D. Polarization approximations for the macroscopic elastic constants of transversely isotropic multicomponent unidirectional fiber composites[J]. *Journal of Composite Materials*, 2015, 49(30).

[11] Tan B D, Xu J S, Sun C X, Jia Y F, Fan X G, A transversely isotropic visco-hyperelastic constitutive model for short fiber reinforced EPDM[J]. *Chinese Journal of Theoretical and Applied Mechanics*, 2017,49(03):677-684.

[12] Rivlin R S. The Hydrodynamics of Non-Newtonian Fluids. I[M]// Collected Papers of R.S. Rivlin. *Springer New York*, 1997:1842-1863.

[13] Reiner M. Elasticity Beyond the Elastic Limit[J]. *American Journal of Mathematics*, 1948, 70(2):433-446.

[14] Li C, Yang G T, Huang Z Z. Nonlinear constitutive equations and potential function of transversely isotropy elastic materials[J]. *Chinese Quarterly of Mechanics*. 2009.30(04):517-522.

[15] Gao H M, Liu H. The Data Analysis of Uniform Design Experiment Based on MATLAB[J]. *Applied Mechanics & Materials*, 2014, 687-691:1544-1547.

IOP Conf. Series: Materials Science and Engineering **281** (2017) 012017 doi:10.1088/1757-899X/281/1/012017

Effect of GNFs/Al Interfacial Characteristics on the Mechanical Properties of Graphene Nanoflakes Reinforced Aluminum Matrix Composites

S Z Zhao, S J Yan *, X Chen and S L Dai

Beijing Institute of Aeronautical Materials, Beijing100095, PR China

* Corresponding author's e-mail address: shaojiuyan@126.com

Abstract. Graphene nanoflakes-reinforced aluminum matrix composites (GNFs/Al) with 1.0 wt.% GNFs were fabricated by a spark plasma sintering (SPS) method under two different condition. The microstructural and interfacial characteristics of the two composites were systematically investigated, and their different tensile mechanical properties were found to depend strongly on the GNFs/Al interfacial characteristics. Prolonged sintering in SPS should be avoided as it promotes obvious interfacial reaction that degrades the mechanical properties. The enhanced mechanical properties of the GNFs/Al composites in this study can be ascribed to their exquisite microstructures, specifically the well-maintained structure of raw reinforcement, the uniform GNF distribution and their parallel alignment with the tensile direction, the absence of mass interfacial reaction, and the good interface bonding.

1. Introduction

Graphene is a perfect two-dimensional (2-D) lattice of carbon atoms [1], which has become the focus of scientific research as its unique structure and superior properties [2-4]. Graphene nanoflakes (GNFs) which compose of a few graphene layers and possess similar properties of single-layer graphene, is much easier to produce and handle. Current researches have applied GNFs to strengthen various polymer, ceramic, and metal matrix composites [5-7]. On the other hand, aluminum and aluminum alloys are among the key materials in aerospace and military vehicles due to their low density; however, their uses are limited by their relatively low strength. Since the graphene-reinforced aluminum composites are expected to be both lighter and stronger, methods to fabricate such composites are greatly desired.

In recent years, some researchers have attempt to fabricated GNFs reinforced aluminum or its alloy composites [8-13]. Their results are quite different. Shin make the composites two times higher than that of monolithic Al with an addition of only 0.7 vol. % GNFs [11], while some results show that GNFs ruin the strength of the matrix [8]. Most of the present researches have been focused on preparation and mechanical properties of GNFs reinforced aluminum composites. However, investigation on interface is not thorough and some results even contradict because of the complexity associated with the interfacial reaction between graphene and Al matrix. As the interface characteristic have crucial effect on properties of composites, then elaborate study on the interface of the composites is essential. As the reactions between graphene and other elements in the alloys could create more complex composites, it is difficult to analyse the strengthening mechanism in composites with aluminum alloy matrices. Therefore, in this study we selected pure aluminum instead of aluminum alloys as the matrix.

2. Materials and experimental procedures

2.1. Fabrication of GNFs/Al composites

Ball milling was employed to disperse 1.0 wt. % GNFs into aluminum powders to obtain GNFs/Al mixtures. First, 5 g of GNFs was dispersed in ethanol by ultrasonication. The solution and 500 g of aluminum powders were poured into a stainless-steel jar, and mixed by ball milling without any dispersive agents. Hardened stainless-steel balls 3 mm in diameter were used in the milling process (100 rpm and 20 h), with the mass ratio between the ball and material being 10:1. Last, the milled slurry was dried in a vacuum furnace at 60 °C for 10 h.

Fig. 1 gives the low- and high-magnification SEM images of the GNFs/Al mixture. Fig. 1 (a) shows that the aluminum powders are flakes with 20 μm in diameter after milling. In Fig. 1(b), the high-magnification image shows that the aluminum powders were wrapped by GNFs, which indicates the even dispersion of GNFs into aluminum powders.

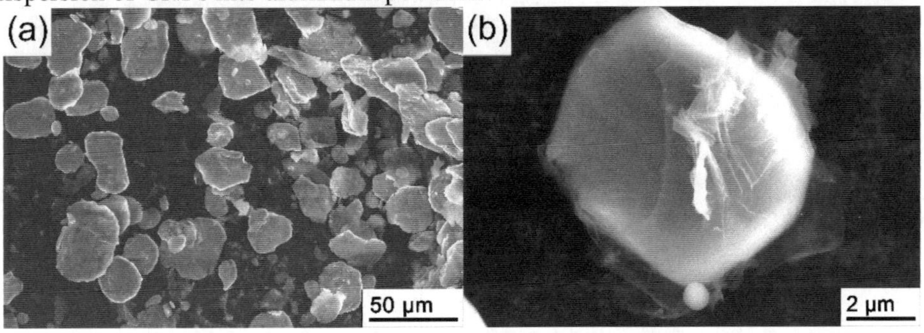

Fig. 1 SEM images of the mixed GNFs/Al powders after milling
(a) low-magnification and (b) high-magnification

SPS was used to fabricate the bulk GNFs/Al composites. The dried GNFs/Al mixture was placed in a graphite die with inner diameter of 30 mm, and then sintered at 560 °C for 4 or 8 min under a maintained pressure of 30 MPa. The sintering time was varied to obtain composites with distinct microstructures and therefore different mechanical properties. The formed bulk GNFs/Al composites were disc-shaped, 30 mm in diameter and 8 mm in thickness. For comparison, pure aluminum specimens were also prepared by the same process.

2.2. Tensile test and microscopy analysis

The density of the bulk materials was measured by Archimedes principle. The tensile mechanical properties of the bulk samples were determined by a testing machine (Instron 5887) at ambient temperature using a displacement rate of 1 mm/min. For minimizing the experimental error, three tensile specimens were cut from each sample along the radial direction with dog bone shape (thickness: 2 mm, gauge length: 10 mm, gauge width: 3 mm). A Hitachi S-4800 SEM system was employed to examine the fracture surface of the specimens after tensile tests. The substructure features and interface characteristics of the composites were observed using a Tecnai G2 F30 transmission electron microscopy (TEM) system. The presence of graphene in the composites was confirmed by Raman spectra (HORIBA LabRAM HR800) using an excitation wavelength of 514 nm.

3. Results and discussion

3.1. Tensile properties

The relative densities of all tested samples, including pure Al and GNFs/Al composites, are all near 100% of the theoretical values. Therefore, the influence of the sintered density on the mechanical properties is neglected. For clarity, the samples are coded according to the sintering time, with the 4-min-sintered bulk materials named "I" and the 8-min-sintered ones named "II".

The tensile test results shown in Table 1 indicate that the sintering time has a distinct influence on the mechanical properties of the GNFs/Al composites, although it has little effect on pure Al. Compared with the tensile strength (111 MPa) and yield strength (61 MPa) of the 4-min-sintered Pure Al I, those of the 4-min-sintered Composite I are enhanced by 42% and 41%, respectively. In terms of ductility, the elongation of Composite I retains the relative high value of around 14.5%, although it is only half that of Pure Al I. On the other hand, compared with the 8-min-sintered Pure Al II, the strength of the 8-min-sintered Composite II changes very little while its elongation declines to just 9.5%. The above results suggest that the fabrication process has an impact on the microstructure and therefore mechanical properties of the GNFs/Al composites.

Table 1 The average tensile values of GNFs/Al composite and pure Al samples with different sintering times.

Material	Tensile Strength (MPa)	Yield Strength (MPa)	Elongation (%)	Sintering time (min)
Composite I	158±4	86±5	14.5±0.5	4
Composite II	107±4	55±5	9.5±0.8	8
Pure Al I	111±2	61±3	26.2±0.3	4
Pure Al II	105±3	56.6±3	27.5±0.5	8

3.2. Fracture surface morphology

Fig. 2 shows the low- and high-magnification SEM images of the tensile fracture surface of Composite I, Composite II, and Pure Al I after tensile tests. As the mechanical properties of Pure Al I and II are almost the same, only Pure Al I was considered in the study of fracture surface morphology as well as the TEM study (to be discussed later). The low-magnification images in Fig. 2 exhibit obviously distinct morphologies among these three fracture surfaces. The fracture surface of Composite I presents an obvious lamellar structure, while those of Composite II and Pure Al I are full of dimple-like structures and dimples, respectively. Besides, some fragments are distributed on the fracture surface of Composite II, which can be speculated to be the Al_4C_3 phase formed in interface reaction according to the TEM observation (to be discussed later). The interface reaction phase can strongly enhance the interface bonding and hence the fracture surface of Composite II is full of dimple-like structure. The high-magnification images in Fig. 2 exhibit lamella structure in both Composite I and II, while that of Pure Al I is full of dimples. In sum, the fracture surface of Composite II is composed of dimple-like structures with some lamellar structures on the dimple edge, showing a transition from the obvious lamellar structure in Composite I to the dimples in Pure Al I.

The high-magnification images in Fig. 2 also show that the GNFs are pulled out along with the aluminum lamellas on the fracture surfaces of Composite I and II. It is also worth noting that the pulled-out GNFs appear smooth, which is distinct from their original wrinkled feature (Fig. 1b). This clearly indicates that the GNFs still endured tensile stress even after the aluminum matrix fractured, leading to their unfolding. In addition, the length of the pulled-out GNFs in Composite II is much shorter than that in Composite I. Previous studies on carbon fiber-reinforced aluminum composites revealed that the fracture behavior of composites depends dramatically on the type of the reinforcement-matrix interface. When the interface between the carbon fiber and matrix is relatively weak, cracks grow at the interface; and when the cracks cross the fiber, long pulled-out carbon fibers appear [14]. When the interface is relatively strong, the propagation of cracks along the interface is reduced, resulting in a shorter pulled-out length of the fibers [14]. When the interface is very strong with the formation of brittle compounds, the fracture surface is flat without the pulled-out fibers [15]. Although the length of the pulled-out GNFs in Composite I is approximately 20–50 nm, which is longer than that in Composite II (less than 20 nm), it is much less than that of the original GNFs (ca. 50 μm) used as raw material in this study. Therefore, the interface between the GNFs and matrix in Composite I is relatively strong. In Composite II, the much shorter pulled-out GNFs, coupled with the dimple-like structure in the fracture surface, indicate that the interface is even stronger than that in Composite I due to the interface reaction. However, although this reaction enhances the interface in Composite II, the elongation is much less than that of pure Al, because Al_4C_3 is brittle.

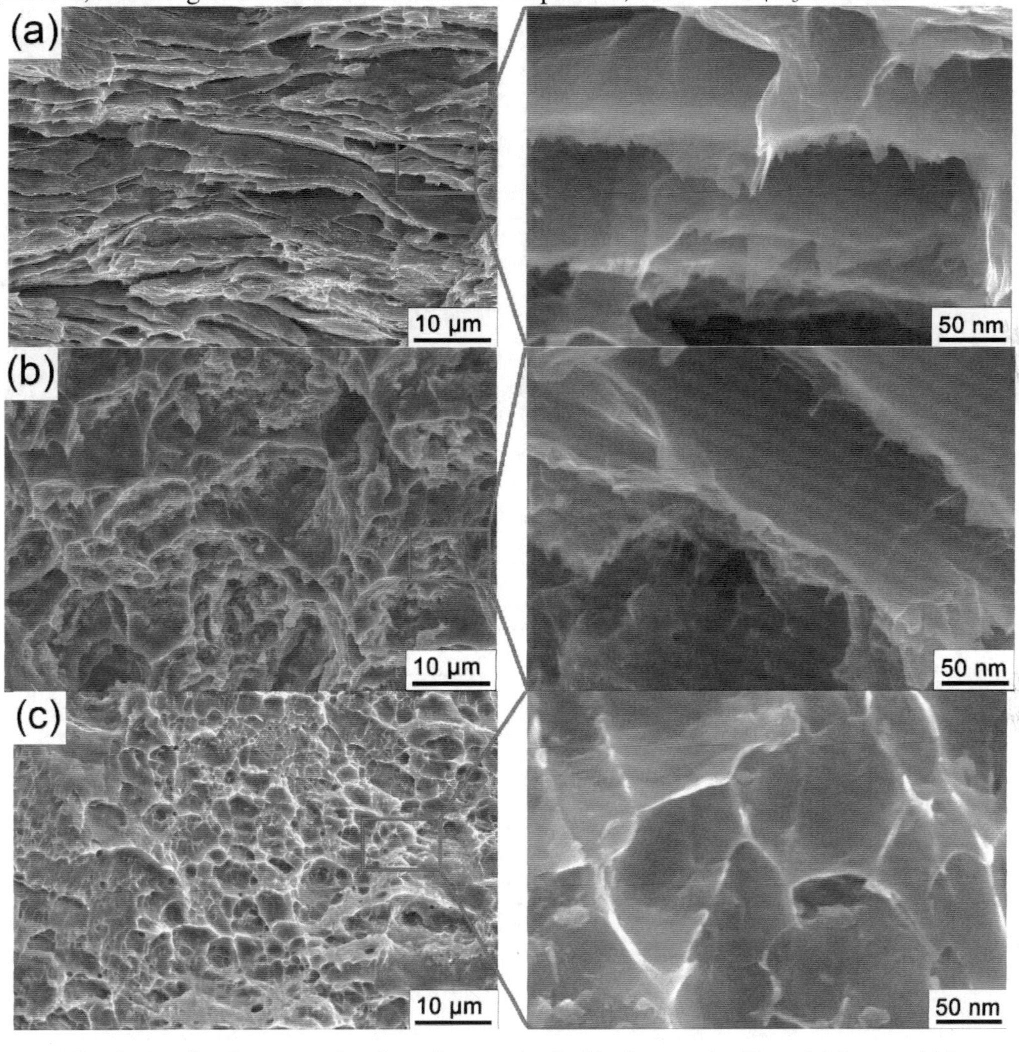

Fig. 2 Tensile fractograph of (a) Composite I, (b) Composite II, and (c) pure Al I

3.3. Microstructures

Fig. 3 exhibits the low- and high-magnification TEM images of Composite I, Composite II, and Pure Al I viewed on the longitudinal section. The low-magnification TEM images show that all three materials possess lamellar structure in the whole, due to the rearrangement of aluminum flakes under the axial pressure applied during the SPS process. It is worth noticing that the thicknesses of aluminum layers in all three samples are almost the same. In addition, the lamella boundary in Composite I is more distinct than those in Composite II and Pure Al I.

The high-magnification TEM images with their selected area electron diffraction (SAED) or high-resolution (HR)-TEM inserted in Fig. 3 exhibit obviously different structures sandwiched between these aluminum lamellas: continuous GNFs dispersed in Composite I (Fig. 3(b)), discontinuous GNFs and an amount of needle-shaped Al_4C_3 immerged in Composite II (Fig. 3(d)), and fractured amorphous Al_2O_3 on Pure Al I (Fig. 3(f)). Masses of needle-shaped Al_4C_3 in Composite II consume a large number of GNFs, thus there is less GNFs between the aluminum lamellas in Composite II than in Composite I, making the lamella boundary less unremarkable in Composite II.

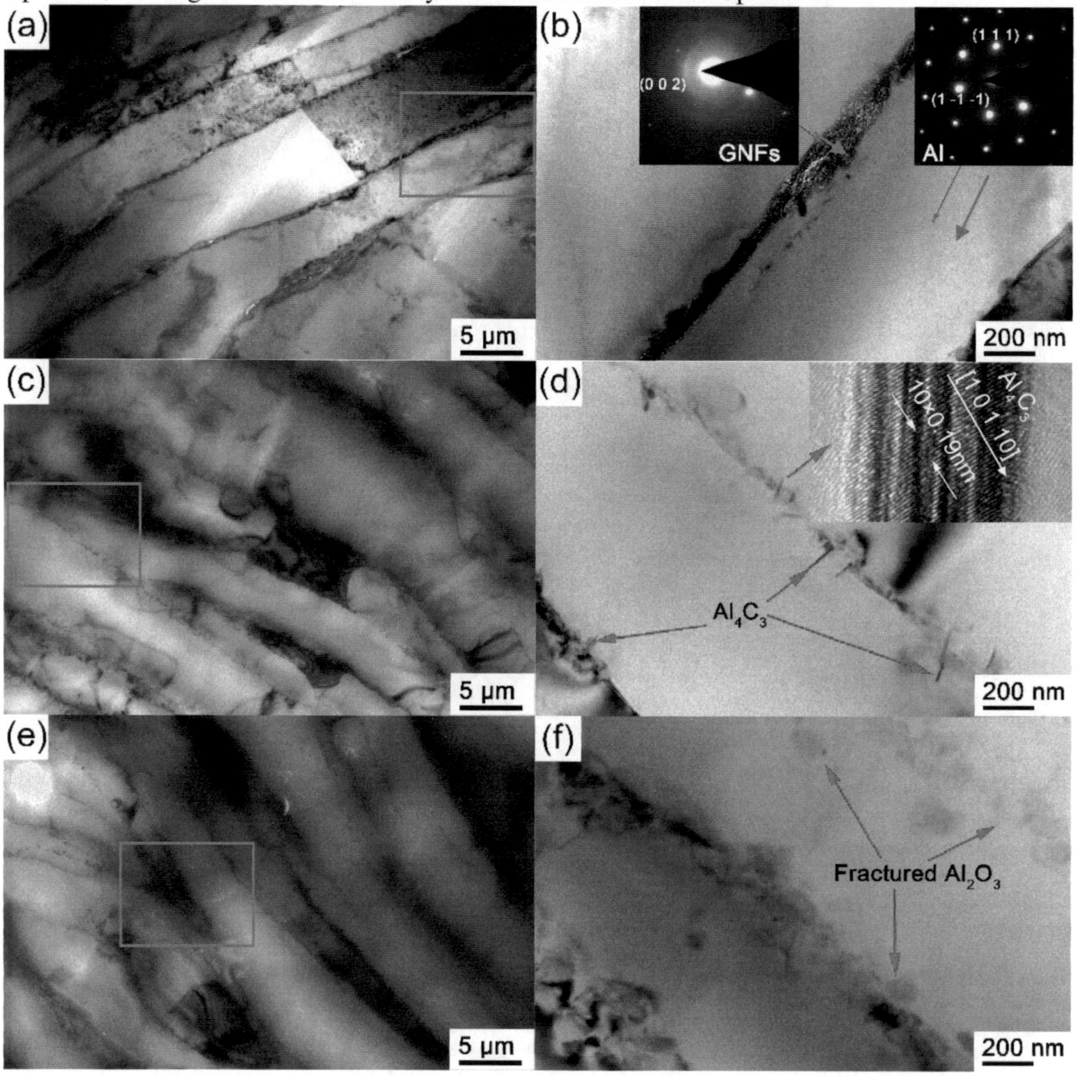

Fig. 3 TEM of (a) Composite I and (b) magnified map of rectangular zone in (a) (SAED of GNFs and Al matrix are inserted); (c) Composite II and (d) magnified map of rectangular zone in (c) (Al_4C_3 are marked by arrows and HRTEM of Al_4C_3 are inserted); (e) Pure Al I and (f) magnified map of rectangular zone in (e) (Al_2O_3 are marked by arrows).

Study on carbon fiber reinforced aluminum composite find that the carbon-aluminum interface reaction generally occurs at temperatures higher than 500 °C [16]. In the current work, the sintering parameters except the sintering time are the same for all materials, and the sintering temperature is higher than 500 °C. Then, the interface reaction is determined by the sintering time. With a shorter sintering time (4 min), the interface reaction is suppressed, while it is fully carried out with an extended sintering time of 8 min. Thus, the existence of sufficient Al_4C_3 phase is the basic distinction between Composite I and II, which further leads to their different mechanical properties. The mechanical test results show that the strength of Composite I is obviously improved compared to its Al matrix, while that of Composite II changes very little and its elongation declines. Hence, the mass interface reaction phase of Al_4C_3 spoiled the mechanical properties of the GNFs/Al composites. To obtain graphene-enhanced aluminum composites, the mass interface reaction should be avoided during the fabrication process.

To understand the strengthening mechanism of GNFs/Al composites, the interface characteristics of Composite I were further studied by HRTEM. The observation demonstrated that two types of GNFs/Al interfaces exist in the composite, as displayed in Fig. 4(a) and (b). The image in Fig. 4(a) shows that the GNFs and the aluminum matrix are separated by an Al_2O_3 layer. The Al_2O_3 layer covers the aluminum particles due to the oxidation of aluminum in air, and it is difficult to remove even during ball milling [17, 18]. The pulse may partially fracture the Al_2O_3 layers during the SPS process [19, 20]. In these fractured areas, direct contact between GNFs and the matrix is feasible, as shown in Fig. 4(b). From Fig. 4(a) and (b), we can see that the interfaces between GNFs and aluminum matrix are tightly bonded on the atomic scale, whether there is an Al_2O_3 layer between them or not. This well-dispersed and compactly bonded interface between GNFs and aluminum matrix facilitates the stress transfer and effectively hinders dislocation movement, which are very important for the strengthening effect of GNFs to be fully realized.

The characteristic of the GNFs in Composite I was also studied by Raman spectroscopy, and compared to the original GNFs powder at the same condition. In the averaged Raman spectra in Fig. 4(c), the D peak at 1350 cm^{-1} indicates the disorder of carbon material, and its intensity ratio to the G peak at 1580 cm^{-1} (I_D/I_G) is a measure of defects in graphene. The obvious D and G bands in the bottom line confirm the existence of GNFs in Composite I, and the lower value of I_D/I_G in the bottom line indicates that the fabrication process did not destroy the structure of GNFs in the aluminum matrix.

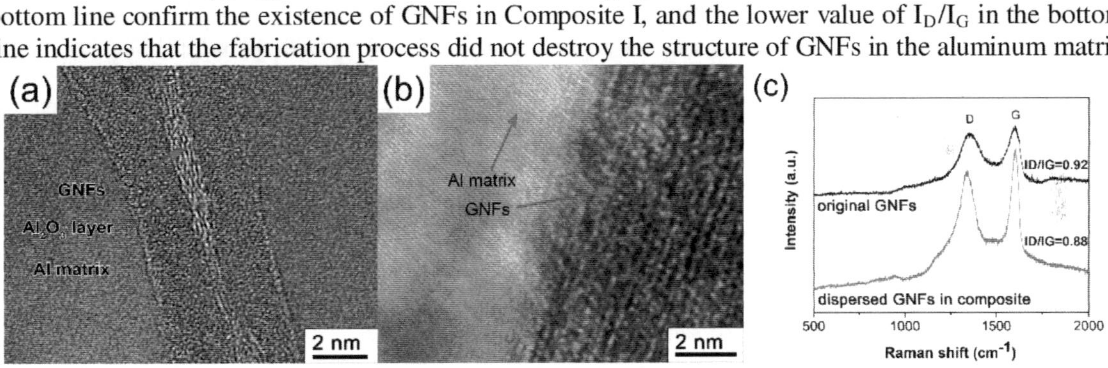

Fig. 4 (a) (b) HRTEM of two kind GNFs/Al interface in Composite I;
(c) Raman spectra of the original GNFs and dispersed GNFs in Composite I.

Based on the above-mentioned fracture surface morphology and microstructure analyses, the structural features of Composite I with enhanced mechanical properties can be described as follows: regular lamellar structure composed of the alternating arrangement of GNFs and aluminum flakes, compact GNFs/Al interface bonded at the atomic scale, and that reinforced phase almost maintains the original structure.

Through the microstructure study, the higher strength in composite I can be attributed to two reasons. First, the reinforcement bears partial stress transferred from the matrix. The dispersed and compactly bonded interface between GNFs and aluminum matrix promotes the stress transfer from the matrix to GNFs. When the GNFs are oriented parallel to the tensile direction, the stress load is aligned with the strongest direction of the GNFs. Therefore, GNFs in the composite with well-maintained raw structure can effectively bear the partial stress transferred from the aluminum matrix. Second, dislocation pile-up, which takes place during the plastic deformation, could be hindered by the reinforcement. The uniform distribution of the GNFs and the compact bonded interface in the composite effectively hinder the dislocation movement. In a word, the GNFs/Al Composite I has remarkably enhanced mechanical properties, because its elaborated structure can bear the partial stress transferred from the matrix and effectively hinder the dislocation movement.

The shorter strain with the steady flow and negligible necking region in Composite I during the tensile process may be attributed to microcracks, which may nucleate for two reasons. Some microcracks are formed within the material due to impurities or voids introduced during the manufacturing process, and others can be caused by stress concentration. As Composite I is dense and uniform, the microcracks mainly stem from stress concentration with severe dislocation pile-up at the GNFs/Al interface, which results in short plastic strain and negligible necking. On the other hand, the good structure of Composite I allows the composite to retain relative high plasticity with a total strain of 14.5%.

4. Conclusions

In the present research, GNFs/Al composites with 1.0 wt.% GNFs, which possess high relative density (near 100%) and homogenous structure, were successfully fabricated by a ball milling and SPS method. The microstructure, interface characteristics, and fracture morphology of two composites with different tensile mechanical properties were systematically investigated. The major conclusions are as follows:

(1) The mechanical properties of GNFs/Al composites depend dramatically on the characteristics of the GNFs-aluminum interface. Mass interfacial reaction reduces the mechanical properties of the composites.

(2) The good mechanical properties of the enhanced GNFs/Al composites can be ascribed to the fact that the structure of the raw reinforcement is well-maintained: the GNFs are distributed uniformly and parallel to the tensile direction, the absence of mass interfacial reaction, and the good interface bonding in the composite. These factors, taken together, allow the high strength and the fine flexibility of the GNFs to be fully utilized to enhance the composites.

(3) SPS parameters have important effects on the interfacial characteristic and mechanical properties of GNFs/Al composites. A long sintering time should be avoided, as it is likely to induce undesirable interface reactions.

Acknowledgments

The authors thank the financial support from Technology Innovation Fund of China Aviation Industry Corporation.

References

[1] Geim A K 2010 *Graphene: Status and Prospects. International of Vacuum Congress* p1530.
[2] Lee C, Wei X, Kysar J W and Hone J 2008 *Science* 321 385
[3] Tsai J L and Tu J F 2010 *Mater. Des.* 31 194
[4] Zheng Q, Geng Y, Wang S, Li Z and Kim J K 2010 *Carbon* 48 4315
[5] Li B and Zhong W H 2011 *J. Mater. Sci.* 46 5595
[6] Walker L S, Marotto V R, Rafiee M A, Koratkar N and Corral E L 2011 *Acs Nano* 5 3182
[7] Chen Q, Zhang L and Chen G 2012 *Anal. Chem.* 84 171
[8] Bartolucci S F, Paras J, Rafiee M A, Rafiee J, Lee S and Kapoor D 2011 *Mater. Sci. Eng.* A 528 7933
[9] Yan S J, Dai S L, Zhang X Y, Yang C, Hong Q H, and Chen J Z 2014 *Mater. Sci. Eng. A* 612 440
[10] Li Z, Guo Q, Li Z, Fan G, Xing D B, and Su Y 2015 *Nano Let.* 15 8077
[11] Shin S E, Choi H J, Shin J H, and Bae D H 2015 *Carbon* 82 143
[12] Liu G, Zhao N, Shi C, Liu E, He F and Ma L 2017 *Mater. Sci. Eng. A* 699 185

[13] Feng S, Guo Q, Li Z, Fan, G, Li, Z and Xiong D B 2017 *Acta Mater.* 125 98
[14] Hajjari E, Divandari M and Mirhabibi AR 2010 *Mater. Des.* 31 2381
[15] Wang X, Chen GQ, Li B, Wu GH and Jiang DM 2009 *J. Mater. Sci.* 44 4303
[16] Lalet G, Kurita H, Miyazaki T, Kawasaki A and Silvain J F 2014 *Mater. Let.* 130 32
[17] Liao J Z, Tan M J and Sridhar I. *Mater. Des.* 31 S96
[18] Lee M, Choi Y, Sugio K, Matsugi K and Sasaki G. 2014 *Compos. Sci. Tech.* 97 1
[19] Lalet G, Kurita H, Miyazaki T, Kawasaki A and Silvain J F. 2014 *Mater. Let. 1*30 32
[20] Xie G, Ohashi O, Chiba K, Yamaguchi N, Song M, Furuya K and Noda T 2003 *Mater. Sci. Eng. A* 359 384

IOP Conf. Series: Materials Science and Engineering **281** (2017) 012018 doi:10.1088/1757-899X/281/1/012018

Research on the Ordered Mesoporous Silica for Tobacco Harm Reduction

Y Wang[1†], Z Y Li[2†], J X Ding[2,3], Z J Hu[1], Z Liu[1], G Zhou[2*], T H Huang [1]

1.Technology Centre, China Tobacco Guangxi Industrial Co., Ltd, Nanning 530001, China
2. Key Laboratory for Biomechanics and Mechanobiology of Ministry of Education, School of Biological Science and Medical Engineering, Beihang University, Beijing 100191, China
3. School of Mechanical-electronic and Automobile Engineering, Beijing University of Civil Engineering And Architecture, Beijing 102616, China

*Corresponding author: Gang Zhou, zhougang@buaa.edu.cn
Key Laboratory for Biomechanics and Mechanobiology of Ministry of Education; School of Biological Science and Medical Engineering, Beihang University . TEL & FAX: Tel:+86-10-82339428,

† The first two authors (Yue Wang and Zhenyu Li) contributed equally to this work.

Abstract. For reducting tobacco harm, this paper prepared an ordered mesoporous silica by using triblock copolymer Pluronic P123 as template. The property of this material was characterized by the X-ray scattering spectrum(XRD), Transmission electron microscopy(TEM), Scanning electron microscopy (SEM) and Nitrogen adsorption/desorption. Then this ordered mesoporous silica was added into the cigarette filter in order to researching its effect of cigarette harm index. The result shows that the feature of SBA-15 was grain morphology, ordered arrangement, tubular porous 2-D hexagonal structure. The application of SBA-15 in cigarette filter can selectively reduce harmful components in cigarette smoke such as crotonaldehyde, hydrogen cyanide, benzo pyrene and tar. The synthesized SBA-15 could properly reduce cigarette harm index.

1. Introduction

The influencing factors of tobacco harm index(H) is related to seven kinds of harmful ingredients in cigarette smoke. There are mainly HCN, NNK, NH3, B[α]P, phenol and crotonaldehyde of the harmful ingredient. At present, the study on cigarette smoke harm index is publicly reported less, mainly including cigarette auxiliary materials design and cigarette paper, etc. Which has become a research hotspot in recent years is to select reducing one or more among seven kinds of harmful ingredients in cigarette smoke, like crotonaldehyde that has intense stimulation to the person's eyes, skin and respiratory tract. The method of selective reduction use mainly traditional adsorbent, such as molecular sieves, chitosan, activated carbon and porous resin, etc. Compared with traditional adsorbent, Ordered mesoporous silica（SBA-15）materials have uniform adjustable large aperture, controllable mesoscopic structure, high specific surface area and functional surface. Silica is an excellent adsorbent and food additives[1]. So far, little researches focus on the effect of SBA-15 on cigarette harm index. Therefore, SBA-15 materials with different aperture are synthesized on a large scale, then detailed characterization of them are performed. And to study the adsorption of harmful ingredients in flue gas and the impact on the H value, they are applied to the cigarette filter as adsorbent.

2. Materials and methods

2.1. Preparing of SBA-15

2kg three block copolymer P123 is added to 15kg deionized water and 60kg 2M hydrochloric acid, then dissolve into solution. By stirring violently,one can get homogeneous solution; By heating, when the temperature reaches 40°C, 4.25g ethyl silicate under the condition of intense stirring is added. After 24h and then transfering into 100°C oven, it occurs to static crystallization; After cooling, by filtering and 100°C drying, and then the sample is placed in muffle furnace, which is heated up to 550°C at a rate of about 1°C /min in air atmosphere. After roasting 6h, one can remove the template agent. After finishing above operations, one can obtain SBA-15.

2.2. Characterization of SBA-15

Small angle X-ray scattering spectrum can be gathered in Nanostar U-type small Angle X-ray scattering (Cu Ka). Then they were observed by SEM and TEM images obtained by SEM. And Nitrogen adsorption/desorption isotherm can be get by using Micromeritics Tristar 3020 adsorption apparatus in 77K [2] .

2.3. Preparation of cigarette samples

Weight ratio about 35% SBA-15 is added into a certain weight of starch adhesive, then is mixed evenly. Add the moderate amount of water under stirring. Through granular molding machine, microwave drying and spheroid granulate machine, the particle is divided into 20~60 mesh by screen. According to the production technology of cigarette, one can roll two kinds of rods respectively into control and test tobacco.

2.4. Analysis of physical and chemical cigarette smoke

After balancing contrast cigarette and test cigarette respectively for 48h in (22±1) ℃, relative humidity conditions (60±2)%. According to the method of YC/T 28 .2-2002, resistance, weight, circumference and rigidity are measured respectively; According to GB/T 19609-2004 and YC/T 156-2001 respectively, smoke tar and release amount of nicotine are measured; According to GB/T 23356-2009, YC/T 254-2008, YC/T 253-2008, YC/T 377-2010, YC/T 255-2008, YQ/T 17-2012 and GB/T 21130-2007 respectively, the release amount of seven kinds of mainstream smoke harmful ingredients including CO, crotonaldehyde, HCN, NH3, phenol, NNK and B[α]P are measured.

3. Results and discussion

3.1. XRD analysis

Characterization of SBA-15 uses the X-ray scattering spectral to investigate mesoscopic order. The result is shown in Figure 1. As can be seen from the Figure 1, SBA-15 within small Angle appear three obvious diffraction peak, whose ratio of interplanar distance (d) is 1:1.73:2, corresponding to typical two-dimensional six-party mesoscopic structure. And space group is p6m.

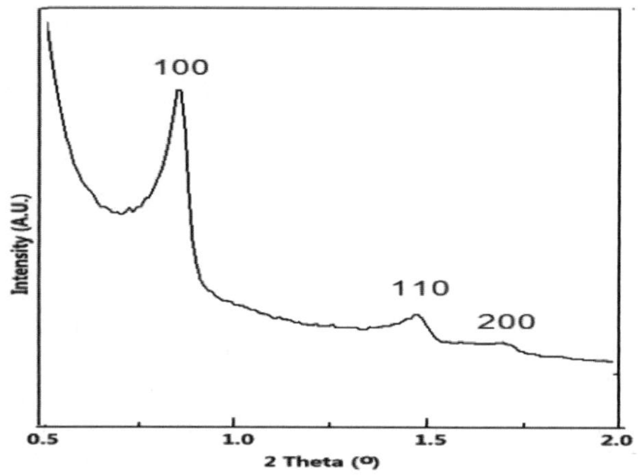

Figure 1. The small angle XRD spectra of SBA-15

3.2. SEM and TEM observation

Scanning electron microscopy (SEM) and transmission electron microscopy (TEM) are carried out on the Characterization of SBA-15, so as to investigate surface structure characteristics and internal fine structure shown in Figure 2.

Figure 2. The pictures of SEM (a,b) and TEM (c,d) of SBA-15

As can be seen from Figure 2, SBA-15 which is observed by SEM shows typical wheat-like morphology. Further amplification can be seen that wheat-like morphology SBA-15 is made of about 1 μm particles (Figure 2a, Figure 2b). TEM pictures shows the tubular channel of orderly arrangement, corresponding to two-dimensional six-party mesoscopic structure and consistent with the results of X-ray scattering spectrum (Figure 2c, Figure 2d).

3.3. Characterization of SBA-15

In order to investigate the nature of the hole of SBA-15, ordered mesoporous materials that are synthesized under the condition of different hydrothermal temperature are characterized with Nitrogen adsorption-desorption. Mesoporous pore size distribution can be calculated by BJH. The results are shown in Table 1.

Table 1. The physical and chemical properties of SBA-15

Samples	S_{BET} /(m^2/g)①	$S_{micropore}$ /(m^2/g)②	V/(cm^3/g)③	bore diameter /nm④	$S_{mesopore}$ /(m^2/g)⑤
SBA-15-HT80	660	107	0.80	7.4	553
SBA-15-HT100	780	89	1.1	9.3	691
SBA-15-HT130	430	34	1.1	12.4	396

Note: (1) BET specific surface area; (2) micro pore specific surface area; (3) pore volume calculated by BET method; (4) mesoporous aperture calculated by BJH; (5) mesoporous specific surface area

According to the Table 1, all Nitrogen isothermal adsorption-desorption of SBA-15 samples shows typical IV curve. The adsorption section shows an obvious capillary condensation phenomenon belong to a type H1 hysteresis loop. That indicates that it has homogeneous mesoporous and tubular mesoporous channel. The aperture size of SBA-15 increased from 7.4 nm to 9.3 nm and 12.4 nm, as the hot water temperature increasing from 80 °C to 80 °C and 100 °C. The specific surface area of the samples that are measured under different hydrothermal temperature are among 430~780 m2/g. Among of those, BET specific surface area of SBA-15 by 100 °C water heat treatment is the largest and is780 m2/g. Therefore, SBA - 15 under 100 °C water heat treatment is chose as cigarette filter additives.

3.4. Application of SBA-15

Through the laboratory screening of SBA-15, SBA-15 under 100°C water heat treatment is chose as cigarette filter additives. In terms of 1.2.3 the flue gas chemical indicators of cigarette samples were shown in Table 2.

Table 2. Flue gas chemical indicators of cigarette samples (mg/branch)

Samples	tar	nicotine	CO	crotonaldehyde	HCN	NH3	phenol	NNK	B[a]P
Control	12.5	1.36	13.3	18.4	117.9	8.4	16.0	4.7	10.8
SBA-15	11.7	1.29	13.4	12.7	114.9	8.4	16.4	5.1	10.5

Table 2 illustrates that compared with the contrast cigarette, H value of test cigarettes adding SBA-15 reduces 0.3/branch, among of which the release quantity of tar reduces 0.8 mg/branch; Nicotine reduces 0.07 mg/branch; HCN reduces 3 μg/branch; B[α]P reduce 0.3 μg/branch; Crotonaldehyde reduces 5.7g/branch, reduced by 30.9%. Experimental results show that adding SBA - 15 can reduce the H value. Meanwhile, the amount of release of tar, nicotine, crotonaldehyde, HCN and B[α]P in cigarette smoke reduces in different degree.

4. Conclusion

Through the synthesis and characterization of SBA-15, the characteristics of the SBA-15 was two-dimensional six-party mesoscopic structure with wheat-like morphology, ordered arrangement, Tubular channel. In the application of SBA-15 in cigarette filter, compared with the contrast cigarette, adding SBA-15 can reduce H value. Meanwhile, the amount of release of tar, nicotine, crotonaldehyde, HCN and B[α]P in cigarette smoke reduce in different degree. Results show that adding SBA-15 can reduce cigarette smoke hazard index and plays an role in reducing tar and decreasing pollution to a certain extent.

References

[1] Wu Z X and Zhao D Y 2011 *Chem Commun*, 47: 3332
[2] Ţălu Ş, Micro and nanoscale characterization of three dimensional surfaces. Basics and applications. Napoca Star Publishing House, Cluj-Napoca,Romania, 2015.

IOP Conf. Series: Materials Science and Engineering **281** (2017) 012019 doi:10.1088/1757-899X/281/1/012019

Numerical Simulations of As-Extruded Mg Matrix Composites Interpenetrated by Metal Reinforcement

H Y Wang , S R Wang*, X F Yang and P Li

School of Mechanical Engineering, University of Jinan, Jinan, 250022, China

Corresponding email: me_wangsr@ujn.edu.cn

Abstract. The interpenetrating magnesium composites reinforced by three-dimensional braided stainless steel wire reinforcement were fabricated and investigated. The extrusion processes of the composites in different conditions were carried out and simulated by finite element method using the DEFORM-3D software. The results show that the matrix and reinforcement of the composites form a good interfacial bonding and the grains were refined by extrusion and the influence of reinforcement, which are in accordance with the enhanced strength and degraded plasticity. The combined quality between the matrix and reinforcement can be strengthened in extrusion chamber where occurred large strain and suffered intense stress, and the effective stress of the material increases continuously with the increase in extrusion ratio and the decrease in extrusion speed until it reaches a stable value.

1. Introduction

Mg and its alloy are the lightest metallurgical based structural materials with high specific strengths, and therefore they are the most attractive materials in many fields where weight reduction is of prime importance, such as automotive, aerospace and mobile electronics and so on. However, owing to the low strength, the Mg alloy was usually reinforced by particles, whiskers and short or long fibers to form magnesium matrix composites to improve the strength and performance of the materials. In recent years, the reinforcement network skeletons of magnesium matrix composites are tended to make from ceramic, glass, cement, intermetallic and so on[1-4], which would lead the composite materials to exhibit some excellent mechanical properties than the traditional magnesium matrix composites[5]. But the defect is that the interpenetrating composites cannot be further deformed to the required shape due to the brittleness of the reinforcement, so their application is still limited in manufacturing the structural components with the cross section shape of fixed type. If the reinforcement of the composites were made from metal that is the metal matrix composites interpenetrated by metal reinforcement (MIMC) would have a good forgeability and machinability which Can be a variety of thermal deformation and Cold deformation. And through the plastic deformation, for example, twin-roll casting (TRC), warm rolling (WR), hot extrusion (HE), torsion straining (TS), reciprocal extrusion (RE) and equal channel angular extrusion (ECAE)[6-8], the combined quality between matrix and reinforcement of the composites can be improved and also the cross-sectional shape of the materials can be changed to meet the requirement . In this study, Mg alloys matrix composites interpenetrated by steel reinforcement (MISC) were manufactured and expected to obtain the excellent mechanical properties attributing to the coordination deformation mechanism of reinforcement with magnesium matrix during deformation[9].

In order to ensure the reliability and safety of subsequent applications, the magnesium matrix composites must be of high quality, especially the combined quality between the reinforcement and matrix of the materials. So the extrusion process of the MISC materials was performed so as to improve the combined quality and the strength of the materials. It is practically impossible to measure stresses and temperatures of the materials in extrusion chamber based on the currently experimental methods. So, in this case, computer simulation based on the finite element method theory (FEM) can play a important role in revealing the evolution and distribution of effective stress, pressure and temperature of the product during the deformation process[10].

In the present works, the fabrication and subsequent extrusion process of the MISC were carried out. The microstructure and texture evolution during deformation was analyzed in detail. And the deformation behaviour of MISC materials during extrusion process were also investigated through finite element methods using a viscoplastic material model. The main objective was to acquired the internal stress of the composites, and the extrusion speed and ratio were used as process variable to evaluate their effects on the extrusion pressures, stresses and temperatures of the extruded product during the extrusion process. The simulation results are looked forward to guide the practical process production and experimentation.

2. Simulation and experimental procedures

2.1. Experimental procedures

The 304 stainless steel (Fe-18Cr-9Ni) wire (diameter as 0.6 mm) was chosen as the reinforcement skeleton materials, and the three dimensional braiding technology was applied to fabricate the skeleton. The AZ31 (Mg-3.31Al-1.05Zn) magnesium alloy were chosen as the matrix alloy. The volume fraction of reinforcement is about 15% which was measured using Archimedes measurements (ASTM C373). The MISC composites were fabricated by using the pressure infiltration technology, and the whole process is protected under argon atmosphere together with the infiltration die. The magnesium alloys were melted at 720°C under an inert atmosphere of CO_2 and SF_6 mixture. After melting, the melt was poured down into the infiltration die where have been placed the reinforcement skeleton in it at the temperatures range of 640-670°C, and then pushing the pressure head. As a result, MISC composites were fabricated successfully. Subsequently the extrusion process was carried out using the extrudate die with a lumen diameter of 40mm, a die half angle of 30° and an extrusion ratio of 16, as shown in Figure 1(a), at a temperature of 350°C under the extrusion speed of 2mm/s. The extrusion billet was machined to a specified diameter of 39 mm before extrusion, the extruded material is quenched to avoid grain growth. Extrusion experiment was performed to illustrate the effects of the compressional stress on the microstructure and texture evolution of MISC materials, and to validate that the material properties can be improved through enhancing the binding force between Mg matrix and the 304 reinforcement.

The microstructure characteristics of MISC material in as-cast and as-extruded state were analyzed by optical microscopy (OM), X-ray diffraction (XRD) and electron back-scatter diffraction (EBSD). The mechanical properties of the MISC materials were tested by tensile tests, and the tensile specimen had a gauge length of 12 mm, a width of 3 mm and a thickness of 2 mm. For comparison, the tests of AZ31 Mg alloy were also carried out. And the tensile tests were conducted at room temperature with a constant speed corresponding to an initial strain rate of $1\times10^{-3}\,s^{-1}$, where the angle between the tensile direction and the extrusion direction was 0 deg.

2.2. Simulation parameters and conditions

The FE simulation of MISC materials extrusion process was conducted using the software of DEFORM-3D, a coupled thermo-mechanical commercial FE code, which can be used for non-isothermal simulation. Thermo-elastic-visco-plastic FE code used in this study have both temperature and displacement as their degrees of freedom and could successfully capture the deformation modes and the specific characteristics of MISC sample extrusion process. In the analysis, a quarter of the geometries of the workpiece and dies were modeled and meshed with tetrahedral elements due to the symmetric boundaery conditions as shown in Figure 1(b). The relative mesh density type was applied for the extrusion head and die which were considered rigid but non-isothermal. However, to improve the accuracy and reliability of FE simulation, the workepiece was meshed using absolute mesh density type where the minimum size of an element was 0.8mm. The total number of elements of the workpiece and tooling were given in table 1 Both of these marerial models neglected the elastic behaviour.

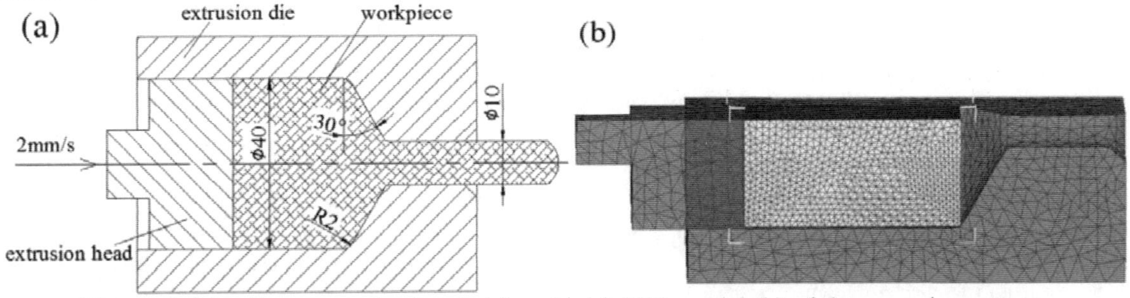

Figure 1. The diagrammatic sketch (a) and initial FE model (b) of the extrusion process

Table 1. The meshing feature of the models

Model	Meshing type	Total number
Extrusion head	relative mesh density type	12000
Extrusion die	relative mesh density type	2500
Workpiece	absolute mesh density type	21660

The material model of AISI-H13 hot-work tool steel in DEFORM-3D software package was chosen for the extrusion head and die. However, a new model should be conducted to match the MISC material because there is no material model employed for it in the software package. The material parameters of MISC as shown in table 2 were obtained from the calculation results based on the parameters of the matrix and reinforcement. The deformation behavior of MISC composites rely on Arrhenius equation,

$$\dot{\varepsilon} = A[\sinh(\alpha\sigma)]^n \exp(-Q/RT) \tag{1}$$

where $\dot{\varepsilon}$ is the strain rate, σ is the flow stress, A and α are the constant related to materials, n is the strain hardening exponent, Q is the activation energy, R is the gas constant and T is the deformed temperature. Arrhenius equation is the sine hyperbolic law which is a comprehensive and perfect equation combining the power law with the exponential law. And the stress-strain constitutive mode of MISC composites which has been validated in the literature is proposed as,

$$\dot{\varepsilon} = 5.047\times10^{13}\left[\sinh\left(0.02063\sigma\right)\right]^{5.355}\exp\left(\frac{-198015}{8.314T}\right) \tag{2}$$

Table 2. The material parameters of MISC

Young's modulus , E / GPa	58.39
Poisson's ratio , υ	0.342
Yield function type	Von Misses
Thermal conductivity λ/ N·(s·°C)$^{-1}$	108.36
Heat capacity C / N·(mm^2·°C)$^{-1}$	1.9876
Thermal expansion α / 10^{-5}	2.016
emissity	0.172

In order to investigate the effects of the extrusion process parameters on the deformation behaviors of the MISC material, the extrusion processes were simulated under different extrusion conditions via changing the extrusion speed (1, 2, 3, 4 and 5mm/s) and the extrusion ratio (4, 9, 16, 25 and 36). The processing temperature of the MISC workpiece was set to 350°C, and the tooling and environment were all set to 300°C owing to the poor plasticity of the magnesium matrix at normal atmospheric temperature. The interface heat transfer coefficient between workpiece and tooling was set to 11 N·(s·mm·°C)$^{-1}$ and between workpiece/tooling and air was set to 0.02 N·(s·mm·°C)$^{-1}$ which assumed to be uniform for the entire surface[11].The shear friction was chosen in this work and chosing 0.3 as the friction factor which was assumed not to vary locally with interface temperature and pressure[12].

3. Results and discussions

3.1. Microstructure characterization

The microstructures and interface characterization of MISC composites are shown in Figure 2. It can be seen that the reinforcement and matrix are anisotropic, interpenetrating and intertwisting. The structure of the reinforcement has a good design ability due to it using the three-dimensional braiding technology. From Figure 2(a), parts of interconnected network structure of reinforcement in longitudinal direction are exposed from the matrix, and only a few of that in transverse direction (black arrow) is exposed., and most parts of it are enwrapped and buried with Mg metal matrix. The matrix and reinforcement form a good interfacial bonding, which is shown in Figure 2(b). The chemical element distributions were examined by Energy Dispersive Spectrometer (EDS) which are shown in Figure 3. The circle reinforcements are Fe which was characterized by blue line, and the Mg matrix was characterized by red line.

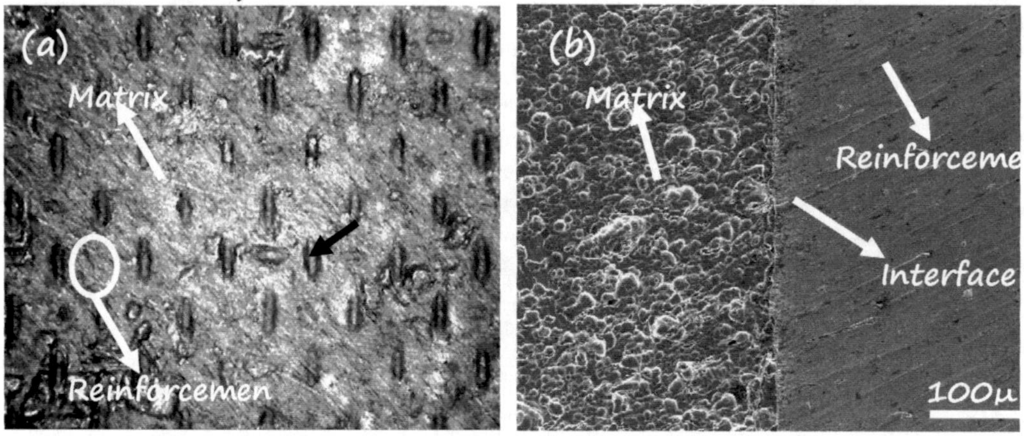

Figure 2. The interpenetrating structure (a) and interface of reinforcement with matrix (b)

Figure 3. The EDS analysis of MISC composites

Figure 4 shows the Optical Micrographs (OM) (Figure 4(a)) and grain size distributions (Figure 4(b)) of Mg matrix in MISC material as-cast state. The grain size distribution is random which does not follow normal distribution. And the area fraction of small grains with less than 50 μm is about 20%. The large grains with greater than 350 μm are about 12%. It is indicated that the area fraction of grains diameter and distribution don't get any influence from the reinforcement materials. That is, in as-cast state, the grain size and grain distribution of Mg matrix in composites has the same characterization with AZ31 Mg alloy. However, as a comparation of the as-cast state, the microstructure of the as-extruded sample has a larger changes in microstructure characteristics. Figure 5 shows the Field Emission Scanning Electron Microscopy (FESEM) and the grain size distributions of magnesium matrix in MISC material as-extrudated state. Due to the accumulated shear strain and dynamic recrystallization during extrusion, the grain size of as-extruded sample became relatively fine (Figure 5(a)), and the grain size distribution follows normal distribution (Figure 5 (b)). The area fraction of small grains with less than 2 μm is about 2%, that of large grains with greater than 30 μm is about 1%. The vast majority of grain sizes are in 10-17μm, and the average grain size is 14 μm.

(a) (b)

Figure 4. OM (a) and the grain size distributions (b) of Mg matrix in MISC material as-cast state

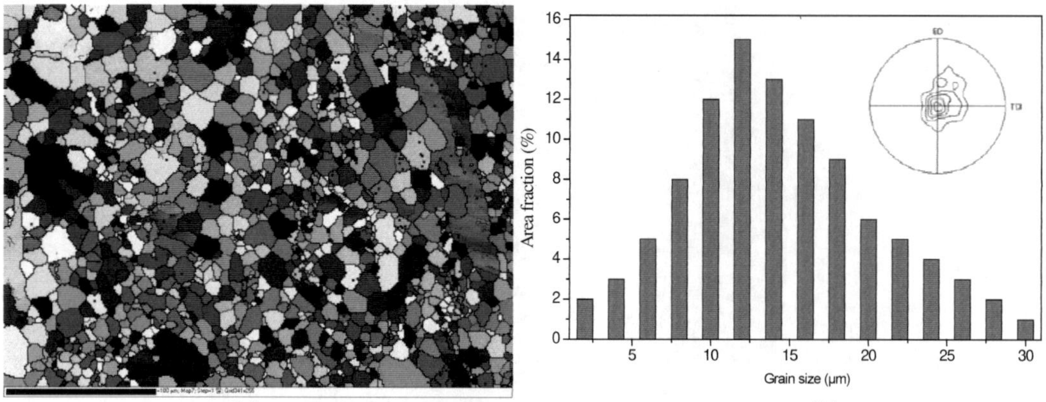

(a) (b)

Figure 5. FESEM (a) and the grain size distributions (b) of magnesium matrix in MISC material as-extrudated state

The difference of grain refinement and sizes distribution can be further demonstrated by the grain orientation changes of Mg matrix. It is indicated that the {0001} basal pole figure for as-cast state shows nearly random distribution which is shown in Figure 4(b). The grain orientation evolution of Mg matrix as extrudated state is revealed by EBSD test data from the {0001} basal pole figure which is set in Figure 5(b), it is shown that the grain orientation forms a strong peak in the centre of the {0001} basal pole figure during extrusion. In MISC material, the stainless steel wires have the highest strength, which cause a larger stress to Mg grains resulting in the basal plane tilted during coordination deformation. Therefore, the dynamical recrystallizations in Mg matrix occur near the strut of the reinforcement. And the room temperature formability of magnesium alloy strongly depends on the influence of the initial crystallographic texture of the basal plane [13]. The basal plane control lead to distinctly improved formability in MISC composites.

3.2. Mechanical properties

Due to the infinluence of reinforcement materials and the extrusion process, MISC composites exhibit different mechanical properties compared with AZ31 magnesium alloy which are shown in Table 3. The ultimate tensile strength (UTS), yield strength (YS) and elongation (δ) of MISC have a significant improvement compared with Mg matrix alloy (AZ31). The improvement can be summarized as follows as (a) high strength and elongation of the reinforcement itself, (b) good interfacial bonding of reinforcement with matrix caused by extrusion, (c) fine and equiaxed grain during coordination deformation of reinforcement and matrix and (d) effective load transfer from matrix to reinforcement.

Table 3. The YS, UTS and δ of AZ31 Mg and MISC meterial

materials	YS (MPa)	UTS (MPa)	δ (%)
AZ31	135	240	7
MISC	245	355	13

3.3. Extrusion process simulations

3.3.1. Deformation behaviors of the MISC workpiece during extrusion.

The mesh characteristics and the accompanying stress state of the MISC workpiece during extrusion process were shown in Figure 6. It can be observed that the flow of the billet through the extrusion die may be divided into three stages: upsetting - filling of the extrusion chamber - continuous extruded. And these stages can also be distinguished in the curve of extrusion pressure corresponging to the extrusion displacement which acquisition from the simulation of the extruded at a discharge rate of 2 mm / s and a extrusion ratio of 16 as shown in Figure 7. The extrusion pressure increased sharply in the filling stage and then remained stable in continuous extruded stage. And the fluction in the pressure-diplacement curve observed was due to the oscillation of nodes in contact with the extrusion die/head. From the distribution of the effective stress (Figure 6), it is noteworthy that the stress is mainly concentrate on the chamber and the outlet of extrusion die where occurred large strain during extrusion. Corresponding the finer meshes in the chamber indicated the material was subjected to strongly compressive force and the elongated meshes in the outlet illustrated the compressive force in radial direction and tensile force along the axial direction for the material. The combined quality between the reinforcement and matrix of MISC composites counld be strengthened in the extrusion chamber where occurred large strain and suffered intense stress, and so as to chieve the interpenetrating and intertwining between them.

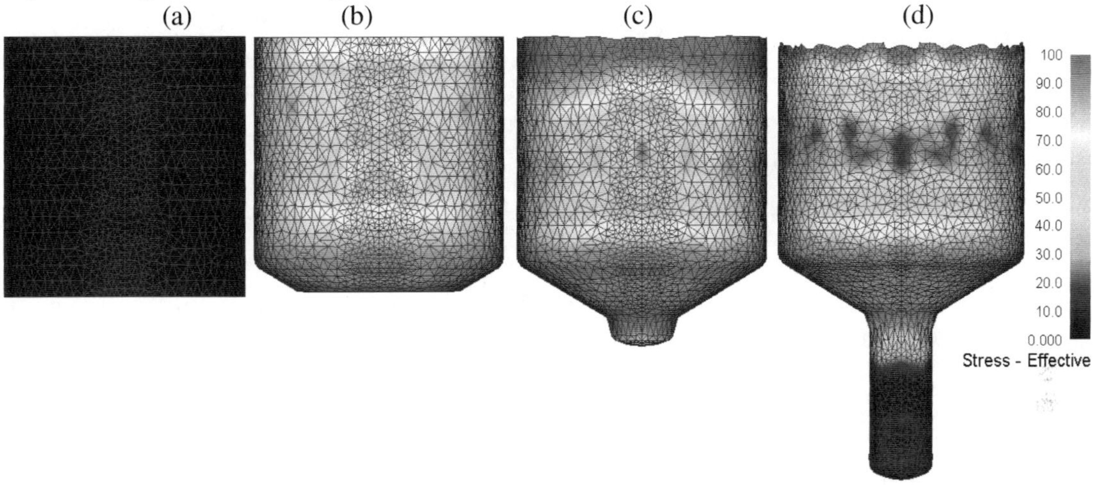

Figure 6. The dynamic extrusion process of MISC: (a)initial state, (b) upsetting, (c) filling of the extrusion chamber and (d) continuous extruded state

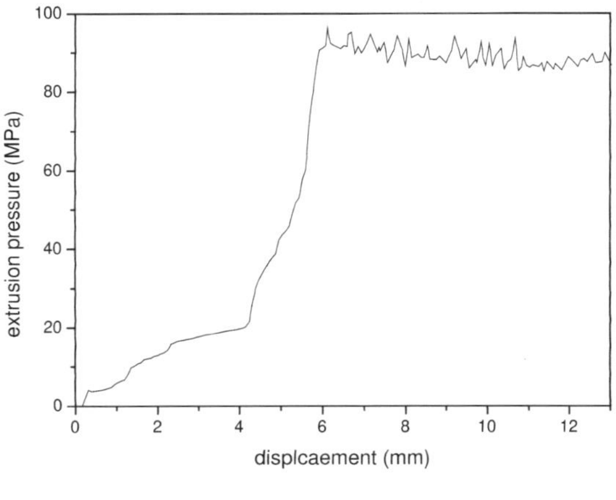

Figure 7. Extrusion pressure evolution obtained from simulation

The deformation characteristics, such as strain distribution, temperature and flow velocity of the MISC workpiece were exhibited in Figure 8 respectively. The material occured large strain in extrusion chamber and channel due to the combined effects of extrusion pressure and friction as show in Figure 8 (a), and which result in a higher temperature in the outlet of extrusion (Figure 8 (b)) since the energy of the plastic deformation was converted to heat and increase the temperature of the workpiece. However, the increase of the temperature would lead to a decrease on effective stress and was not conducive to the bonding between the matrix and the reinforcement of MISC composites. The metal flow velocity distribution as shown in Figure 8(c) indicate that the dead metal zone does not obviously in the corners of die due to the chamferings, and it can ensure the smooth and continuous flow of material in extrusion. It also can be seen that the metal in the outlet zone flows much faster than other zones which illustrate the larger strain and stress have been suffered in this zone.

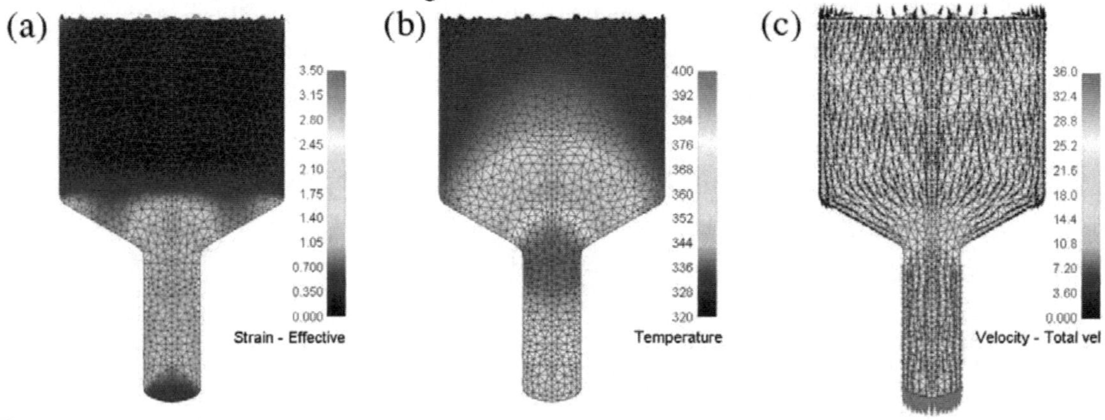

Figure 8. The deformation behaviors of the workpiece during extrusion (a) strain distribution; (b) temperature distribution; (c) metal flow velocity distribution

3.3.2. The influence of the extrusion conditions to the deformation chatacteristics of MISC material.
The bonding strength between the matrix and reinforcement depends on the value of the effective stress of MISC composites suffered in the deformation region which significantly influenced by extrusion conditions, such as extrusion ratio, speed and temperatures. In this study, all the simulations of extrusion process were carried out under high temperatures due to the poor plasticity of magnesium matrix at room temperature, so only the extrusion speed and ratio were selected as the variable process parameters. In order to observe the difference of the deformation characteristics of composites under different conditions, a series of points in deformation region of material as shown in Figure 9 was tracked. And due to the sizes of the die with different extrusion ratio are different, the tracked points are not the same under different conditions, but always throughout the whole extrusion chamber.

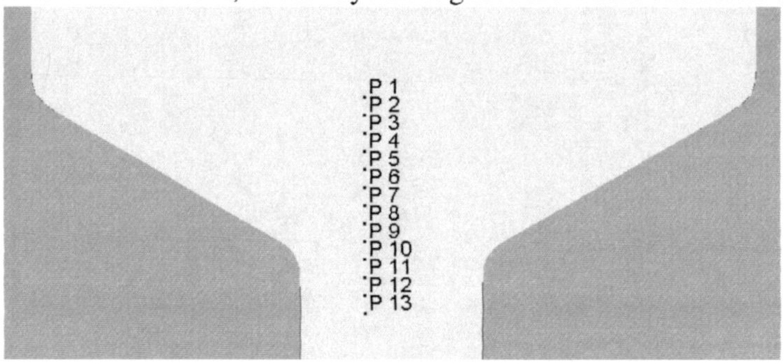

Figure 9. Deformation zone and the points for extracting deformation values ($d_{1\text{-}13}$=12mm)

The temperatures of the tracked points in deformation region of MISC composites under different conditions were exhibited in Figure 10. It can be seen that the temperatures of material increase significantly with the increase of extrusion ratio and speed and reach a peak near the outlet, i.e. the position of maximum strain. And the temperatures increase slowly along the line from point 1 to point 12 with a relative small extrusion ratio (r<16) and this also illustrated the uniform distribution of temperature of the workpiece. However, when extrusion ratio turnes larger, the temperatures rise rapidly along the line (as shown in Figure 10(a)) due to the sharply increase of the effective strain

IOP Conf. Series: Materials Science and Engineering **281** (2017) 012019 doi:10.1088/1757-899X/281/1/012019

which will convert to heat. It can be noted the temperatures only have a little increase when the extrusion speed reaches a certain value (v > 3mm/s) as shown in Figure 10(b). Figure 11 shows the simulated strain rate of material in deformation zone along the extrusion direction. It can be seen that the strain rates also increase with extrusion speed and ratio which are similar to the variation of temperature.

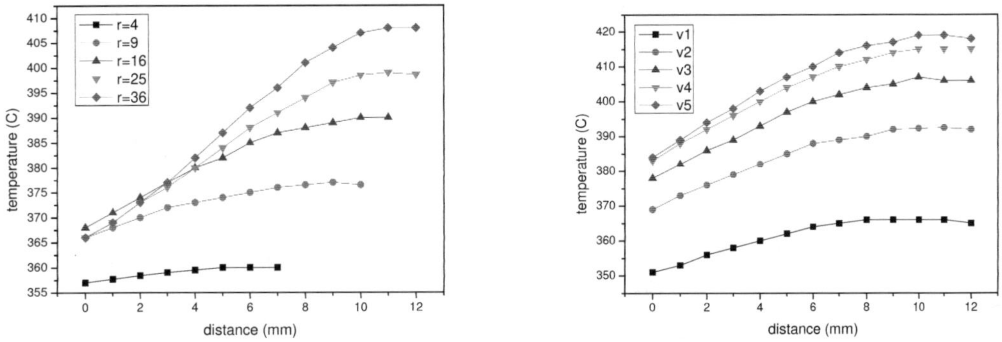

(a) (b)

Figure 10. Simulated temperatures in deformation zone along the extrusion direction (a) at different extrusion speeds; (b) at different extrusion ratio

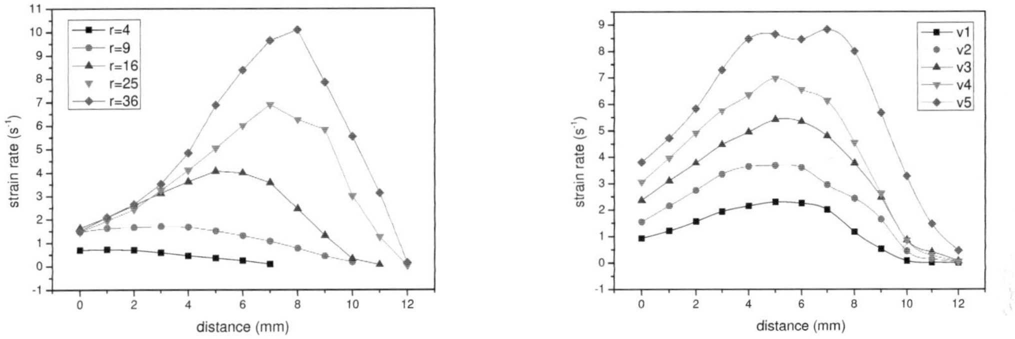

(a) (b)

Figure 11. Simulated strain rate in deformation zone along the extrusion direction (a) at different extrusion speeds; (b) at different extrusion ratio

And the effective stresses of MISC material in deformation zone are directly related to the vatiations of its temperature and strain rate according to the flow stress equation as shown in Figure 12. The effective stress of the MISC workpiece increases constinuously with the increase in extrusion ratio until it reaches a stable value when the extrusion ratio is too large (r>16, as shown in Figure 12(a)). In contrast, the effective stress decreases with the increase in extrusion speed and reaches a stable value when v>2mm/s (as shown in Figure 12(b)). In conclusion, the effective stress would be influenced directly by the deformation temperatures and strain rate of material which affected by extrusion conditions, i.e. extrusion speed and ratio, and it increase with the increase in strain rate and the decrease in extrusion temperature.

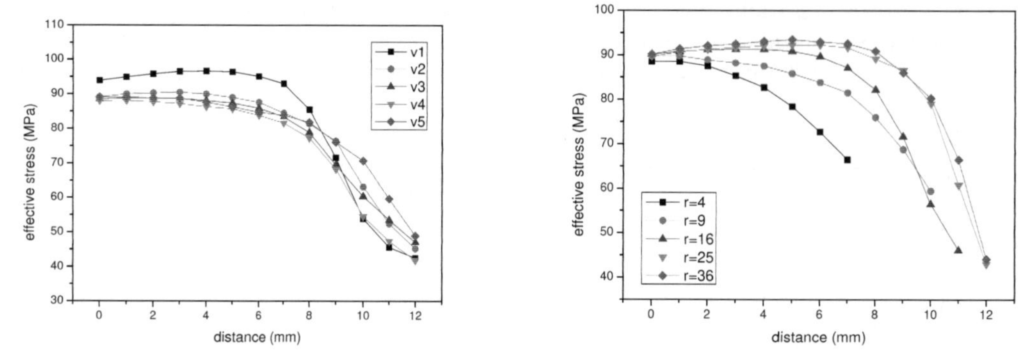

(a) (b)

Figure 12. Simulated effective stress in deformation zone along the extrusion direction (a) at different ram speeds; (b) at different extrusion ratio

3.3.3. Extrusion pressure in relation with extrusion conditions.

Extrusion pressure changes with the displacement of extrusion head at different extrusion conditions are exhibited in Figure 13. It is observed that as a constant extrusion speed of 2mm/s, extrusion pressure is significantly increased with the increase of extrusion ratio(Figure 13(a)). So considering the workability and strength of the MISC material, especially the plasticity of magnesium matrix and the operability of the extrusion experiment, the extrusion ratio should not be choosed too large despite it can introduce greater stress. Generally the extrusion ratio can be selected as 16 to 25. Notably, the extrusion pressure is not affected obviously by extrusion speed when the extrusion ratio keep to a constant value of 16 as shown in Figure 13(b). So based on the previous analysis, a smaller extrusion speed can be choosed to enhance the effective stress when processing. And to ensure the efficiency, the extrusion speed should not be too small.

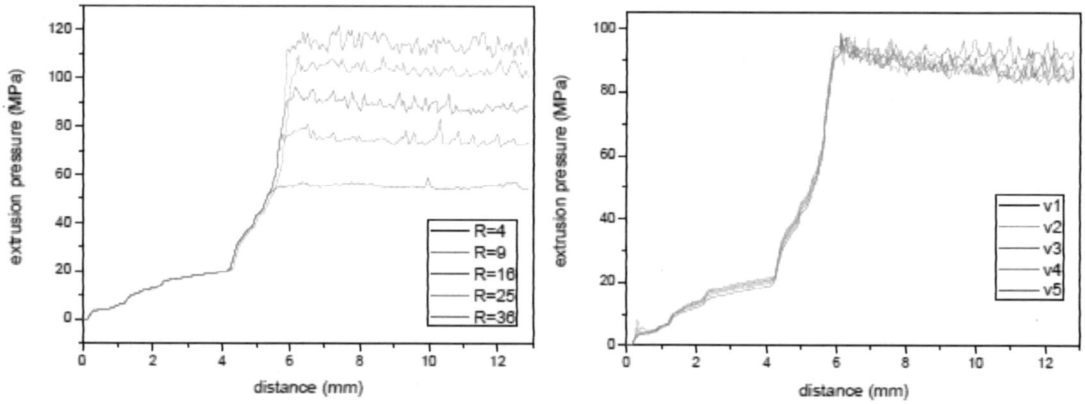

(a) (b)

Figure 13. Extrusion pressure evolutions (a) at different extrusion ratio; (b) at different extrusion speed

4. Conclusions

1. One kind of interpenetrating magnesium composite reinforced by three-dimensional braided steel reinforcement (MISC composites) were constructed, and the microstructure characteristics of it in as-cast and as-extruded state were analyzed. The MISC composites in extruded state have much finer grains compared to the as-cast state and an improvement in mechanical properties compared to the Mg matrix material due to the effective load transfer from matrix to reinforcement and good interfacial bonding quality caused by extrusion.

2. The extrusion processes of MISC material were simulated by FE method. The flow of blank through the extrusion die can be divided into three steps: upsetting - filling of the extrusion chamber - continuous extruded according to the mesh characteristic and the evolution of extrusion pressure. And the combined quality between the reinforcement and matrix of MISC composites counld be strengthened in the extrusion chamber where occurred large strain and suffered intense stress.
3. The effective stress of the MISC workpiece increases continuously with the increase in extrusion ratio and the decrease in extrusion speed until it reaches a stable value. And in order to acquire the larger effective stress so as to strengthen the bonding quality between Mg matrix and steel reforcement, a appropriately high extrusion ratio and a low extrusion speed should be selected.

Acknowledgements
This work was supported by the National Natural Science Foundation of China (51372101) and the S&T Developing Program of Shandong Province, China (2016GGX103008, 2015GSF118016).

References
[1] Wang S R, Song L H and Kang S K: Deformation behavior and microstructure evolution of wrought magnesium alloys, *Chinese Journal of Mechanical Engineering*, Vol. 26-3 (2013), p.437
[2] Fletcher M, Siebert-Timmer A, Bichler L and Sediako D, Evolution of Strain and Microstructure During Creep of Wrought AE42 and ZE10 Magnesium Alloys, *Transactions of the Indian Institute of Metals* , Vol. 66-2(2013), p. 133
[3] Alan A. Luo and Anil K. Sachdev: Development of a New Wrought Magnesium-Aluminum-Manganese Alloy AM30, *Metallurgical and Materials Transactions A* ,Vol. 38-6 (2007), p. 1184
[4] Zhang X L , Chen G and Du Y L : Synthesis of Plastic Mg-Based Bulk-Metallic-Glass Matrix Composites by Bridgman Solidification, *Metallurgical and Materials Transactions A*,Vol. 43-8(2002), p. 2604
[5] Varin R A: Intermetallic-reinforced light-metal matrix in-situ composites, *Metallurgical and Materials Transactions A,* Vol. 33-1 (2002), p. 193
[6] Wang S R, Guo P Q, Yang L Y and Wang Y J: Microstructure and mechanical properties of AZ91 alloys by addition of rare earth yttrium, *Journal of materials engineering and performance,* Vol. 6 (2008), p.13
[7] Wang S R, Wang Y, Li C C, Chi Q and Fei Z Y: The dry sliding wear behavior of interpenetrating titanium trialuminide /Aluminium composites, *Applied Composite Materials,* Vol. 14 (2007), p. 129
[8] Wang S R, Kang S B, Cho J H and Wang Y J: Microstructure evolution of ZK60 magnesium alloy at CC/TRC/TRCHC processing following by annealing, *Transactions of Nonferrous Metals Society of China,* Vol. 20-5(2010),p.763
[9] Shouren WANG, Wang M, Ma R, Wang Y and Wang Y J: Microstructure and hot compression behavior of twin-roll-casting AZ41M magnesium alloy, *Rare metals,* Vol. 29-4 (2010), p. 396
[10] Palani swamy Hariharasudhan, Gracious Ngaile, Taylan Altan: Finite element simulation of magnesium alloy sheet forming at elevated temperatures, *Journal of Materials Processing Technology;* Vol. 146(2001), p.52
[11] Wang S R, Guo P Q and Yang L Y: Deformation Behavior and Failure Analysis of Wrought Magnesium During Twin Rolling Casting and Hot Compression, *Journal of Engineering Materials and Technology,* Vol. 134(2012),p. 041002
[12] Ma R, Wang S R, Wang Y and Yang L Y: Microstructure evolution of ZK60M twin-roll-casted magnesium alloy during hot compression and annealing, *Advance in materials research,* Vol. 284-286 (2011),p. 1502
[13] Wang S R, Yang L Y, Ma R and Guo P Q: Fracture failure analysis of twin rolling casting magnesium by hot compression, *Advanced Materials Research* Vols. 123-125 (2010). P. 547

IOP Conf. Series: Materials Science and Engineering **281** (2017) 012020 doi:10.1088/1757-899X/281/1/012020

Modification of Wood Fiber for Use in Cement Board

F Q Han[1], X Tan[1] and F Q Zhao[1,2,*]
[1]Hebei University of Science & Technology, Shijiazhuang 050018, PR China
[2]Hebei Engineering Research Center of Solid Wastes Utilization

Email:zhaofq3366@126.com

Abstract. When ordinary Portland cement is used for wood fiber cement (WFC) board, the setting time is too long, even hard to solidify. Three methods can be used for wood fiber modification, i.e., soaking in water, treated with alkali solution and coated with some substances on the fiber surface. The results show that the proper water-cement ratio of WFC paste is 1:1.3 in the case of wood cement ratio being 1:1. The WFC board from modified wood fiber and cement is better than the control samples, in which the combined treatment, i.e. soaking in hot water and then coating with alkali-BFS-EVA slurry, behaves best. It is proved that ordinary Portland cement can be used to produce WFC board, with the modified wood fiber, which can greatly reduce production costs.

1. Introduction

Since thousands years ago, natural fibers have been used as reinforced inorganic materials such as straw and reeds for brick and mortar. Among these fibers, wood fiber is strong, lightweight, abundant, non-hazardous and relatively inexpensive. Therefore, WFC composites have been used in the fabrication of building materials for more than 60 years, has been more popular recently [1,2].

Cement wood board is a kind of cementinous composites, with mostly rapid hardening cement as binder, wood fiber as lightweight strengthening materials[3]. WFC board is an environmentally friendly building material, with energy-saving, sound insulation, anti-fire, anti-freeze-thaw, and other advantages [4]. It is generally used for sound-absorbing, insulation, and concrete decoration materials in the construction and transportation sector. WFC board has become one of the most popular building materials on the market.

Considering the cost of rapid hardening cements are relatively high, ordinary Portland cement can be used instead. However, that imposed a serious problem: when wood fiber is mixed with OPC, the setting time becomes longer, which influence the production of WFC board. It is known that wood generally contains cellulose, hemicellulose and lignin composition [5]. Hemicellulose is mostly composed of glucose, mannose, galactose, xylose and arabinose and some water-soluble monosaccharides, which tends to form calglucon with calcium ions resulting from the hydration of cement, and covered in the cement surface, blocking the cement and water. As the reaction proceeds, the film slowly becomes thicker, the hydration reaction rate of the cement slows down, and the hardening rate of the cement is gradually prolonged [6].

The key of production of WFC board using OPC to replace rapid hardening cement is to solve the problem of the retarding effect of cement when mixed with wood fiber.

IOP Conf. Series: Materials Science and Engineering **281** (2017) 012020 doi:10.1088/1757-899X/281/1/012020

2. Modification method for wood fiber

Various treatments can be used to solve the problem of retarding effect, most of which are based on soaking in water, alkaline hydrolysis and retention of sugar and hemicelluloses[7]. These treatments will effectively improve the cement curing properties of the WFC paste, to some extent.

The first method is extraction with water to remove the water-soluble compounds which have retarding effect on cement hydration. The wood fibre is socked in hot water for some time; the monosaccharide leach out from wood dissolved in hot water. This is suitable for extracting the water-soluble ingredients in the wood; the leaching efficiency is relatively low.

The second method is alkaline hydrolysis. Wood fibre is socked in a certain concentration of NaOH solution, stirring, the monosaccharide dissolves out. In the present of NaOH, hemicelluloses and sugar decomposed into non-retarding substances. This is the most common means in practice. It is simple and easy to operate, but will be destroy celluloses, which contributes the strength of WFC board.

The third method is the retention of sugar and hemicelluloses. In this process, a thin coating on the surface of wood fibre forms, inhibiting the material to be released into the curing medium. It is the commonly used approach for WFC board production. A protective film on the surface of wood fibre provides a barrier between the retarding agents and cement, removing retardation effect and prolonging the setting time of paste. Some mineralisers as blast-furnace slag, fly ash and steel slag can be used for this purpose [8,9]. And the mineralisers in the coating layer not only have strong effect on early strength of cement board, but also improve durability and water resistance and other properties.

Wood fibre modification might reduce the final mechanical properties of WFC board, but the compensation obtained is the reduction of settling time of WFC paste. The combined use of different methods may bring the synergistic effect into play and obtain the optimal results.

3. Experimental

3.1. Materials

The wood fiber, granulated blast-furnace slag(BFS) and Portland cement are provided by Shijiahzuang Changanjiayi construction material Co. ltd. Caustic soda and EVA are available on market.

3.2. Methods

Water-cement ratio. According to practice and initial test, the ratio of water and cement varies from 1:1 to 1:1.5. With fixed mass ratio of wood fiber and cement 1:4, three parallel tests are designed to obtain the proper water-cement ratio.

Wood fiber modification. Three modification methods are tested. The first is to treat wood fiber in hot water. The second is to treat wood fiber in alkali solution. The third is combined treatment of soaking in hot water and coating with alkali-BFS-EVA slurry.

(1) Soaking in hot Water. Take some wood fiber, soak it in hot water of 80°C for 1.5 h, filtering and washing, air dried.

(2) Soaking in alkali solution. Take some wood fiber, soak it in alkali solution at room temperature for 1.5 h, filtering and washing, air dried.

(3) Combined treatment. Take some wood fiber, the first soaking in water for 5 minutes, and then mixed with the mixture of NaOH solution, BFS and VAE emulsion, forming a thin layer film on the surface of the wood fiber, air dried.

Preparation of samples. The modified wood fiber is mixed with cement and water. Samples with the dimensions of 70×70×210 mm are molded, standard curing, and de-molded after 5 hours, as shown in Figure. 1.

Figure 1 Test block sample

3.3.Results and discussion

3.3.1 Water-cement ratio Water-cement ratio is an important factor, which influence the de-molding and mechanical strength of WFC board. Table 1 shows the effect of water-cement ratio on WFC board samples after curing for 5 hours.

Table 1 Effect of water-cement ratio on hardness of the specimen

Experiment number	Wood cement ratio	Water-cement ratio	Hardness of the specimen
1	1:4	1:0.8	Soft
2	1:4	1:1	Soft
3	1:4	1:1.3	Hard
4	1:4	1:1.5	Hard+

From the Table 1 we can see that water-cement ratio has remarked effect on the hardness of WFC paste. When water-cement ratio is at and above 1:1, the specimen is soft. The proper amount of water not only affects the hydration of cement strength, but also makes cement and wood fiber mixed evenly; but excessive water will make the influence the hardening of the paste. The results show that the water-cement ratio of 1:1.3 is appropriate.

3.3.2 Modification of wood fiber Three modification methods were tested and compared with the blank. See Table 2.

Table 2 Comparison of various modification methods for wood fiber

Method	Hardness of the specimen	Cost estimation
blank	Soft	Low
Soaking in hot Water	Hard	Low
Soaking in alkali solution	Hard	Moderate
Combined treatment	Hard+	Moderate

As can be seen from the Table 2, the hardness of WFC specimen with modified fiber is higher than the blank, in which the combined treatment behaves best. For the blank sample, suppression of substances contained in wood fibers, which tends to form calglucon with calcium ions resulted from the hydration of cement, and covered in the cement surface, blocking the cement and water. Therefore, the hydration of cement is retarded. In the case of the combined treatment, alkali-BFS-VAE slurry formed a layer of dense protective film, covering the surface of the wood fiber, blocking the cement and the inhibitor. In this way, the retarding effect of wood fiber on cement is removed, obtaining normal setting time for WFC board. Besides, the cost is moderate. Based on the above factors, the method of combined treatment is suitable for WFC board.

IOP Conf. Series: Materials Science and Engineering **281** (2017) 012020 doi:10.1088/1757-899X/281/1/012020

4. Conclusions

(1) Some organic materials in wood fiber inhibit the setting of the cement and reduce its suitability as WFC board. When wood fiber is modified, ordinary Portland cement can be used to produce WFC board instead of rapid hardening cement, thus reducing production cost.

(2) Water-cement ratio is an important factor in WFC production. In the case of wood cement ratio being 1:1, as water-cement ratio is at and above 1:1, the paste is difficult to harden. The proper water-cement ratio is 1:1.3.

(3) Three methods for wood fiber modification, soaking in water, alkaline hydrolysis treatment and the combined approach of soaking in hot water and then coating with alkali-BFS-EVA slurry, can be used to modify wood fiber. Among these, the combined method behaves best and the cost is moderate.

References

[1] Mahzabin M S, Hamid R, Badaruzzaman W H W. Evaluation of Chemicals Incorporated Wood Fibre Cement Matrix Properties [J]. *Journal of Engineering Science & Technology*, 2013, **8**(4).

[2] Xuedong X U, Duan X, Zhao S, et al. Experimental study on the build of eco and energy-saving rural housing with wood wool cement board [J]. *New Building Materials* [In Chinese], 2010.

[3] Zhong S Y. Large Wood Welded Cement Wallboard and Its Application [J]. *New building materials* [In Chinese], 2016, **43**(8):100-105.

[4] Ling C. Wooden cement board [J]. *Concrete world* [In Chinese], 2007(3): 64-66.

[5] Zhang S Y. Chemical Components Effect on Mechanical Properties of Wood Cell Wall [D]. *Chinese Academy of Forestry Sciences*, 2011.

[6] Tu P T. Cement wood board and its production technology [J]. *China Forest Products Industry*, 2007, **34**(2):000034-37.

[7] Quiroga A, Marzocchi V, Rintoul I. Influence of wood treatments on mechanical properties of wood-cement composites and of Populus Euroamericana wood fibers [J]. *Composites Part B Engineering*, 2016, **84**.

[8] Zhao F Q , Ni W , Wang H J , Liu H J. Activated fly ash/slag blended cement [J]. *Resources Conservation & Recycling"* , 2008 (2) :303-313

[9] Zhao F Q , Liu H J , Hao L X , Li Q . Water resistant block from desulfurization gypsum. *Construction & Building Materials* , 2012 (1) :531-533

IOP Conf. Series: Materials Science and Engineering **281** (2017) 012021 doi:10.1088/1757-899X/281/1/012021

On-line Monitoring Device for High-voltage Switch Cabinet Partial Discharge Based on Pulse Current Method

S Y Tao[*1,2], X Z Zhang[1], H W Cai[1,2], P Li[1,2], Y Feng[1,2], T C Zhang[1,2], J Li[1,2], W S Wang[1,2] and X K Zhang[2]

[1] Power Research Institute of State Grid Beijing Electric Power Company, Beijing, 100075, China

[2] Beijing Dingcheng Heng'an Technology Development Co., Ltd., Beijing, 100075, China

Email: 563450993@qq.com

Abstract. The pulse current method for partial discharge detection is generally applied in type testing and other off-line tests of electrical equipment at delivery. After intensive analysis of the present situation and existing problems of partial discharge detection in switch cabinets, this paper designed the circuit principle and signal extraction method for partial discharge on-line detection based on a high-voltage presence indicating systems (VPIS), established a high voltage switch cabinet partial discharge on-line detection circuit based on the pulse current method, developed background software integrated with real-time monitoring, judging and analyzing functions, carried out a real discharge simulation test on a real-type partial discharge defect simulation platform of a 10KV switch cabinet, and verified the sensitivity and validity of the high-voltage switch cabinet partial discharge on-line monitoring device based on the pulse current method. The study presented in this paper is of great significance for switch cabinet maintenance and theoretical study on pulse current method on-line detection, and has provided a good implementation method for partial discharge on-line monitoring devices for 10KV distribution network equipment.

1. Introduction

High voltage switch cabinets are critical electrical equipment in power system distribution networks. The safe operation of high-voltage switch cabinets is of great significance for grid security. Defects or deterioration of the internal insulation components of high voltage switch cabinets or poor contact at conductive connection parts will threaten their safe operation and directly affect the safe and reliable operation of the power grid. Relevant data show that the failures caused by insulation and current-carrying errors account for 40% of the total 6-10kV switch cabinet accidents in China's power system. Among these, accidents arising from insulation flashovers account for 79% of the total number of accidents caused by insulation failure. The status of the insulation within the switch cabinet can be analyzed based on relevant information obtained through partial discharge on-line monitoring since partial discharge may be present during the accident incubation period[1-8].

2. Significance of high voltage switch cabinet partial discharge

The internal components, supporting insulator and busbar are installed into the switch cabinet after arrival at the site, and can only be tested after on site assembly. However, carrying out high-voltage withstand testing or partial discharge testing on site is much more difficult than in the plant. Generally, high voltage equipment presents higher integration. Once an accident happens, the hazard rating will be much high than that of previous discrete and open-type equipment. The failure of one equipment component will endanger the adjacent components, and the maintenance cycle will be much longer. Furthermore, switch gear products are becoming increasingly compact, which leads to a corresponding reduction in insulation margin. Especially in South China, the hot and humid summer air is prone to causing surface creepage of internal insulation components which can hardly be detected outside the cabinet. In recent years, many surface creepage failures have been detected. Therefore, it is necessary to monitor the insulation state of high-voltage switch cabinets [9-10].

The most common high voltage switch cabinet partial discharge detection method is the transient earth voltage (TEV) method which is adopted for routing inspection. However, quantitative analysis cannot be carried out while using this method for partial discharge detection. Furthermore, it shows poor technical and economic performance since it can be easily disturbed by ultrahigh frequency (UHF) or acoustic emission (AE), and requires a lot of equipment if electro-acoustic methods are adopted. Long-term on-line monitoring and partial discharge source localization cannot be widely used for switch cabinets with this method[11-15]. Consequently, a more efficient high-voltage switch cabinet partial discharge on-line monitoring system is needed.

3. Pulse current detection principle for VPIS

The pulse current method is the first developed and most widely used method for detection at present. The International Electrotechnical Commission (IEC) formulated a specific detection standard for it in 2000. Each partial discharge will lead to the neutralization of positive and negative charge, accompanied by a sharp current pulse. Thus, the pulse current method is detecting the occurrence of partial discharge by measuring this pulse current. This method is generally applied in type testing and other off-line tests of electrical equipment at delivery. In spite of the high sensitivity of off-line measurement, it is prone to be affected by external noise on site and shows poor immunity from interference[16-20].

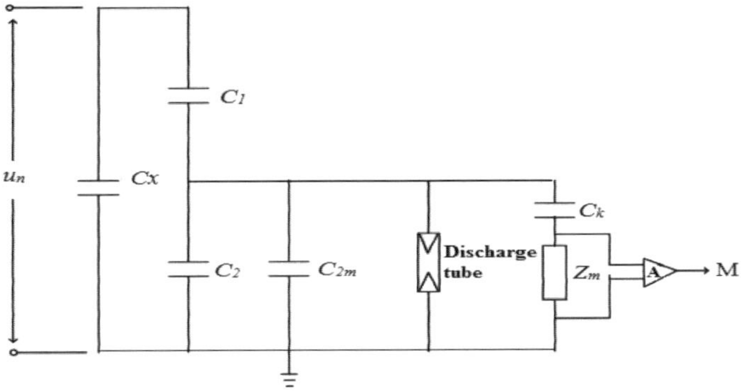

Figure 1. VPIS Partial of discharge detection circuit

According to the characteristics of the pulse current method, this paper improves the traditional IEC partial discharge pulse current test method by utilizing the partial voltage principle of high-voltage presence indicating systems (VPIS). Coupling capacitance and detection impedance are added according to the partial voltage principle of insulator sensors and VPIS, and external interfering noise is efficiently eliminated by combining the method with three-phase common mode noise inhibition, amplification and detection. As a result, the partial discharge monitoring device can be applied on site efficiently, breaking through the previous bottleneck where the pulse current method could only be used in a laboratory for high voltage switch cabinet application. Figure 1 above is the VPIS partial discharge detection circuit .

As shown in Figure 1, u_n is the rated voltage of the switch cabinet busbar; C_x is the capacitance of the test sample; C_1 is the level 1 capacitance of the insulator; C_2 is the level 2 capacitance of the insulator; C_{2m} is the matching capacity of the VPIS; C_k is the coupling capacitance of the PD; Z_m is the detecting impedance; A is the amplifier; and M is the test system.

Detecting impedance Z_m is obtaining the high-frequency pulse signal generated by partial discharge. Since the signal amplitude is small and requires amplification with an amplifier, Z_m and the amplifier must match perfectly, which determines detection sensitivity and pulse resolution.

Figure 1 shows that the total impedance Z_0 of C_1, C_{2l}, C_{2m}, C_{2k} and Z_m *is:*

$$C_2 = C_{2l}+C_{2m} \tag{1}$$

$$Z_0 = \frac{1-jwC_kZ_m}{w^2C_kC_2Z_m - jw(C_k+C_2)} - j\frac{1}{wC_1} \tag{2}$$

The signal collected by the test system is a partial discharge pulse voltage signal, the voltage amplitude of which is u_m. The known rated operating voltage of the primary busbar of the switch cabinet is u_n. The transitive relation between u_m and u_n can be derived via formula (1) and (2):

$$u_m = \frac{u_nC_1Z_m(jwC_k + w^2C_k{}^2)}{C_1 - C_2 - C_k - jwZ_m(2C_kC_2 + C_k{}^2) + w^2C_k{}^2Z_m{}^2(C_1+C_2)} \tag{3}$$

The partial discharge pulse detected by the test system is in proportion to the actual partial discharge quantity in electric charge. The specific proportional relation is related to the measuring circuit, amplifier etc. It is very difficult to calculate the quantity of electric charge from the indicated value. Instead, it is determined via experiment. That is, the measuring device for the pulse current method must be calibrated. The calibration method involves applying a pulse of a certain discharge capacity which simulates a partial discharge, and observing if the partial discharge amount displayed on the measuring device is consistent with that applied on the test sample. In case of any inconformity, the measuring device is adjusted with the calibration coefficient k until the inconformity is eliminated. The measurement and calibration with this technology can be performed when the high voltage switch cabinet is online, which is more advanced than the traditional off-line calibration. Online calibration positions and pictures are as shown in Figure 2.

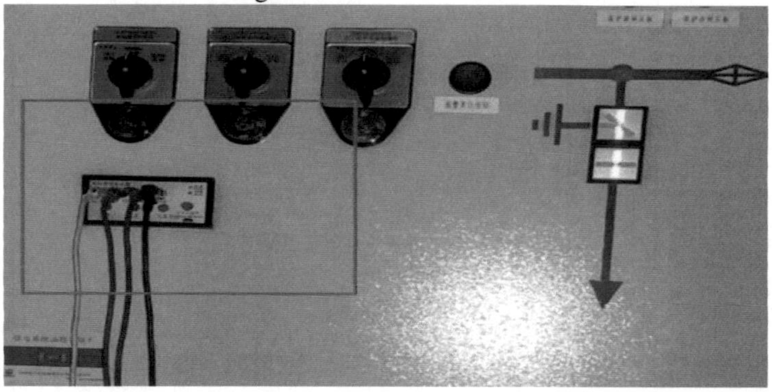

Figure 2. On-line calibration wiring diagram

During the partial discharge calibration process, if a pulse signal with a discharge capacity of q_1 is injected at both ends at the matching capacitance C_{2m} of the VPIS, the relationship between the actually measured voltage amplitude and injected signal can be obtained as shown below:

$$q_1 = k\int_0^{t_0} u'mdt \tag{4}$$

Where in Formula (4), k is the calibration coefficient, t_0 is voltage duration of the partial discharge, and $u'm$ is the actually measured voltage amplitude.

4. Device configuration and implementation

4.1 Compositions and installation

The system established in this paper mainly includes the VPIS, amplifier, collection and display unit, analysis software and local storage unit and the background summary display unit. A system composition block diagram and connection relationships are shown in Figure 3. In the stable operation of the system, the system parameter table is shown in table 1.

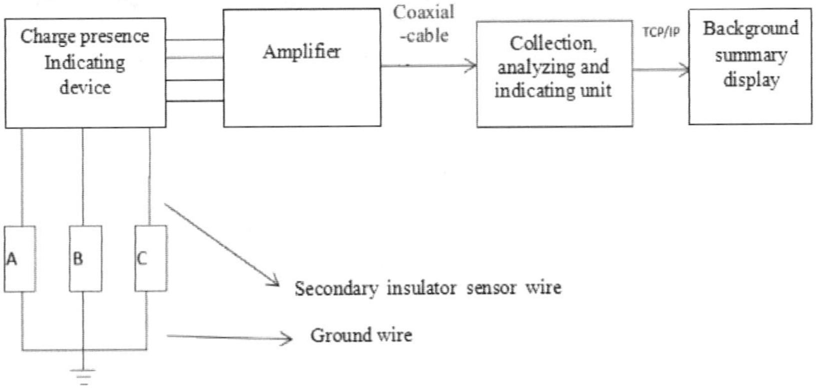

Figure 3. Block diagram of VPIS partial discharge measurement system.

Table 1. System parameter table

System project	Parameter
Primary system voltage	10kV
Detection voltage	AC 100±5V
Auxiliary power	DC 110V
Power supply	AC 220V
Detection sensitivity	≤50pC
Alarm function	Available
Continuous detection function	Available
Phase analysis function	Available

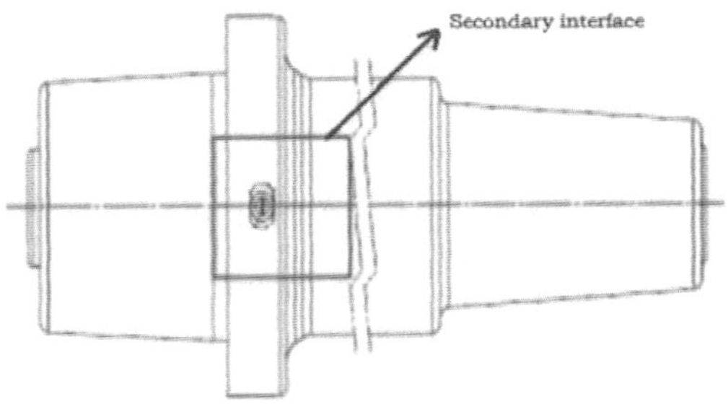

Figure 4. Secondary interface insulator sensor location diagram

This equipment and the switch cabinet indicating system share one set of insulator type capacitance sensors. The connection position is as shown in the red blocks of Figure 4. The PD signal amplifier can be connected to the three-phase interface on the back plate of the charge display device. The PD signal is transferred to the analyzing host for data analysis after being amplified with a signal amplifier.

4.2. Noise suppression

The device can only be powered on for debugging after being installed properly. After power on, the analysis and display unit will display ambient noise. After judging the noise size, adjust the appropriate noise suppression device via analysis software.

4.3. Calibration setting

Since the device is based on the pulse current method, calibration setting shall be carried out before monitoring. The calibrator is specifically used for on-line calibration and measurement accuracy. Specifically, the on-line calibration method is as follows:

- Check whether the calibrated switch cabinet and adjacent ones are equipped with electrophase-checking interfaces, and are normally energized.
- Connect the partial discharge calibrator and the electrophase-checking interface of the calibrated switch cabinet and adjacent ones. The positive pole of the calibrator signal output wire is connected to any phase of the three-phase electrophase-checking interface. The negative pole is connected to the grounding interface to ensure reliable connection of the calibrated signal wire and electrophase-checking interface.
- The capacitance of the capacitance sensor of the running switch cabinet can be tested online by inputting the grid operating voltage to the calibrator. Rotate the voltage adjusting knob to emit a corresponding partial discharge pulse. Then equip the analyzing host with partial discharge analysis software, and adjust the power value as indicated by the analysis software to be consistent with the calibrator to complete calibration.

4.4. Software analysis

As shown in Figure 5, we can process and analyze the collected data via analysis software, and check the partial discharge phase diagram and discharged power of each channel. Identify the switch cabinet with partial discharge through comparing the size of partial discharge of each switch cabinet.

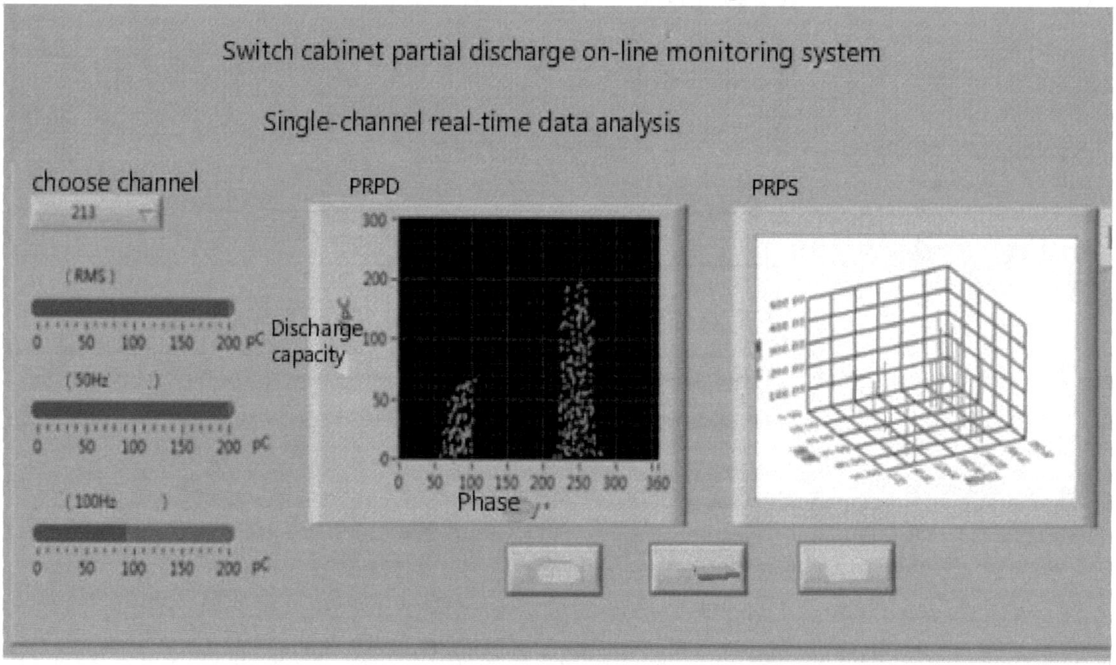

Figure 5. Analysis software interface of switch cabinet partial

Figure 5. Analysis software interface of switch cabinet partial discharge on-line monitoring system

5. Verification test

In Dec. 2016, the device was installed on the partial discharge defect simulation platform of a 10KV switch cabinet belonging to the State Grid Beijing Electric Power Research Institute to verify the defect test effects, including PD monitoring effect, sensor sensitivity, repeat ability, partial discharge mode recognition etc.

The test established an air gap insulator defect with low power discharge in the 222 switch cabinet as shown in Figure 6.

Figure 6. Defect insulator within the 222 switch cabinet

After applying voltage to the switch cabinet, the device was subject to on-line calibration by utilizing the partial discharge calibrator. During the test process, a certain grade of voltage was applied to the 222 switch cabinet until an obvious PD signal was obtained. The device was calibrated by utilizing the PD calibrator and performing partial discharge monitoring. The voltage u_n applied to the 222 switch cabinet was10KV. Testing with the device presented in this paper was carried out. The test results are as shown in Figure 7 and 8. When the switch cabinet was under 50pC, the internal defect discharge was 121pC. The phase of the discharge spectrogram was narrow. While compared to the voltage phase level applied in switch cabinet A, the discharge phase was distributed between 270 and 300°. Due to the 50pC background interference of the switch cabinet simulation platform, the discharge spectrogram complied with the phase distribution features of phase 1 and 3. Therefore, it can be determined that the discharge is from phase A of the switch cabinet and is in line with the actual defect setting.

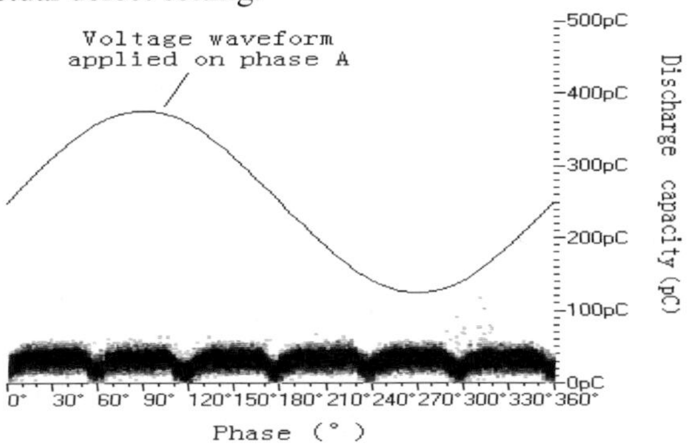

Figure 7. PRPD spectrogram of partial discharge signal.

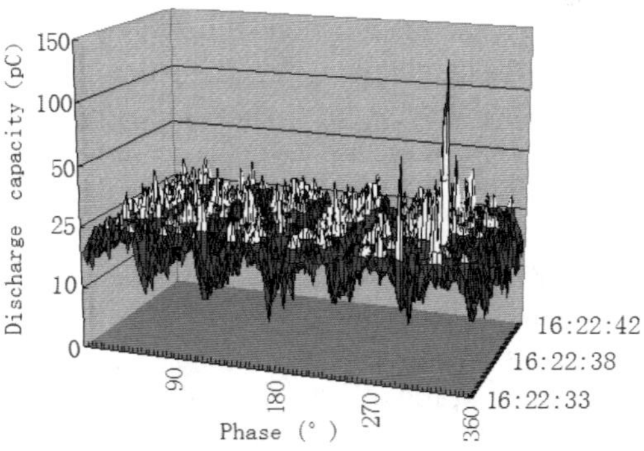

Figure 8. PRPD 3D spectrogram of partial discharge signal (20ms).

IOP Conf. Series: Materials Science and Engineering **281** (2017) 012021 doi:10.1088/1757-899X/281/1/012021

6. Conclusions

The simulation status within the switch cabinet can be reflected effectively and intuitively via a VPIS partial discharge monitoring device which can not only effectively prevent a catastrophic failure, but also optimally control the O&M cost of the entire system. This paper obtained the following conclusions through establishing the monitoring circuit, formula, device development and platform verification:

- The detection circuit has a partial discharge signal calibration and detection function through the designed pulse current partial discharge detection circuit and formula. This method is consistent with the traditional pulse current partial discharge measurement.
- The device can effectively monitor the actual partial discharge signal in the switch cabinet as verified by the simulation platform of the State Grid Beijing Electric Power Research Institute.
- The three-phase common-mode suppression circuit of the device amplifier can effectively filter external interference signals. The actual detection sensitivity is no greater than 50 pC.
- The background analysis software developed in this paper can perform phase spectrogram analysis. The discharge source phase can be determined through comparing with the actual operating phase.

References

[1] Yu Y, Li R P, Huang C, et al. Study on TEV distribution characteristic of switchgear surface based on finite-difference time-domain method [J].Insulating Materials,2015,48(4);45-51

[2] Guan Y G and Qian J L, 2001 Research on the Application of the RF Method on High Voltage Switch Cabinet PD Monitoring [J] *High-Voltage Electrical Appliances.* 5

[3] Shang Pe, Wang Y F and Zhao L, 2013 Detection and Analysis of High Voltage Switch Cabinet Partial Discharge [J] *Electronics World.* 22

[4] Cui Z Y, Jiang C Y and Piao J J, 2007 UHF Partial Discharge On-line Real-time Monitoring System Preventing Failure of Gas Insulation Switch Devices [J] *Power System Technology.* 4

[5] Tian Y and Tian J L, 1996 Statistical Accident Analysis and Improvement Suggestions for 6-10kV Switch Cabinets [J] *Northeast Electric Power Technology.* 8

[6] Cheng S Y, Lu Fa C and Xie Q, 2012 Study of Partial Discharge Locating Method in Transformers Based on Transient Earth Voltage and Ultrasonic Array Signals [J] *Transactions of China Electrotechnical Society.* 27(**4**): 255-262

[7] Chen X, 2012 A Thesis Submitted in Partial Fulfillment of the Requirements for the Degree of Master of Engineering [D] *Huazhong University of Science & Technology*

[8] Wang Y Y, Li Y W and Lu G J, 2011 Simulation of Transient Earth Voltages .Aroused by Partial Discharge in Switch gears [J] *High Voltage Engineering.* 07:1683-1688

[9] Ren M, Peng H D, Chen X Q and Dong M, 2010 Application of the Transient Ground Voltage Method to Comprehensive Detection of Partial Discharge in Switchgear [J]*HighVoltage Engineering.* 10:2460-2466

[10] Wang L H and Lv H, 2014 Application Study on Detection Technology of Switch gear Partial Discharge [J] *High Voltage Apparatus.* 03:62-67

[11] Han T, 2012 Switch gear Partial Discharge Detection and Pattern Recognition Based on TEV [D] *Tianjin University*

[12] Li Y, Wang Y and Lu G 2010 Simulation of transient earth voltages aroused by partial discharge in switchgears[C]//*High Voltage Engineering and Application (ICHVE) 2010 Conference on. IEEE.* 2010: 309-312

[13] Ye H F, QianY and Man Y Y, 2013 Simulation Analysis of Transient Earth Voltages Aroused by Partial Discharge in Switchgear [J] *High Voltage Apparatus.* 07:13-17+24

[14] Chen P, 2014 Research on Partial Discharge Monitoring and Fault Pattern Recognition in Switchgear Based on the Characteristics of the Signal Energy [D] *Chongqing University*

IOP Conf. Series: Materials Science and Engineering **281** (2017) 012021 doi:10.1088/1757-899X/281/1/012021

[15] Xianzhi Y and Jian Z 2011 Application of 10 kV Switch gear Partial Discharge Detecting Based on Ultrasonic and TEV Technology in the Power Distribution Network [J] *Electric Power Science and Engineering*. 6: 018

[16] Feng B, Xiao D B and Li Y, 2015 Detection of Partial Discharge in Switchgear Under Different Humidity Using Transient Earth Voltage and Ultrasonic Method [J] *High Votage Apparatus*. 04:67-71

[17] Yao M 2010 The application of temporary earth voltage (TEV) measurement in the online monitoring of the partial discharge of HV switch cabinet [C]//Electricity (CICED), China *International Conference on IEEE 1-6*

[18] Li P, Huang D C, Ruan J J, et al.Electromagnetic interference on secondary smart devices caused by breaking 10 k V switch cabinet[J].*Power System Technology*,2015,39(1);110-117

[19] *IEC270* 2000 High-voltage Test Techniques-partial Discharge Measurements

[20] Xin Y C, Cui J D, Zhou C,et al.Fault classification for smart switchgear based on multivariate multiscale cloud sample entropy and fuzzy support vector machine[J]. *Power System Technology*,2016,40(11);3597-3603

IOP Conf. Series: Materials Science and Engineering **281** (2017) 012022 doi:10.1088/1757-899X/281/1/012022

Exploration on Wire Discharge Machining Added Powder for Metal-Based Diamond Grinding Wheel on Wire EDM Dressing and Truing of Grinding Tungsten Carbide Material

H M Chow[1*], L D Yang[2], Y C Lin[1] and C L Lin [3]

[1]Department of Mechanical Engineering, Nan Kai University of Technology, Nantou 54243, Taiwan
[2] Department of Industrial Management, Nan Kai University of Technology, Nantou 54243, Taiwan
[3] School of Creative Design, City College of Dong-guan University of Technology, Dong-guan, China.

Email: hmchow@nkut.edu.tw

Abstract. In this paper, the effects of material removal rate and abrasive grain protrusion on the metal-based diamond grinding wheel were studied to find the optimal parameters for adding powder and wire discharge. In addition, this kind of electric discharge method to add powder on the metal-based diamond grinding wheel on line after dressing and truing will be applied on tungsten carbide to study the grinding material removal rate, grinding wheel wear, surface roughness, and surface micro-hardness.

1. Introduction

It is important that grinding wheel surface abrasive to maintain the appropriate amount of protrusion in order to have a good processing quality. Grinding process will produce high temperature, adhesion and other effects and abrasive will wear, loss of the original geometry and size; due to high-speed grinding of the grinding debris is extremely small and easy to block the surface of the grinding wheel. In order to make the diamond grinding wheel to maintain a good grinding state, it is necessary for grinding wheel dressing and truing after the abrasive wear. The shaping and repairing principle of the grinding wheel is to remove the metal bonding agent from the surface of the grinding wheel, so that the abrasive grains are prominent to increase the grinding performance. In order to make the diamond grinding wheel to maintain a good grinding state, it is necessary for grinding wheel dressing and truing after the abrasive wear. The shaping and repairing principle of the grinding wheel is to remove the metal bonding agent from the surface of the grinding wheel, so that the abrasive grains are prominent to increase the grinding performance. Wire EDM is often used conductive material as a binding agent for grinding wheel dressing and truing which is employing a forming conductive electrode to grind wheel for dressing and sharpening [1-3]. Diamond grinding wheel hardness, wear resistance, compressive strength and thermal conductivity and other characteristics, known as super abrasive, when the processing components required ultra-precision, diamond grinding wheel has gradually replaced the silicon carbide wheel, alumina wheel [4]. The metal method is a powder metallurgy with a metal powder as a binder. Metal-based diamond grinding wheel has good abrasive adhesion、thermal conductivity, and excellent abrasion resistance, can maintain a sharp shape after grinding. It can be applied to the ultra-precision grinding like mirror grinding, non-ferrous metal or non-metallic cutting difficult materials, high-performance ceramic high-speed grinding and forming grinding [5].

2. Experimental method

In this study, the electric discharge method to add powder on the metal-based diamond grinding wheel on line after dressing and truing will be applied on tungsten carbide to study the grinding material removal rate, grinding wheel wear, surface roughness, and surface micro-hardness. Experimental procedure is shown in Figure 1. Taguchi method is applied to analyse and conduct the experiment by assigning the grinding parameters shown Figure 1, then L18 experimental array will be used to conduct the experiments and the optimal level combination will be found out.

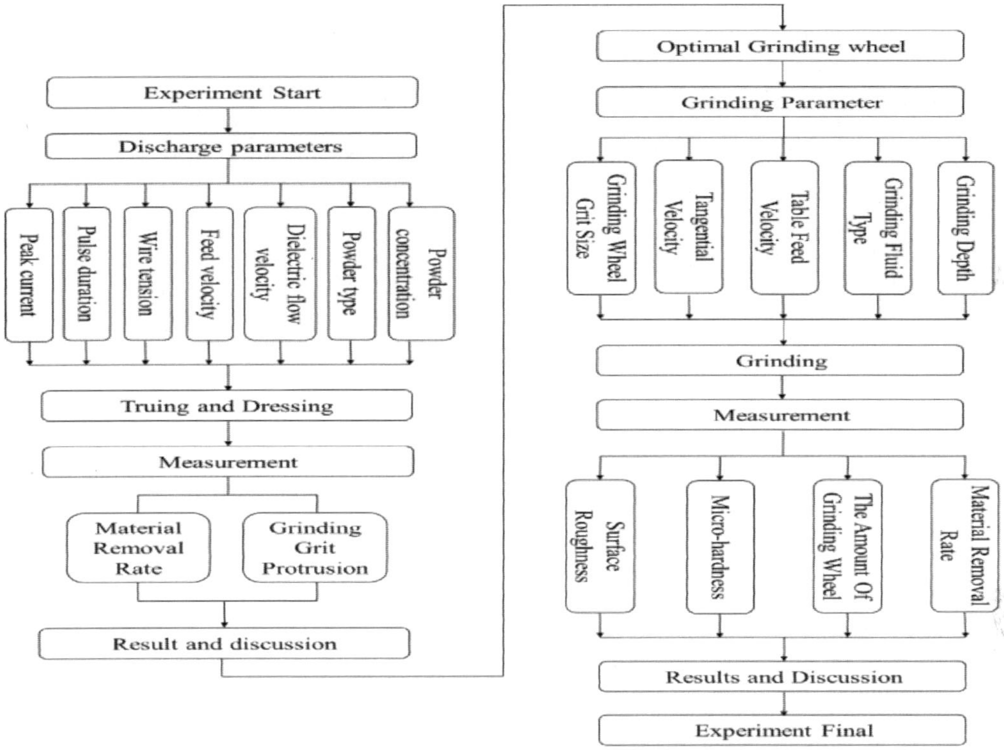

Figure 1. Experimental architecture flowchart

2.1 Experimental materials

Table 1 Powder characteristics

	Aluminum	Silicon carbide	Chromium
Particle size	1~3μm	1~3μm	1~5μm
Bulk density g/cm^3	1.143	1.254	1.775
Purity	99.5%	99.5%	99%
1.75cm^3/L	2.000g	2.194g	3.106g
3.5cm^3/L	4.000g	4.389g	6.212g
7cm^3 /L	8.000g	8.778g	12.425g

The consumable equipment used in this study contains metal-based diamond wheels, copper electrodes, aluminum powder, silicon carbide powder, chromium powder, and tungsten carbide blocks shown in Table 1.

2.2 Adding powder to the EDM experiments

The parameters are discharge current, pulse time, feed rate, wire tension, dielectric fluid flow rate and the concentration of three kinds of powder processing fluid shown in Table 2. The 18 sets of experimental models are established to obtain the optimal combination of discharge parameters shown in Table 3. In this study, there are two characteristic points observed, large the better (LTB) material removal rate (MRR) and nominal the better (NTB) abrasive grain (Grinding Grit Protrusion, GGP). As the diamond abrasive particles in the binder is irregularly distributed and dispersed, so each combination before the experiment carbon steel blocks must be grounded and the new datum plane can be defined in the modification of the experiment to avoid interference between the experiments.

In this study, the removal rate and the amount of protrusion were calculated by the average roughness (Ra) and ten point roughness (Rz), respectively. From the Taguchi experiment, a better set of discharge parameters and factor responses are obtained. Then the confirmation of the experiments will be conducted. To prove that this experiment is suitable for the diamond wheel of each particle size, the # 100 diamond grinding wheel with a particle size of about 150 and # 1000 diamond grinding wheel particle size is about 13 were chosen. And then to explore the different diameter of the grinding wheel processing methods and parameters. The center line as a benchmark to the bottom of the curve refraction shown in Figure 2, and then calculate the center line above the full curve after the curve covered by the area, and then divided by the measured length, the resulting value in μm units, that is, the length of the processing surface within the measurement center line average surface roughness value shown in Figure 3.

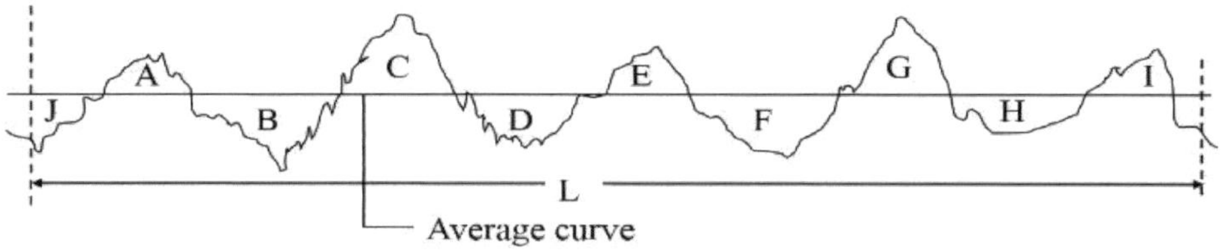

Figure 2 Curve average line

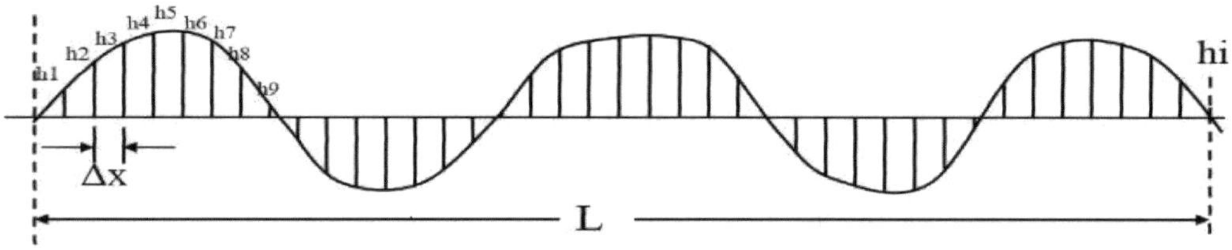

Figure 3 Central line average surface roughness (Ra)

Table 2 Discharge experimental factors and levels

Factor / Level	A	B	C	D	E	F	G
	Peak current (Amp)	Pulse time (μs)	Feed rate (mm/min)	Line tension (kgf)	Dielectric flow rate(mm/s)	Powder type	Powder Concentration (g/l)
I	1	10	0.5	1	0.05	AL	1.75
II	3	20	1	3	0.10	SiC	3.5
III	5	30	2	5	0.15	Cr	7

Table 3 $L_{18}(3^7)$ experiment parameters and levels

Exp	A Peak current (Amp)	B Pulse time (μs)	C Feed rate (mm/min)	D Line tension (kgf)	E Dielectric flow rate (mm/s)	F Powder type	G Powder concentration
1	1	10	0.5	1	0.05	Al	1.75
2	1	20	1	3	0.1	SiC	3.5
3	1	30	2	5	0.15	Cr	7
4	1	10	0.5	1	0.1	Cr	7
5	1	20	1	3	0.15	Al	1.75
6	1	30	2	5	0.05	SiC	3.5
7	3	10	1	3	0.05	SiC	7
8	3	20	2	5	0.1	Cr	1.75
9	3	30	0.5	1	0.15	Al	3.5
10	3	10	1	5	0.15	SiC	1.75
11	3	20	2	1	0.05	Cr	3.5
12	3	30	0.5	3	0.1	Al	7
13	5	10	2	3	0.15	Cr	3.5
14	5	20	0.5	5	0.05	Al	7
15	5	30	1	1	0.1	SiC	1.75
16	5	10	2	5	0.1	Al	3.5
17	5	20	0.5	1	0.15	SiC	7
18	5	30	1	3	0.05	Cr	1.75

2.3Tungsten carbide grinding experiment

In order to obtain the optimal parameter combination, the parameters for the wheel size, tangent speed, bed feed, cutting fluid and grinding depth are selected to conduct experiments shown in Table 4. There are three types of quality characteristics such as small the better (STB), large the better (LTB), and nominal the better (NTB). In this experiment, four observations were roughness (STB), micro hardness (STB), abrasion (STB), and material removal rate (LTB). The Taguchi experiment array is shown in Table 5.

Table 4 Grinding experiment factor and level

Factor / Level	A Grain size	B Grinding Wheel rate(m/s)	C Feed rate(mm/s)	D Cutting fluid	E Grinding depth(mm)
I	100#	27.1	12	water	0.005
II	1000#	31.6	24	Emulsifier	0.010
III	None	36.1	36	oil	0.015

Table 5 $L_{18}(3^5)$ experimental parameter array

Exp	A Grain size	B Grinding wheel rate(m/s)	C Feed rate(mm/s)	D Cutting fluid	E Grinding depth(mm)
1	100	27.1	12	water	0.01
2	100	27.1	24	Emulsifier	0.015
3	100	27.1	36	Oil	0.02
4	100	31.6	12	Water	0.015
5	100	31.6	24	Emulsifier	0.02
6	100	31.6	36	Oil	0.01
7	100	36.1	12	Emulsifier	0.01
8	100	36.1	24	Oil	0.015
9	100	36.1	36	Water	0.02
10	1000	27.1	12	Oil	0.02
11	1000	27.1	24	Water	0.01
12	1000	27.1	36	Emulsifier	0.015
13	1000	31.6	12	Emulsifier	0.02
14	1000	31.6	24	Oil	0.01
15	1000	31.6	36	Water	0.015
16	1000	36.1	12	Oil	0.015
17	1000	36.1	24	water	0.02
18	1000	36.1	36	Emulsifier	0.01

3. Result and discussion

3.1 Large particle size diamond grinding wheel dressing

The results showed that the experimental combination with the optimal material removal rate was 17, the removal rate was 5.921 mm3 / min, and the optimal removal rate of Taguchi experiment was 5.76 mm3 / min. The experimental combination with the optimal protrusion is the 14th combination and the abrasive grain protrusion is 74.711 μm. The experimental results are 75.478 m shown in Table 6. Figure 4 shows the large diameter diamond wheel before and after dressing from SEM photos.

Table 6 Removal rate of abrasive material and abrasive grain protrusion

EXP (100)	A Peak current (Amp)	B Pulse time (μs)	C Feed rate (mm/min)	D Line tension (kgf)	E Dielectric flow rate (mm/s)	F Powder type	G Powder concentration	GGP (μm)	MMR (mm^3/min)
L1	1	10	0.5	1	40	Al	1.75	68.716	4.757
L2	1	20	1	3	50	SiC	3.5	66.675	5.062
L3	1	30	2	5	60	Cr	7	62.358	5.310
L4	3	10	0.5	3	50	Cr	7	65.434	5.075
L5	3	20	1	5	60	Al	1.75	75.779	4.407
L6	3	30	2	1	40	SiC	3.5	68.304	5.776
L7	5	10	1	1	60	SiC	7	71.337	5.563
L8	5	20	2	3	40	Cr	1.75	68.429	5.059
L9	5	30	0.5	5	50	Al	3.5	77.929	5.162
L10	1	10	2	5	50	SiC	1.75	66.573	4.798
L11	1	20	0.5	1	60	Cr	3.5	69.769	5.167
L12	1	30	1	3	40	Al	7	70.245	4.836
L13	3	10	1	5	40	Cr	3.5	65.984	5.194
L14	3	20	2	1	50	Al	7	74.711	5.153
L15	3	30	0.5	3	60	SiC	1.75	71.859	4.817
L16	5	10	2	3	60	Al	3.5	78.833	5.733
L17	5	20	0.5	5	40	SiC	7	69.154	5.921
L18	5	30	1	1	50	Cr	1.75	70.077	4.871
Best GGP	3	20	1	1	60	A1	7	75.478	
Best MMR	5	10	2	1	40	SiC	3.5		5.760

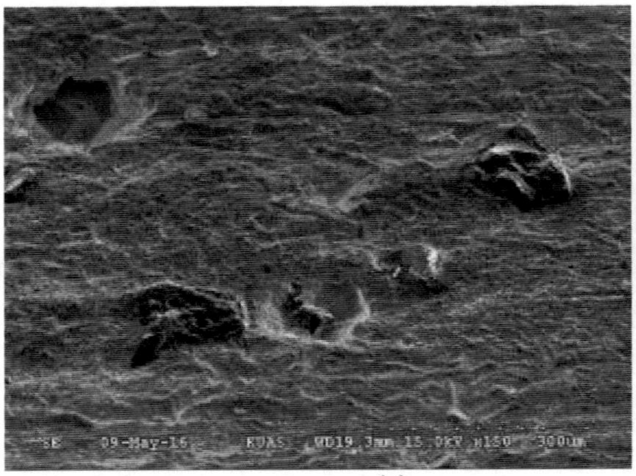

(a)

(b)

Figure 4 Large diameter diamond wheel (a) before dressing (b) after dressing

3.2 Small particle size diamond grinding wheel dressing

The results showed that the experimental combination with the optimal material removal rate was the seventh combination with the removal rate of 3.357 mm3 / min, and the optimal removal rate of Taguchi was 3.208 mm3 / min. The optimal experimental parameter combination is the fourth group, the protrusion is 8.621 μm, and the optimal experimental value of Taguchi is 8.478μm in this study shown in Table 7. The SEM photos of small diameter diamond wheel before and after dressing are in the Figure 5.

Table 7 Removal rate and protrusion of small particle size wheel

EXP (1000)	A	B	C	D	E	F	G	GGP (μm)	MMR (mm^3/min)
	Peak current (Amp)	Pulse time (μs)	Feed rate (mm/min)	Line tension (kgf)	Dielectric flow rate (mm/s)	Powder type	Powder concentrat ion		
L1	1	10	0.5	1	40	Al	1.75	11.743	2.985
L2	1	20	1	3	50	SiC	3.5	10.624	3.017
L3	1	30	2	5	60	Cr	7	9.213	2.737
L4	3	10	0.5	3	50	Cr	7	8.621	2.804
L5	3	20	1	5	60	Al	1.75	11.570	2.864
L6	3	30	2	1	40	SiC	3.5	10.097	3.173
L7	5	10	1	1	60	SiC	7	13.123	3.357
L8	5	20	2	3	40	Cr	1.75	11.795	3.006
L9	5	30	0.5	5	50	Al	3.5	16.218	2.958
L10	1	10	2	5	50	SiC	1.75	9.845	3.120
L11	1	20	0.5	1	60	Cr	3.5	9.232	2.831
L12	1	30	1	3	40	Al	7	10.538	2.858
L13	3	10	1	5	40	Cr	3.5	10.753	2.901
L14	3	20	2	1	50	Al	7	11.623	2.985
L15	3	30	0.5	3	60	SiC	1.75	11.295	2.732
L16	5	10	2	3	60	Al	3.5	16.328	2.842
L17	5	20	0.5	5	40	SiC	7	11.661	3.130
L18	5	30	1	1	50	Cr	1.75	12.960	2.953
Best GGP	1	20	0.5	3	40	Cr	7	8.478	
Best MMR	5	10	2	1	40	SiC	7		3.208

Figure 5 Small diameter diamond wheel (a) before the dressing (b) after finishing

3.3 Grinding results

The grinding results show that the roughness of the machined surface is optimal in the 14th combination, the roughness is 0.184 um, the optimal roughness of Taguchi is 0.175 um. The experimental combination with the optimal micro hardness of the machined surface is the fifth combination with the hardness value of 2046Hv and the optimal hardness of Taguchi is 1980Hv.

The optimal experimental combination of diamond grinding wheel wear is the seventh combination, the wear amount is 0.020 mm and the Tanaka is 0.020 mm. The optimal level for removal rate was 3, the removal rate was 3.1589.

3.4 Discussions

Large diameter diamond grinding wheel dressing in the addition of powder caused by excessive conductive particles easily interfere with the discharge phenomenon, so that the discharge is unstable and refer to surface roughness deterioration, simultaneously the material removal rate is low. After adding the powder, the amount of abrasive grains is increased because the conductive particles increase in the processing fluid, and both the discharge gap and the processing depth increases.

Small diameter diamond grinding wheel in the addition of powder due to the increase in conductivity which is easy to cause the phenomenon of bridge, the discharge gap increases, so that the processing depth increases, so the material removal rate increased. After adding the powder, the amount of protrusion is too large to cause the abrasive particles to fall off.

4. Conclusions

This study has the following four conclusions

(1) The results show that the optimal parameters of the surface roughness experiment are combination 14, the grinding wheel has a grain size of 1000, a tangent speed of 31.6 m / s, a bed feed of 24 mm / s, a cutting fluid of oil, a grinding depth of 0.01 mm, and a surface roughness of0.184 um. The optimal Taguchi parameters are grinding wheel size 1000, tangential speed 31.6 m / s, bed feed 36 mm / s, cutting fluid is oil, grinding depth 0.02 mm, roughness 0.175 um.

(2) The results also show that the optimum parameters for micro hardness experiment are combination 5, the grinding wheel has a grain size of 100, a tangent speed of 31.6 m / s, a bed feed of 24 mm / s, a cutting fluid of water + oil, a grinding depth of 0.02 mm, a micro hardness of 2046 Hv. The optimal Taguchi parameters include a grinding wheel size of 1000, a tangent speed of 31.6 m / s, a bed feed of 24 mm / s, a cutting fluid of water + oil, a grinding depth of 0.02 mm and a micro hardness of 1980 hv.

(3) Then, the optimal parameters for abrasive test are combination 7, the grinding wheel has a particle size of 100, a tangent speed of 36.1 m / s, a bed feed of 12 mm / s, a cutting fluid of water + oil, a grinding depth of 0.01 mm, and a wear of 0.020 Mm. The optimal parameters of Taguchi are wheel size 100, tangent speed 31.6 m / s, bed feed 12 mm / s, cutting fluid is water + oil, grinding depth 0.01 mm.

(4) Finally, the best parameters for material removal rate was the third combination, the grinding wheel size was 100, the tangent speed was 27.1 m / s, the bed feed with was 36 mm / s, the cutting fluid was oil, the grinding depth was 0.02 mm, the material removal rate (MRR) of 3.15887. The optimal Taguchi parameters are wheel size 100,tangent speed 31.6 m / s, bed feed 36 mm / s, cutting fluid is water + oil, grinding depth 0.02 mm, material removal rate is 4.0034.

Acknowledgements

Thanks to the Ministry of Science and Technology 104-2221-E-252-008-MY3 program subsidy to make this study more successful. This study successfully established the wire discharge machining method for metal-based diamond grinding wheel precision dressing and truing technical ability. And to understand the effect of fine powder shaping on the surface roughness and material removal rate of the metal-based diamond grinding wheel by adding the powder to the wire discharge machining method.

References

[1] Wang X, Ying B, Liu W, "EDM dressing of fine grain super abrasive grinding wheel." *Journal of Materials Processing Technology* Vol.62pp.299-302, 1996.

[2] Eduardo Weingärtner & Sascha Jaumann &Friedrich Kuster & Konrad Wegener (2010) On-machine wire electrical discharge dressing (WEDD) of metal-bonded grinding wheel

[3] Rhoney B K, Shih A J, Scattergood RO, Akemon JL, Gust DJ, Grant MB (2002) WireElectrical Discharge Machining of Metal Bond Diamond Wheels for Ceramic Grinding. International *Journal of Machine Tools & Manufacture* 42:1355-1362.

[4] Rhoney B K, Shih A J, Scattergood RO, Akemon JL, Gust DJ, Grant MB (2002) Wire Electrica l Discharge Machining of Metal Bond Diamond Wheels for Ceramic Grinding. International *Journal of Machine Tools & Manufacture* 42:1355-1362.

[5] Wang X Y, Wu Y B, Wang J, Xu W J and Kato M, 2005 J Mater Process Technol, 164-165 1128.

IOP Conf. Series: Materials Science and Engineering **281** (2017) 012023 doi:10.1088/1757-899X/281/1/012023

Study on Single-yarn Pullout Test of Ballistic Resistant Fabric under Different Preloads

Q C Fang, Z K Lei*, F Y Qin, W K Li and R X Bai

State Key Laboratory of Structural Analysis for Industrial Equipment, Dalian University of Technology, Dalian 116024, China

*E-mail: leizk@163.com

Abstract. During bullet penetrating fabric, the pull-out force of yarn in fabric is related to the impact resistance of fabric when the yarn is pulled out from the fabric. The complex uncrimping and friction slip behavior occur during the yarn pullout process, which is critical to learn the impact resistance of fabric. Based on digital image correlation technique, the deformation behavior of Kevlar 49 fabric subjected to preload during the single-yarn pullout process was studied in this paper. The pullout force and displacement curve shows a straight rise and an oscillated decrease. In the linear rise stage, the yarn uncrimping causes a static friction effect. The maximum of the pullout force is not linearly increased with the preload. In the oscillating descending stage, the local descent of the pullout force indicates that the yarn end is gradually withdrawn from the fabric, and the local rise indicates that the yarn end moves to the next weft/warp interaction until the yarn is completely pulled out. The shear deformation of fabric corresponds to the single-yarn pullout process.

1. Introduction

Aramid fiber Kevlar made by DuPont company can be used as a kind of bulletproof textile fiber with excellent performance, such as low fiber density, high crystalline, high temperature resistance, chemical stability, high tensile modulus, good energy absorption and anti-shearing ability. In recent years, many scholars have studied the behavior of fabric by combining experiment with finite element method [1, 2], Bilisik et al. [3] aimed to study the pullout properties by changing the yarn linear density, demonstrating that the pullout force increased with the yarn linear density. Chen et al. [4] reported on an analytical model of ballistic impact of multi-layer woven fabrics, finding that during the ballistic impact of multi-layer woven fabrics, shear failure occurs before tensile failure for the front layers, preventing the high strength polymer fibers reaching full energy absorption potential.

In the process of bullet penetrating fabric, the yarns are pulled out from fabric under the action of friction, so the pullout force of the yarns is related to the impact resistance. As a complex process of fabric impact, the impacting damage is not only related to fabric weaving ways, yarn material properties, yarn friction coefficient, shearing contact area, but also bullet shape and speed [5]. Yarn tensile fracture and shear tearing are two common failure forms. The crimp yarns are stretched up to fracture and the adjacent yarns are squeezed each other to produce a frictional shear effect. The single-yarn pullout test is used to study the energy-absorption behavior of fabric under impact. Tapie et al. [6] studied the response of plain weaving fabric during in-plane and out-of-plane pullouts and the energy dissipation in yarns was mainly caused by frictional force. Duan et al. [7, 8] studied the role of the friction on the ballistic impact behavior of fabric with the method of experiment and simulation, and the results showed that the friction contributed to increasing the fabric energy absorption not only through frictional sliding dissipated energy but also by increasing yarn strain energy and yarn kinetic energy, which indicated an indirect effect on the fabric energy absorption. Zhou et al. [9] introduced yarn gripping as a key parameter to learn the mechanisms of the energy absorption, concluding that resistance against the pullout force of fabric increased with the enhanced yarn gripping.

The uncrimping and frictional slipping behavior is complex during the yarn pullout, which is important to investigate the energy absorption behavior. Many methods have been proposed to investigate the interactions. King et al. [10] established a continuum constitutive model to characterize the process of the yarn pullout with the introduction of the anisotropic shell; they also extended the model to three dimensions. Bazhenov [11] intended to learn the friction of yarns during pullout, characterizing energy transferred to different fabric layers and noting that the friction improved energy dissipation.

In this paper, the deformation behavior of Kevlar 49 fabric subjected to preload during the single-yarn pullout process was studied by digital image correlation (DIC) technique. The influence of preloads on the maximum pullout force and the shear deformation was studied. The study is helpful to understand the energy absorption mechanism of fabric under impact.

2. Testing process

2.1. Kevlar 49 plain fabric parameters

The basic properties of the Kevlar 49 plain fabric are shown in Table 1.

Table 1. Material properties of fabric

Name	Density (g/cm3)	Fiber diameter (μm)	Fiber number per yarn
Kevlar 49	1.44	12	1000

3. Experiments

The fabric specimen has 65 pairs of intersections in the warp and weft direction, as shown in Figure 1(a). Before the testing, both sides of the specimen are clamped in a stress-free state by the fixture. The preload can be applied through adjusting the slider on the rail.

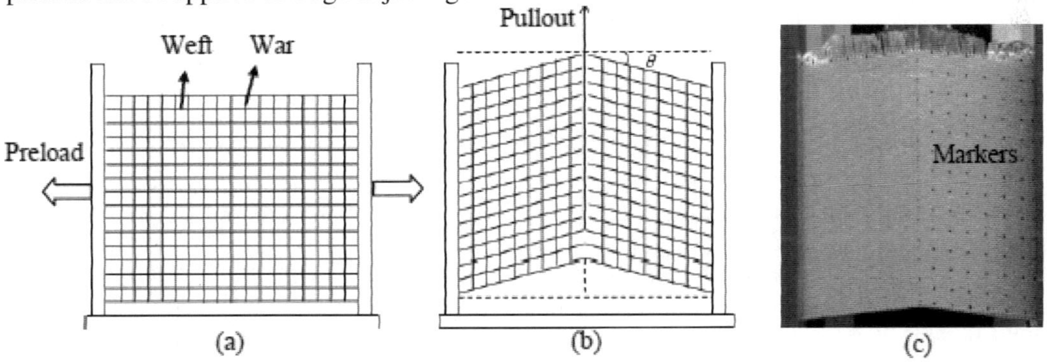

Figure 1. (a) Plain fabric specimen, **(b)** schematic of single yarn pullout test and **(c)** schematic of black markers

A single warp yarn in the middle of the specimen was clamped into the upper fixture of the tensile machine (Instron 3345) and was pulled out, as shown in Figure 1(b). The preloads of 25N, 65N and 100N were applied, respectively. A displacement loading rate was 100 mm/ min.

In order to identify the fabric deformation, the black markers (Fig. 1c) with equal intervals were made and identified by the DIC method which used the local matching algorithm to reveal the full-field deformations with accuracy. The deformed images were captured by a camera (Guppy F080b) with a image resolution of 1024 × 768 pixels and the sampling frequency was 5 fps.

4. Results and discussion

4.1. Pullout force - displacement curve

The curves of pullout force and displacement in single-yarn pullout tests under different preloads are shown in Figure 2. There are two parts: a linear ascending phase and an oscillating descending phase, which correspond to an uncrimping process and a sliding process of the pulled yarn, respectively.

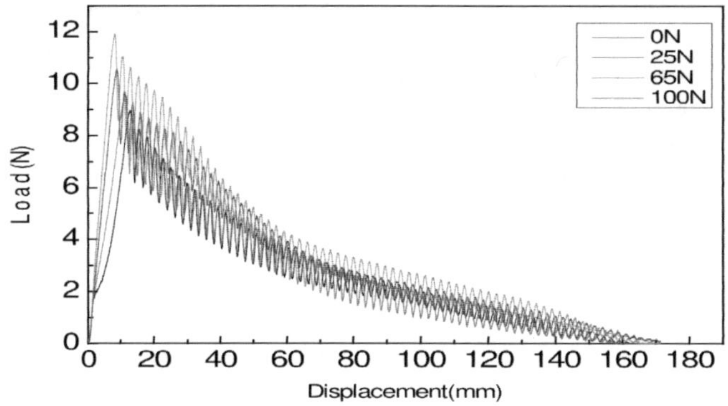

Figure 2. Typical curves of pullout force and displacement in single-yarn pullout test under different transverse preloads

(1) Linear ascending phase

At the beginning of the experiment, the pullout force increases linearly with the displacement and the pulled warp yarn is gradually stretched from the initial crimping state, and the pullout load is transferred by the friction between the yarns. Due to the uncrimping of the pulled yarn end, the cross-section of the yarn is reduced and the contact surface of the warp and weft yarns decreased. The uncrimping portion of the drawn warp yarn has begun to slide before reaching the maximum pullout force. In contrast, the remaining crimp part of the warp is still in the static friction stage, at this time the pullout force is shared together by the dynamic friction and the static friction. It can be seen that the linear ascending phase is not only the process of uncrimping of the pulled yarn, but also the transformation process from the static friction to the dynamic friction.

(2) Oscillating descending phase

When the maximum pullout force is reached, the pulled warp yarn has been completely uncrimped and then slipped; the pullout force is entirely shared by the dynamic friction in the oscillating descending stage. The decrease in pullout force indicates that the end of the warp is pulled out from the last intersection, and the rise in pullout force indicates that the warp end moves to the next intersection. It results in a drop of the pullout force in the sliding friction phase. Thus, the "slip-out and slip-in" process at the end of the pulled yarn means a pair of warp/ weft intersections is crossed, producing a "wave" of pullout force. As the yarn is gradually pulled out, the intersections gradually decrease until the yarn is completely pulled out, and the load is reduced to zero.

4.2. The effects of the preloads

In order to analyze the effects of different transverse preloads, the maximum pullout force and its displacement data are extracted from Figure 2, as listed in Table 2. Obviously, the maximum pullout force is increased with the increase in preload and the spent time is shorter, because the weft yarn deformation under preload expands the contact area between the weft and warp yarns. It is noted that the contact pressure at the intersection of the warp and weft yarns is increased as the weft yarn under preload, so the frictional force at the intersections is also increased.

Table 2. Effects of preload on pullout parameters

Preload (N)	Time (s)	Max load (N)	Displacement (mm)	Slope (N/mm)
0	7.6	8.933	12.985	0.688
25	6.9	9.533	11.475	0.724
65	5.4	10.539	8.967	1.175
100	5.0	11.905	8.333	1.429

It can also be seen from Table 2 that the maximum pullout force increases nonlinearly with the preload. The preload of 25N is taken as an example, when the preload force increased to 260% and 400%, the corresponding maximum pullout force increased by about 110.6% and 124.9% respectively. It indicates that the increase in preload will inhibit the fabric shear deformation. According to time data of the maximum pullout force in Table 2, it can be seen that the presence of the preload causes the single-yarn to be pulled early into the oscillating descent stage. The increased preload changed the initial buckling state of the weft yarns and increased the contact pressure. According to the curves in Figure 2, the slopes of the linear ascending phase are listed in Table 2, it can be seen that the slope also increases with the preload, indicating that the yarn is more difficult to be pulled out.

4.3. Fabric energy absorption

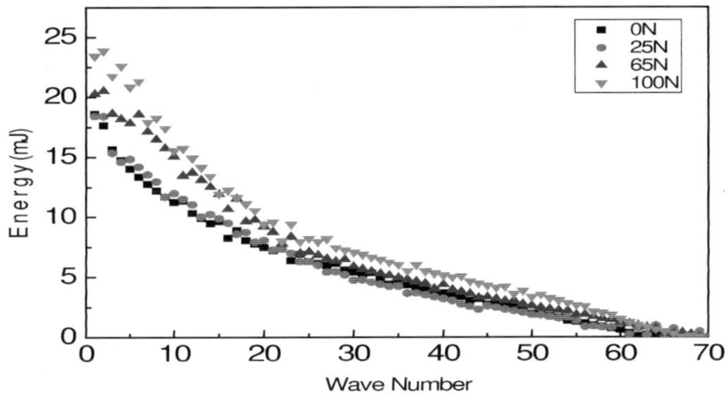

Figure 3. Absorbed energy curves with wave number under different preloads

In the yarn pullout force and displacement curve, the area under the curve is regarded as the total work of the pullout force. The area under the pullout peaks was integrated and the relationship between the wave number and the absorption energy was obtained, as is shown in Figure 3. It can be seen that the yarn intersections is decreased in the yarn pullout process, and the overall energy absorption capacity of the fabric is also decreased. In the initial dynamic friction stage, the fabric energy absorption ability was exponentially decayed; after the yarn was pulled out to half of the intersections, the yarn energy absorption rate decreased and the energy absorption capacity was weakened. It indicates that the effect of preload on the fabric deformation is mainly reflected in the forward stage in single-yarn pullout test. Comparing the energy absorption characteristics under different preloads, it can be found that the stronger the preloads, the stronger the energy absorption effect of the fabric, which indicates that the preload expanded the energy absorption potential of the fabric and improved the ability of the fabric to resist the deformation. Moreover, the energy absorption efficiency is not the same, which was better under the preload of 25N or 65N. At last, four curves in Figure 3 tended to coincide after the half intersections pulled out, then the preload no longer played a major role of limiting the fabric deformation.

4.4. Fabric shear deformation

Displacement measurement of markers on the fabric surface (Figure 1c) was achieved by using the DIC method. A correlation function was used to measure the similarity of the markers before and after the deformation, and the displacements generated by the markers were obtained. In order to obtain subpixel accuracy, a bilinear interpolation method was used to the image before and after the deformation.

Since the pulled yarn is located in the middle of the fabric, which is symmetrical in both geometric and boundary conditions, thus only the right side of the fabric is used to study on the shear deformation. The horizontal angle of the weft yarn on the right side of the fabric is defined as the shear angle θ (Figure 1b), and the shear angle without fabric deformation is 0°. The markers on the top weft line of the fabric are regarded as the row 1, then the row 2 is followed and so on (Figure 1c).

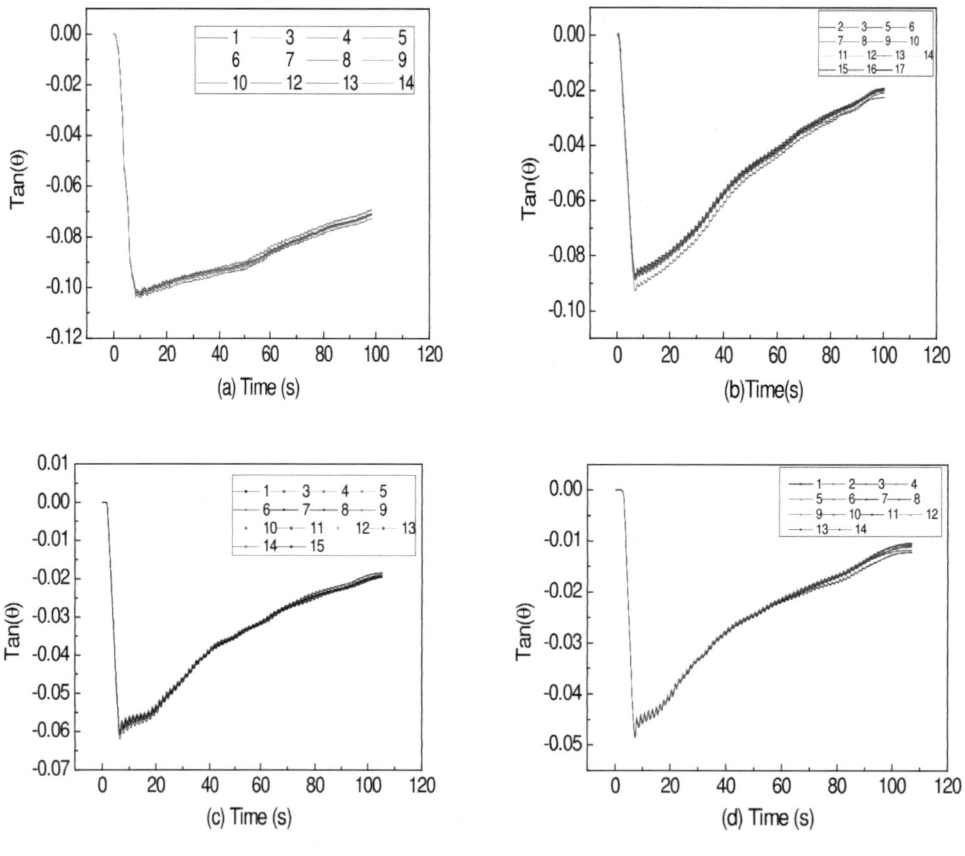

Figure 4. Tan(θ) with time curve for different weft lines under preload of (a) 0N, (b) 25N, (c) 65N and (d)100N

The yarn pullout process is accompanied by the overall deformation of the fabric, which can be analyzed by the DIC method. The tangential value of the shear angle for each weft row with time is shown in Figure 4. It can be seen that there are similar changes of shear angles for different weft lines. In the initial stage, the absolute shear angle increased linearly and rapidly and reached the maximum at the moment of the largest static friction. After entering the dynamic friction stage, the absolute shear angle was gradually reduced until the yarn was completely pulled out. The shear angle was remained at last, which indicates that the fabric remains larger deformation when the yarn was pulled out completely. In the dynamic friction stage, the absolute shear angle decreases oscillatingly, which corresponds to the oscillation of the curve of pullout force and displacement. It shows that the decrease of the shear angle of the fabric is closely related to the pullout force.

Table 3. Shear angles under different preloads

Preloads (N)	Range of shear-angle tangents	Range of residual shear-angle tangents
0	[-0.10401, -0.10174]	[-0.07275,-0.06903]
25	[-0.08879,-0.08698]	[-0.02057, -0.01897]
65	[-0.06196, -0.05994]	[-0.02016,-0.01843]
100	[-0.04865, -0.04834]	[-0.01222, -0.01037]

The minimum shear angle tangent is used to characterize the maximum deformation of the weft yarns, meanwhile, the residual shear angle is used to characterize the residual deformation after the weft is fully pulled out, which are summarized in Table 3. It can be seen that these tangent values increases with the preload, which indicates that the increase of the preload not only restrains the fabric shear deformation, but also increases the ability of the fabric to restore deformation. The greater the preload, the better the effect of fabric deformation recovery, which is due to the warp pullout causing the vertical deformation of the weft yarns.

5. Conclusions

By conducting the single-yarn pullout test under different preloads, the influence of preload on the deformation parameters such as the pullout force, absorbing energy and shear angle was studied. The static friction plays a major role in the yarn uncrimping elongation stage. Then, the sliding friction takes the main role in the yarn slip stage. The maximum pullout force increases non-linearly with the preload so that the pulled warp enters the dynamic friction stage early. The fabric energy absorption decays exponentially, and the effect of preload on the fabric deformation is mainly reflected in the stage of the first half intersections of the yarn. The fabric shear deformation corresponds to the yarn pullout process. The absolute shear angle of the fabric increases linearly in the yarn uncrimping stage. In the dynamic friction stage, the absolute shear angle oscillates down to the residual value, which indicates that the fabric still remains residual deformation after the warp pulled out. The greater the preload, the smaller the fabric residual deformation.

Acknowledgments
The authors thank the National Natural Science Foundation of China (Nos. 11472070, 11772081, 11572070), the National Basic Research Program of China (2014CB046506) and the Natural Science Foundation of Liaoning Province of China (No. 2015020145)

References

[1] Dong Z and Sun C T. 2009 Testing and modeling of yarn pull-out in plain woven Kevlar fabrics *COMPOS PART A* **40** 1863
[2] Bilisik K 2011, Properties of yarn pull-out in para-aramid fabric structure and analysis by statistical model *COMPOS PART A* **42** 1930
[3] Bilisik K and Yolacan G, 2011 Single and multiple yarn pull-out on E-glass woven fabric structures *TEXT RES J* **81** 2043
[4] Chen X, Zhu F and Wells G, 2013 An analytical model for ballistic impact on textile based body armour *COMPOS PART B-ENG* **45** 1508
[5] Rao MP, Duan Y, Keefe M, Powers BM and Bogetti TA 2009 Modeling the effects of yarn material properties and friction on the ballistic impact of a plain-weave fabric *COMPOS STRUCT* **89** 556
[6] Tapie E, Y.B. Guo and Shim V.P.W 2016 Yarn mobility in woven fabrics - a computational and experimental study *INT J SOLIDS STRUCT* **80** 212
[7] Duan Y, Keefe M, Bogetti TA and Cheeseman B.A 2005 Modeling friction effects on the ballistic impact behavior of a single-ply high-strength fabric *INT J IMPACT ENG* **31** 996

IOP Conf. Series: Materials Science and Engineering **281** (2017) 012023 doi:10.1088/1757-899X/281/1/012023

[8] Duan Y, Keefe M, Bogetti TA, Cheeseman B.A and Powers B 2006 A numerical investigation of the influence of friction on energy absorption by a high-strength fabric subjected to ballistic impact *INT J IMPACT ENG* **32** 1299

[9] Zhou Y, Chen X and Wells G 2014 Influence of yarn gripping on the ballistic performance of woven fabrics from ultra-high molecular weight polyethylene fibre *COMPOS PART B-ENG* **62** 198

[10] King MJ, Jearanaisilawong P and Socrate S 2005 A continuum constitutive model for the mechanical behavior of woven fabrics *INT J SOLIDS STRUCT* **42** 3867

[11] Bazhenov S 1997 Dissipation of energy by bulletproof aramid fabric *J MATER SCI* **32** 4167

IOP Conf. Series: Materials Science and Engineering **281** (2017) 012024 doi:10.1088/1757-899X/281/1/012024

Effect of α-Platelet Thickness on the Mechanical Properties of Ti-6Al-4V Alloy with Lamellar Microstructure

Y Ren[1,*], S M Zhou[2], Z Y Xue[1], W B Luo[3], Y J Ren[1] and Y J Zhang[1]

[1] School of energy, power and mechanical engineering, North China Electric Power University, Beijing 102206, PR China
[2] No. 52 Institute of China Ordnance Industries, Yantai, Shandong 264003, China
[3] School of Materials Science and Engineering, University of Science and Technology Beijing, Beijing 100083, PR China

[*]Email: renyuljz@163.com

Abstract. In this paper, lamellar microstructures with average α-platelet thicknesses ranging from 0.55 μm to 1.98 μm were obtained in Ti-6Al-4V (Ti64) alloy by solution-treated above β-transus temperature and followed cooling through different methods. Then effects of the average thickness of plate-like α phase on the mechanical properties of this alloy were investigated. The results show that the yield stress and/or the flow stress of Ti64 decrease with increased average α-platelet thickness during compression under various strain-rate and temperature conditions. Conversely, the ductility increases with increased average α-platelet thickness. The greater plate thickness facilitates slip and avoids severe pile-up of dislocations in α phase, resulting in the augmentation of the fracture strain as well as decrease of the strength of Ti64. Finally, the variation of yield stress with average α-platelet thickness satisfies Hall-Petch relationship at both quasi-static and dynamic loading strain-rates.

1. Introduction

Ti-6Al-4V alloy (hereafter denoted as Ti64) has been widely used in the aerospace and military industries due to its attractive specific strength, low density and corrosion resistance. Ti64 is a typical two-phase alloy and consists of hexagonal-close-packed (hcp) α and body-centered-cubic (bcc) β phases [1]. This alloy contains about 4 wt.% of the β-stabilizing element vanadium so as to preserve part of β phases at room temperature. Different types of microstructures, such as equiaxed, bimodal or lamellar microstructure, can be obtained in Ti64 through heat treatments conducted under various combinations of temperatures, holding times and cooling-rates [2]. The lamellar (Widmanstatten) microstructure, which can be obtained by means of solid solution at temperatures above beta-transus point in Ti64, is often adopted in the aerospace industry due to its preferable specific strength, fracture toughness (damage tolerance) and creep resistance [3, 4].

Expectedly, the initial microstructural characteristics, such as the morphology, size and volume fraction of equiaxed (primary) α-phase, α-platelet thickness, prior β grain size, etc., will strongly affect the mechanical properties of Ti64 at different temperatures, stress states and strain-rates. Semiatin and Bieler [5] reported that the peak flow stresses of Ti64 with lamellar microstructures during hot compression at 1088–1228 K follow a Hall–Petch dependence on α-platelet thickness. Zhang et al. [6] found that Ti64 with bimodal microstructure has higher fracture strain but lower flow stress than that with Widmanstatten microstructure at both quasi-static and dynamic strain-rates. After a systematic analysis of the literature data, Wu et al. [7] found that the high cycle fatigue (HCF) strengths of Ti64 decrease in the order of bimodal, lamellar and equiaxed microstructure and microstructure details also has significant influence on the HCF strength of this alloy. Besides, the formability of Ti64 under high temperature conditions depend dramatically on the microstructural characteristics as well [8-10]. However, the effect of microstructure details, such as average α-platelet thickness, on the mechanical properties of Ti64 at broader strain-rates and temperatures has not yet been studied in detail.

In this work, lamellar microstructures with different average α-platelet thicknesses are obtained through heat treatments. Then the effect of the average thickness of the plate-like α-phase on the mechanical properties of this alloy is investigated under various strain-rate and temperature conditions. These findings aid in understanding the deformation behavior of titanium alloys.

2. Materials and experimental procedures

The Ti64 alloy used was a 30 mm-diameter forging bar stocks. The chemical compositions of Ti64 are listed in Table 1. The beta-transus temperature T_β of this alloy determined by differential scanning calorimetry is 1266 K. In order to acquire lamellar microstructures with different α-platelet thicknesses in Ti64, the forging bars were solution-treated at 1286 K, above T_β, for 30 min followed by water quenching (WQ), air cooling (AC) and furnace cooling (FC), respectively [2]. Then these solution-treated materials were annealed at 873 K for 2 h followed by air cooling to relieve the internal stress generated during the cooling process of the solution treatment.

Table 1. Chemical compositions of Ti64 (wt.%).

Al	V	Fe	O	C	H	N	Ti
5.82	3.84	0.16	0.08	0.024	0.0014	0.023	Balance

The materials were sectioned from the center area of the heat-treated forging bars into several compression cylinders for quasi-static (4 mm in diameter and 8 mm in thickness) and dynamic (4 mm in diameter and 4 mm thick) compression tests. Quasi-static compression tests were performed at both room and elevated (473 K) temperatures under a strain rate of 10^{-3} s^{-1} using an MTS 810 hydraulic servo machine. Tests were continuously conducted until the cylinders fractured. Elevated temperature compression tests were carried out in a high-temperature environmental chamber. Samples were held for 120 s at 473 K before compression. Dynamic compression tests were carried out at room temperature under a strain rate of about 3000 s^{-1} utilizing a split Hopkinson pressure bar (SHPB) apparatus. A more complete description of the SHPB test can be found in Refs. [11, 12].

Optical microscope (OM) and transmission electron microscope (TEM) were employed to determine the microstructural characteristics of the heat-treated forging bars. Specimens for OM studies were prepared by electrochemical polishing in a solution of 95% ethanoic acid + 5% perchloric acid, followed by etching in a solution of 2% HF + 10% HNO_3 + 88% H_2O at room temperature [11]. The metallographical characteristics of Ti64 after heat treatment were examined using a ZEISS Axio Observer A1m optical microscope. TEM analyses were conducted using a JEOL JEM-2100 system at an accelerating voltage of 200 kV. For TEM observations, thin samples with an initial thickness of 0.5 mm were cut from the blocks after OM examinations, reduced to less than 40 μm thick by mechanical means, then punched into several standard 3 mm-diameter TEM discs, and finally thinned by ion milling.

3. Results and discussion

3.1. Microstructural characteristics

Figure 1 shows the optical microstructures of Ti64 after solution-treated and cooled using different methods. The microstructures of the heat-treated Ti64 all show typical lamellar characteristics and consist of acicular or plate-like α phases in equiaxed β matrix. However, the thickness of α-platelet is

IOP Conf. Series: Materials Science and Engineering **281** (2017) 012024 doi:10.1088/1757-899X/281/1/012024

significantly distinct in this alloy at different cooling-rates. The mean grain diameter of the equiaxed β matrix is about 698±33 μm and barely varies with decreased cooling-rate, indicating that the prior β grain size is determined more by the solution temperature and the holding time (which are kept constant in this work) rather than the cooling-rate. Besides, some intergranular α with thickness of about 5.7–16.4 μm (indicated by the arrows in Figure 1(c)) can be found in the grain boundaries of the FC material. As shown in Figure 1(a) and (b), when the cooling-rate is very fast (WQ and AC), the thickness of those α-platelets is unable to distinguish accurately due to an insufficient resolution of OM. Hence, more in-depth substructure analyses using TEM must be carried out to further inspect the microstructure details of Ti64 heat-treated at different cooling-rates.

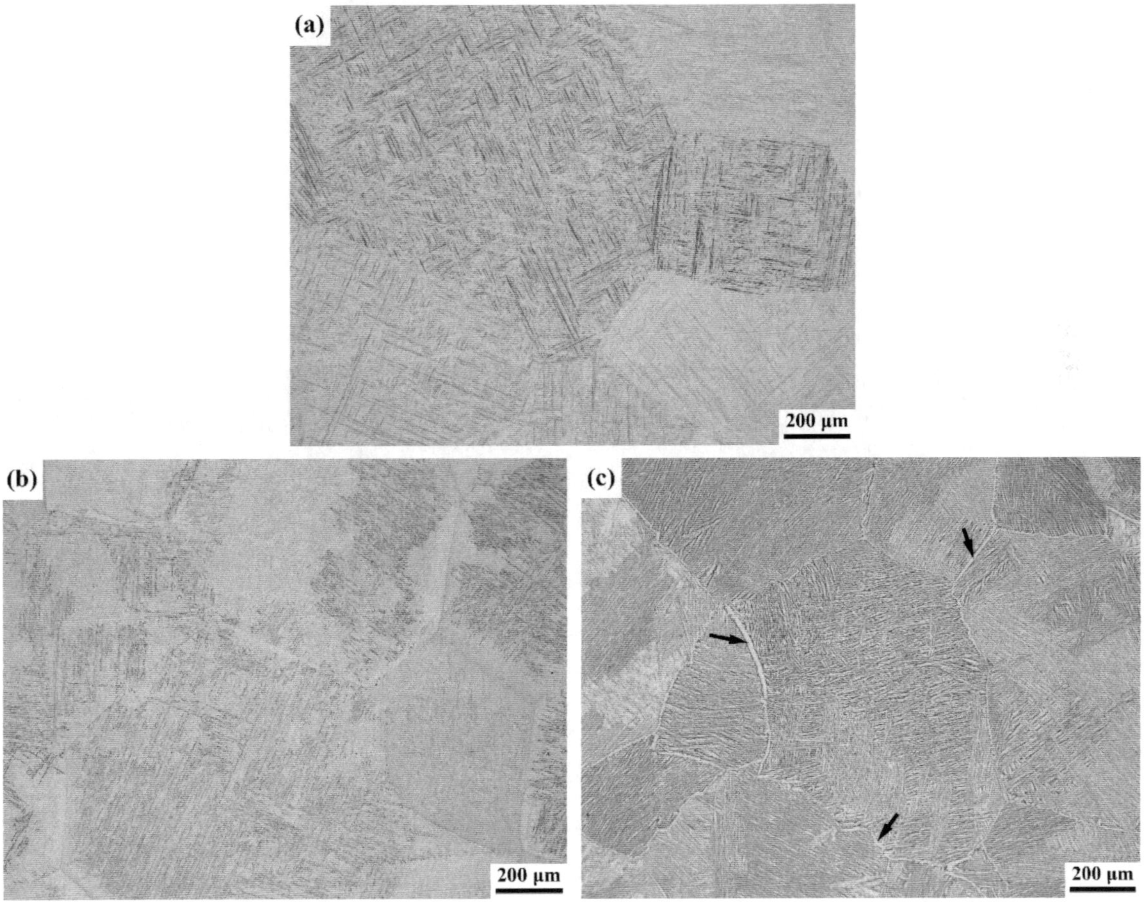

Figure 1. Optical micrographs of Ti64 alloy solution-treated above T_β and then followed by (a) water quenching, (b) air cooling and (c) furnace cooling.

Figure 2 displays the corresponding TEM images of Ti64 with different α-platelet thickness. One can find that the plate-like α phase widens with decreased cooling-rate (from WQ to FC). The distance between two α-platelets increases slightly with decreased cooling-rate. These TEM images also indicate a low dislocation density and the absence of ω phase in alloys before compression testing.

To obtain the precise thickness of α-platelets generated using different cooling methods, statistics were carried out over 20 TEM pictures. The average α-platelet thickness was ascertained using the statistical method for determining average grain size [13]. The result displays that the average α-platelet thickness of Ti64 cooled using WQ, AC and FC methods are 0.55±0.05 μm, 0.86±0.05 μm and 1.98±0.09 μm, respectively. When the temperature declines below T_β during the cooling process of the solution treatment, α phase precipitates from the prior β matrix. The precipitation of α platelets involves the nucleation and growth processes of nuclei. The number of nuclei and growth rate are dependent on the composition and the cooling-rate [2]. When the cooling-rate declines, crystal nucleation process is suppressed due to a lower degree of supercooling. Consequently, fewer nuclei make α-platelets have plenty of space to grow, resulting in the formation of wider plates at a lower cooling-rate.

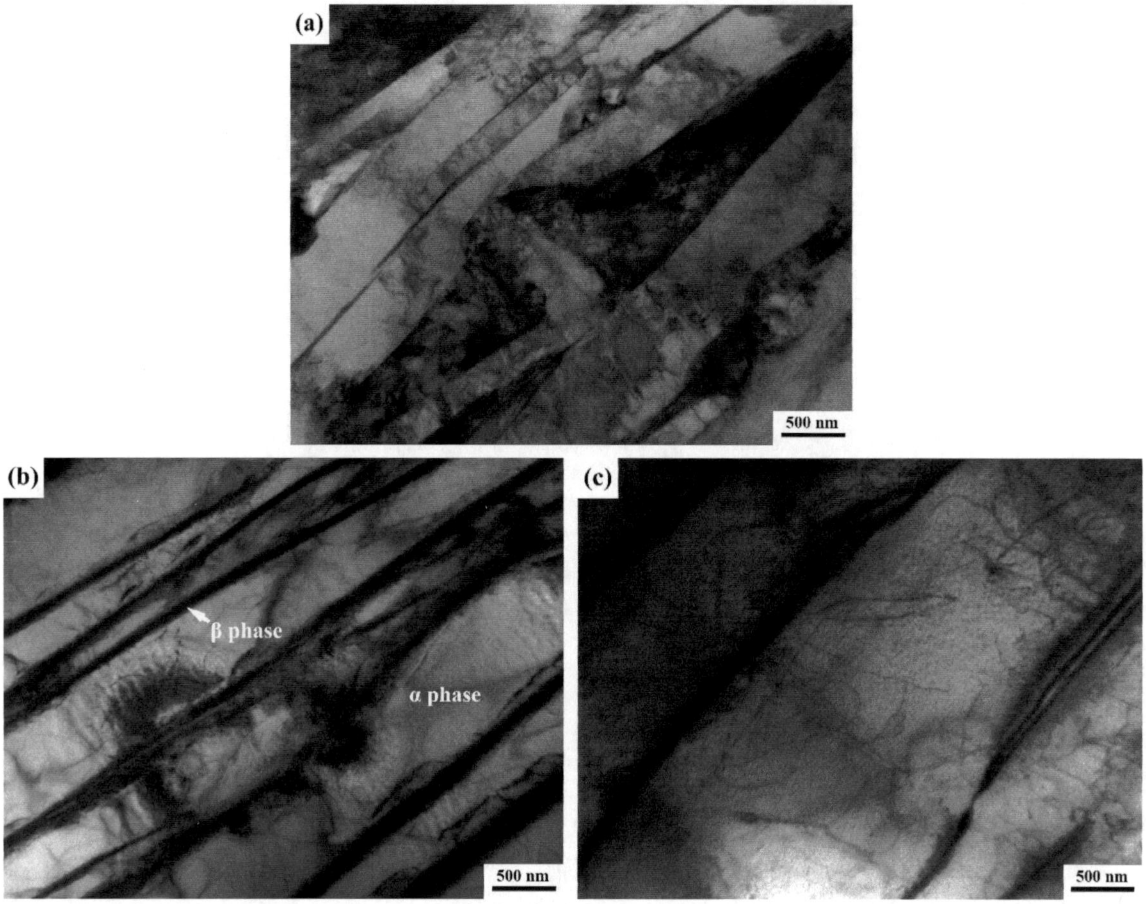

Figure 2. Bright field TEM images of Ti64 alloy solution-treated above T_β and then followed by (a) water quenching, (b) air cooling and (c) furnace cooling.

3.2. Mechanical properties

3.2.1. At quasi-static strain-rate.
Figure 3 presents the compression stress–strain curves and strength variation of Ti64 with different average α-platelet thicknesses at both room temperature and 473 K and at a strain rate of 10^{-3} s^{-1}. At room temperature, as shown in Figure 3(a), the flow stress of Ti64 decreases with increased average α-platelet thickness, although the shape of these curves is similar regardless of the thickness of the plate-like α phase. The strain-hardening rate (SHR) of this alloy hardly changes when the plastic deformation proceeds, implying that the α-platelet thickness has no effect on the SHR of Ti64. On the other hand, the fracture strain increases to some extent (from 0.30 to 0.32) with increased average α-platelet thickness, suggesting the benefit of α-phase broadening on the ductility of Ti64. The fracture strain is defined as the maximum strain that the specimen endures before it fractures during the

compression test. The greater plate thickness facilitates slip and avoids severe pile-up of dislocations in α phase, resulting in the augmentation of the fracture strain as well as decrease of the strength of Ti64.

When the temperature elevates to 473 K (Figure 3(b)), the Ti64 exhibits a certain degree of thermal softening during compression process. Namely, the flow stress of materials decreases with increased plastic strain, i.e. they have a negative SHR during high-temperature compression. The fracture strain also increases with increased average α-platelet thickness. The augmentation of fracture strain is more obvious when the average thickness of plate-like α phase exceeds 1 μm. Figure 3(c) shows the quasi-static compression yield stress as a function of average α-platelet thickness of Ti64 at various temperatures. The yield stress of Ti64 has roughly the same change tendency at both room and elevated temperatures and decreases with increased average α-platelet thickness. All yield stress values at 473 K are about 270 MPa lower than those at room temperature, indicating non-influence of the α-platelet thickness on the decreasing degree of yield stress in Ti64. However, the negative SHR of Ti64 with the α-platelet thickness of 0.86 μm is less evident than the other two (Figure 3(b)), suggesting that AC microstructure has a higher resistance to thermal softening.

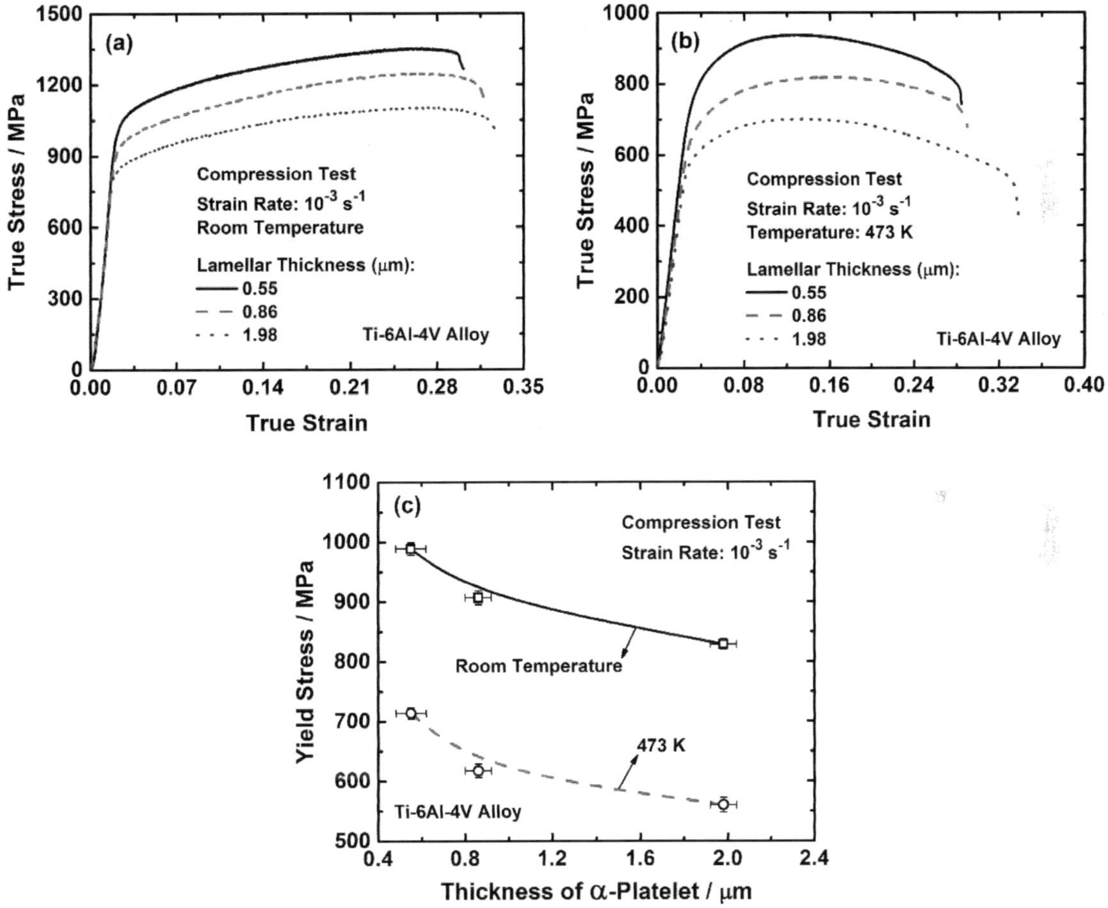

Figure 3. Typical stress–strain curves of Ti64 with different average α-platelet thicknesses quasi-statically compressed at (a) room temperature and (b) 473 K. (c) Yield stress at both room and elevated temperature vs. average α-platelet thickness for Ti64 with lamellar microstructure. Each of the strength values is the average of three measurements.

3.2.2. At dynamic strain-rate.

Figure 4 shows the compression stress–strain curves and strength variation of Ti64 with different average α-platelet thicknesses at room temperature and at a strain rate of about 3000 s^{-1}. Compared with the flow response at quasi-static strain-rate (Figure 3), the α-platelet thickness has a significant impact on the ductility of Ti64 under dynamic strain-rate loading conditions, as shown in Figure 4(a). The fracture strain increases substantially from 0.15 to 0.40 as the average α-platelet thickness increases from 0.55 μm to 1.98 μm. Loading at higher strain-rates leads to more dislocations generated during plastic deformation [12, 14]. So dislocations are more likely to tangle and block at high strain-rate. The greater plate thickness facilitates slip and avoids severe pile-up of dislocations in α phase, resulting in the augmentation of the plastic strain as well as decrease of the strength of Ti64. Besides, the fracture strain of Ti64 with 0.55 μm lamellar thickness in dynamic compression test is less than that in quasi-static compression test, while the opposite performance occurs in the material with 1.98 μm lamellar thickness. This interesting phenomenon may be related to the dynamic fracture behavior (such as nucleation and propagation of adiabatic shear bands) of Ti64 under high-strain-rate loading conditions. Further in-depth studies have been on the way to verify this conjecture.

The α-platelet thickness also has a more pronounced effect on the SHR and further the thermal softening behavior of Ti64 during dynamic compression process. The SHR gradually reduces with increased average α-platelet thickness, implying a more obvious thermal softening of Ti64 with the wider lamellar microstructure at high strain-rates. The high strain-rate loading process is in an adiabatic state. Thus, the rate of heat generation is greater than the rate of heat loss during deformation at high strain-rates, which increases the temperature of the material. A continuous rise in temperature results in the simultaneous lowering of the flow stress as the strain proceeds [15]. Therefore, greater fracture strain means longer deformation time and more heat accumulation, resulting in a more significant thermal softening effect in Ti64 having a broader lamellar microstructure during dynamic compression. Besides, as shown in Figure 4(b), the yield stress of Ti64 at the strain rate of 3000 s^{-1} also decreases with increased average α-platelet thickness and has a similar change tendency compared to that at quasi-static strain-rate (Figure 3(c)).

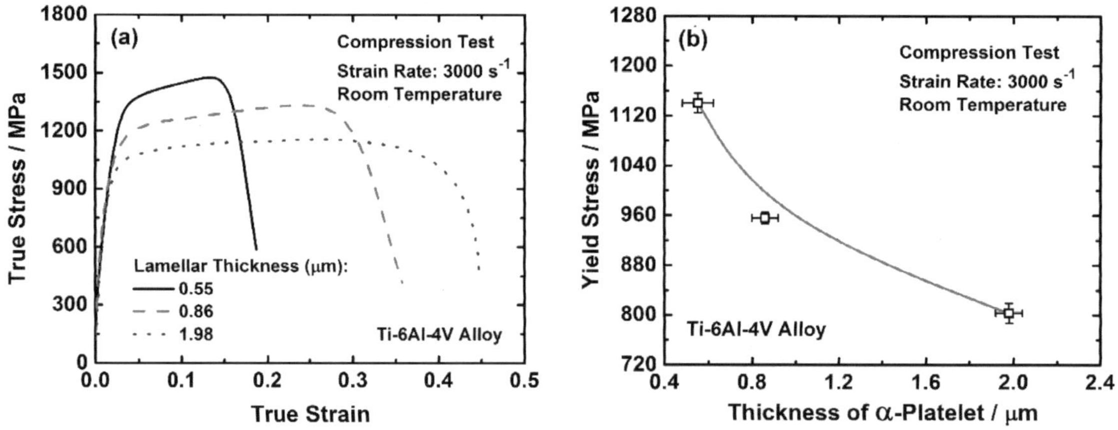

Figure 4. (a) Typical compression stress–strain curves and (b) yield stress vs. average α-platelet thickness for Ti64 with lamellar microstructure at room temperature under a strain rate of 3000 s^{-1}. Each of the strength values is the average of three measurements.

3.3. Hall-Petch behavior

The strengthening behavior generated due to the existence of grain boundaries in most polycrystalline metallic materials can be described through the following formulae known as the Hall–Petch relationship [16]:

$$\sigma = \sigma_0 + k_S \cdot l^{-1/2} \tag{1}$$

where σ is the strength value (yield stress, flow stress or peak stress, etc.), l is the factor quantifying the microstructural scale (grain diameter or α-platelet thickness, etc.), σ_0 and k_S are constants. The Hall–Petch relationship has been successfully applied to explain the deformation behavior of Ti64 at temperatures higher than 1073 K [5, 17]. Hence, this theory is predictably able to be used to describe the deformation behavior of Ti64 under broader strain-rate and temperature conditions. Figure 5 shows the variation of yield stress with the inverse square root of the average α-platelet thickness ($l^{-1/2}$) for Ti64 under different strain-rate and temperature loading conditions. One can find that the yield stress of Ti64 increases linearly with $l^{-1/2}$ in all cases, indicating that the change of yield stress with the average α-platelet thickness of Ti64 obeys the Hall–Petch relationship under a wide range of strain-rates and temperatures. Therefore, linear fitting of these data points according to Equation (1) are carried out. The results are shown in Figure 5 and listed in Table 2.

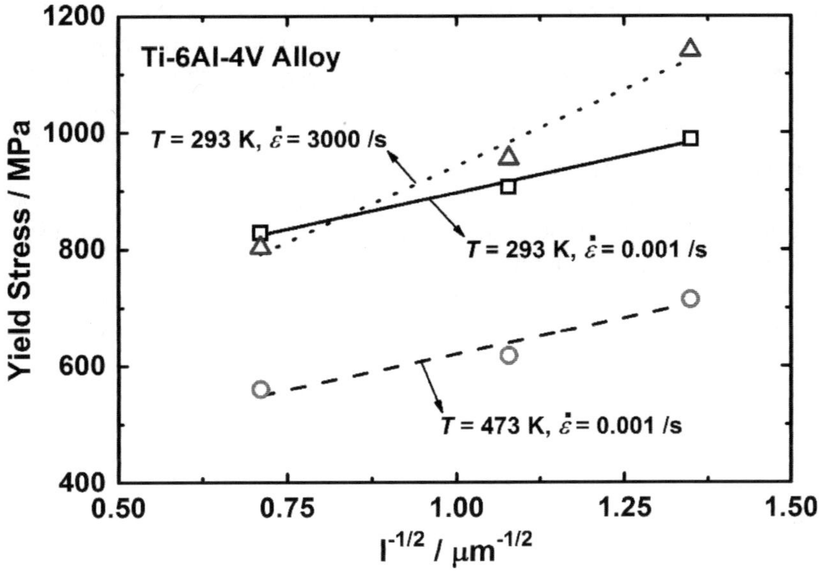

Figure 5. Yield stress vs. the inverse square root of the average α-platelet thickness ($l^{-1/2}$) for Ti64 at various strain-rates and temperatures.

As shown in Table 2, the k_S value significantly increases with increased strain-rate, however, it declines slightly with increased deformation temperature. The constant k_S can be calculated by [18]:

$$k_S = \sqrt{4\sigma^* Gb / \alpha\pi} \tag{2}$$

where σ^* is the barrier strength, G is the shear modulus, b is the burger's vector, and α is a constant. For the same material (Ti64), the value of k_S is only related to σ^*. σ^* value may be a function of strain rate and temperature [5] and increases with increased strain-rate. Hence, at high strain-rate, the augment of yield stress caused by the variation of α-platelet thickness is more obvious in Ti64. The temperature difference in this work is not great, leading to a smaller gap of k_S at quasi-static strain-rate, similar to the behavior of Ti64 at higher temperatures [5]. In addition, the fact that the σ_0 value of Ti64 at dynamic strain-rate is lower than that at quasi-static strain-rate is somewhat surprising. This may be due to the more evident thermal softening in Ti64 with lamellar microstructure (Figure 4(a)).

Table 2. Hall–Petch constants (σ_0 and k_S) based on yield stress data for Ti64 with lamellar microstructures.

Strain-rate (s^{-1})	Temperature (K)	Hall–Petch constants	
		σ_0 (MPa)	k_S (MPa·$\mu m^{-1/2}$)
0.001	Room temperature	647.49	249.19
0.001	473	374.13	246.23
3000	Room temperature	427.47	527.88

4. Conclusions

Solution treatments with different cooling-rates were carried out on Ti64 to obtain the lamellar microstructures with different average α-platelet thicknesses. Then a series of compression tests were performed on this alloy under various strain-rate and temperature conditions to investigate the effect of the average α-platelet thickness on the mechanical properties of this alloy. The major conclusions are as follows:

(1) The average α-platelet thicknesses of 0.55 μm, 0.86 μm and 1.98 μm are obtained in Ti64 through solution treatment and subsequently cooled using WQ, AC and FC methods, respectively.

(2) The yield stress and/or the flow stress of Ti64 decrease with increased average α-platelet thickness during compression under various strain-rate and temperature conditions. Conversely, the ductility increases with increased average α-platelet thickness.

(3) The greater plate thickness facilitates slip and avoids severe pile-up of dislocations in α phase, resulting in the augmentation of the fracture strain as well as decrease of the strength of Ti64.

(4) The strengthening effect caused by the α-platelet refinement obeys the Hall–Petch relationship in Ti64 deformed at both quasi-static and dynamic loading strain-rates.

Acknowledgments

This work was supported by the Young Scientists Fund of the National Natural Science Foundation of China (grant number: 51501064).

References

[1] Joshi V A 2006 *Titanium Alloys: An Atlas of Structures and Fracture Features* (Boca Raton: Taylor & Francis Group) p 9
[2] Lütjering G and Williams J C 2007 *Titanium* (*2nd edition*) (Heidelberg: Springer-Verlag Berlin)
[3] Leyens C and Peters M 2003 *Titanium and Titanium Alloys* (Weinheim: WILEY-VCH Verlag GmbH & Co. KGaA) p 33
[4] Huo D, Li S and Fan Q 2013 *Rare Metal Mater. Eng.* **42** 0457
[5] Semiatin S L and Bieler T R 2001 *Acta mater.* **49** 3565
[6] Zhang J, Tan C, Ren Y, Wang F and Cai H 2011 *Trans. Nonferrous Met. Soc. China* **21** 39
[7] Wu G Q, Shi C L, Sha W, Sha A X and Jiang H R 2013 *Mater. Des.* **46** 668
[8] Guan R G, Je Y T, Zhao Z Y and Lee C S 2012 *Mater. Des.* **36** 796
[9] Wang Y C and Langdon T G 2013 *Mater. Sci. Eng. A* **559** 861
[10] Fu J, Ding H, Huang Y, Zhang W and Langdon T G 2015 *J. Mater. Res. Tech.* **4** 2
[11] Liu X Q, Tan C W, Zhang J, Wang F C and Cai H N 2009 *Int. J. Impact Eng.* **36** 1143
[12] Meyers MA 1994 *Dynamic Behavior of Materials* (New York: John Wiley & Sons, Inc.)
[13] Bhattacharjee A, Varma V K, Kamat S V, Gogia A K and Bhargava S 2006 *Metall. Mater. Trans.* **37A** 1423
[14] Asay J R and Shahinpoor M 1992 *High-pressure shock compression of solids* (New York: Springer-Verlag, Inc.)

IOP Conf. Series: Materials Science and Engineering **281** (2017) 012024 doi:10.1088/1757-899X/281/1/012024

[15] Nemat-Nasser S and Guo W G 2003 *Mech. Mater.* **35** 1023
[16] Armstrong R, Codd I, Douthwaite R M and Petch N J 1962 Philos. Mag. **7** 45
[17] Semiatin S L and Bieler T R 2001 *Metall. Mater. Trans. A* **32A** 1787
[18] Eshelby J D 1963 *Phys. Status Solidi* **3** 2057

IOP Conf. Series: Materials Science and Engineering **281** (2017) 012025 doi:10.1088/1757-899X/281/1/012025

Researches Regarding the Main Chain Vibrations Study of a Shaper

D G Băgnaru[1], M M Stănescu[2], C M Miriţoiu[1*], D Bolcu[1] and V Ionică[1]

[1] Faculty of Mechanics, University of Craiova, Calea Bucureştu Street, No. 165, 200620, Craiova, Dolj, Romania
[2]Applied Mathematics Department, University of Craiova, 13 A I Cuza, 200396, Craiova, Dolj, Romania

Email: miritoiucosmin@yahoo.com

Abstract. In this paper we highlight the importance of applying the integral transform for determining the partial dynamic response of the assembly tool – tool holder following the vibrations that appear during the working time.

1. Introduction

The paper [1] presents a mathematical model which allows the simulation of three dimensional behaviour of planetary/epicyclic and helical gears. The motion equations are solved by using a time-integration scheme and a contact algorithm simultaneously for all the networking.

A mechanical and mathematical model for a tool and the support unit of a milling device is presented in the paper [2]. In this way, a mechanical model with three freedom degrees was created for which there was applied the Langrage formalism leading to a system with three differential equations of second degree. By using the integral transform Laplace, the mathematical model was reduced to a polynomial equation which was easily resolved.

Based also on the Laplace transform, unilaterally in relation with time, in [3] there were obtained the transversal displacements for kinematical elements that are in vibratory motion.

In [4] the longitudinal and transversal displacements fields were determined for the linear elastic rod of an R mechanism (RRT) by using an iterative method. In the paper [5] there were obtained the same elements, but for a linear elastic connecting rod, part of an parallelogram mechanism.

The dynamic behaviour modeling of a mechanical system is essential if it is desired to obtain vibration and acoustical acceptable levels. In this context, in [6], the authors affirm that there is a very large consensus in the researches domain to take into consideration the transmission errors fluctuations. Here, the authors develop the main techniques used and the hypotheses associated with these ones.

The essential problems of structural systems analysis subjected to time analysis were presented in [7]. More precisely, there were determined the solutions of these problems by using a method in which there were idealized these systems by using an assembly of discrete structural elements and by reaching to equations sets solved conveniently matrix algebraically operations.

The importance of vibrations study in the mechanical systems design processes or in the manufacturing processes is studied in [8] through different methods, by insisting over the noises occurrence as an immediate consequence of the vibrations.

The vibrations influence over the errors that may appear in the motion transmission process of the mechanical systems is studied in [9].

In [10] it is developed an analytical model of mechanical systems vibrations for motion transmission. This is used to investigate the natural frequencies and vibrations modes. The vibrations modes are classified in rotation and translation modes. The unique characteristics of each mode type are detailed investigated analytically.

Many of the calculus methods used are based on a matrix formalism easily implemented on computers. A good part of differential equations solutions, which describe, for example, the material systems behaviour at the internal and external factor influence that act upon them, can be obtained by using the computer.

In order to validate the essential mechanical models are the dynamic tests. In this way, an important role is played by the modal experimental analysis problems and the way in which the correspondence between the calculus – tests is made, but also the model actualization. Largely used in mechanics, particularly in vibration mechanics, are the inverse problems. Through the applications and methods used we can cite the sources inverse problems, shape optimization, devices active control for fluid transports or, in the end, the integral equations methods largely used in mechanics and acoustics. The present work can be enclosed on this direction.

2. Mechanical model. Mathematical model

Vibrations mechanical model of the main kinematical chain from a shaper should contain the tool holder masses, its support, the rotating device and of headwork, all considered as non-deformable solids. But, the basis vibrations elements analysis allows the elimination of partial systems with small influence over the elastic system dynamic behaviour in its assembly, reason for which there can be studied the vibratory partial system of tool holder assembly (2-1), like the one from the figure 1, of which its m_s mass is concentrated in its mass centre C [11].

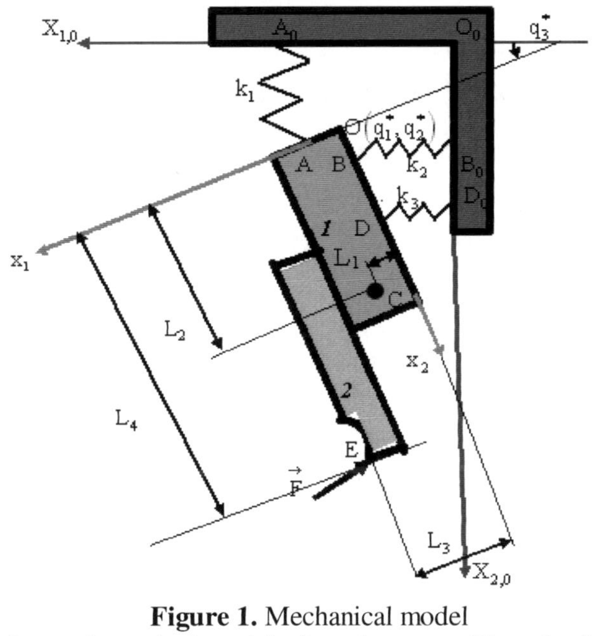

Figure 1. Mechanical model

In order to describe the mathematical model of motion, we will apply the Lagrange formalism by neglecting the resisting forces that are non-elastically, or, by other words, neglecting the damping and also the tool holder gravity force.

We insert the generalized coordinates (1), where $q_{i0}^*, i=\overline{1,3}$ give the stable static equilibrium position of the vibratory system when it is loaded with the stationary chipping force $\overrightarrow{F_0}=-F_{10}\overrightarrow{i_{10}}-F_{20}\overrightarrow{i_{20}}$, in order to have a vibration around a stable equilibrium.

$$q_i=\Delta q_i^*=q_i^*-q_{i0}^*, i=\overline{1,3}, \tag{1}$$

The relations of moving the coordinates for some important points of the vibratory system, depending on the generalized coordinates, to the inherent coordinate system $T(X_1OX_2X_3)$, with the orthonormal base $(\vec{i_1},\vec{i_2},\vec{i_3})$, to the fixed coordinate system $T_0(X_{1,0}O_0X_{2,0}X_{3,0})$, with the orthonormal base $(\overrightarrow{i_{1,0}},\overrightarrow{i_{2,0}},\overrightarrow{i_{3,0}})$, in a matrix writing are given by (2), where by accepting the angular coordinate q_3 as being small, so with $\cos q_3\cong 1$ and $\sin q_3\cong q_3$, we obtain (3).

$$\begin{Bmatrix} X_{1,A}\\ X_{2,A}\end{Bmatrix}=\begin{Bmatrix} q_1^*\\ q_2^*\end{Bmatrix}+[M]\begin{Bmatrix} l_1\\ 0\end{Bmatrix};\begin{Bmatrix} X_{1,B}\\ X_{2,B}\end{Bmatrix}=\begin{Bmatrix} q_1^*\\ q_2^*\end{Bmatrix}+[M]\begin{Bmatrix} 0\\ l_2\end{Bmatrix};$$
$$\begin{Bmatrix} X_{1,D}\\ X_{2,D}\end{Bmatrix}=\begin{Bmatrix} q_1^*\\ q_2^*\end{Bmatrix}+[M]\begin{Bmatrix} 0\\ l_3\end{Bmatrix};\begin{Bmatrix} X_{1,C}\\ X_{2,C}\end{Bmatrix}=\begin{Bmatrix} q_1^*\\ q_2^*\end{Bmatrix}+[M]\begin{Bmatrix} L_1\\ L_2\end{Bmatrix}; \tag{2}$$
$$\begin{Bmatrix} X_{1,E}\\ X_{2,E}\end{Bmatrix}=\begin{Bmatrix} q_1^*\\ q_2^*\end{Bmatrix}+[M]\begin{Bmatrix} L_3\\ L_4\end{Bmatrix};[M]=\begin{bmatrix}\cos q_3^* & -\sin q_3^*\\ \sin q_3^* & \cos q_3^*\end{bmatrix},$$

$$\sin q_3^*=\sin(q_3+q_{30}^*)=\sin q_3\cos q_{30}^*+\cos q_3\sin q_{30}^*\cong q_3\cos q_{30}^*+\sin q_{30}^*,$$
$$\cos q_3^*=\cos(q_3+q_{30}^*)=\cos q_3\cos q_{30}^*-\sin q_3\sin q_{30}^*\cong \cos q_{30}^*-q_3\sin q_{30}^*, \tag{3}$$
$$l_1=OA; l_2=OB; l_3=OC.$$

The kinetic energy of the tool holder assembly is given by (4).

$$T=\frac{1}{2}m_S\vec{v}_c^{\,2}+\frac{1}{2}J_{33}\omega^2=\frac{1}{2}m_S\vec{v}_c^{\,2}+\frac{1}{2}J_C\vec{\omega}^{\,2}. \tag{4}$$

In addition, the relations (5) and (6) are verified.

$$\dot{q}_i^*=\dot{q}_i+\dot{q}_{i0}^*=\dot{q}_i, i=1,2, \frac{d}{dt}(\sin q_{30}^*)=0, \frac{d}{dt}(\cos q_{30}^*)=0, \vec{\omega}=\dot{q}_3^*\vec{i}_3=\dot{q}_3\vec{i}_3, \tag{5}$$

$$\vec{v}_c=(\vec{i}_{1,0},\vec{i}_{2,0})\begin{Bmatrix}\dot{X}_{1,C}\\ \dot{X}_{2,C}\end{Bmatrix}=(\vec{i}_{1,0},\vec{i}_{2,0})\left(\begin{Bmatrix}\dot{q}_1\\ \dot{q}_2\end{Bmatrix}+\begin{bmatrix}-\dot{q}_3\sin q_{30}^* & -\dot{q}_3\cos q_{30}^*\\ \dot{q}_3\cos q_{30}^* & -\dot{q}_3\sin q_{30}^*\end{bmatrix}\begin{Bmatrix}L_1\\ L_2\end{Bmatrix}\right)\Rightarrow$$
$$\vec{v}_c=(\vec{i}_{1,0},\vec{i}_{2,0})\begin{Bmatrix}\dot{q}_1-\dot{q}_3(L_1\sin q_{30}^*+L_2\cos q_{30}^*)\\ \dot{q}_1+\dot{q}_3(L_1\cos q_{30}^*-L_2\sin q_{30}^*)\end{Bmatrix}. \tag{6}$$

By inserting the kinetic energy relation (4) in (5) and (6), there results the relation (4[1]) for the kinetic energy of tool holder assembly.

$$T=\frac{1}{2}m_S\left\{\left[\dot{q}_1-\dot{q}_3(L_1\sin q_{30}^*+L_2\cos q_{30}^*)\right]^2+\left[\dot{q}_2+\dot{q}_3(L_1\cos q_{30}^*-L_2\sin q_{30}^*)\right]^2\right\}+$$
$$+\frac{1}{2}J_{33}\dot{q}_3^2\Rightarrow$$

$$T=\frac{1}{2}m_S\dot{q}_1^2+\frac{1}{2}m_S\dot{q}_2^2+\frac{1}{2}\left[m_S\left(L_1^2+L_2^2\right)+J_c\right]\dot{q}_3^2-$$
$$-m_S\dot{q}_1\dot{q}_3\left(L_1\sin q_{30}^*+L_2\cos q_{30}^*\right)+m_S\dot{q}_2\dot{q}_3\left(L_1\cos q_{30}^*-L_2\sin q_{30}^*\right)\qquad(4^1)$$

The generalized forces are given by the relations (7).

$$Q_j=Q_{j,c}+Q_{j,e},\ j=\overline{1,3},\qquad(7)$$

In (7), $Q_{j,c}$ are the conservative generalized forces and $Q_{j,e}$ are the active generalized forces given by (8) and (9), where the chipping force $\vec{F}$ is given by (10).

$$Q_{j,c}=-\frac{\partial V}{\partial q_j},\ j=\overline{1,3},\qquad(8)$$

$$Q_{j,e}=\vec{F}\frac{\partial\vec{r}}{\partial q_j}=\vec{F}\frac{\partial\overrightarrow{O_0E}}{\partial q_j},\ j=\overline{1,3},\qquad(9)$$

$$\vec{F}=-F_{10}\sin(\omega t)\vec{i}_{10}-F_{20}\sin(\omega t)\vec{i}_{20}.\qquad(10)$$

We mark with Δ the variation of a value between an instantaneous position and the static equilibrium position. By developing in Taylor series limited to the linear terms, we can write (11).

$$\Delta\left(\sin q_3^*\right)=\sin q_3^*-\sin q_{30}^*\cong q_3\cos q_{30}^*,\ \Delta\left(\cos q_3^*\right)\cong -q_3\sin q_{30}^*.\qquad(11)$$

The elastic potential can be calculated with (12).

$$V=\frac{k_1}{2}\left(\Delta\overrightarrow{AA_0}\right)^2+\frac{k_2}{2}\left(\Delta\overrightarrow{BB_0}\right)^2+\frac{k_3}{2}\left(\Delta\overrightarrow{DD_0}\right)^2\Rightarrow$$
$$V=\frac{k_1}{2}\left[q_1^2+q_2^2+l_1^2q_3^2+2l_1q_3\left(q_2\cos q_{30}^*-q_1\sin q_{30}^*\right)\right]+$$
$$+\frac{k_2}{2}\left[q_1^2+q_2^2+l_2^2q_3^2-2l_2q_3\left(q_1\sin q_{30}^*+q_2\cos q_{30}^*\right)\right]+\qquad(12)$$
$$+\frac{k_3}{2}\left[q_1^2+q_2^2+l_3^2q_3^2-2l_3q_3\left(q_1\sin q_{30}^*+q_2\cos q_{30}^*\right)\right]$$

By using (8), (9), (10), (11) and (12) in (7) we obtain the generalized forces (7^1), (7^2).

$$Q_1=Q_{1,c}+Q_{1,e}=-\frac{\partial V}{\partial q_1}+\vec{F}\frac{\partial\overrightarrow{O_0E}}{\partial q_1}=-\left[(k_1+k_2+k_3)q_1-(k_1l_1+k_2l_2+k_3l_3)q_3\sin q_{30}^*\right]+$$
$$+\left[-F_{10}\sin(\omega t)\vec{i}_{10}-F_{20}\sin(\omega t)\vec{i}_{20}\right]\cdot\frac{\partial}{\partial q_1}\left\{\left[q_1+q_{10}^*+L_3\left(\cos q_{30}^*-q_3\sin q_{30}^*\right)-\right.\right.$$
$$-L_4\left(\sin q_{30}^*+q_3\cos q_{30}^*\right)\Big]\vec{i}_{10}+\Big[q_2+q_{20}^*+L_3\left(\sin q_{30}^*+q_3\cos q_{30}^*\right)+$$
$$\left.+L_4\left(\cos q_{30}^*-q_3\sin q_{30}^*\right)\Big]\vec{i}_{20}\right\}\Rightarrow$$
$$Q_1=-\left[(k_1+k_2+k_3)q_1-(k_1l_1+k_2l_2+k_3l_3)q_3\sin q_{30}^*\right]-F_{10}\sin(\omega t),\qquad(7^1)$$

$$Q_2=Q_{2,c}+Q_{2,e}=-\frac{\partial V}{\partial q_2}+\vec{F}\frac{\partial\overrightarrow{O_0E}}{\partial q_2}=-\left[(k_1+k_2+k_3)q_2-(k_1l_1+k_2l_2+k_3l_3)q_3\sin q_{30}^*\right]+$$
$$+\left[-F_{10}\sin(\omega t)\vec{i}_{10}-F_{20}\sin(\omega t)\vec{i}_{20}\right]\cdot\frac{\partial}{\partial q_2}\left\{\left[q_1+q_{10}^*+L_3\left(\cos q_{30}^*-q_3\sin q_{30}^*\right)-\right.\right.$$
$$-L_4\left(\sin q_{30}^*+q_3\cos q_{30}^*\right)\Big]\vec{i}_{10}+\Big[q_2+q_{20}^*+L_3\left(\sin q_{30}^*+q_3\cos q_{30}^*\right)+$$
$$\left.+L_4\left(\cos q_{30}^*-q_3\sin q_{30}^*\right)\Big]\vec{i}_{20}\right\}\Rightarrow$$
$$Q_2=-\left[(k_1+k_2+k_3)q_2+(k_1l_1-k_2l_2-k_3l_3)q_3\sin q_{30}^*\right]-F_{20}\sin(\omega t),\qquad(7^2)$$

$$Q_3 = Q_{3,c} + Q_{3,e} = -\frac{\partial V}{\partial q_3} + \vec{F}\frac{\partial \overrightarrow{O_0E}}{\partial q_3} =$$
$$= -\left[\left(k_1l_1^2 + k_2l_2^2 + k_3l_3^2\right)q_3 + \left(k_1l_1 - k_2l_2 - k_3l_3\right)q_2 \sin q_{30}^* - \left(k_1l_1 + k_2l_2 + k_3l_3\right)q_1 \sin q_{30}^*\right] +$$
$$+\left[-F_{10}\sin(\omega t)\vec{i}_{10} - F_{20}\sin(\omega t)\vec{i}_{20}\right]\cdot\frac{\partial}{\partial q_3}\left\{\left[q_1 + q_{10}^* + L_3\left(\cos q_{30}^* - q_3 \sin q_{30}^*\right) -\right.\right.$$
$$\left.- L_4\left(\sin q_{30}^* + q_3\cos q_{30}^*\right)\right]\vec{i}_{10} + \left[q_2 + q_{20}^* + L_3\left(\sin q_{30}^* + q_3\cos q_{30}^*\right) +\right.$$
$$\left.\left.+ L_4\left(\cos q_{30}^* - q_3 \sin q_{30}^*\right)\right]\vec{i}_{20}\right\} \Rightarrow$$
$$Q_3 = -\left[\left(k_1l_1^2 + k_2l_2^2 + k_3l_3^2\right)q_3 + \left(k_1l_1 - k_2l_2 - k_3l_3\right)q_2 \cos q_{30}^* -\right.$$
$$\left.- \left(k_1l_1 + k_2l_2 + k_3l_3\right)q_1 \sin q_{30}^*\right] + F_{10}\left(L_3 \sin q_{30}^* + L_4 \cos q_{30}^*\right)\sin(\omega t) - \tag{7^3}$$
$$- F_{20}\left(L_3 \cos q_{30}^* - L_4 \sin q_{30}^*\right)\sin(\omega t).$$

Because the links are totally homonymic, the Lagrange equations are given by (13).

$$\frac{d}{dt}\left(\frac{\partial T}{\partial \dot{q}_j}\right) - \frac{\partial T}{\partial q_j} = Q_j,\ j = \overline{1,3}. \tag{13}$$

By replacing in the equation system (13) the kinetic energy given by (4^1) and the generalized forces given by (7^1), (7^2) and (7^3), there results the linear differential equations system of second degree order, which is the motion mathematical model (which was similarly obtained in [11]), under the matrix form (13^1).

$$[A]\left\{\ddot{q}\right\} + [K]\{q\} = \{Q_e\}, \tag{13^1}$$

$$[A] = \begin{bmatrix} a_{11} & 0 & a_{13} \\ 0 & a_{22} & a_{23} \\ a_{13} & a_{23} & a_{33} \end{bmatrix};\ [K] = \begin{bmatrix} k_{11} & 0 & k_{13} \\ 0 & k_{22} & k_{23} \\ k_{13} & k_{23} & k_{33} \end{bmatrix};\ \{Q_e\} = \{Q_{1,e}, Q_{2,e}, Q_{3,e}\}^T;$$

$$\left\{\ddot{q}\right\} = \left\{\ddot{q}_1, \ddot{q}_2, \ddot{q}_3\right\}^T;\ \{q\} = \{q_1, q_2, q_3\}^T;\ Q_{1,e} = -F_{10}\sin(\omega t);\ Q_{2,e} = -F_{20}\sin(\omega t);$$

$$Q_{3,e} = \left[F_{10}\left(L_3 \sin q_{30}^* + L_4 \cos q_{30}^*\right) - F_{20}\left(L_3 \cos q_{30}^* - L_4 \sin q_{30}^*\right)\right]\sin(\omega t);$$

$$a_{11} = a_{22} = m_S;\ a_{33} = m_S\left(L_1^2 + L_2^2\right) + J_c;$$

$$a_{13} = -m_S\left(L_1 \sin q_{30}^* + L_2 \cos q_{30}^*\right);\ a_{23} = m_S\left(L_1 \cos q_{30}^* - L_2 \sin q_{30}^*\right);$$

$$k_{11} = k_{22} = k_1 + k_2 + k_3;\ k_{33} = k_1l_1^2 + k_2l_2^2 + k_3l_3^2;$$

$$k_{13} = -\left(k_1l_1 + k_2l_2 + k_3l_3\right)\sin q_{30}^*;\ k_{23} = \left(k_1l_1 - k_2l_2 - k_3l_3\right)\cos q_{30}^*.$$

By applying the (13^1) system Laplace transform in relation with time and by accepting initial homogenous conditions, in a matrix writing, there results the algebraic system (13^2).

$$\left([K] + s^2[A]\right)\{\tilde{q}(s)\} = \left\{\tilde{Q}_e(s)\right\},\ \{\tilde{q}(s)\} = \{\tilde{q}_1(s), \tilde{q}_2(s), \tilde{q}_3(s)\}^T;\ \left\{\tilde{Q}_e(s)\right\} = \left\{\tilde{Q}_{1,e}(s), \tilde{Q}_{2,e}(s), \tilde{Q}_{3,e}(s)\right\}^T; \tag{13^2}$$

$$\tilde{Q}_{1,e}(s) = -F_{10}\frac{\omega}{s^2 + \omega^2},\ \tilde{Q}_{2,e}(s) = -F_{20}\frac{\omega}{s^2 + \omega^2},$$

$$\tilde{Q}_{3,e}(s) = \left[F_{10}\left(L_3 \sin q_{30}^* + L_4 \cos q_{30}^*\right) - F_{20}\left(L_3 \cos q_{30}^* - L_4 \sin q_{30}^*\right)\right]\frac{\omega}{s^2 + \omega^2}.$$

The solution of (13^2) system is (14).

$$\tilde{q}_1(s)=\frac{P_1(s)}{P(s)},\ \tilde{q}_2(s)=\frac{P_2(s)}{P(s)},\ \tilde{q}_3(s)=\frac{P_3(s)}{P(s)}, \tag{14}$$

$$P(s)=\left(s^2+\omega^2\right)\left(a_{11}s^2+k_{11}\right)\cdot$$
$$\cdot\left[\left(-a_{13}^2-a_{23}^2+a_{11}a_{33}\right)s^4+\left(a_{33}k_{11}-2a_{13}k_{13}-2a_{23}k_{23}+a_{11}k_{33}\right)s^2-k_{13}^2-k_{23}^2+k_{11}k_{33}\right],$$
$$Q_1=-F_{10},\ Q_2=-F_{20},\ Q_3=\left[F_{10}\left(L_3\sin q_{30}^*+L_4\cos q_{30}^*\right)-F_{20}\left(L_3\cos q_{30}^*-L_4\sin q_{30}^*\right)\right],$$
$$P_1(s)=-\omega\left(a_{23}^2Q_1-a_{11}a_{33}Q_1-a_{13}a_{23}Q_2+a_{11}a_{13}Q_3\right)s^4-$$
$$-\omega\left(-a_{33}k_{11}Q_1+a_{13}k_{11}Q_3-a_{23}k_{13}Q_2+a_{11}k_{13}Q_3+2a_{23}k_{23}Q_1-a_{13}k_{23}Q_2-a_{11}k_{33}Q_1\right)s^2-$$
$$-\omega\left(k_{11}k_{13}Q_3-k_{13}k_{23}Q_2+k_{23}^2Q_1-k_{11}k_{33}Q_1\right),$$
$$P_2(s)=\omega\left(a_{13}a_{23}Q_1-a_{13}^2Q_2+a_{11}a_{23}Q_3\right)s^4+$$
$$+\omega\left(a_{33}k_{11}Q_2-a_{23}k_{11}Q_3+a_{23}k_{13}Q_1-2a_{13}k_{13}Q_2+a_{13}k_{23}Q_1-a_{11}k_{23}Q_3+a_{11}k_{33}Q_2\right)s^2-$$
$$-\omega\left(-k_{11}k_{23}Q_3+k_{13}^2Q_2+k_{13}k_{23}Q_1+k_{11}k_{33}Q_2\right),$$
$$P_3(s)=\left(a_{11}s^2+k_{11}\right)\left[\omega\left(a_{13}Q_1+a_{23}Q_2-a_{11}Q_3\right)s^2-\omega\left(k_{11}Q_3-k_{13}Q_1-k_{23}Q_2\right)\right]$$

3. Dynamic response

By applying the Laplace transform inverse in relations (14), there results the vibratory system dynamic response with the form of time – functions $q_i=q_i(t),\ i=\overline{1,3}$, as written in (15).

$$q_1(t)=\frac{\sqrt{2}}{4a_1^2\sqrt{\frac{c_1}{a_1}\left(b_1^2-4a_1c_1\right)}}\left(b_1^2d_1-2a_1c_1d_1-b_1d_1\sqrt{b_1^2-4a_1c_1}+a_1b_1d_2-a_1\sqrt{b_1^2-4a_1c_1}\,d_2-2a_1^2d_3\right)\cdot$$
$$\cdot\left[\sqrt{\frac{-b_1-\sqrt{b_1^2-4a_1c_1}}{a_1}}\sinh\left(\sqrt{\frac{-b_1+\sqrt{b_1^2-4a_1c_1}}{2a_1}}t\right)+\sqrt{\frac{-b_1+\sqrt{b_1^2-4a_1c_1}}{a_1}}\sinh\left(\sqrt{-\frac{b_1+\sqrt{b_1^2-4a_1c_1}}{2a_1}}t\right)\right],$$

$$q_2(t)=\frac{\sqrt{2}}{4a_1^2\sqrt{\frac{c_1}{a_1}\left(b_1^2-4a_1c_1\right)}}\left(b_1^2D_1-2a_1c_1D_1-b_1D_1\sqrt{b_1^2-4a_1c_1}+a_1b_1D_2-a_1\sqrt{b_1^2-4a_1c_1}\,D_2-2a_1^2D_3\right)\cdot$$
$$\cdot\left[\sqrt{\frac{-b_1-\sqrt{b_1^2-4a_1c_1}}{a_1}}\sinh\left(\sqrt{\frac{-b_1+\sqrt{b_1^2-4a_1c_1}}{2a_1}}t\right)+\sqrt{\frac{-b_1+\sqrt{b_1^2-4a_1c_1}}{a_1}}\sinh\left(\sqrt{-\frac{b_1+\sqrt{b_1^2-4a_1c_1}}{2a_1}}t\right)\right],$$

$$a_1=-a_{13}^2-a_{23}^2+a_{11}a_{33};\ b_1=a_{33}k_{11}-2a_{13}k_{13}-2a_{23}k_{23}+a_{11}k_{33};\ c_1=-k_{13}^2-k_{23}^2+k_{11}k_{33};$$
$$A_1=\omega a_{13}F_1+\omega a_{23}F_2-\omega a_{11}F_3;\ A_2=-\omega F_3k_{13}+\omega F_1k_{13}+\omega F_2k_{23};$$
$$d_1=-\omega\left(a_{23}^2Q_1-a_{11}a_{33}Q_1-a_{11}a_{23}Q_2+a_{11}a_{33}Q_3\right);$$
$$d_2=\omega\left(-a_{33}k_{11}Q_1+a_{13}k_{11}Q_3-a_{23}k_{13}Q_2+a_{11}k_{13}Q_3+2a_{23}k_{23}Q_1-a_{13}k_{23}Q_2-a_{11}k_{33}Q_1\right);$$
$$d_3=\omega\left(k_{11}k_{23}Q_3-k_{13}k_{23}Q_2+k_{23}^2Q_1-k_{11}k_{13}Q_1\right);\ D_1=\omega\left(a_{11}a_{33}Q_1-a_{13}^2Q_2+a_{11}a_{33}Q_2-a_{11}a_{23}Q_3\right);$$

$$q_3(t)=\frac{1}{4\omega c_1\left(\omega^4 a_1-\omega^2 b_1+c_1\right)\sqrt{b_1^2-4a_1c_1}}\left\{-\omega A_1 c_1\left[b_1\sqrt{2}\Omega_1\sin(\Omega_1 t)-\sqrt{2}\Omega_2\sin(\Omega_2 t)\right]-\right.$$
$$-2\sqrt{2}\omega^2 a_1\left[-\sqrt{2}\Omega_1\sin(\Omega_1 t)+\sqrt{2}\Omega_2\sin(\Omega_2 t)\right]-$$
$$-\sqrt{b_1^2-4a_1c_1}\left[4\omega\sin(\omega t)+2\Omega_1\sin(\Omega_1 t)+2\Omega_2\sin(\Omega_2 t)\right]+ \tag{15}$$
$$+4A_2c_1\sin(\omega t)\sqrt{b_1^2-4a_1c_1}+\sqrt{2}\omega A_2 a_1\left[-\sqrt{2}\Omega_1\sin(\Omega_1 t)+\sqrt{2}\Omega_2\sin(\Omega_2 t)\right]-$$
$$-b_1\sqrt{2}\omega\left(\omega^2 a_1-b_1\right)\left[\sqrt{2}\Omega_1\sin(\Omega_1 t)-\sqrt{2}\Omega_2\sin(\Omega_2 t)\right]+$$
$$\left.+\sqrt{2}\omega\left(\omega^2 a_1-b_1\right)\sqrt{b_1^2-4a_1c_1}\left[\sqrt{2}\Omega_1\sin(\Omega_1 t)+\sqrt{2}\Omega_2\sin(\Omega_2 t)\right]\right\}$$

$$\Omega_1=\sqrt{\frac{\sqrt{b_1^2-4a_1c_1}-b_1}{2a_1}};\ \Omega_2=\sqrt{\frac{b_1+\sqrt{b_1^2-4a_1c_1}}{2a_1}};\ D_3=\omega\left(-k_{11}k_{23}Q_3+k_{13}^2Q_2+k_{13}k_{23}Q_1+k_{11}k_{33}Q_2\right)$$

$$D_2=\omega\left(-k_{11}a_{33}Q_2+a_{23}k_{11}Q_3-a_{23}k_{13}Q_1+2a_{13}k_{13}Q_2-a_{13}k_{23}Q_1+a_{11}k_{23}Q_3-a_{11}k_{33}Q_2\right);$$

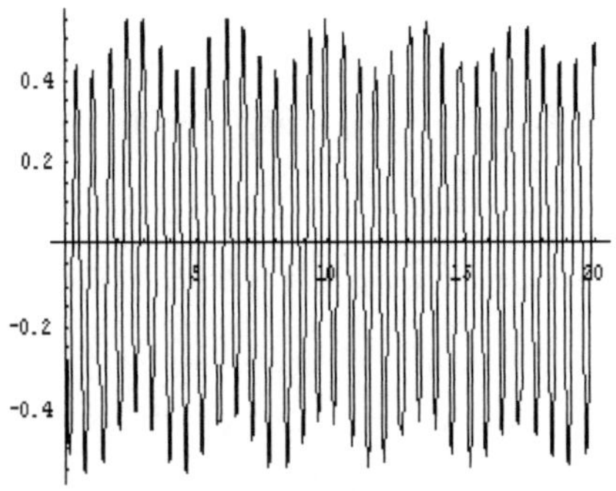

Figure 2. $q_1=q_1(t)$.

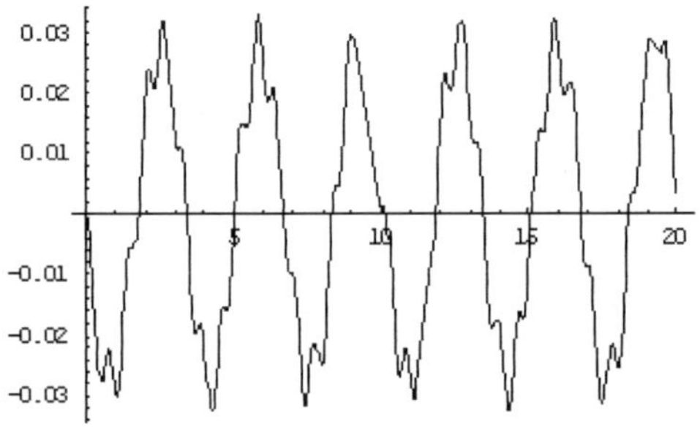

Figure 3. $q_2=q_2(t)$

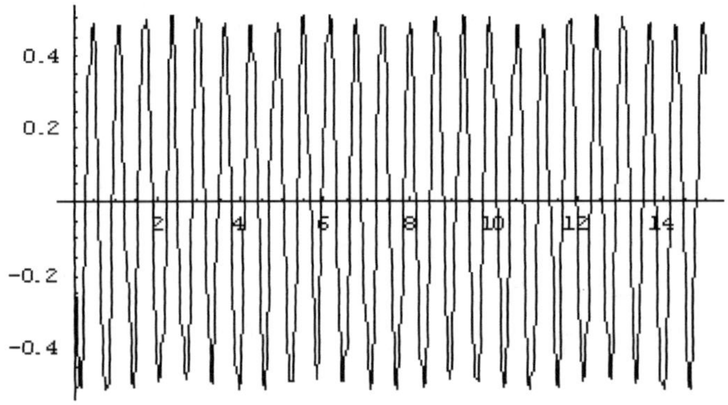

Figure 4. $q_3=q_3(t)$

In the case of a numerical application, in which are given the next initial data: $m_S=0.056\left[\frac{N\cdot s^2}{cm}\right]$, $l_1=20\,[mm]$, $L_4=250\,[mm]$, $l_2=30\,[mm]$, $J_c=2.04\,\left[N\cdot cm\cdot s^2\right]$, $L_1=37\,[mm]$, $L_2=168\,[mm]$, $L_3=70\,[mm]$, $k_1=28\cdot 10^4\left[\frac{N}{cm}\right]$, $l_3=50\,[mm]$, $k_3=11\cdot 10^6\left[\frac{N}{cm}\right]$,

$k_2 = 28 \cdot 10^4 \left[\frac{N}{cm}\right]$, we obtain the graphics of time – functions $q_i = q_i(t), i = \overline{1,3}$, from the 2, 3 and 4 figures.

4. Conclusions

The dynamic behaviour of mechanical systems is often affected by the non-linear effects, which, in many cases, are situated at the interfaces levels of coupling between different parts (dynamic excitations generated by the contact between the movements transmitting parts and the efforts, the instability related to the friction in mechanical systems) which may lead to unwanted effects (noises) or can harm the mechanism functioning (excessive wear).

In our days, the research activities are focused on the experimental parts development, theoretically and numerically multidisciplinary inserted when the structures, machine parts or, generally the mechanical systems, are designed. The research purpose is to improve the knowledge regarding the materials and structures behaviour, to develop models and instruments useful in the design process of machines and structures and to exploit a technical culture regarding the analysis, conception and fabrication analysis.

In this context, from the design stage, there must known the vibrations influence in the products manufacturing processes, and, in definitive, the machine dynamic behaviour, fact that is high lightened by the above study.

References

[1] Abousleiman V and Velex P 2006 *A hybrid 3d finite element/lumped parameter model for quasi-static and dynamic analyses of planetary/epicyclic gear sets* (Mechanism and Machine Theory 41(6)), pp 725–748

[2] Băgnaru D Gh Stănescu M M Cuţă P 2008 *Vibrations of the tool and tool holder unit for a shaper device* (8th International Conference "Research and Development in Mechanical Industry" RaDMI 2008, 14-17 September 2008, Užice, Serbia) pp 202-205

[3] Băgnaru D Gh 1997 *Sur l'analyse des vibrations des barres viscoelsatique, les elements constitutifs des mecanismes plans* (SYROM 97 Bucureşti) pp 25-30.

[4] Băgnaru D Gh Hadăr A Bolcu D Stănescu M M Cuţă 2009 P *The vibrations influence on the field of accelerations of the linear-elastic connecting rod of a mechanism connecting rod lug* (Annals of DAAAM for 2009 & Proceedings 20(1)) pp 833-835.

[5] Băgnaru D Gh Stănescu M M Bolcu D Cuţă P 2008 *Approximate method in order to determine the displacement fields for the kinematic elements standing in vibration* (Second international congres automotive, safety and environment, SMAT, Craiova, 23-26 October) pp 9-12.

[6] Bard C 1995 *Modelisation du comportement dynamique des transmissions par engrenages* (L'institut National des Sciences Appliquees de Lyon)

[7] Craig R R Bampton M C C 1968 *Coupling of substructures for dynamics analysis* (AIAA Journal, 6(7)) pp 1313–1319

[8] Gafiţanu M Focşa V Merticaru V Biborosch L 1980 *Vibrations and noises* (Junimea Publishing, Iaşi).

[9] Kahraman A 1994 *Planetary gear train dynamics* (Journal of mechanical design, 116(3)) pp 713–720

[10] Lin J Parker RG 1999 Analytical characterization of the unique properties of planetary gear free vibration (*Journal of vibration and acoustics*, 121(3)) pp 316–321.

[11] Băgnaru D Gh Stănescu M M 2017 *Aplicaţii ale transformatei Laplace în abordare teoriilor moderne în domeniul vibraţiilor unor siteme mecanice* (Universitaria Publishing Craiova ISBN 978-606-14-1230-3)

IOP Conf. Series: Materials Science and Engineering **281** (2017) 012026 doi:10.1088/1757-899X/281/1/012026

Creep Tests and Modeling Based on Continuum Damage Mechanics for T91 and T92 Steels

J P Pan[1*], S H Tu[1], X W Zhu[1], L J Tan[2], B Hu[3] and Q Wang[4]

1 Jiaxing Special Equipment Inspection Institute, Jiaxing 314051, China
2 Shanghai Jiao Tong University, Shanghai 200240, China
3 China Special Equipment Inspection and Research Institute, Beijing 100029, China
4 China Jiliang University, Hangzhou 310018, China

* Email: jxlpjp@foxmail.com.

Abstract. 9-11%Cr ferritic steels play an important role in high-temperature and high-pressure boilers of advanced power plants. In this paper, a continuum damage mechanics (CDM)-based creep model was proposed to study the creep behavior of T91 and T92 steels at high temperatures. Long-time creep tests were performed for both steels under different conditions. The creep rupture data and creep curves obtained from creep tests were captured well by theoretical calculation based on the CDM model over a long creep time. It is shown that the developed model is able to predict creep data for the two ferritic steels accurately up to tens of thousands of hours.

1. Introduction

Construction of advanced power plants requires materials with improved high-temperature strength and superior resistance to high-temperature corrosion. Enhancing the operating temperature and pressure of boilers has led to the development of creep-resistant steels [1,2]. During the past decades, continuous efforts have been devoted to the developing creep-resistant 9-11%Cr ferritic steels. The complex microstructure of 9-11%Cr steels has hindered the establishment of a convincing and comprehensive model for creep deformation of these alloys. This in turn has caused problems with reliable extrapolation of creep behavior from short-term tests [3-5]. The MIKORA model considered subgrain walls to be the main source of strengthening, and their effect was modeled by considering them to be hard regions surrounded by a soft ferritic matrix [6]. Another model by Semba et al. was on the basis of continuum damage mechanics (CDM) and a large amount of work in this area over the past years [7,8]. Semba et al. regarded nanosized MX carbonitride particles as the obstacles to dislocation motion, whose spacing was used to define the reference stress. The back stress was assumed to arise as a result of stress redistribution between hard and soft regions in the alloy [9].

Notwithstanding the lack of complete knowledge of the various microstructural features, there have been attempts to construct CDM-based creep models for bainitic steels [10]. In the present study, the CDM framework was used to develop a creep model for 9-11%Cr ferritic steels. The calculated results based on the creep model were compared with the experimental results for T91 and T92 steels. The creep tests lasted for a period of over 20,000 h.

2. Experimental

2.1. Materials

The T91 and T92 steels sampled from boilers of a power station were studied. The chemical compositions of the two steels were determined using an Optical Emission Spectrometer (ARL 4460, Thermo Scientific), and the results are listed in Table 1.

Table 1. Chemical compositions of T91 and T92 steels (%).

Steel	C	Si	Mn	P	S	Cr	Mo	Nb	V	N	Ni	W	Ti
T91	0.083	0.314	0.403	0.009	0.002	8.59	0.918	0.074	0.202	/	/	/	/
T92	0.09	0.378	0.4	0.012	0.003	8.68	0.374	0.049	0.19	0.048	0.187	1.54	0.003

2.2. Creep tests

Tensile creep tests in air at various temperatures with preset load were conducted using an electronic high temperature creep & rupture testing machine (GWT2105, MTS) to determine the creep rupture times. Standard cylindrical samples according to DIN50125 B 4×20 were used.

3. Results and discussion

3.1. Creep model description

The most important property of heat-resistant steels is the creep rupture strength. Therefore, a set of CDM equations utilized for the T91 and T92 steels studied here is as follows.

Master equation: $$\dot{\varepsilon} = \dot{\varepsilon}_0' \exp(-Q_C/RT) \sinh[\frac{\sigma(1-H^*(1-D_S))}{\sigma_o(1-D_P)}] \quad (1)$$

Primary creep: $$\dot{\sigma}_o = K_1(1-\sigma_o/K_2)\dot{\varepsilon} \quad (2)$$

Subgrain evolution: $$\dot{D}_S = \frac{\dot{\varepsilon}}{S_i}(K_{S1} + K_{S2}\exp\left(-\frac{Q_S}{RT}\right)) \times (1-D_S)^2 \quad (3)$$

MX evolution: $$\dot{D}_P = \frac{K_P}{P_i^2}\exp\left(-\frac{Q_P}{RT}\right)(1-D_P)^4 \quad (4)$$

The corresponding parameters for the set of equations are listed in Table 2.

Table 2. CDM parameters for the T91 and T92 steels

Parameter	Units	T91	T92
$\dot{\varepsilon}_0'$	s^{-1}	7.23×10^{6}	1.06×10^{7}
K_1	MPa	5.81×10^{3}	6.67×10^{3}
K_2	MPa	11.7	11.8
H^*	None	0.45	0.45
Q_c	kJ mol^{-1}	360	360
Q_p	kJ mol^{-1}	270	286
Q_S	kJ mol^{-1}	282	294
K_p	$m^3\ s^{-1}$	3.4×10^{-16}	4.5×10^{-17}
K_{S1}	m s^{-1}	4.10×10^{-6}	3.95×10^{-6}
K_{S2}	m s^{-1}	2.97×10^{12}	5.9×10^{12}
P_i	nm	7.22	3.89
S_i	nm	245	263

3.2. Creep rupture analysis

CDM prediction curves as well as the experimental rupture data at four specific temperatures for T91 and T92 steels are shown in Figure 1. For both steels, the creep rupture stress decreased with increasing temperature, as the creep rupture becomes severer at higher temperatures. Subjected to higher loads, it took shorter time periods for the two steels to rupture, as expected. In addition, the T92 steel is more creep resistant than T91 steel, since the creep rupture stress of the T92 steels was higher than that of the T91 steel under the same conditions. The experimental rupture data were captured well by the predicted curves based on the CDM model. It is thus possible to describe most of the features of the creep curves of ferritic steels with the proposed CDM model.

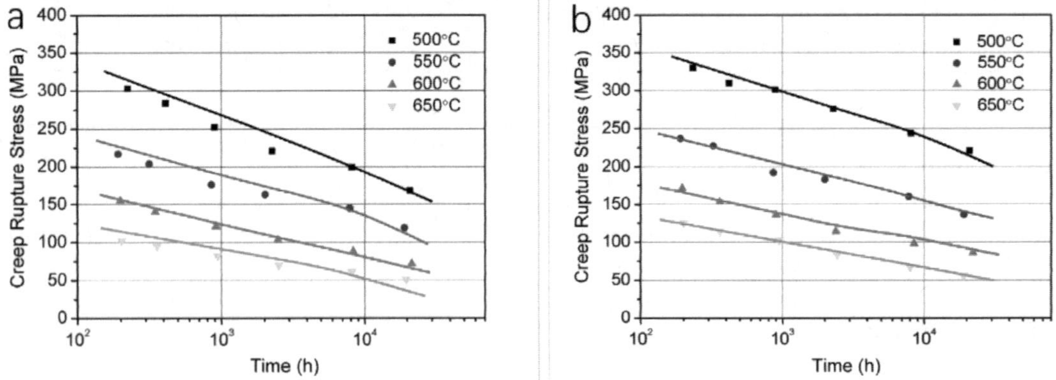

Figure 1. Creep test results (scatters) along with CDM predictions for (a) T91 and (b) T92 at different temperatures. The curves are obtained by CDM equations.

3.3. Creep curves
Figure 2 shows experimental creep curves as well as predicted creep curves based on the CDM model for T91 and T92 steels under different conditions. For the T92 steel, a pronounced primary creep can be found in the curves with similar magnitudes, despite the significantly different stresses and temperatures applied. The CDM curves agreed perfectly with the experimental data, especially at the creep time shorter than 15000 h. In the case of T91 steel, the experimental data were also predicted well by the CDM model in the data range, further indicating the reliability of the proposed model. Besides, the T91 steel seemed to behave much differently under the two conditions, implying that the creep behavior of this ferritic steel is affected greatly by temperature and loading.

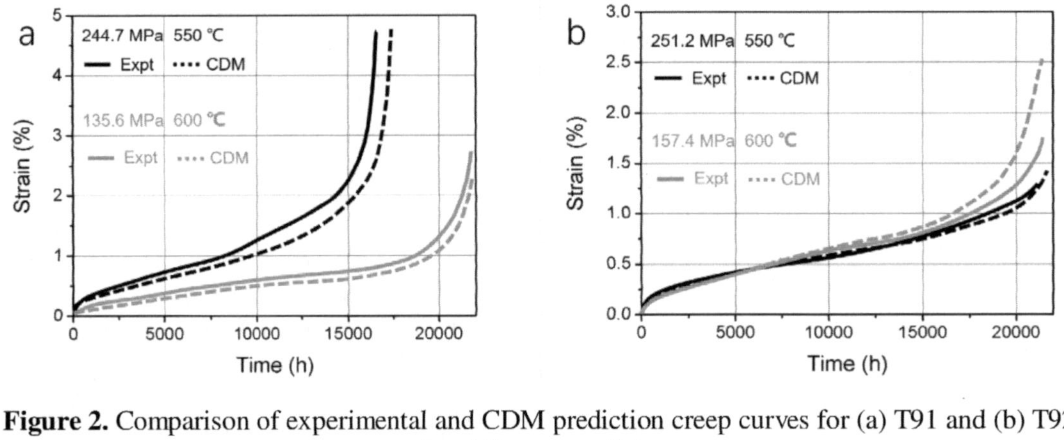

Figure 2. Comparison of experimental and CDM prediction creep curves for (a) T91 and (b) T92 under different conditions.

4. Conclusions
In conclusion, we have successfully established a CDM-based creep model for typical 9-11%Cr ferritic steels, i.e., T91 and T92 steels. Creep tests were performed for the T91 and T92 steels, the

IOP Conf. Series: Materials Science and Engineering **281** (2017) 012026 doi:10.1088/1757-899X/281/1/012026

results of which indicate that the set of equations works well in describing and predicting high-temperature creep behavior for both T91 and T92 steels. The soundness and reliability of the proposed CDM model have been confirmed.

Acknowledgment
This work is financially supported by the Quality and Technology Supervision of Zhejiang Province (Grant No. 20170251), the AQSIQ Science and Technology Project (Grant No. 2017QK037) and the National Key Research and Development Program of China (Grant No. 2017YFF0210702).

References

[1] Kaupilla P, Kouhia R, Ojanper¨J, Saksala T and Sorjonen T 2016 A continuum damage model for creep fracture and fatigue analyses. *Procedia Structural Integrity* **2** 887.
[2] Auerkari P, Holmstrom S, Veivo J and Salonen J 2007 Creep damage and expected creep life for welded 9-11% Cr steels. *Int. J. Pres. Ves. Pip.* **84** 69.
[3] Kimura K, Kushima H and Sawada K 2003 In: Proceedings of the 6th int Charles Parsons turbine conference p 443.
[4] Kimura K 2005 *Creep and fracture in high temperature components-design and life assessment issues*. ed I A Shibli, S R Holdsworth and G Merckling (Lancaster, PA: Des Tech Publications) p 1009.
[5] Auerkari P, Holmstrom S, Veivo J and Salonen J 2007 Creep damage and expected creep life for welded 9-11% Cr steels. *Int. J. Pres. Ves. Pip.* **84** 69.
[6] Polcik P, Straub S, Henes D and Blum W 1998 *Microstructural stability of creep resistant alloys for high temperature plant applications*. ed A Strang, J Cawley and G W Greenwood (London: Institute of Materials) p 405.
[7] Semba H, Dyson B F and McLean M 2005 *Creep and fracture in high temperature components-design and life assessment issues*. ed I A Shibli and S R Holdsworth (Lancaster, PA: Des Tech Publications) p 419.
[8] Dyson B F 1988 Creep and fracture of metals: mechanisms and mechanics. *Rev. Phys. Appl.* **23** 605.
[9] Oruganti R, Karadge M and Swaminathan S 2011 Damage mechanics-based creep model for 9-10%Cr ferritic steels. *Acta Mater.* **59** 2145.
[10] Mustata R and Hayhurst D R 2005 Creep constitutive equations for a 0.5Cr 0.5 Mo 0.25V ferritic steel in the temperature range 565 °C-675 °C. *Int. J. Pres. Ves. Pip.* **82** 363.

Experimental Studies on the Synthesis and Performance of Boron-containing High Temperature Resistant Resin Modified by Hydroxylated Tung Oil

J X Zhang[1]*, Z Y Ren[1], G Zheng[1], H F Wang[1], L Jiang[1], Y Fu[1], W Q Yang[2] and H H He[2]

[1]High & New Technology Research Center of Henan Academy of Sciences, Zhengzhou, China
[2] School of Materials Science and Engineering, Zhengzhou University, Zhengzhou, China

Email:zjx865185@163.com

Abstract: In this work, hydroxylated tung oil (HTO) modified high temperature resistant resin containing boron and benzoxazine was synthesized. HTO and ethylenediamine was used to toughen the boron phenolic resin with specific reaction. The structure of product was studied by Fourier-transform infrared spectroscopy(FTIR), and the heat resistance was tested by Differential Scanning Calorimetry (DSC) and Thermogravimetric Analysis(TGA). The results indicated that the conjugated triene structure of HTO was involved in the crosslinking of the heating curing progress, and in addition, the open-loop polymerization reaction of benzoxazine resin during heating can effectively reduce the curing temperature of the resin and reduce the release of small molecule volatiles, which is advantageous to follow-up processing. DSC data showed that the initial decomposition temperature of the resin is 350-400 °C, the carbon residue rate under 800 °C was 65%. It indicated that the resin has better heat resistance than normal boron phenolic resin. The resin can be used as an excellent ablative material and anti-friction material and has a huge application market in many fields.

1. Introduction

High temperature resistant resin has been widely used in the fields of refractory materials, super hard materials, abrasives, friction materials, aerospace materials and adiabatic materials[1,2]. The existing high temperature resistant materials include polyimide resin, polysulfone resin, polyether resin, polyphthalocyanine resin, phenolic resin and so on[3-5]. Phenolic resin was the first synthetic resin which was discovered and industrialized in the world, and it has a long history so far. Due to the cheap reactants, simple production process and low capital investment, high mechanical strength and excellent heat resistant performance, phenolic resin has been widely used in many fields [6]. The improvement of phenolic resin is one of the important topics in the field of the high temperature resistant resin [7,8].

Main methods of improving thermal stability of phenolic resin include alkyl phenol or phenyl phenol modification, vegetable oil modification, rare elements (such as molybdenum, tungsten and organic boron) modification, organic silicon modification, and other impurity atoms of polymer materials (such as polyimide and polyamide) copolymerization modification and blending modification, etc[9-12]. Above all, organic boron modification is one of the most effective and economic methods. Due to the introduction of boron atoms, C-O-B bong and C-O→B ligand bond were formed in the molecules of resin. Since the B-O bond enthalpy is 523 kJ/mol and it is higher than that of 347 kJ/mol of C-O bond enthalpy, the heat-resistant performance of boron-containing resin is excellent, and the thermal decomposition temperature is significantly improved[13,14]. Additionally, the introduction of vegetable oil into the phenolic resin could also increase the thermal stability and toughness of the resin[15,16]. As the existing boron phenolic resin was brittle and easy to absorb moisture, the curing temperature was high and the release of small molecule volatiles in curing was large, it is necessary to improve the performance of resin continually.

Therefore, this work adopted HTO through derivative processing. The modified high temperature resistant resin containing boron and benzoxazine was synthesized with specific reaction. The HTO and ethylenediamine were used to toughen the boron phenolic resin. Through researching the effects of reaction conditions on the properties of modified resin, the optimum process conditions of high temperature resistant resin was selected to meet the requirement of market. It would provide effective reference basis for the production and application of novel high temperature resistant resin.

2. Experimental

Certain amount of boric acid, phenol and HTO were added in 500 mL of four-necked flask with electric mixer. Esterification was carried out when oil bath was heated highter than 140 °C, and water was steamed out. Then the system was cooled to 60 °C, a certain amount of paraformaldehyde and ethylenediamine were added. The material was heated to 105 ~ 110 °C and refluxed 1h.
Vacuum-dewatering process was conducted up to the required viscosity at the temperature of 80 °C, and then the product was discharged. Figure 1 shows the pathway for the preparation of boron phenolic resin.

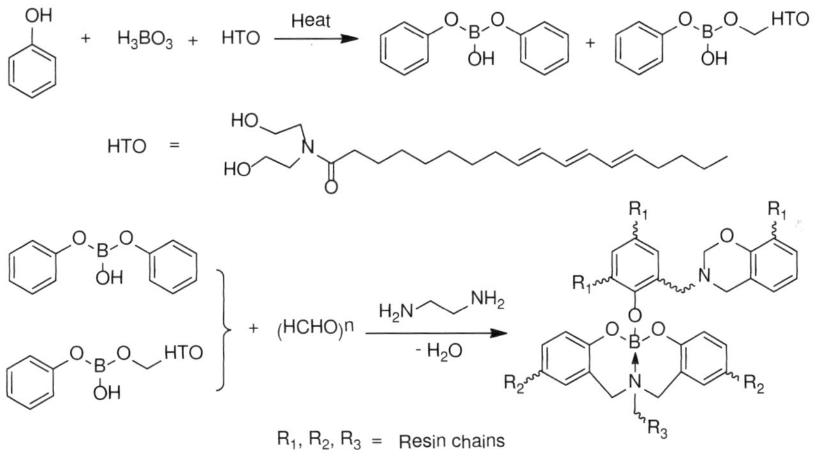

Figure 1. Synthesis routes of boron phenolic resin modified by HTO

3. Results and discussion

In this work, the reaction conditions were optimized by many experiments. The suitable ratio of the raw materials is n(phenol): n(boric acid): n(paraformaldehyde)=1:0.3:1.3. Ethylenediamine was used as alkaline catalyst, and it can also participate in the reaction to generate the benzoxazine structure. Since the vacant orbital of the boron atom can coordinate with the lewis base, the coordination of nitrogen atoms and boron atoms could enhance the water resistance of the resin. The boron phenolic resin was modified by HTO, which was better than the direct use of tung oil. In the synthesis process, esterification reaction went ahead between the hydroxyl of HTO and boric acid. The retained conjugate triene structure was cross-linked in the process of resin curing, and it was beneficial to improve the heat resistance of resin.

IOP Conf. Series: Materials Science and Engineering **281** (2017) 012027 doi:10.1088/1757-899X/281/1/012027

3.1. Structural characterization

Figure 2 is the FTIR spectrum of HTO modified boron phenolic resin (HTO-BPF). It can be observed that the characteristic bands of boron phenolic resin are very obvious, including the *ν*OH band at 3354cm^{-1}, the amide band at 1595cm^{-1}, the *ν*BO band at 1388cm^{-1}, the *ν*Ph-O band at 1226cm^{-1}, as well as the *ν*C-B-N band at 1080cm^{-1}. It is also seen that there is a small band at around 990cm^{-1} [17,18], showing the triene structure from HTO remains in boron phenolic resin.

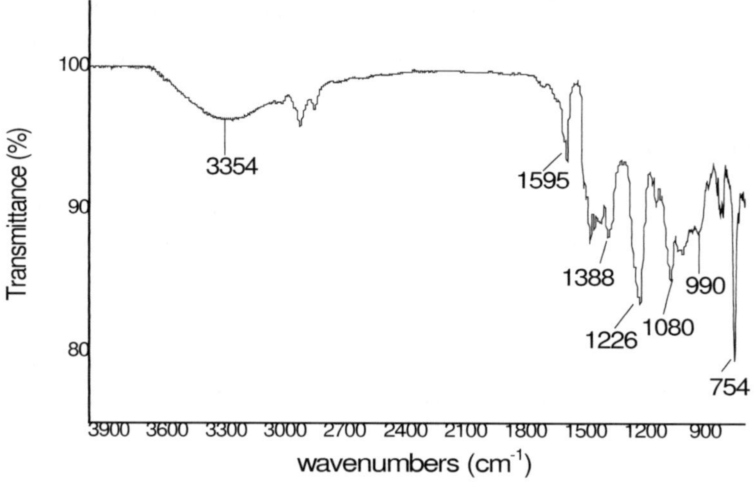

Figure 2. FTIR spectrum of HTO modified boron phenolic resin (HTO-BPF)

3.2. Heat-resistant properties

Figure 3 was the TG spectra of HTO modified boron phenolic resin (BPF) (here referred to as HTO-BPF) and common BPF. It can be seen that although the initial heat resistance of HTO-BPF (146 °C, 98%) is lower than that of common BPF (196 °C, 98%), the heat resistance in the high temperature zone of HTO-BPF is excellent. The char yield of HTO-BPF is 86 % at 500 °C, and the ratio is 65% at 800 °C. The results indicated that the conjugated triene structure of HTO was involved in the crosslinking of the heating curing progress and it may improve the heat resistance of the resin in the high temperature zone. In addition, the open-loop polymerization reaction of benzoxazine structure in the resin during heating could effectively reduce the curing temperature of the HTO-BPF.

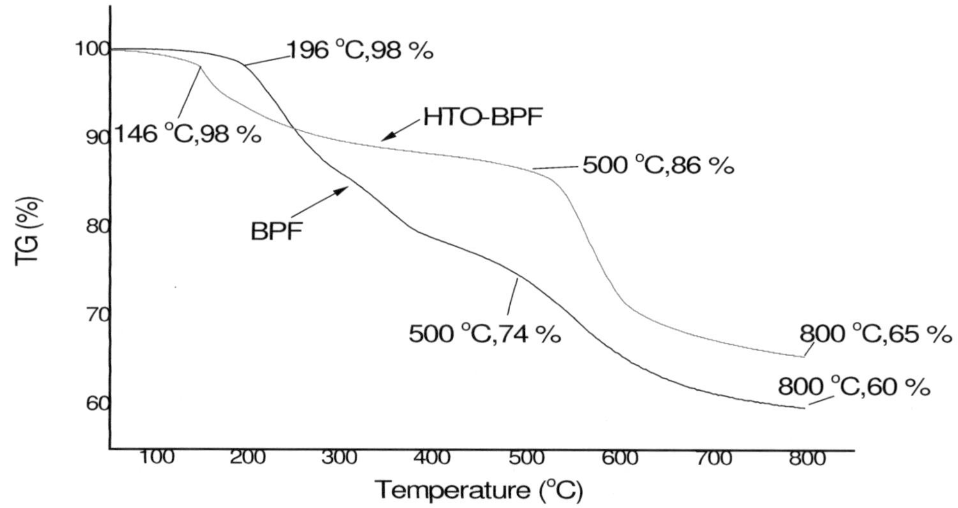

Figure 3. Comparative TG curves between HTO modified boron phenolic resin (HTO-BPF) and common boron phenolic resin(BPF)

4. Conclusions

In summary, HTO modified high temperature resistant resin containing boron and benzoxazine was synthesized. The structure of product was studied by FTIR, and the heat resistance was tested by DSC and TGA. The data indicated that the conjugated triene structure of HTO was involved in the product, and the benzoxazine structure was also formed. DSC data showed that the initial decomposition temperature of the resin is 350-400 °C, the carbon residue rate under 800 °C was 65%. It indicated that the resin has better heat resistance than normal boron phenolic resin. The results can provide a good theoretical basis for the preparation of novel high temperature resistant resin.

Acknowledgements

This study has been supported by the Henan Science and Technology Project Foundation (the basic and frontier research project) (162300410023).

References

[1] Chen X X, Liang G Z, Gu A J, Yuan L. Flame retarding cyanate ester resin with low curing temperature, high thermal resistance, outstanding dielectric property, and low water absorption for high frequency and high speed printed circuit broads. 2015, *Ind. Eng. Chem. Res.*, **54** (6),1806-1815.

[2] Miao J T, Yuan L, Guan Q B, Liang G Z, Gu A J. Biobased heat resistant epoxy resin with extremely high biomass content from 2,5-furandicarboxylic acid and eugenol. 2017, *ACS Sustainable Chem. Eng.*, Articles ASAP.

[3] Williams J C, Nguyen B N, McCorkle L, Scheiman D, Griffin J S, Steiner III S A, Meador M B. Highly porous, pigid-pod polyamide aerogels with superior mechanical properties and unusually high thermal conductivity. 2017, *ACS Applied Materials & Interfaces.*, **9** (2), 1801-1809.

[4] Podgórski M, Wang C, Yuan Y, Konetski D, Smalyukh I, Bowman C N. Pristine polysulfone networks as a class of polysulfide-derived high-performance functional materials. 2016, *Chem. Mater.* ,**14**, 5102-5109.

[5] Nazari B, Rhoades A M, Schaake R P, Colby R H. Flow-Induced crystallization of PEEK: isothermal crystallization kinetics and lifetime of flow-induced precursors during isothermal annealing. 2016, *ACS Macro Lett.* , **5** (7), 849-853.

[6] Pilato L. Phenolic resins: 100 Years and still going strong. 2013, *Reactive & Functional Polymers.* **73**, 270-277.

[7] Monni J, Alvila L, Pakkanen T T. Structural and physical changes in phenol-formaldehyde resol resin as a function of the degree of condensation of the resol solution. 2007, *Ind. Eng. Chem. Res.* **46**, 6916-6924.

[8] Streckova M, Sopcak T, Medvecky´ L', Bures R, Faberova M, Batko I, Briancin J. Preparation, chemical and mechanical properties of microcomposite materials based on Fe powder and phenol-formaldehyde resin. 2012, *Chemical Engineering Journal.* **180**, 343-353.

[9] Liu Y H, Jing X L. Pyrolysis and structure of hyperbranched polyborate modified phenolic resins. 2007, *Carbon.,* **45**(10), 1965-1971.

[10] Wang D C, Chang G W, Chen Y. Preparation and thermal stability of boron-containing phenolic resin/clay nanocomposites. 2008, *Polymer Degradation and Stability.*, **93**, 125-133.

[11] Liu L, Ye Z P. Effects of modified multi-walled carbon nanotubes on the curing behavior and thermal stability of boron phenolic resin. 2009, *Polymer Degradation and Stability.*, **94**:1972-1978.

[12] Li Y U, Lin H E, Yan J W. Research on the Heat-Stability of the Nano-Particle SiC Modified Phenolic Resin. 2007, *Polymer Materials Science & Engineering.*, **23**(3), 148-151.

[13] Zhang W F, Liu C L, Ying Y G, et a1. The preparation and characterization of boron-containing phenolic fibers. 2010, *Materials Chemistry and Physics.*, **121**(1-2), 89-94.

[14] Zhu M M, Wang R M, Wei X Y. Research progress of synthesis and modification for boron phenolic resin. 2011, *China Adhesives.,* **20**(6), 60-63.

[15] Vinay S, Kund P. Addition polymers from natural oils-A review. 2006, *Prog. Polym. Sci.*, **31**, 983-1008.
[16] Biermann U , Butte W, Holtgrefe R , et al. Esters of calendula oil and tung oil as reactive diluents for alkyd resins. 2010, *Eur. J. Lipid. Sci. Techno.*, **112**, 103-109.
[17] Zhao W, Wang H F, Ren Z Y, et al. Structural confirmation of a new hydroxylated tung oil compound by NMR. 2015, *Magn. Reson. Chem.*, **53**(2), 160-162.
[18] Ren Z Y, Liu L, Wang H F, Fu Y, Jiang L, Ren B X. Novel Amphoteric Polyurethane Dispersion with Postpolymerization Crosslinking Function Derived from Hydroxylated Tung Oil: Synthesis and Properties. 2015, *RSC Advances,* **5**(35), 27717-27721.

IOP Conf. Series: Materials Science and Engineering **281** (2017) 012028 doi:10.1088/1757-899X/281/1/012028

High Performance Resistive Switching Characteristics of SiN Films with a Cu/Ta/SiN/Cu/SiN/TiN Multilayer Structure

D Kumar and T Y Tseng*

Department of Electronics Engineering and Institute of Electronics, National Chiao Tung University, Hsinchu 30010, Taiwan

*Corresponding author email: tseng@cc.nctu.edu.tw

Abstract. The bipolar resistive switching properties of SiN based conductive bridge random access memory (CBRAM) device are investigated for non-volatile memory applications in a Cu/Ta/SiN/Cu/SiN/TiN multilayer structure. The device shows good switching characteristics with set voltages between 0.8 V and 1.3 V and reset voltages between -0.3 V and - 0.7 V with a variation of less than 0.1 V. The Cu/Ta/SiN/Cu/SiN/TiN multilayer CBRAM device exhibits excellent memory performance, such as long stable endurance cycles (> $4.5x10^3$) during the test without any degradation, good retention ability (>10^4 s) at a temperature of 120 ^{0}C with more than 10^2 on/off resistance ratio.

1. Introduction

Resistive random access memory (RRAM) has been considered to be a promising candidate to replace NAND FLASH memory because of its high scalability, high retention, large memory window, fast speed and low power consumption [1-3]. Among the several kinds of RRAMs, the CBRAM has profound interest due to its excellent scaling, large non-volatile memory window, low voltage, low current applications, reconfigurable logic circuits, and CMOS compatibility [4,5]. Several kinds of switching materials have been used to investigate the CBRAM characteristics such as oxide materials like Al_2O_3, ZrO_2, HfO_2, SiO_2, TiO_2 and non-oxide materials, SiCN, Si_3N_4, a-Si_3N_4 [6-14]. The resistive switching (RS) phenomenon in Ox-RAM device is based on the formation and rupture of the conductive filament due to the oxygen vacancies. Furthermore, the reduction of metal (Cu^+, Ag^+) ions are responsible for the RS in the CBRAM devices instead of oxygen vacancies by the application of the external bias [2,5,14]. The transition metal oxides (TMO) have shown advantages than other materials in their excellent process compatibility [15,16]. However, several problems such as large reset currents, small on/off ratio, long reset time and large variations in both resistances states [17]. Nitride-based CBRAM shows relatively excellent RS characteristics such as high on/off ratio, low voltage, superior endurance, retention, and high switching speed [14,18]. Silicon Nitride and Cu metal are widely used in current CMOS technology as insulating layer and interconnect materials, respectively. Sun et al. have reported that an Ag/Si_3N_4/Pt bipolar resistive switching (BRS) device working based on the Ag filament model has shown good retention characteristics [16]. Kim et al. have also reported bipolar resistive switching in Ag/Si_3N_4/Al with the on/off resistance ratio more than 10^7 but the device has shown poor endurance and high forming voltage [18]. Still, these devices using Ag metal electrodes may possibly limit the applicability to CMOS manufacturing technology. However, the natural random formation and rupture of conductive filament (CF) is serious problem in the CBRAM devices, which shows a huge variation of switching voltages and resistance states [13,16,19].

To reduce these problems, we have investigated a Cu/Ta/SiN/Cu/SiN/TiN multilayer structure with the insertion of thin Cu metal layer between the SiN layers. This multilayer device has shown good and highly stable switching characteristics with minimum dispersion in the memory switching parameters like very low set/reset voltages, high endurance and long retention time, showing its applications in real time devices. In this study, we have investigated a SiN-based CBRAM device with the Cu/Ta/SiN/Cu/SiN/TiN multilayer structure, and systematically investigate resistive switching characteristics of this device. The device shows that the low forming voltage, narrow fluctuations in the operating voltages and good stable endurance.

2. Experimental Details

In the experiment, the Cu/Ta/SiN/Cu/SiN/TiN device was fabricated as follows. Firstly, a 200 nm thick SiO_2 layer was deposited on Si substrate by using Plasma Enhanced Chemical Vapor Deposition (PECVD). Secondly,30 nm thick TiN was deposited on the SiO_2/Si substrate by Atomic Layer Deposition (ALD). After that, a 5 nm thick SiN layer was deposited on the TiN bottom electrode (BE) by electron beam evaporation system. Then, a 2 nm thick Cu film was deposited on the SiN layer by RF sputtering. After that, again a 5 nm thick SiN layer was deposited on Cu layer. Finally, a 4 nm Ta adhesion barrier layer and 250 nm Cu top electrode (TE) were deposited by RF magnetron sputtering to form Cu/Ta/SiN/Cu/SiN/TiN multilayer CBRAM device. The Agilent B1500A semiconductor parameter analyzer was used to measure the resistance switching characteristics of the present device, in which the bias voltage was applied to the Cu top electrode with the TiN bottom electrode grounded.

3. Results and Discussion

The schematic diagram of the Cu/Ta/SiN/Cu/SiN/TiN multilayer structure is illustrated in Figure 1. The schematic diagram shows the clear and distinguished presence of various layers within the device. The multilayer device consists of Cu top electrode, 4 nm Ta adhesion barrier layer, 6 nm SiN films are on both sides of 2 nm Cu layer, and 30 nm TiN bottom electrode.

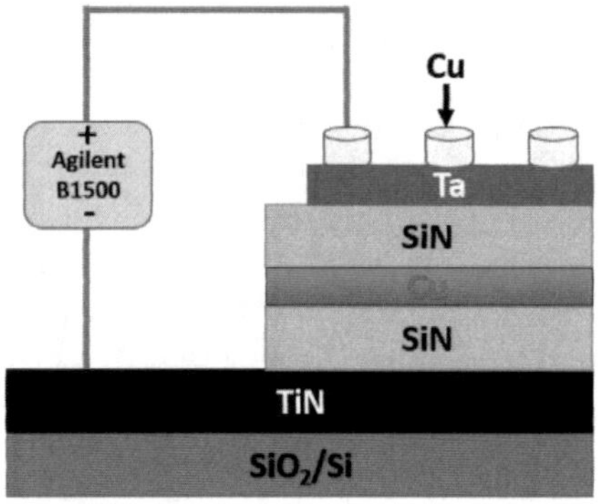

Figure 1. A schematic configuration of the Cu/Ta/SiN/Cu/SiN/TiN multilayer device

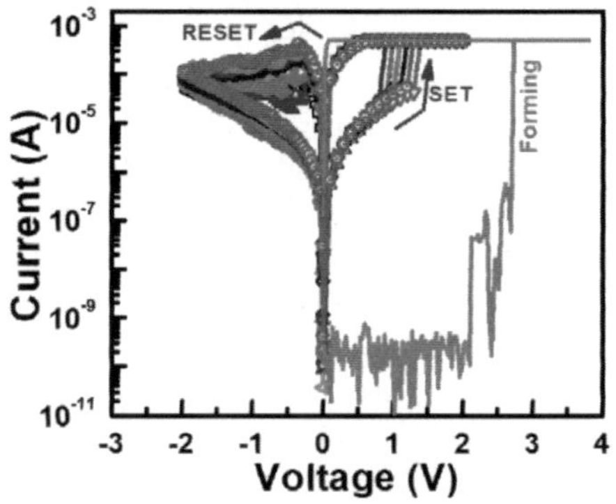

Figure 2. Forming process and I-V resistive switching curves of the multilayer device

Figure 2 shows the typical forming behavior of the Cu/Ta/SiN/Cu/SiN/TiN multilayer device, where compliance current (CC) of 500 μ A is used to protect the device. The bipolar resistive switching current – voltage (I–V) of the multilayer device is revealed in Figure 2. During the forming process, the device is switched from the initial state to the low resistance state (LRS), with the application of positive voltage on the Cu top electrode. By sweeping the negative voltage from 0 V to – 2 V on Cu top electrode, the device is switched from the LRS to the high resistance state (HRS), which is called as the reset process. Then, the device is swept by positive voltage from 0 V to +2 V on the Cu top electrode to switch the device back to the LRS, which is called the set process. The device has shown LRS/HRS ratio of about 10^2 for multilayer device at the read voltage of 0.2 V, which is useful for non-volatile memory applications.

The resistive switching characteristics of the Cu/Ta/SiN/Cu/SiN/TiN multilayer device are very similar to that of transition oxide based CBRAM devices [14, 19-20]. The resistive switching phenomenon of the multilayer device is based on formation and rupture of Cu conductive filament. When the positive bias is applied on the Cu top electrode, the Cu atoms get oxidized and become Cu ions. The electric field generated by the applied positive voltage between top and bottom electrodes drifts the ionized Cu atoms from the top electrode towards the TiN bottom electrode, resulting in the formation of Cu filaments, called set process.

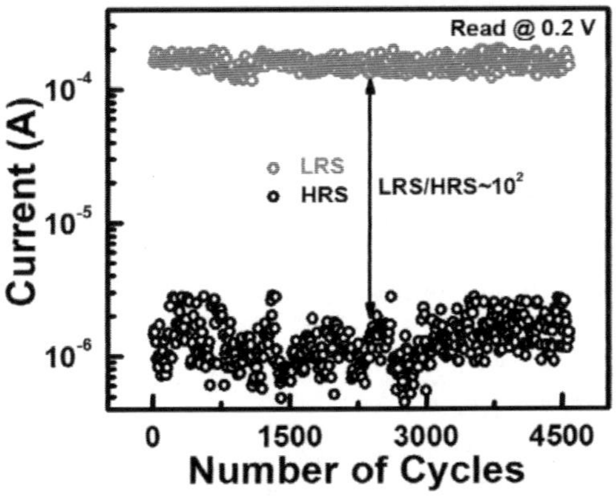

Figure 3. Endurance test for multilayer device measured at 120 ^{0}C temperature

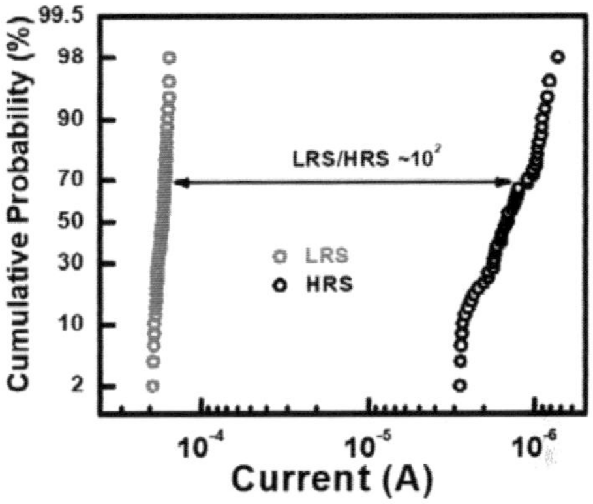

Figure 4. Resistance distribution of the device at both HRS and LRS conditions

Conversely, the negative bias applied on the Cu top electrode causes the electrochemical reaction, and the Cu atoms becomes Cu ions and drift back to the TE due to the applied electric field, resulting rupture of the Cu filaments, called reset process. Figure 3 shows the endurance characteristics the Cu/Ta/SiN/Cu/SiN/TiN multilayer structure at the read voltage of 0.2 V.

The device is maintained its both resistance (LRS/HRS) states up to 4500 cycles without showing any degradation. During the resistive switching cycles, the resistance ratio of HRS to LRS is maintained in the range of 10^2. The cycle to cycle variation of the device in both LRS/HRS resistance states are detected at the read voltage of 0.2 V during the 100 continuous switching cycles as shown in Figure 4. As shown in the resistance evolution during these given cycles, there is a minimum fluctuation in HRS and LRS. The device shows sharp distribution in LRS and narrow distribution in HRS with maintained the resistance ratio around 10^2.

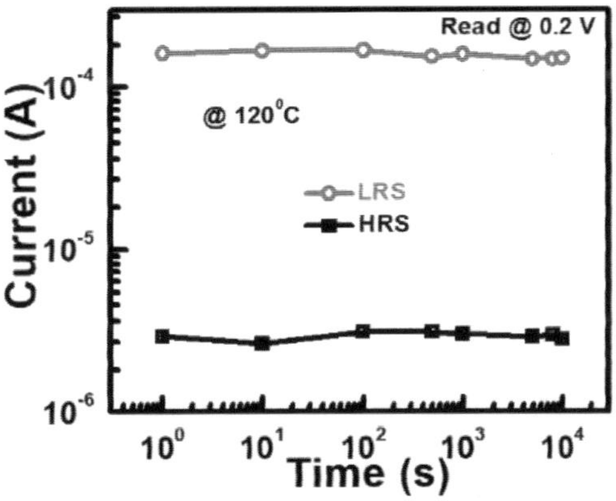

Figure 5. Retention behavior is measured at 120 ^{0}C for multilayer device

The high temperature retention characteristic is also measured for the approval of memory performance of the device with the read voltage of 0.2 V, at temperature of 120 ^{0}C as shown in Figure 5. It is observed that the device can well maintain its both LRS/HRS states for more than 10^4 s without any degradation.

4. Conclusions

The switching characteristics and uniformity of the Cu/Ta/SiN/Cu/SiN/TiN multilayer device has been investigated. The device shows that the LRS/HRS resistance ratio around 10^2 and endurance more than 4500 cycles are achieved without any degradation during the continuous resistive switching cycles. The device also possesses good retention property up to 10^4 s at 120 ^{0}C. These results indicate that the multilayer structure has high potential for non-volatile memory applications

References

[1] Tseng T Y and Sze S M 2012 *An introduction to nonvolatile memories," in Nonvolatile Memories: Devices, and Applications* vol 1, ed T Y Tseng and S M Sze (American Scientific Publishers, CA, USA, 2012) p 1–9.
[2] Yang J J, Strukov D B, and Stewart D R 2013 *Nature Nanotechnology* **8** 13–24
[3] Waser R and Aono M 2007 *Nat. Mater.* **6** 833
[4] Suri M, Bichler O, Querlioz D, Palma G, Vianello E, Vuillaume D, Gamrat C and Salvo B D 2012 IEEE *Int. Electron Devices Meeting (IEDM)*, p 10.3.1–10.3.4,
[5] Chand U, Huang C-Y, Kumar D, and Tseng T-Y 2015 *Appl. Phys. Lett.***107** 203502
[6] Liu Q, Long S, Wang W, Zuo Q, Zhang S, Chen J and Liu M 2009 *IEEE Electron Device Lett.* **30** 1335
[7] Yang Y C, Pan F, Liu Q, Liu M and Zeng F, 2009 *Nano Letters* **9** 1636
[8] Belmonte A, Kim W, Heylen N, Fantini A, Houssa M, Jurczak M and Goux L 2013 *IEEE Tran. Electron Device* **60** 3690
[9] Schindler C, Thermadam S P, Waser R and Kozicki M *N* 2007 *IEEE Tran. Electron Device* **54** 2762
10] Lv H, Xu X, Liu C H, Liu R, Liu Q, Banerjee W, Sun H, Long S, Lee L and Liu M 2017 *Scientific Rep.* **5** 7764
[11] Kim K M, Choi B J, and Hwang C S, 2007 Appl. Phys. Lett. **90** 242906
[12] Chand U, Huang K -C, Huang C –Y, Ho C -H, Lin C-H, and Tseng T –Y 2015 *J. Appl. Phys.* **117** 184105
[13] Chand U, Huang C -Y and Tseng T -Y 2014 *IEEE Electron Device Lett.* **35** 1019
[14] Kumar D, Aluguri R, Chand U and Tseng T -Y 2017 *Appl. Phys. Lett.* **110** 203102
[15] Kim K -H, Gaba S, Wheeler D, Cruz-Albrecht J M, Hussain T, Srinivasa N and Lu W 2012, *Nano Lett.* **12** 389
[16] Sun B, Liu L F, Wang Y, Han D D, Liu X Y, Han R Q and Kang F 2008 *9th Int. Conf. on Solid-State and Integrated-Circuit Technology* (Beijing) p925
[17] Banno N, Sakamoto Iguchi T N, Sunamura H, Terabe K, Hasegawa T and Aono M 2008 IEEE Trans. Electron. Device **55** 3283
[18] Kim H -D, An H -M, Kim K C, Seo Y, Nam K -H, Chung H -B, Lee E B and Kim T G 2010 *Semicond. Sci. Technol.* **25** 065002
[19] Wang Y, Liu Q, Lv H -B, Long S -B, Zhang S, Li Y -T, Lian W -T, Yang J –H, Liu M 2011 *Chin. Phys. Lett.* **28** 077201
[20] Wang Y, Liu Q, Long S, Wang W, Wang Q, Zhang M, Zhang S, Li Y, Zuo Q, Yang J and Liu M 2010 *Nanotechnology* **21** 045202

A Flexible Metamaterial Terahertz Perfect Absorber

X R Chen[1,2], Y W Zheng[1,2], L M Qin[1,2], G C Wei[1,2], Z P Qin[1,2], N G Zhang[1,2],K Liu[1,2], S Z Li[1,2] and S X Wang[1,2,*]

[1] School of Electronic and Electrical Engineering, Wuhan Textile University, Wuhan, Hubei, 430073 People's Republic of China

[2] State Key Laboratory for Hubei New Textile Materials and Advanced Processing Te chnology, Wuhan Textile University, Wuhan, 430020, P. R. China

[*]Email: shxwang@wtu.edu.cn

Abstract. We designed a THz matematerial absorber using metallic wires (MWs) and split resonant rings (SRRs). This matematerial absorber exhibits perfect absorption which up to 96% at 4.03 THz and is capable of wrapped around objects because of flexible polyimide dielectric substrate.

1. Introduction

The metamaterial consists of artificially designed cell units are capable of exhibiting exotic properties that natural materials cannot achieve at operating wavelengths that are much smaller than the unit structure dimension. This unit cells resemble the molecules of natural materials and behave different properties by adjusting the properties of the unit cell. With the rapid progress of metamaterial research, the practical application of metamateials in from microwave, infrared to optical frequency regions has attracted enormous interest [1-3].

Terahertz range is a particularly interesting band which refers to the frequency in the range of from 0.1 THz to 10 THz. Due to the lack of high efficient terahertz source and detection equipment, the development of THz applications is almost blank and commonly known as "THz gap". Terahertz wave exhibits many unique properties[4, 5]. A large amount of nonpolar molecules cannot absorb terahertz waves, so terahertz waves have strong penetrability for many materials and can be used at security field. In addition, the terahertz wave photon energy is so less than that of the traditional X-ray so that the damage of terahertz wave to the detected object is much smaller relatively. The highly flexible polyimide substrate can be used on non-planar devices because it can be easily wrapped around objects with a diameter of millimeters[6].

The metamaterial structure based on a polyimide substrate with nearly perfect absorption characteristics in terahertz frequency regime can be applied to non-planar device and these results serve as an important step forward in constructing a functional THz cloak[7-9]. In this paper, we designed a metamaterial based on polyimide flexible substrate with perfect absorption which up to 96% at 4.03 THz.

2. Design and simulation

We designed a compact multi-layer absorber which is illustrated in Figure1. The unit cell of the metamaterial consists of four alternating metal structure and dielectric layer. The two-layer intermediate dielectric layer is composed of polyimide and the dielectric constant of which is $\varepsilon = 2.4 + i0.005$ in the interested frequency[10]. Patterned metallic structure is a combination of metallic wire and split resonant rings and it is made of Au. The upper and lower layers split resonant rings is mirror symmetrical along the horizontal axis. The repeat period metamaterial parameters is illustrated in the inset of Figure 1 in top-right corner and specific dimension is Py=125 um, Px=106 um, Ry=28 um, Rx=27.5 um, R=19.5 um, r=14.5 um, d=5 um, w=10 um, L=110 um, l=100 um. The thickness of polyimide dielectric layer and pattened metal structure are 5 um and 500 nm, respectively.

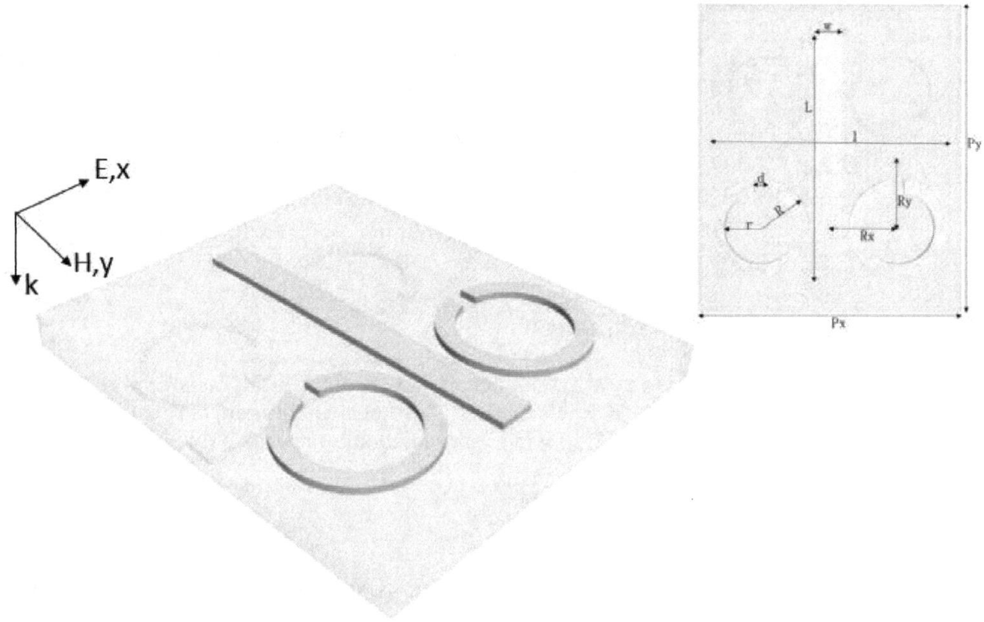

Figure 1. The unit schematic of the designed metamaterial structure and the inset in top-right shows the structure parameters of the designed unit cell of the metamaterial.

We implemented the simulation through finite element calculation, which incident plane wave from port is perpendicular to the structure plane. The direction of electric component of incident electromagnetic wave is parallel to x-axis as that of the magnetic component parallel to y-axis. The transmission parameter S_{21} and the reflection parameter S_{11} are used to describe the transmission and reflection characteristics of the entire model structure. The absorption of the terahertz electromagnetic wave is given by $A = 1 - |T| - |R|$.

On the other hand, we have simulated the absorption of the different components of the metamaterial interacting with the terahertz electromagnetic wave. As it shows in Figure 2, the dominant absorptivity of the structure lies in the split resonant rings of which the absorption curve is highly coincident with the unit structure.

3. Discussion

We designed the metamaterial structure with perfect absorption characteristics in terahertz region. The absorpiton spectra in Figure 3 reveal resonance peak with the absorption efficiency up to 96% at the frequency of 4.03 THz, which could be potentially applied for the subsequent development of metamaterial devices. It's worth noting that Figure 3 also illustrate two drastic changing absorption peak at 3.26 THz and 4.44 THz. To figure out the nature of these two peaks, we extract the distribution of the electric field in the x-direction and shown in Figure 4. It can be seen that the enhancement effect of the electric field mainly concentrates on the gaps of the split of the SRRs. It is well known that the split resonant ring structure can be equivalent to an inductance-capacitance (LC) circuit, and the equivalent inductance and capacitance have a great influence on the resonant frequency.

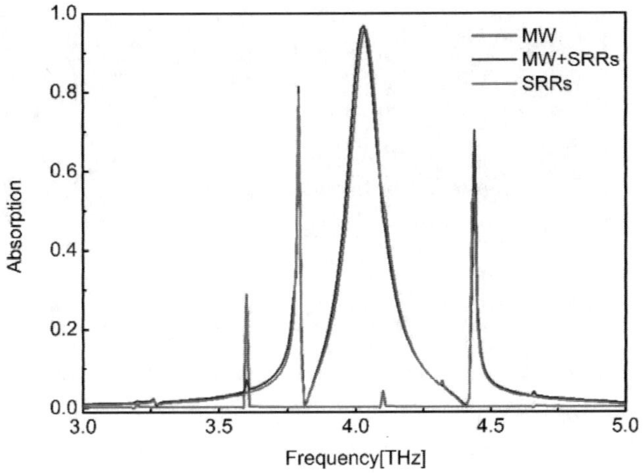

Figure 2. The absorption of different components of the unit matamaterial structure

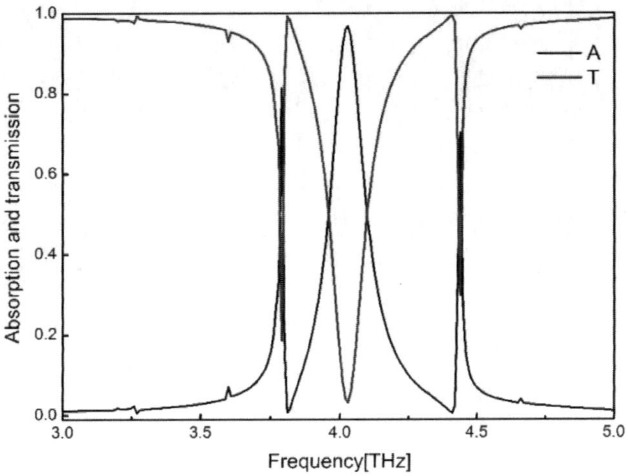

Figure 3. The absorption and transmission curve of the unit structure reacting with terahertz electromagnetic wave.

The distribution of current density inside the split resonant ring at 4.03 and 4.44 THz are displayed in Figure 5. The relatively weak current density portion of SRRs accompanied with enhanced electric field in adjacent dielectric is shown in Figure 5. It is easy to make it clear that the absorption peak at 4.03 THz is due to resonance and the absorption peaks at 4.44 THz and 3.26 THz have the same mechanism. It is remarkable that the fluctuation of the charges density distribution leads to the generation of singular absorption peaks and its physical mechanism is consistent with the main peak. On the other hand, the full width at half maximum（FWHM）of this metamaterial device is 0.15 THz which is adequate to practical application.

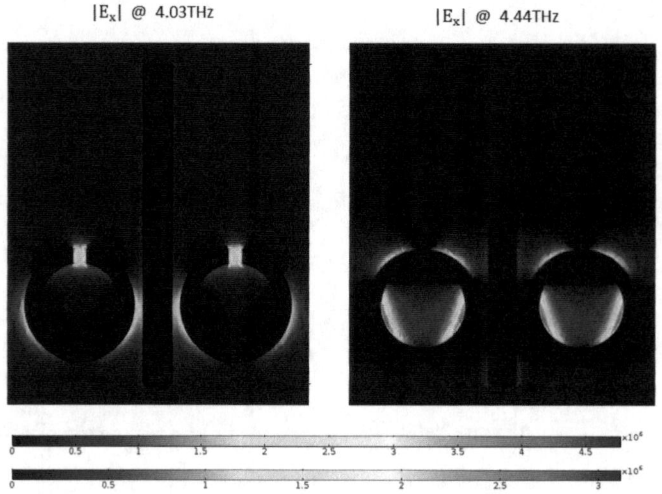

Figure 4. The distribution of electric field component and the upper legend corresponds to the electric field distribution at 4.03THz.

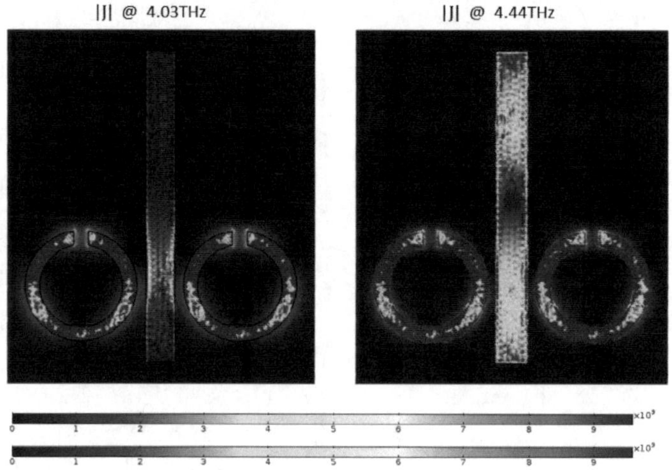

Figure 5. The distribution of current density and the upper legend corresponds to the current density at 4.03THz.

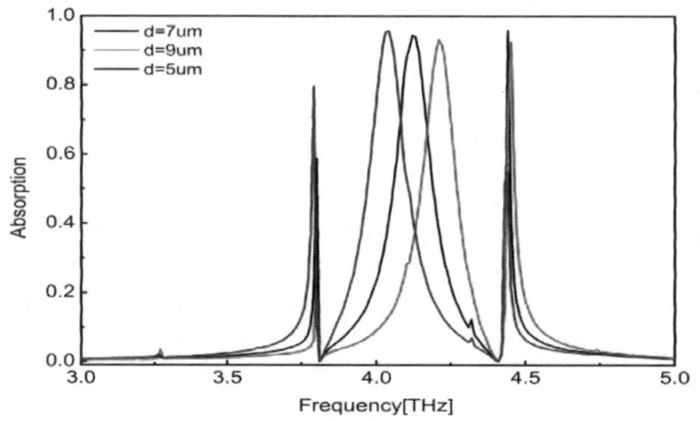

Figure 6.The influence of different split width of SRRs on the absorptivity of the unit metamaterial structure

As mentioned, the split resonant rings has a decisive effect on the absorption of electromagnetic waves. To further study the absorption characteristics of the split resonant rings, we simulate the effect of changing the width of the split of SRRs. The results are shown in Figure 6. As the split width increased, the resonant absorption becomes smaller and blue shift of resonant frequency occurs. The peak value of the singular peak on the left side descends and the peak value of the singular peak on the right increases. The split width of split resonant rings can be equivalent to the capacitance of the LC circuit as described above, so the change of split width affect the capacity of equivalent capacitor and the blue-shift occurs in addition with the maximum of absorption efficiency declining. However, another method with the same effect is to alter the inner and outer diameters of the split resonant rings[11].

4. Conclusions

In conclusion, we have proposed a THz metamaterial absorber could be applied for perfect absorption in THz region and be capable of wrapped around objects due to flexible polyimide dielectric substrates. The absorption intensity reaches 96% at 4.03 THz. We further explored the effects of the different components of the metal structure on the absorption and found that the metal split resonant rings had a decisive influence on the absorption. The distributions of electric field component are used to illustrate the nature of singular peak at 3.26 THz and 4.44 THz. The absorption peak can be adjusted by changing the structure of the metamaterial. This perfect flexible metamaterial absorber will be applied as candidate materials for a functional THz thermal detector or imaging application [12-15].

Acknowledgements

This study was supported by National Natural Science Foundation of China (51302196) and State Key Laboratory for Hubei New Textile Materials and Advanced Processing Technology (ZDSYS201711).

References

[1] Bingham C M, Tao H, Landy N I, Averitt R D, Padilla W J and Zhang X 2008 A metamaterial absorber for the terahertz regime: Design, fabrication and characterization *Opt. Express* 16 7181.
[2] Grant J, Escorcia-Carranza I, Li C, Mccrindle I J H, Gough J and Cumming D R S 2013 A monolithic resonant terahertz sensor element comprising a metamaterial absorber and micro - bolometer *Laser & Photonics Reviews* **7** 1043-8.
[3] Cheng Y, Gong R and Cheng Z 2016 A photoexcited broadband switchable metamaterial absorber with polarization-insensitive and wide-angle absorption for terahertz waves *Opt. Commun.* **361** 41-6.
[4] Li E P, Lim H C, Federici J F, Hor Y L and Szabó Z 2010 Terahertz response of microfluidic-jetted three-dimensional flexible metamaterials *Appl. Opt.* **49** 1179.
[5] Tao H, Bingham C M, Pilon D, Fan K, Strikwerda A C, Shrekenhamer D, Padilla W J, Zhang X and Averitt R D 2010 A dual band terahertz metamaterial absorber *Journal of Physics D Applied Physics* **43** 225102.
[6] Tao H, Bingham C M, Strikwerda A C, Pilon D, Shrekenhamer D, Y N I L, Fan K, Zhang X, Padilla W J and Averitt R D 2008 Highly flexible wide angle of incidence terahertz metamaterial absorber: Design, fabrication, and characterization *Physics* **78** 1879-82.
[7] Chen H, Wu B I, Zhang B and Kong J A 2007 Electromagnetic wave interactions with a metamaterial cloak *Phys. Rev. Lett.* **99** 063903.
[8] Wang B X, Wang L L, Wang G Z, Huang W Q, Li X F and Zhai X 2014 A broadband, polarisation-insensitive and wide-angle coplanar terahertz metamaterial absorber *The European Physical Journal B* **87**.
[9] Yao J Q 2017 Solid analyte and aqueous solutions sensing based on a flexible terahertz dual-band metamaterial absorber *Optical Engineering* **56** 027104.

IOP Conf. Series: Materials Science and Engineering **281** (2017) 012029 doi:10.1088/1757-899X/281/1/012029

[10] Wang B-X, Wang L-L, Wang G-Z, Huang W-Q, Li X-F and Zhai X 2013 A simple design of ultra-broadband and polarization insensitive terahertz metamaterial absorber *Appl. Phys. A* **115** 1187-92.

[11] Ye Q, Liu Y, Lin H, Li M and Yang H 2012 Multi-band metamaterial absorber made of multi-gap SRRs structure *Appl. Phys. A* **107** 155-60.

[12] Grant J, Ma Y, Saha S, Khalid A and Cumming D R 2011 Polarization insensitive, broadband terahertz metamaterial absorber *Opt. Lett.* **36** 3476-8.

[13] Ma Y, Chen Q, Grant J, Saha S C, Khalid A and Cumming D R S 2011 A terahertz polarization insensitive dual band metamaterial absorber *Opt. Lett.* **36** 945.

[14] Withayachumnankul W, Shah C M, Fumeaux C, Ung B S Y, Padilla W J, Bhaskaran M, Abbott D and Sriram S 2014 Plasmonic Resonance toward Terahertz Perfect Absorbers *Acs Photonics* **1** 625-30.

[15] Cong L, Tan S, Yahiaoui R, Yan F, Zhang W and Singh R 2015 Experimental demonstration of ultrasensitive sensing with terahertz metamaterial absorbers: A comparison with the metasurfaces *Appl. Phys. Lett.* **106** 26

IOP Conf. Series: Materials Science and Engineering **281** (2017) 012030 doi:10.1088/1757-899X/281/1/012030

Machining of Molybdenum by EDM-EP and EDC Processes

K L Wu[1,*], H J Chen[2], H M Lee[3] and J S Lo[4]

[1]Department of Mechanical Engineering, Tungnan University, New Taipei City 222, Taiwan.
[2]Department of Mechanical Engineering, National Central University, Jhong-Li City 320, Taiwan.
[3] Department of Mechanical Engineering, Army Academy, Taoyuan City 32093, Taiwan.
[4]Department of Mechanical Engineering ,National United University, Miaoli 36003, Taiwan.

E-mail: klwu@mail.tnu.edu.tw

Abstract. Molybdenum metal (Mo) can be machined with conventional tools and equipment, however, its refractory propertytends to chip when being machined. In this study, the nonconventional processes of electrical discharge machining (EDM) and electro-polishing (EP) have been conducted to investigate the machining of Mo metal and fabrication of Mo grid. Satisfactory surface quality was obtained using appropriate EDM parameters of $I_p \leqq 3A$ and $T_{on} < 80\mu s$ at a constant pulse interval of 100μs. The finished Mometal has accomplished by selecting appropriate EP parameters such as electrolyte flow rate of 0.42m/s under EP voltage of 50V and flush time of 20 sec to remove the recast layer and craters on the surface of Mo metal. The surface roughness of machined Mo metal can be improved from Ra of 0.93μm (R_{max} = 8.51μm) to 0.23μm (R_{max}=1.48μm).Machined Mo metal surface, when used as grid component in electron gun, needs to be modified by coating materials with high work function, such as silicon carbide (SiC). The main purpose of this study is to explore the electrical discharge coating (EDC) process for coating the SiC layer on EDMed Mo metal. Experimental results proved that the appropriate parameters of I_p = 5A and T_{on}=50μs at T_{off} =10μs can obtain the deposit with about 60μm thickness. The major phase of deposit on machined Mo surface was SiC ceramic, while the minor phases included $MoSi_2$ and/orSiO_2 with the presence of free Si due to improper discharging parameters and the use of silicone oil as the dielectric fluid.

1. Introduction

Molybdenum (Mo) metal is a refractory metal with characteristics of high melting point, great strength, good thermal/electrical conductivity and low thermal expansion coefficient. It has been widely applied in electrical and electronic devices, material processing, glass manufacturing, high-temperature furnaces, aerospace and defence applications. Although Mo metal and its alloys can be machined by common metal removal processes, its superior mechanical properties pose difficulty to high-quality surface polishing due to the very high temperature reached by the cutting edges of machining tool during processing.

Electrical discharge machining (EDM) is an electro-thermal process which removes materials by melting and vaporization without direct contact with the workpiece surface. Hence, EDM can be employed to process difficult-to-machine materials, high-strength temperature-resistant alloys with complicated geometrics [1-3]. However, care must be taken when EDMing molybdenumand its alloys because the surface zone frequently contains a recast layer. This structure is susceptible tomicro-cracking and needs improving by mechanical or chemical polishing prior to placing the part in service [4]. A technique of electro-chemical machining (ECM) named electro-polishing Electro-polishing (EP) is one of the surface treatment processes which can be employed to finish the machined Mo metal.

The EP process involves an anodic dissolution or electrochemical reaction between workpiece and electrode. It improves the machined surface to microscale or nanoscale smoothness through leveling and brightening. The proposed mechanisms include mass transfer mechanism, current distribution, shape control, and passive oxide film behaviours [5]. According to the partial inter-granular corrosion theory, Lee et al. investigated the corrosion resistance properties of 316L stainless steel and proved that EP can enhance corrosion resistance significantly [6]. Hocheng et al. examined EP parameters using different electrode feeding rates, rotating electrodes and disks of various shapes, and explored new applications of EP [7-8]. For decades, machined Mo metal has been applied in vacuum devices [9]. It can lead to hot electron emission from the Mo grid and destroy the grid cut-off characteristics of the advanced microwave tube, rendering it unstable [10]. To alleviate this undesirable effect, the Mo grid needs to be coated with materials of high work function, such as silicon carbide (SiC) (φ=4.4eV) [11-12]. The objective of this study is to investigate the surface quality of Mo metal by using EDM, EP and EDC process separately and to obtain the appropriate combination of their working parameters.

2. Experimental setup and procedure

The EDM/EDC processes were carried out using the conventional Die-EDM equipment shown in Figure 1(a) In EDM, a pure copper electrode was employed to machine 99.9% Mo metal in kerosene; while in EDC, an artificial graphite electrode (IG-11, Toyo Tanso, Japan) was utilized to coat the Mo metal surface in silicone oil.

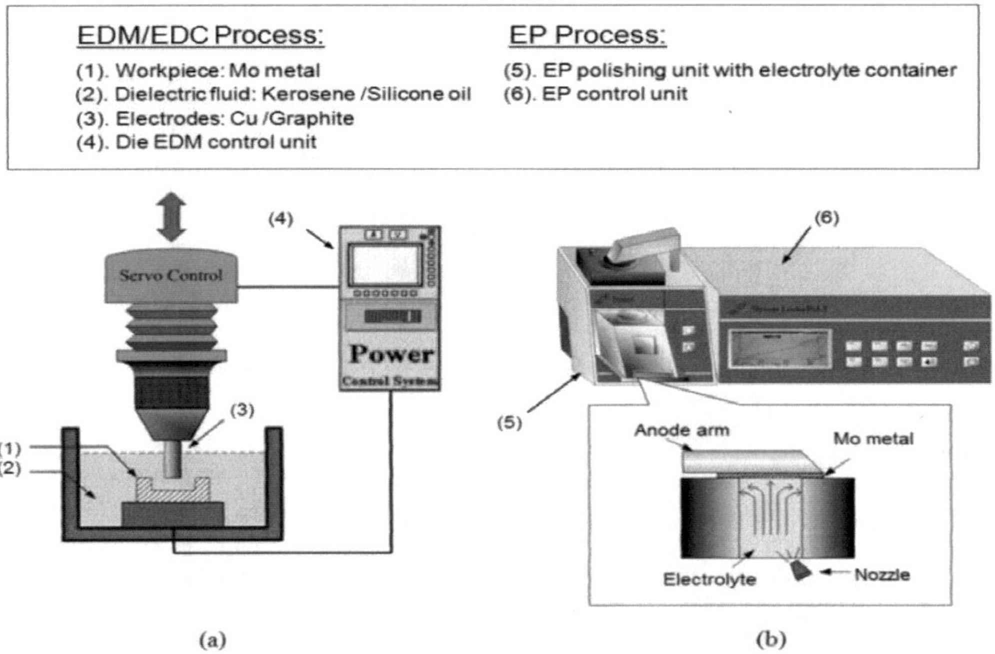

Figure 1 Schematics of (a) EDM/EDC and (b) EP devices.

Table 1 lists the properties of the materials used in this study. The material removal rate (MRR) of the Mo metal and the tool wear rate (TWR) of electrodes before and after processing were measured using a fine balance (accuracy:0.0001g).A mixed signal oscilloscope (DL-1200A, Yokogawa, Japan) captured the discharge waveforms during the EDM process. The EP treatment was operated in a commercial machine (LectroPol-5, Struers™, Denmark) as shown in Figure 1(b). The commercial electrolyte (A3,Struers™) made up of methanol (60.36vol%), 2-butoxyethanol (55.49vol%), perchloric acid (4.51 vol%) and H_2O (3.00 vol%) was used. During polishing, the electrolyte is pumped onto the surface of the workpiece at a specific flow rate and voltage. The anodic polarization (V-I) curve was monitored by a potentiostat (Parstat 2263). The morphologies and elemental distributions of the machined surfaces were examined by SEM/EDS (JSM-6360LV, Hitachi, Japan). The deposit on Mo metal wasanalyzed by an X-ray diffractometer (XRD-6000 Shimadzu, Japan) to verify the phases under various treatment conditions. The surface roughness(Ra) of machined surface was measured by using a contact type stylus based surface roughness tester, 2D Surfcoder (E35A, Tokyo Seimitsu,Japan).

Table 1 Properties of materials used in EDM/EDC processes.

Material	Density (g/cm³)	Melting Point (℃)	Boiling Point (℃)	Thermal Conductivity (W/m°K)	Specific Heat (J/kg°K)	Tensile Strength (MPa)	Young's Modulus (GPa)
Mo	10.22	2617	4612	138	255	415	330
Cu	8.96	1083	2567	391	385	215	110
Graphite	1.90	3852	4250	139	713	29.4	12
Dielectric Fluid	**Molecular Formula**	**Density (g/cm³)**	**Flash Point (℃)**	**Thermal Conductivity (W/m°K)**	**Dielectric Breakdown (MV/m)**	**Viscosity (mm²/s)**	**Surface Tension (mN/m)**
Kerosene	$C_{12}H_{26}$	0.78~0.81	101	0.15	20	2.4	28
Silicone Oil	$(C_2H_6OSi)_n$	0.96	316	0.10	10~15	1000	21.2

3. Result and discussion

3.1. EDM of Mo metal

The electrode polarity, discharge current (I_p) and pulse duration (T_{on}) of EDM parameters affect the machining quality soundly. Figure 2 displays the effect of polarity of electrode on MRR and TWR of Mo metal under constant I_p=0.4A and T_{on}= 100μs. As seen in Figure 2(a), the copper electrode with positive polarity (i.e., acathodicworkpiece) achieved high MRR while maintaining low TWR. In other words, when using an anodic electrode, the discharge energy can be effectively employed for removal of Mo metal, while the carbon coating on the electrode surface offers protection against rapid electrode wear [4]. However, the opposite was observed when using an electrode with negative polarity. As seen in Figure 2(b), the MRR of workpiece was much reduced as a result of the carbon layer protection, while the TWR of electrode increased more than three folds as compared with that for an anodic electrode, shown in Figure 2(a). Sufficient discharge energy favored the dissociation of kerosene to produce cathodic carbon ions. Thus, anodic workpiece would attract the carbon ion to form carbon layer on the surface of workpiece during the electrical discharge. Figure 3 compares surface roughness on EDMed Mo metal obtained using electrodes of positive and negative polarity. As can be seen, lower surface roughness of machined Mo metal was obtained by a negative electrode (i.e., an anodic workpiece) while a coarse machined Mo surface was obtained by a positive electrode (i.e., a cathodicworkpiece).

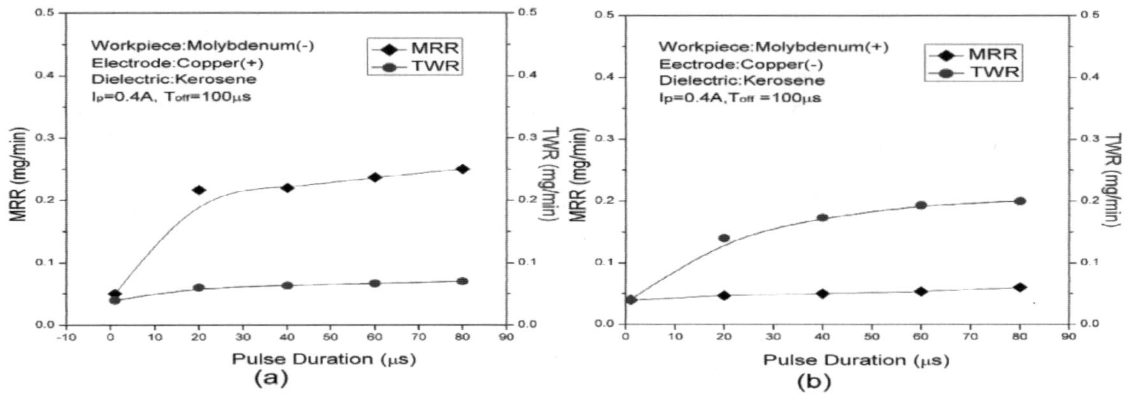

Figure 2 Effect of (a) positive and (b) negative polarity of electrode on MRR/TWR of Mo metal.

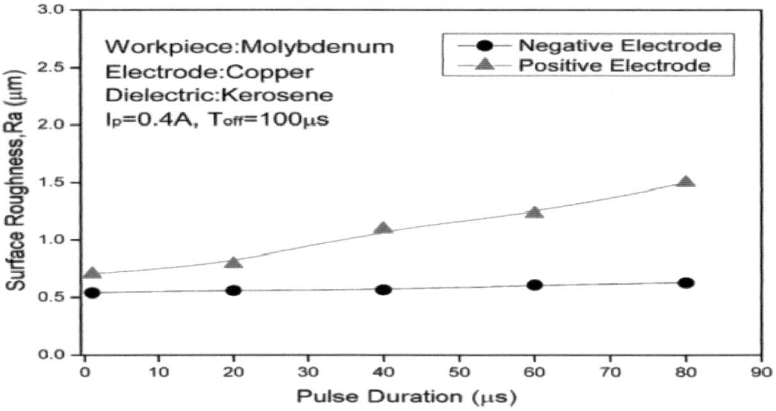

Figure 3 Effect of electrode polarity on surface roughness of machined Mo metal.

In order to obtain an acceptable surface quality of Mo metal via EDM process, assigned the polarity of electrode as negative. However, in such case, the discharge parameters need to be modulated to reduce electrode wear. Figure 4 displays the MRR and TWR obtained under various discharge currents (I_p= 0.4-9A) and pulse durations (T_{on}=1-240μs). As can be seen, when I_p<6A, MRR was always lower than TWR (cf. Figure 4 (a) and (b)), revealing that low discharge energy yielded poor material efficiency but caused rapid wearing of the electrode. It was ascribed to the high melting point of Mo metal protected by carbon layer. When $I_p \geq 6$A and in the case of $T_{on} \leq 80$μs, MRR and TWR showed identical trend of increase. In this stage, each discharge column continued expanding and exploding periodicly between electrode and workpiece. The weak bonding protective carbon layer on anodic workpiece was easily destroyed as a result of severe electron bombardment and thus resulted in the similar weight loss. On the other hand, when Ton>80μs, MRR declined slowly while TWR remained unchanged. Such changes can be attributed to the lower discharge energy density under long pulse duration which caused inefficient machining of Mo metal.

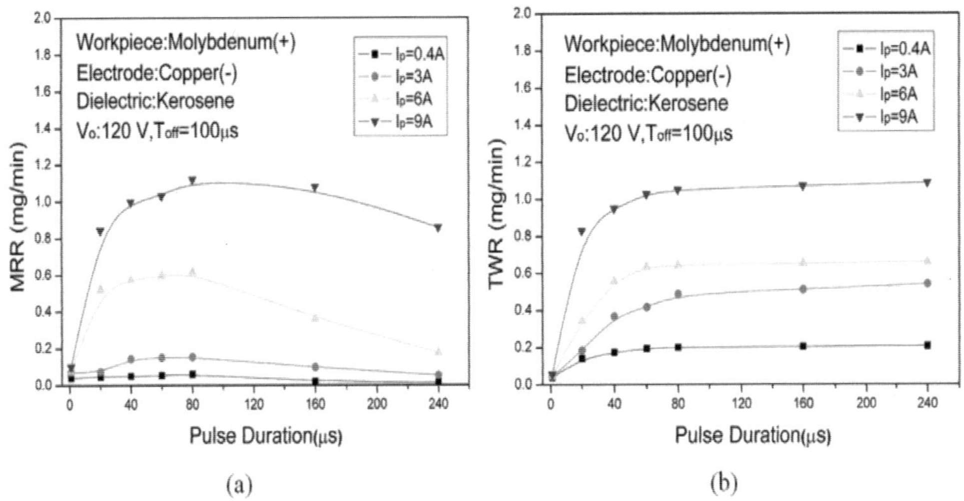

Figure 4 Effects of discharge current and pulse duration on MRR and TWR of Mo metal.

Figure 5 presents the variations of surface roughness of EDMed Mo metal under different discharge currents (Ip) and pulse durations (Ton). Although Ra at Ip = 0.4A was kept small for all Ton, the material removal efficiency was also low. On the other hand, high Ip (> 0.4A) yielded rough surface and long Ton (> 20 μ s) deteriorated the surface quality. Considering the machining time of refractory Mo metal with an acceptable surface quality, the EDM parameters combination of Ip≦3A and Ton = 20 μ s was adopted in the subsequent experiments.

Figure 6 displays the SEM images of Mo metal surface before and after EDM processes (Ip = 3A, Ton = 20 μ s). As can be seen, the surface roughness after EDM increased from Ra = 0.17 μ m to Ra = 0.93 μ m due to formation of the recast layer, craters and micro-cracks on the EDMed Mo metal, thus necessitating surface finishing.

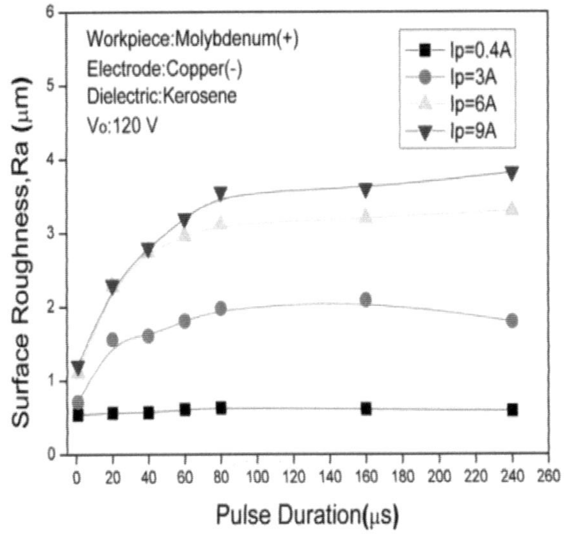

Figure 5 Variations in surface roughness of Mo EDMed under different I_p and T_{on}

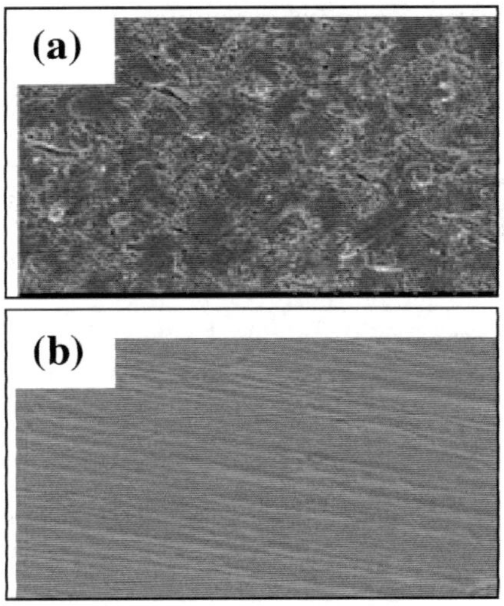

Figure 6 SEM images of Mo metal surface (a) before and (b) after EDM.

3.2. EP of EDMed Mo metal

Figure 7 shows how surface roughness variations of machined Mo metal by EP voltage. As can be seen, surface roughness of Mo metal EDMed was apparently improved from Ra = 0.93μm to a minimum value of0.36μmwhen EP voltage=50V. Meanwhile, it showed that R_{max} on EPed Mo metal dropped rapidly from 8.5μm to 5.3μm when EP voltage =10V. During EP, discharging will occur at the sharpest point of crater on the EDMed surface, which results in a levelling effect. As a result, the rugged surface then becomes smooth and the dissociation of material will slow down. Therefore, the minimum Ra of EMDed Mo metal under flush time of 10 sec and electrolyte flow rate of 0.42 m/s can be obtained when EP voltage = 50V. During EP process, bubbles formed at the anode due to dissociation, which increased the resistance of electrolyte, resulting in pitting on the workpiece surface. The stagnant electrolyte will further accelerate this phenomenon. Therefore, flowing of electrolyte is essential for achieving a good-quality EP surface. In this study, the flowing rate of electrolyte ranged from 0.42 to 1.05m/s.

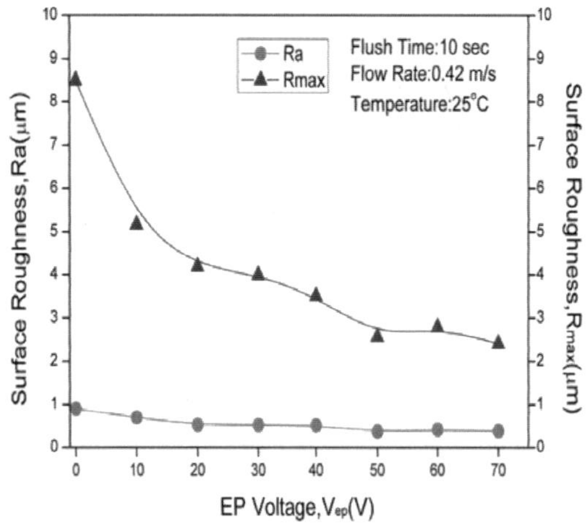

Figure 7 Effect of EP voltage on EDMed Mo metal surface roughness.

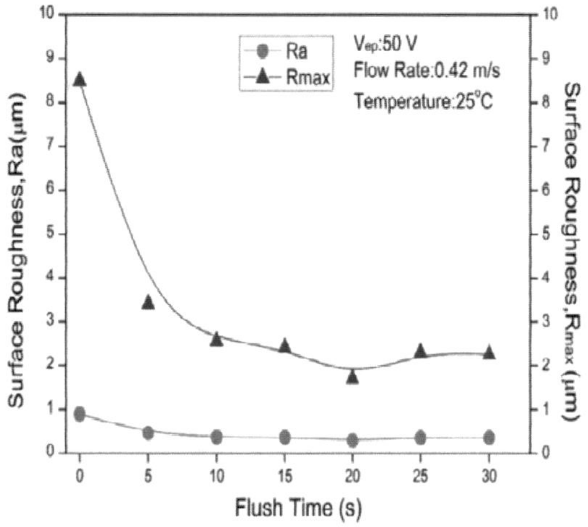

Figure 8 Effect of EP flush time on surface roughness of EDMed Mo metal.

Figure 8 illustrates the improvement in surface roughness under EP with different flush durations. As can be seen, the average surface roughness was reduced to 0.46μm (R_{max} = 3.04μm) after 5sec and approached the minimum value of 0.29μm (R_{max} = 1.70μm) when flush time reached 20sec. As the EDC process would further modify the EPed Mo, the interactive influences of EP voltage, flush time and electrolyte flow rate were ignored in this stage and hence chose flush time of 20sec under EP voltage = 50V and electrolyte flow rate = 0.42m/s for further investigation.

3.3. EDC of Mo metal

The graphite and silicone oil adopted instead of copper and kerosene, respectively because their ion sources of carbon and silicon can confer a SiC coating on the EDMed Mo metal surface. In addition, during the EDM/EDC process, the higher the discharge current, the larger the discharge energy and explosive force will be, thus yielding a coarse machined surface. Hence, to explore the effect of discharge current and pulse duration, surface roughness of the coated Mo metal obtained under EDC parameters I_p=1-13A, T_{on} = 1-200μs, T_{off}=10 and 50μs were examined.

Figure 9 shows the MRR of Mo workpiece and TWR of graphite electrode under EDC with different discharge currents and pulse durations. When I_p>1A and T_{on}> 10μs, the MRR and TWR were negative, indicating that both workpiece and electrode gained, instead of lost, weight when discharge energy increased. The increase in their net weights is not only due to the polarity effect

(TWR>MRR), but also contributed by the layer of surface deposit made up of ions from graphite, silicone oil and Mo metal during electrical discharge.

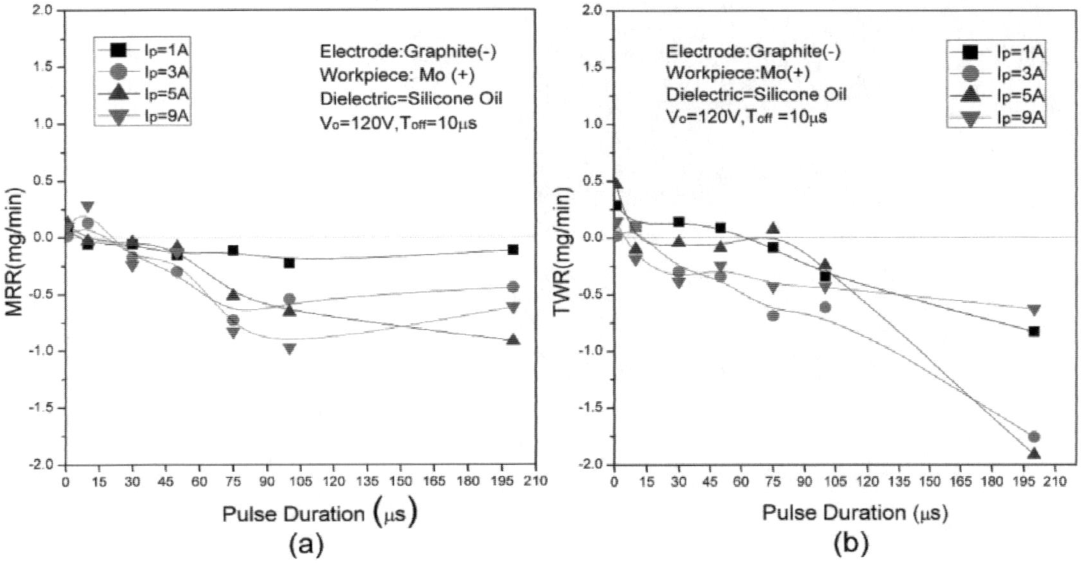

Figure 9 Effects of Ip and Ton on MRR and TWR under EDC.

Figure 10 shows the surface roughness of various EDCed Mo metal. As can be seen, surface roughness on EDCed Mo varied with discharge current and pulse duration. The higher the I_p and T_{on} of EDC process, the coarser the EDCed Mo metal. Between 50μs and 100 μs, the Ra reached the maximum for different discharge currents. Thereafter, Ra declined with further increase in energy density. According to these results, appropriate EDC parameters should be low I_p and long T_{on}.

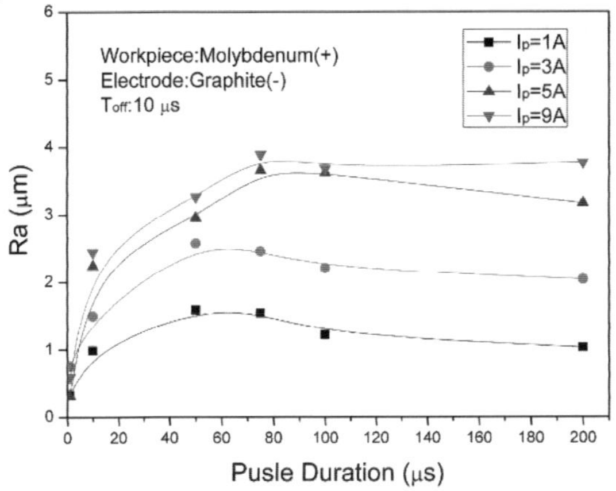

Figure 10 Effects of I_p and T_{on} on surface roughness of EDCed Mo metal.

Figure 11 SEM images of Mometal after EDC process.

Figure 11 displays the SEM images of Mo surface after EDC process (I_p=5A, T_{on}=200μs,T_{off}=10μs). As can be seen, the EDCed Mo metal surface has a fine grain texture with average surface roughness of 3.17 μm (R_{max}=4.16μm). XRD proves that the layer is SiC phase with minor silica. Figure 12 demonstrates the cross-sectional view of EDCed Mo metal. The thickness was approximately 60μm as shown in Figure 12(a). Results of EDS analyses on the elemental distribution of the coating deposited on Mo metal, as indicated in Figure 12(b), confirmed that the coating comprised mainly Si and C with few oxygen ion soriginated from graphite and silicone oil, respectively. Nevertheless, Mo did not found on the surface of the coating layer.

IOP Conf. Series: Materials Science and Engineering **281** (2017) 012030 doi:10.1088/1757-899X/281/1/012030

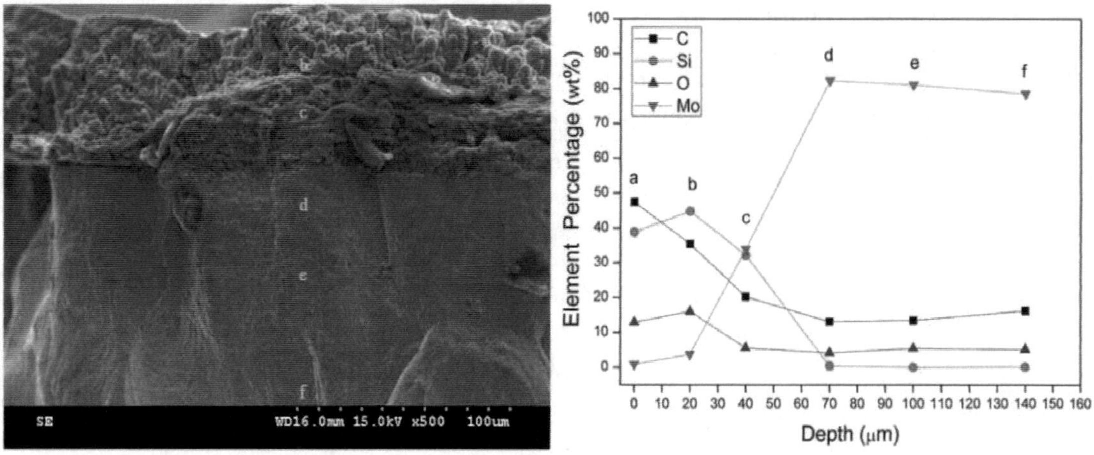

(a) (b)

Figure 12 (a) SEM image of cross-sectional view of EDCed Mo metal and

(b) elemental distribution by EDS analyses

4. Conclusions

The present study demonstrates that the refractory Mo metal has successfully machined by EDM, finished by EP and surface modified by EDC processes. According to the results obtained, the following conclusions include as follows:

1) Machining of Mo requires high energy density due to its high melting point. However, the polarity of electrode and workpiece can alter the quality of machining. For better-machined surface roughness and lower electrode wear rate, workpiece of positive polarity (+) and low discharge current should be used.
2) The appropriate EDM discharge current (I_p) and pulse duration (T_{on}) for machining anodic Mo metal should be ≦3 A and <80μs, respectively at T_{off} =100μs.
3) EP of EDMed Mo metal undervoltage=50V, flush time=20sec and electrolyte flowing rate=0.42m/s improved surface roughness from Ra =0.93μm (R_{max}= 8.51μm) to Ra = 0.23μm (R_{max}= 1.48μm), evidencing the effectiveness of EP in enhancing surface quality.
4) The EDMed Mo metal surface has successfully modified by EDC process using graphite as electrode and silicone oil as dielectric fluid. The appropriate parameters are I_p =5A and T_{on} =50μs at T_{off} =10μs. The deposit on the EDCed surface comprised mainly the SiC ceramic phase.

References

[1] SortinoM, TotisG andProsperiF2013 *J. Mater. Process. Technol.* 2131179
[2] OjhaK, GargRK and SinghKK2010*J. Miner. Mater.Charact.Eng.*9(8)709
[3] AbbasMM, SolomonD Gand BahariMF 2007 *Int. J. Mach.Tool. Manuf.*47 (7-8)1214
[4] Demellayer R and RichardJ2013*Procedia CIRP*689
[5] Rokicki R and Hryniewicz T 2012*Trans. Inst. Met. Finish.*90(4)188
[6] Lee SJ andLaiJJ2003*J. Mater. Process. Technol.*140206
[7] Hocheng H and PaPS2003*J. Mater. Process. Technol.*120 6
[8] Hocheng Hand PaPS.2003*J. Mater. Process. Technol.* 142 203
[9] Lawrence R. 2004*IEEE Trans. Plasma Sci.*32(3)1277
[10] JiangJ, JiangB, Ren C, Zhang F, Feng T and Wang X 2006*Vac.*80537
[11] Tang H Tan S, Huang Z, Dong S and Jiang D2005*Surf. Coat. Technol.* 197161
[12] Wang J, Wu XM and Zhuge LJ 2007*Vac.*81890

IOP Conf. Series: Materials Science and Engineering **281** (2017) 012031 doi:10.1088/1757-899X/281/1/012031

The Effect of High Frequency Pulse on the Discharge Probability in Micro EDM

Y Liu[1], Y Qu[1], W Zhang[1], F Ma[1], Z Sha[1], Y Wang[2], B Rolfe[2] and S Zhang[1]*

1School of Mechanical Engineering, Dalian Jiaotong University, Dalian 116028, China
2School of Engineering, Deakin University, Geelong VIC3220, Australia

Email:zsf@djtu.edu.cn

Abstract. High frequency pulse improves the machining efficiency of micro electric discharge machining (micro EDM), while it also brings some changes in micro EDM process. This paper focuses on the influence of skin-effect under the high frequency pulse on energy distribution and transmission in micro EDM, based on which, the rules of discharge probability of electrode end face are also analysed. On the basis of the electrical discharge process under the condition of high frequency pulse in micro EDM, COMSOL Multiphysics software is used to establish energy transmission model in micro electrode. The discharge energy distribution and transmission within tool electrode under different pulse frequencies, electrical currents, and permeability situation are studied in order to get the distribution pattern of current density and electric field intensity in the electrode end face under the influence of electrical parameters change. The electric field intensity distribution is regarded as the influencing parameter of discharge probability on the electrode end. Finally, MATLAB is used to fit the curve and obtain the distribution of discharge probability of electrode end face.

1. Introduction

Micro EDM introduces the principle of conventional EDM to the field of micromachining by reducing the energy of discharge machining and servo control. Short pulse duration and high frequency pulse power supply is used in micro EDM to ensure a small amount of energy of single pulse processing to achieve the machining accuracy and in the meanwhile improve the efficiency of processing. As micro EDM is significantly influenced by the electrical parameters, the process of machining is bound to be influenced by the impact mechanism from the high frequency pulse power supply. M T Yan et al.[1] presented the development of a high frequency power supply for surface quality improvement of wire electrical discharge machining (wire-EDM). A novel fixed pulse-width modulation pulse control method was proposed to generate high frequency and short-duration pulse control signals. X Li et al.[2] studied the wear mechanism of Ni-TiN/Cu composite electrode in the case of high frequency pulse current, and the influence of the fluctuation frequency of discharge current on electrode wear in micro-EDM was found out. D K Chung et al.[3] developed a new high frequency bipolar pulse generator for micro EDM in order to prevent electrolytic corrosion. Using the new pulse generator, micro holes without electrolytic corrosion were successfully fabricated in deionized water and tap water. R Casanueva et al.[4] found a new application of the resonant converter in EDM power supply, which is designed for the purpose of developing small size EDM systems. By tuning the conversion frequency to the natural resonant frequency an optimum and stable operation using tap water as dielectric fluid was achieved preventing the generation of undesired impulses and keeping the distance between the electrode and the workpiece within the optimum stable range. Q Liu et al.[5] discussed the influences of tool electrode diameter on the micro-EDM process based on the skin-effect and area effect. It is demonstrated that the machining speed, electrode wear, and taper rate are different with the increase of tool electrode diameter. Y W Yong et al.[6] experimentally verified the electromagnetic properties of plasma under the condition of skin-effect, which was used to study the effect of electromagnetic field on machining mechanism in discharge channel of EDM. B Schacht et al.[7] studied the effect of pulse frequency on resistance of the wire electrode under the skin-effect of wired EDM and explained the importance of the skin-effect to high frequency wired EDM. Y Liu et al.[8] based on the theory of skin-effect studied the principle of the electrode energy distribution of RC circuit and the variation law of discharge location, and achieved the EDM machining of freeform surfaces.

In this paper, the effect of pulse frequency, electrical current and permeability on the skin-effect under high frequency pulse and the skin-effect on discharge probability in the process of micro EDM are explored. The principle of skin-effect in high frequency electromagnetic field is applied to the analysis of micro EDM discharge process.

2. The skin-effect in high frequency pulse EDM

The skin-effect is a phenomenon that affects the energy distribution of electromagnetic fields caused by the frequency variation of the electromagnetic field, which is more significant in high frequency conditions. It is found that the strength of the skin-effect can be reflected by the skin depth. The expression for skin depth δ is:

$$\delta=\sqrt{\frac{2}{f\mu\sigma}} \tag{1}$$

where, f is frequency of pulse, μ is magnetic conductivity, and σ is electrical conductivity.

The influence of the skin-effect on micro EDM is mainly to change the distribution of electric energy within the electrode. Due to that the energy distribution directly affects the discharge probability of electrode discharge location, so the skin-effect indirectly affects the electrode discharge location. Skin-effect will promote the location of discharge distributing on the electrode edges, which leads to a fast consumption of the electrode edges in the machining process.

3. The micro electrode energy transfer model under high frequency pulse

3.1. Establishment of mathematical model

When the electrode is applied to a high frequency pulse current and the spark discharge occurs, a changing magnetic field whose direction is perpendicular to the current is generated inside and outside of the electrode. That the electric field and the magnetic field excite each other can be expressed by the Maxwell equation of the time-varying field [9].

The relationship between electric field strength E and electric potential V is:

$$E=-\nabla V \tag{2}$$

The relationship between transient current density J and electric field E is:

$$J=\sigma E+\varepsilon_0\varepsilon\frac{\partial E}{\partial t} \tag{3}$$

where, ε_0 is vacuum dielectric constant. The instantaneous variational current can generate induced magnetic field, which is shown as:

$$\begin{cases}\sigma\dfrac{\partial A}{\partial t}+\nabla\cdot(\mu^{-1}B)=J\\ B=\nabla\cdot A\end{cases} \tag{4}$$

where, A is the magnetic field vector. The strength of the induced electric field E_i is calculated according to the variation of magnetic field as:

$$E_i = -\frac{\partial A}{\partial t} \tag{5}$$

3.2. Establishment of the micro electrode energy transfer simulation model
The simulation software COMSOL Multiphysics is used to simulate the discharge process of EDM at a high frequency pulse. The simulation method of magnetic field and electric field coupling is used when modelling, and then the 3D steady state and frequency domain simulation is performed.

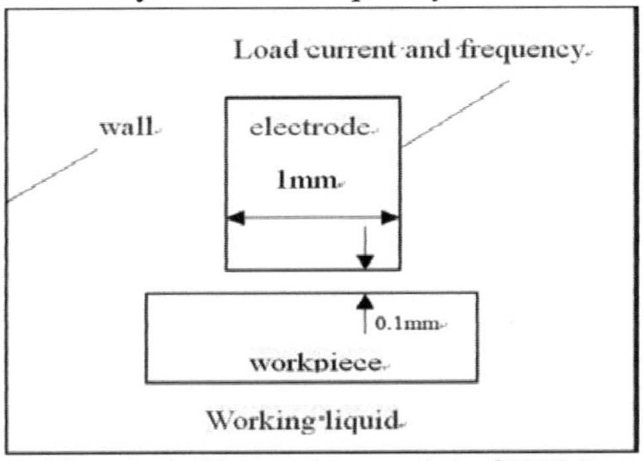

Figure 1. Simulation model of EDM

The micro electrode energy transfer model includes a 1mm diameter tool electrode, workpiece and the surrounding insulation working fluid, as shown in Figure 1. After loading pulse and current on the whole electrode, the electric field intensity between the electrode and the workpiece can be up to 1.2 x 10^7 V/m, which is pretty consistent with the actual EDM.

4. Simulation and result analysis
The simulation of energy distribution is carried out under different pulse frequency, electrical current and magnetic permeability with COMSOL Multiphysics simulation software. The simulation parameters in three groups are shown in Table 1.

Table 1. The simulation parameter list

	material	Pulse frequency(kHz)	electrical current(A)	Magnetic permeability(H/m)
1	copper	500,1000	20	1
2	copper	500	20,40	1
3	copper, iron	500	20	1,400

4.1. The simulation results and analysis of different pulse frequencies
Figure 2 shows the current density distribution of the copper electrode under the conditions as I = 20 A, f = 500 kHz and μ = 1 H/m. As shown in Figure 2, the current density distributes non-uniformly within the electrode. The electrode surface has the highest current density value, and current density value is the lowest in the electrode center. This is due to the skin-effect promoting the electrical energy transferring through electrode surface rather than the center.

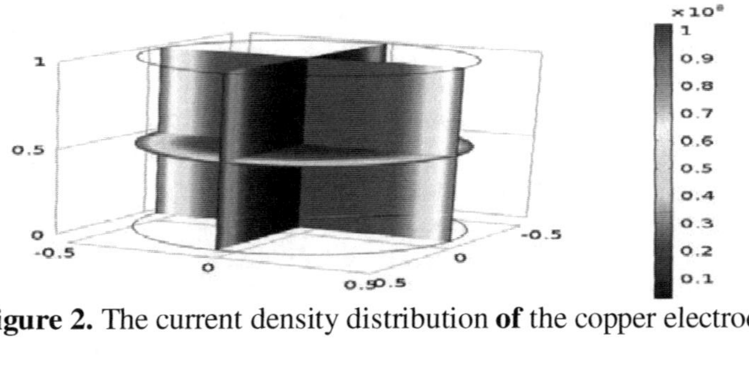

Figure 2. The current density distribution **of** the copper electrode

The current density value is derived from the simulation results based on different pulse frequency as 500 kHz and 1000 kHz, the comparison of current density distribution is shown in Figure 3. From the Equation (3), the simulation result can be post-reprocessed, and the electric field intensity distribution can be obtained under different pulse frequencies, as shown in Figure 4.

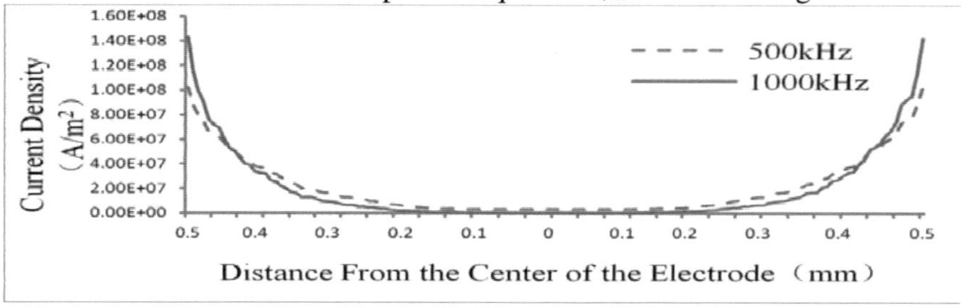

Figure 3. Current density comparison chart of the copper electrode at different pulse frequencies

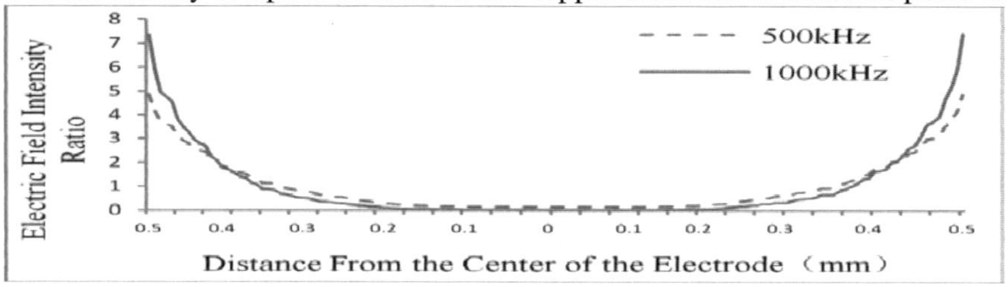

Figure 4. Electric field intensity comparison chart of the copper electrode at different pulse frequencies

From Figure 3 and Figure 4, it can be observed that the two curves of the graph resemble the shape of bathtub under the influence of the skin-effect, so they are named as "bathtub curve". As shown in Figure 3 and Figure 4, with the pulse frequency increasing, the electric field intensity and current density are more concentrated on the surface of the electrode, which causes the electric field intensity and current density of the interior of electrode reduced accordingly. Compared the two figures, it is found that there exist intersections of the two curves when the pulse frequency changes. Within a radius of 0.43mm, the current density and electric field intensity is lower when the pulse frequency is 1000kHz. While, in other regions the current density and electric field intensity is higher when the pulse frequency is 1000 kHz. The reason is that the higher the pulse frequency is, the more obvious the skin-effect phenomenon is, so the "bathtub curve" is steeper. Finally, it is concluded that as the pulse frequency increases, the distribution of current density and electric field intensity is more concentrated on the electrode surface. At the same time, the skin depth becomes small, which makes the skin-effect phenomenon more obvious.

4.2. The simulation results and analysis of different electrical current

The curves in Figure 5 show that with the double electrical current, the corresponding current density of both the electrode surface and the center is about two times that of the original, though the distribution pattern of the current density is unchanged. The reason is that the current has no effect on the distribution of current density in the electrode. But the larger the current is, the more energy is transmitted in the electrode. Therefore, the current density on the electrode surface is increased.

From Figure 6, it can be seen that when the current changes the ratio of electric field intensity at each point to the average electric field intensity doesn't change and the two curves are identical. This proves once again that the current change has no effect on the distribution of electric field intensity. But the Figure 5 shows that the current density value of the electrode surface is increased, which suggests that although the increase of current does not directly affect the ratio of the electric field intensity distribution, the increase of current density on electrode surface is bound to lead the phenomenon of skin-effect to be more apparent, which will affect the electrode wear of EDM.

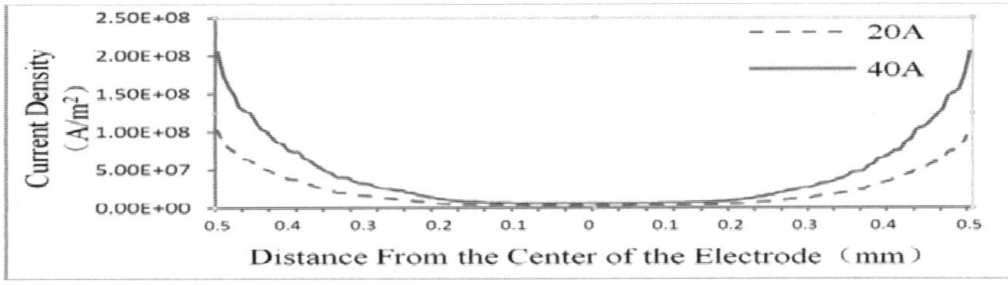

Figure 5. Current density comparison chart of the copper electrode at different electrical current

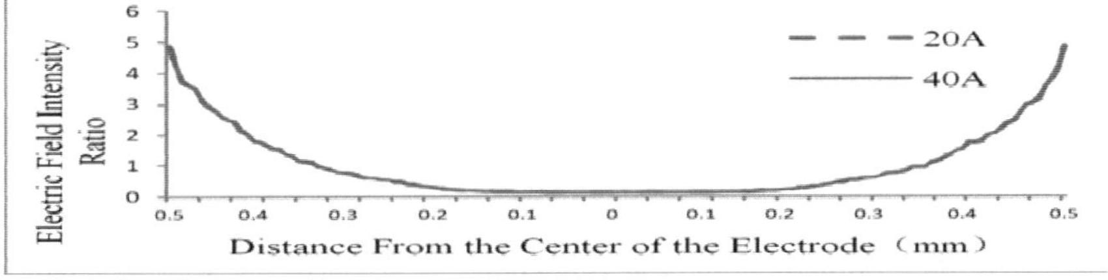

Figure 6. Electric field intensity comparison chart of the copper electrode at different electrical current

4.3. The simulation results and analysis of different magnetic permeability

By observing the Figure 7, the skin depth of the iron electrode is very small compared with the copper electrode. The reason is that the magnetic permeability of iron is 400 times more than that of copper, and the skin depth reduces with the increase of magnetic permeability. As a result, the skin depth shown in Figure 7 is much smaller than that in Figure 2.

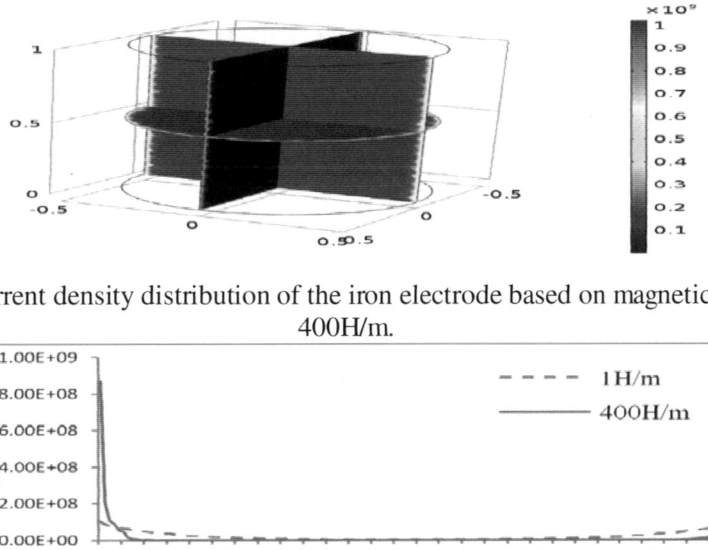

Figure 7. The current density distribution of the iron electrode based on magnetic permeability is 400H/m.

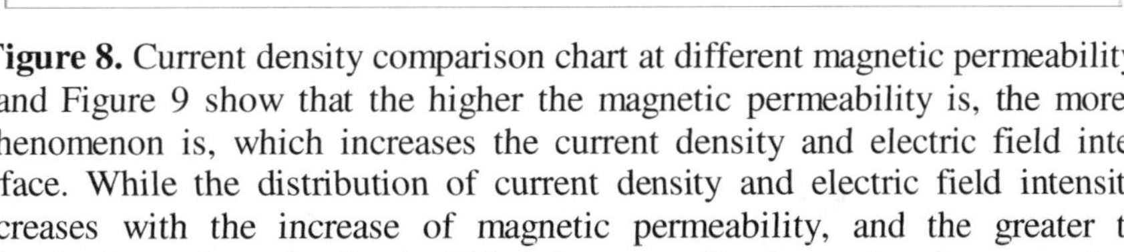

Figure 8. Current density comparison chart at different magnetic permeability

Figure 8 and Figure 9 show that the higher the magnetic permeability is, the more obvious the skin-effect phenomenon is, which increases the current density and electric field intensity on the electrode surface. While the distribution of current density and electric field intensity within the electrode decreases with the increase of magnetic permeability, and the greater the magnetic permeability is, the faster the reduction is. This is because after increasing the magnetic conductivity, under the skin-effect, the current is almost all gathered near the surface of the electrode, which causes the current density to decrease inside of the electrode. Meanwhile, the electric field intensity on the electrode surface increases with the sharp increase of current density on the surface of the electrode. It is concluded that as the magnetic permeability increases, skin-effect phenomenon of the electrode is strengthened obviously, which increases the current density and electric field intensity on the electrode surface, meanwhile decreases the current density and electric field intensity within the electrode.

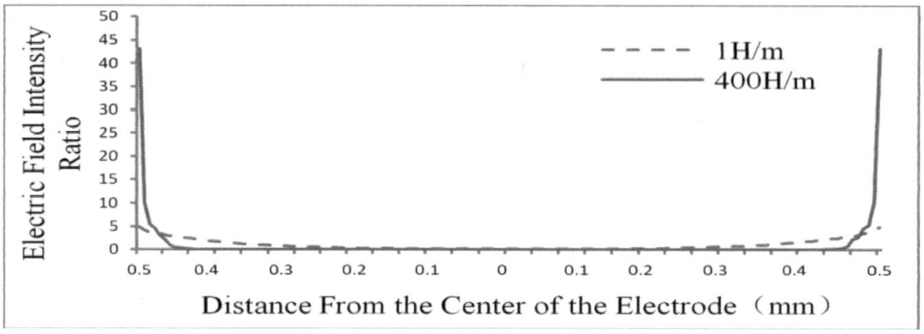

Figure 9. Electric field intensity comparison chart at different magnetic permeability

5. Calculation of the probability of electrode end face

5.1. The fitting of the electric field intensity distribution curve

The method of exponential function to curve fitting is used in MATLAB software to fit the left half of the electric field intensity distribution curve, which is obtained in the case of $I = 20$ A and $f = 500$ kHz with the copper electrode. Then the fitting curve and the curve equation can be obtained. The fitting equation of the two terms exponential function [10] is expressed as follows:

$$f(x) = A \cdot e^{Bx} + C \cdot e^{Dx} \tag{6}$$

where, A, B, C and D are coefficients. When A, B, C and D is 0.01272, -9.807, 0.03827 and 5.313 respectively, the fitting degree of is the highest.

After calculation, it can be found that the degree of fit is above 0.99. Therefore, the fitting curve can be used to represent the distribution of the electric field intensity for relevant analysis.

5.2. The normalization of the fitting curve

In order to obtain the function of discharge probability, the fitting curve should be normalized. In this paper, the fitting curve is treated by standard normalization method and the fitting curve of discharge probability is obtained as shown in Figure 10.

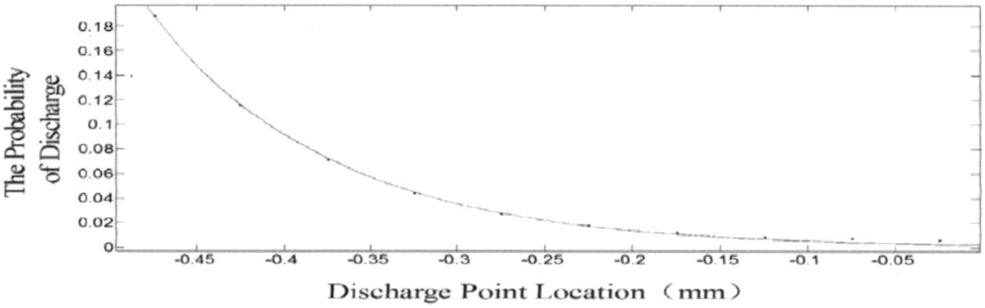

Figure 10. The graph of fitting curve of discharge probability

The equation of the discharge probability curve for the whole electrode is:

$$f(x) = 0.001779 \times e^{9.809|x|} + 0.004294 \times e^{-4.186|x|} (-0.5 \le x \le 0.5) \tag{7}$$

Discharge probability can be used for simulating random discharge of electrode end face. Combining with the electrode material removal in volume and multiple discharge iterations, the laws of electrode wear and shape change in EDM machining can be revealed.

IOP Conf. Series: Materials Science and Engineering **281** (2017) 012031 doi:10.1088/1757-899X/281/1/012031

6. Conclusions

In this paper, the distribution laws of electric field intensity and current density on the electrode end face are analyzed under the condition of different pulse frequency, electrical current and magnetic permeability. Based on the distribution curve of electric field intensity, the fitting curve and the curve equation in the direction of the diameter of electrode end face is obtained. Finally, the fitting curve is normalized and the probability of discharge is obtained. The conclusions are as follows:

(1) With the increasing of pulse frequency and magnetic permeability of electrode, the skin-effect phenomenon of the electrode is more obvious, which results in the current density and the electric field intensity more concentrated on electrode surface, also decrease the current density and electric field intensity inside of the electrode. With the increasing of electrical current, there are no effect on the distribution of current density and electric field intensity, but the values of current density and electric field intensity in the whole electrode are increased, which affects the discharge distribution on the surface of electrode end.

(2) The discharge probability at each point along the direction of diameter is symmetric about the center of electrode, and with the change of location, the discharge probability can be expressed by exponential function. The fitting curve and the curve equation of the electric field intensity distribution curve can be used to represent the actual distribution state of the electric field intensity, which can provide the theoretical basis for the further study of the electrode wear law.

Acknowledgments

The financial support from the National Natural Science Foundation of China under grant number 51405058, Scientific Research Platform Foundation of Liaoning Province under grant number JDL2016006 and Talent Special Foundation of Dalian City under grant number 2016RQ054 are acknowledged.

References

[1] Yan M T, and Liu Y T 2009 Design, analysis and experimental study of a high-frequency power supply for finish cut of wire-EDM J. *International Journal of Machine Tools & Manufacture* 49(10) pp 793-796
[2] Li X P, Wang Y G, Zhao F L, Hua M,W U and Liu Y 2014 Influence of high frequency pulse on electrode wear in micro-EDM J. *Defence Technology* 10(3) pp 316-320
[3] Chung D K, Hong S S, Min S P and Chong N C 2011 Machining characteristics of micro EDM in water using high frequency bipolar pulse J. *International Journal of Precision Engineering & Manufacturing* 12(2) pp 195-201
[4] Casanueva R, Azcondo F J, Branas C and Bracho S 2005 Analysis, design and experimental results of a high-frequency power supply for spark erosion J. *IEEE Transactions on Power Electronics* 20(2) pp 361-369
[5] Liu Q, Zhang Q, Zhu G, Wang K, Zhang J and Dong C 2016 Effect of Electrode Size on the Performances of Micro-EDM J. *Materials & Manufacturing Processes* 31(4) pp 391-396
[6] Yong Y W, Guo C N and Yang L 2015 Study on EDM Machining Characteristics Based on Skin Effect and Magnetic Field J. *Machinery Design & Manufacture* (in Chinese)
[7] Schacht B, Kruth J P, Lauwers B and Vanherck P 2004 The skin-effect in ferromagnetic electrodes for wire-EDM J. *The International Journal of Advanced Manufacturing Technology* 23(11) pp 794-799
[8] Liu Y, Sundaram M M, Zhao F L and Rajurkar K P 2011 Study on manufacturing free form surface parts in micro EDM based on skin effect theory J. *High Technology Letters* 17(4) pp 439-445
[9] Shi S, Zhang L and Li J 2007 Time-Varying Electromagnetic Fields and Maxwell's Equations M. Basic Electromagnetism and Materials. *Springer New York pp* 157-171
[10] Du J, Song M, Hu N N, Cao K N, Huang W T and Xu P 2014 A Double Exponential Function Fitting Algorithm for Optimize Parameter of μ H Curve J. *Advanced Materials Research* 960-961 pp 1146-1150

Prediction of Welding Deformation and Residual Stress of Stiffened Plates Based on Experiments

R X Bai, Z F Guo and Z K Lei *

State Key Laboratory of Structural Analysis for Industrial Equipment, Dalian University of Technology, Dalian 116024, China

*E-mail: leizk@163.com

Abstract. Thermo-elastic-plastic (TEP) method is a method that can accurately predict welding deformation and residual stresses, but the premise is to select the appropriate heat source parameters. Aiming at the two welded joints in the stiffened plate studied in this paper, the welding experiments of simple components were carried out respectively, and the corresponding welding deformation and residual stresses were measured. Based on the welding experiment, the corresponding TEP model was established, and the corresponding heat source parameters were obtained according to the experimental data. The comparison between the experimental results and the numerical results shows that the obtained heat source parameters can well predict the welding deformation and residual stress of the welded structure. And then, the obtained heat source parameters were applied to the TEP model of the stiffened plate. The prediction results show that the T-type fillet welds of the stiffened plate can reduce the angular deformation caused by the butt welds to a certain extent. In addition, we can also find that the heat of the subsequent welds can reduce the residual stresses at the completed welds. This method not only can save a lot of experimental costs and time, but also can accurately predict the welding deformation and residual stresses.

1. Introduction

The stiffened plate is often used in the structure of the ship to improve the carrying capacity of the panel and the whole structure. Welding is a commonly used metal connection process with high production efficiency and low production costs. However, the welding process also has unavoidable shortcomings [1] due to the high concentration of heat near the weld and rapid cooling of the cooling process, making the welding structure produce a certain welding deformation and residual stresses.

With the rapid development of computer performance and the wide application of finite element method, numerical simulation has become an effective and reliable method to predict the welding deformation and residual stress. Lee et al. [2] investigated the temperature field and residual stress of the dissimilar steel welding process. The simulation results show that the magnitude and distribution of the welding residual stresses of the dissimilar welds is different from the corresponding butt welds. Murakawa et al. [3] based on the inherent deformation method and applied the interface element to simulate the welding deformation of the thin plate structure. The numerical results agree well with the experimental results.

TEP method considering welding conditions and welding sequence in detail has been widely used in the prediction of welding deformation and residual stresses, which can more accurately predict the welding deformation and residual stresses compared to other methods such as inherent strain method. T. Schenk [4] studied the effect of support distance on the welding deformation of T-joints. Wei[5] presents a validated TEP model to investigate the effect of the welding sequence on weld induced residual stress distributions in a multi-pass welded piping branch junction.

In this paper, we propose a method to obtain the appropriate heat source parameters based on the simple structure welding experiment and apply the parameters to the prediction of welding deformation and residual stresses of large and complex structures. Based on this method, T-beam welding and plate-butt welding are carried out for the two welds in the stiffened plate, and the corresponding welding deformation and residual stresses are measured. Based on the welding experiment, the corresponding TEP model was established to obtain the appropriate heat source parameters and then the parameters was applied to the welding analysis of the stiffened plate.

2. Welding experiment

2.1. T-beam welding experiment

For the fillet weld in the stiffened plate, the T-beam is welded with a wing size of 500x200x10mm and a web size of 500x380x6mm, as shown in figure 1. The welding method was manual arc welding (MAW) by a Panasonic YD-400SS3HGE welding machine and the welding conditions are given in Table 1. The material is 1045 steel, and the material composition and mechanical properties are shown in Table 2. The in-plane displacement of the web was measured using 2D-DIC, and the Y-displacement shift on a1 and a2 along Line2 was measured using a displacement meter with the precision of 0.05mm. The vernier caliper with precision of 0.01mm was used to obtain the Z-displacement on b1 and b2.

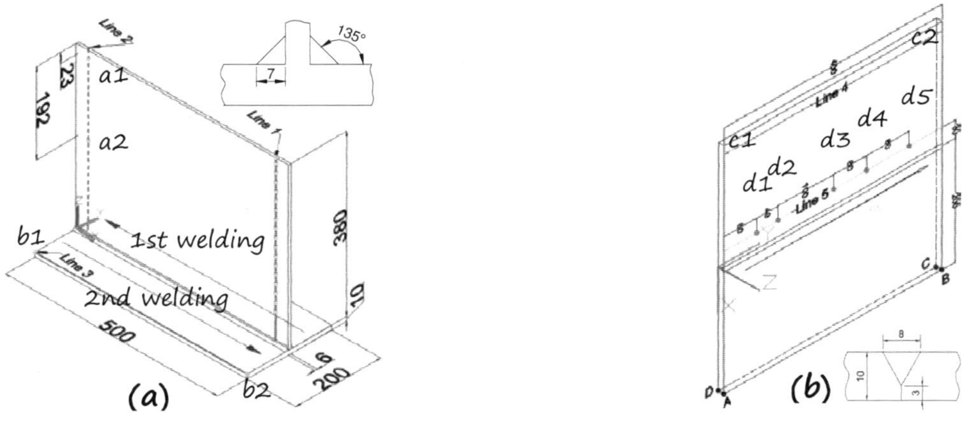

Figure 1. Geometric model and measuring points of (a) T-Beam (b)butt welding plate

Table 1. Welding conditions

Seam Type	Welding method	Wire	Pass number	Welding voltage (V)	Welding current (A)	Welding speed (mm/s)	Welding thermal efficiency
T-fillet welds	MAW	J507	1,2	64	130	4	0.6
Butt weld	MAW	J507	1	64	130	2.797	0.6

2.2. Plate butt welding experiment

In order to obtain the heat source parameters during butt welding, the plate butt welding experiment was carried on. Two 1045 steel plates with a length of 400 mm, a width of 200 mm and a thickness of 10 mm were used for butt welding, as shown in figure 1, the weld size was also shown. The Z-displacement of c1 and c2 was measured by YHD-50 displacement meter with the precision of 0.05mm.

Table 2. 1045 steel material composition and mechanical properties

Chemical composition (%)						Mechanical properties	
C	Si	Mn	P	S	Cr	Yield Strength (MPa)	Tensile Strength (MPa)
0.46	0.24	0.59	0.016	0.005	0.01	366	642

According to GB / T 31310-2014 standard, the strain release coefficient A=-4.587×10^{-7}mm^2/N and B=-7.04×10^{-7}mm^2/N were obtained by calibration experiment. And the release strain ε_0, ε_{45} and ε_{90} at d1-d5 were obtained by drilling method, as shown in figure 1. Thereby the transverse residual stresses and the longitudinal residual stresses can be obtained at d1-d5.

3. Validation of FEM model

3.1. Finite element simulation

The TEP analysis is a sequentially coupled finite element method that the nodal temperatures captured in the thermal analysis were applied as loads in the mechanical analysis. In the thermal analysis, two kinds of boundary conditions, convective heat transfer and radiant heat transfer are considered, and they are expressed by the composite heat release coefficient (CHRC), as shown in equation (1) and Table 3. In the mechanical analysis, the boundary condition is only to avoid the occurrence of rigid body displacement. Thermal analysis and mechanical analysis use the same element meshes, only the different types of element: Solid70 and Solid185 were used respectively. We assume that the material at the weld is the same as the base metal, and the changes in material properties with temperature were taken into account in the numerical simulations [6], as shown in figure 2.

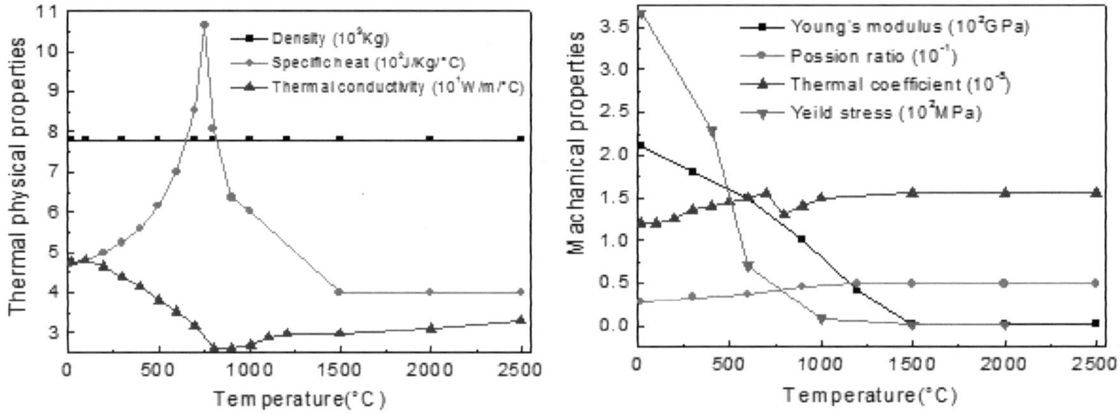

(a) (b)

Figure 2. Temperature-dependent(a) thermal properties and (b) mechanical properties

$$q_h = \alpha_h (T - T_0) \tag{1}$$

Where q_h is heat loss due to convection and radiation, α_h is used to represent the composite heat dissipation coefficient.

Table 3. Temperature-dependent CHRC

Temperature (°C)	20	100	500	750	1000	1500	2000
CHRC (W·m^{-2}/°C)	1.336	6.68	33.4	91.15	148.9	264.4	379.9

For T-type fillet welds, the combined heat source model can more accurately describe the distribution of welding heat [7]. In this paper, a volumetric heat source and surface heat source with uniform density are used to express droplet heat and arc heat respectively, as shown in equation (2) and (3). And for the plate butt welds, the moving volumetric heat source is used.

$$Q_V = \frac{\beta\eta UI}{V} \tag{2}$$

Where U and I are the welding voltage and current, V is the heat source volume. η is the welding thermal efficiency. The coefficient β is the proportion coefficient between the homogeneous surface heat source and the total welding heat. Generally, $\beta = 0.6$ [8].

$$Q_A = \frac{(1-\beta)\eta UI}{A} \tag{3}$$

Where A is the arc surface area.

The finite element model of T-beam and plate butt welding is shown in figure 3. In order to take into account the computational efficiency and accuracy, a dense mesh is used near the weld, and the element size increases as the distance from the weld increases. In the finite element model of T-beam, the length of each element in the welding direction is 10 mm, and the smallest element in the finite element models are 1 mm×1mm×10 mm. And in the finite element model of plate butt welding, the element size is the same as above, except for the length in the welding direction, which is 5 mm in length.

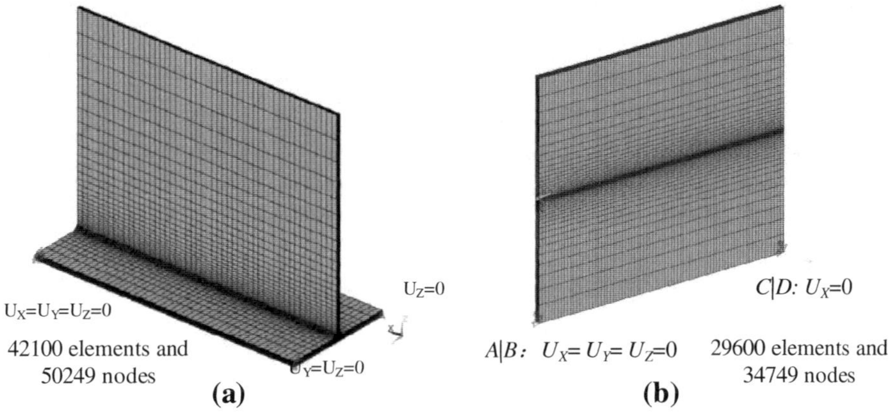

Figure 3. Finite element model of (a)T-Beam (b) butt welding plate

3.2. Comparison between experiment and numerical simulation

Figure 4 compares the welding deformation of T-beam welding experiments and numerical simulations. We can see that the numerical simulation results are in good agreement with the experimental results, not only accurately reflect the welding deformation trend, but also accurately predict the welding deformation value.

Figure 5 shows the results of the plate butt welding experiments and the numerical simulation of the welding deformation and residual stresses. For the most important welding deformation, ie, the angular deformation, the experimental results agree well with the numerical results. For the residual stresses, the minimum error is the transverse residual stress at the Y coordinate of 0.34 m, which is 1.13 MPa. And the maximum error is the longitudinal residual stress at the Y coordinate of 0.26 m, which is 156.82 MPa. The reason is that, in the actual welding process, the two flat plates are welded together with the formation of the weld pool, while in the finite element simulation, prior to the start of welding, the two plates are bonded together, causing the error of finite element results with the experimental results, and this effect is more pronounced in the longitudinal direction. So the numerical results of the transverse residual stresses agree well with experimental results. In general, the numerical simulation results accurately describe its trend, and the error between the two is very weak except for the individual measurement points.

From the above analysis we can see that the finite element model and the selected heat source parameters can accurately predict the welding deformation and residual stresses.

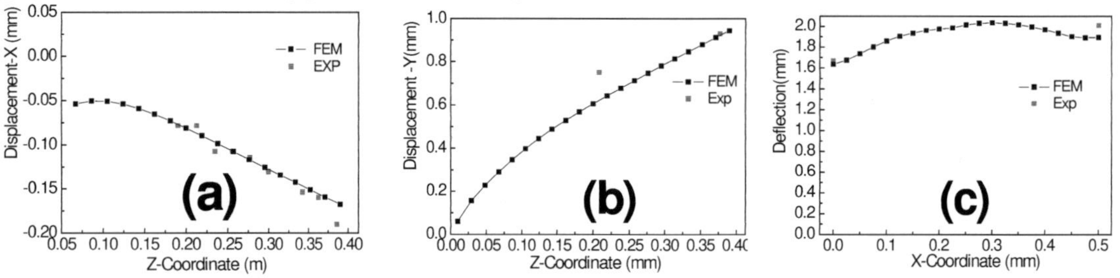

Figure 4. Comparison of T-beam welding experiment and numerical simulation: (a) X-displacement along Line 1 (b) Y-displacement along Line 2 (c) Z- displacement along Line 3

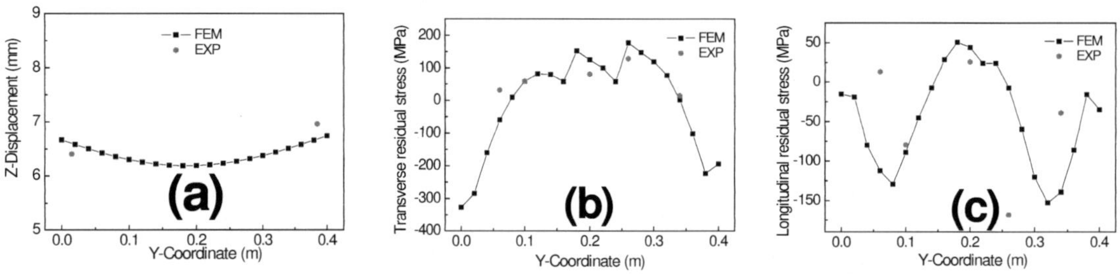

Figure 5. Comparison of plate butt welding experiment and numerical simulation: (a) Z-displacement along Line 4, (b) Transverse residual stresses along Line 5, (c) Longitudinal residual stresses along Line 5.

4. Prediction of welding deformation and residual stress of stiffened plates

The heat source parameters obtained above were applied to thermal analysis to simulate the temperature field. And then, the temperatures obtained in the thermal analysis were applied as a load in the mechanical analysis to solve the welding deformation and residual stresses.

Figure 6 shows the geometric model and dimensions of the stiffened plate. The dimensions of the two welds are shown in figure 1. Using the same geometric size and element size of the finite element model of plate butt welding, the finite element model was established, as shown in figure 7, with a total of 86880 elements, 97767 nodes. And the welding sequence is also shown in figure 6, that is, the butt weld is first welded, then the left fillet weld, and finally the right fillet weld.

Figure 8 shows the total displacement of the stiffened plate. We can see that the main deformation mode is transverse bending, and the displacement at the left bar is the largest, while the longitudinal bending is almost negligible. This is due to the large thickness of the stiffened plate. At the same time, we can find that the deformation at the end of the weld is larger than the beginning of the weld. The reason is that the base metal is more resistant to the welding deformation due to the low temperature of the structure at the beginning of the welding, while the base metal is less resistant to the welding deformation due to the higher temperature of the structure at the end of the welding.

Figure 9 demonstrates the Y-direction displacement along Line 6, and the curve is symmetrical about the butt weld. We can see that in the vicinity of the butt weld, the trend of increasing deformation is gradually reduced, and the deformation is reduced near the butt weld, which is caused by the angular deformation of the T-type fillet welds on both sides.

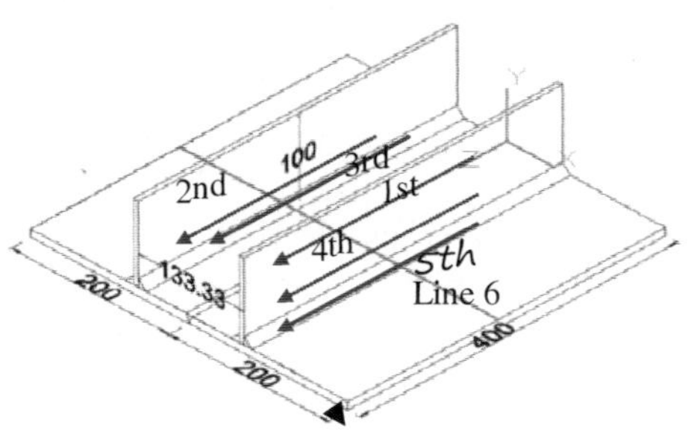

Figure 6. Geometrical model of stiffened plate

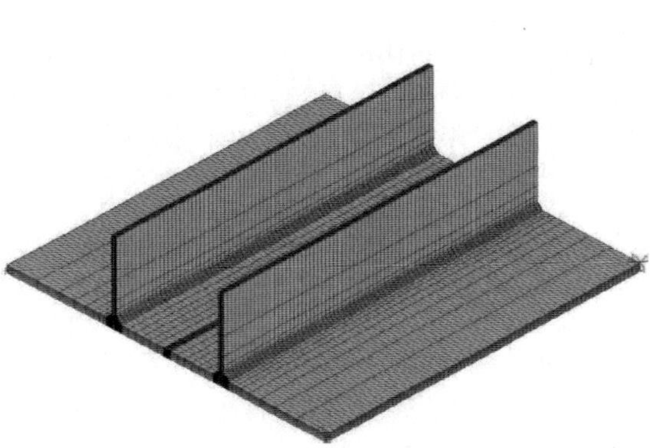

Figure 7. Finite element model of stiffened plate

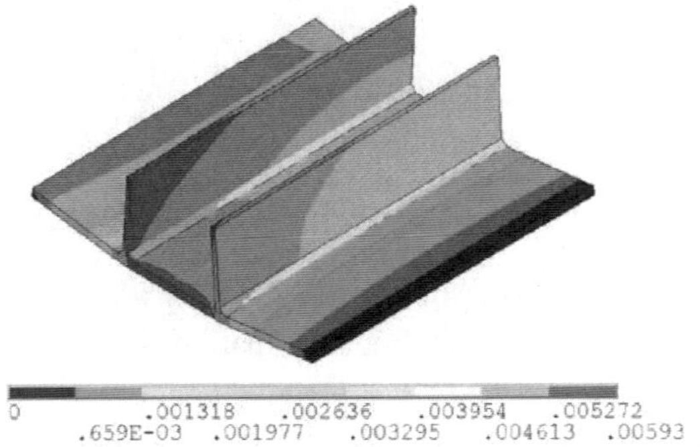

Figure 8. Total displacement of the stiffened plate

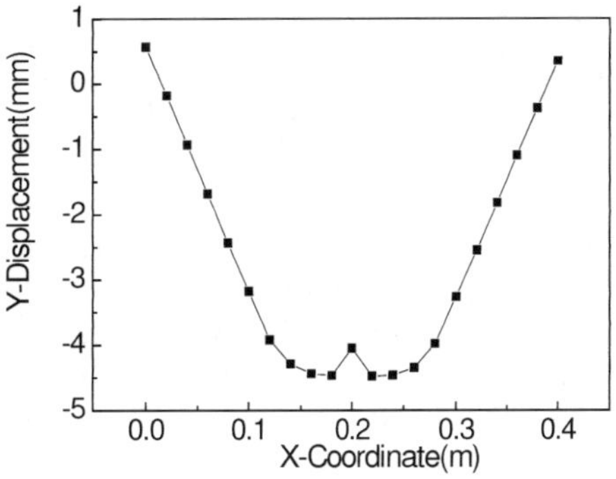

Figure 9. Y-Displacement along Line 6

Figure 10 shows the equivalent von Mises stress of the stiffened plate with greater stresses at the weld and less stresses away from the weld zone, and the largest residual stress occurred in the weld start / end. Figure 11 shows the magnitude and distribution of the Mises stress along Line 6. We can see that the maximum stress is close to the yield stress of the 1045 steel, which has a negative effect on the service performance of the stiffened plate. Through the careful observation we can find that the residual stresses at the butt weld is less than that of the left T-type fillet, while the left T-shaped fillet is less than the right. This is due to the fact that the completed welds surf from thermal effect from the lateral welds, so that the residual stresses is released.

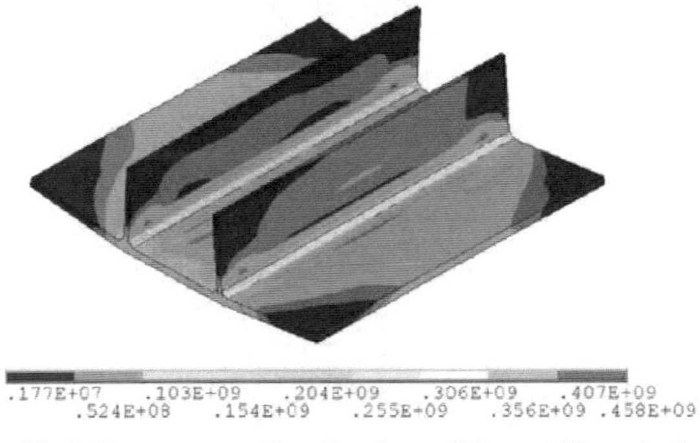

Figure 10. Mises stress distribution of the stiffened plate

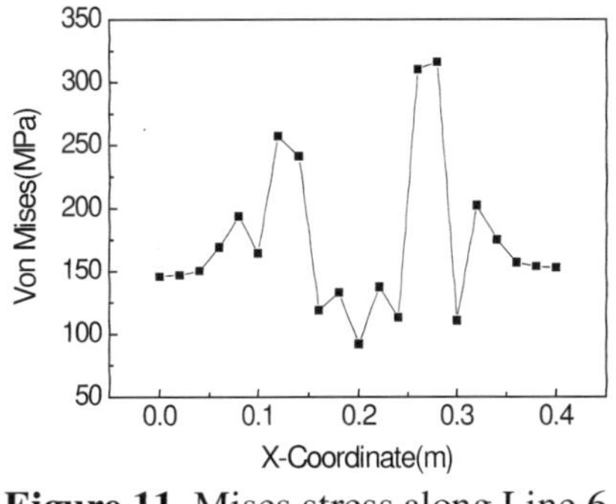

Figure 11. Mises stress along Line 6

5. Conclusion

In this paper, the corresponding TEP model is established according to the simple welding experiments, and the appropriate heat source parameters are obtained. The numerical results agree well with the experimental results. Applying of the heat source parameters, the welding deformation and residual stresses of the stiffened plates are predicted, and the following conclusions were drawn.

(1) This paper presents a method to obtain accurate heat source parameters based on simple welding structures and to apply them to the prediction of welding deformation and residual stresses of large welded structures.
(2) The T-type fillet welds of the stiffened plate reduce the angular deformation caused by the butt welds to a certain extent.
(3) The equivalent Mises stress at the welds is very large and close to the yield stress.
(4) The heat of the subsequent welds has a thermal effect on the completed welds, which can reduce the residual stresses at the completed welds.

Acknowledgement
This research was financially supported by the National Natural Science Foundation of China (Nos. 11472070, 11772081, 11572070), the Natural Science Foundation of Liaoning Province of China (No.2015020145).

References
[1] Lee C H and Chang K H, 2012, *Appl Therm Eng.*, Temperature fields and residual stress distributions in dissimilar steel butt welds between carbon and stainless steels, Vol **45-46**, pp. 33-41.
[2] Deng D, 2009, *Mater Des.*, FEM prediction of welding residual stress and distortion in carbon steel considering phase transformation effects, Vol **30**, pp. 359-66.
[3] Murakawa H, Deng D A, Ma N S and Wang J C, 2011, *Comput. Mater. Sci.*, Applications of inherent strain and interface element to simulation of weldingdeformation in thin plate structures, Vol **51**, pp. 43-52.
[4] Schenk T, Doig M, Esser G and Richardson I M, 2010, *Sci. Technol. Weld. Join.*, Influence of clamping support distance on distortion of welded T joints, Vol **15**:7, pp. 575-82.
[5] Wei J and Yahiaoui K, 2012, *Int J Press Vessels Piping.*, Effect of welding sequence on residual stress distribution in a multipass welded piping branch junction, Vol **95**, pp. 39-47.
[6] Gao K, Qin X P, Wang Z, Chen H, Zhu S X, Liu Y X and Song Y L, 2014, *J. Mater. Process. Technol.*, Numerical and experimental analysis of 3D spot induction hardening of AISI 1045 steel, Vol **214**, pp. 2425–33.
[7] Deng D A, Liang W and Murakawa H, 2007, *J. Mater. Process. Technol.*, Determination of welding deformation in fillet-welded joint by means of numerical simulation and comparison with experimental measurements, Vol **183**, pp. 219–25.
[8] Pardo E and Weckman D C, 1989, *Metall. Trans. B*, Prediction of weld pool and reinforcement dimensions of GMA welds using a finite element model, Vol **20B**, pp. 937–47.

IOP Conf. Series: Materials Science and Engineering **281** (2017) 012033 doi:10.1088/1757-899X/281/1/012033

Preparation and Characteristics of Porous Ceramics by a foaming Technology at Low Temperature

H Q Zhang[*1,2,3], S P Wang[2], J Wen[1,2,*], N WU[2] and S H Xu[2]

[1]State Key Laboratory of Silicate Materials for Architectures (Wuhan University of Technology), Wuhan 430070, China
[2]School of Materials Science and Engineering, Wuhan University of Technology, Wuhan 430070,China
[3]Zhongshan Advanced Engineering Technology Research Institute, Wuhan University of Technology, Zhongshan 528400, China

E-mail: zhhquan@whut.edu.cn

Abstract. Recycling and converting coal gangue and red mud into porous ceramics with good performance is a feasible disposal route. In this present work, porous foam ceramics was prepared using coal gangue and red mud as main raw materials at low sintering temperature, The amount of coal gangue and red mud were up to 70 wt%. To regulate the forming and sintering performance of the product, quartz sands and clay material were added to the formula. The green body was formed by a foaming technology using aluminum powders as foaming agents at room temperature. After foamed, the specimens were dried at 60-80 °C, and then calcined at 1060℃. Effects of concentration of NaOH and amount of aluminum powders on the phase, mechanical properties and microstructure were investigated here. Such study is expected to provide a new utilization route of the coal gangue and red mud, and brings both intensive environmental and economic benefits.

1. Introduction

With the rapid development of industrial technology, all kinds of industrial solid wastes also increase year by year. In China, red mud and coal gangue are the prominent solid waste produced by the aluminum processing and coal mining and processing. Simply stockpile and stacking treatment continue to still be used now, which not only occupy too much land, also give rise to serious environmental problems and huge waste of resources[1, 2]. Therefore, it is vital to develop a comprehensive utilization technology that consumes red mud and coal gangue or converts them into a secondary resource. Up to now, scientists have carried out research projects to explore disposal and utilization routes of these wastes. According to their unique physical and chemical properties, it has been found that they could be utilized to produce environmental protection product for adsorption to purify water[3], light weight aggregate concrete and cementitious materials[4-6], traditional clay brick and ceramic products[4, 7-8]. Nevertheless, their use dosage is limited in a low percentage, especially for red mud, and their comprehensive utilization rate in China was less than 4% for red mud and 60% for coal gangue, respectively [1,9].

Porous ceramic materials have several unique properties such as low density, low thermal conductivity, high surface area, high permeability, etc., and widely used as filters, catalyst supports, refractory linings for furnaces, or sound proof materials. There are various processing methods for fabricating the porous ceramics, such as: burn-out of organic pore-formers, replication of polymeric sponges, foaming of suspensions, sintering of spherical particles and freeze casting [10]. These methods allow obtaining materials of different porosity, poring size and poring geometry. Unfortunately, although considerable attention is paid on obtaining a porous ceramic materials, there are few reports about the preparation of porous ceramic products mainly using coal gangue and red mud by a foaming techniques at a low sintering temperature so far. Therefore, in the present work, effects of different pore creators and pore stabilizers on the mechanical property, porosity and pore microstructure of the foam ceramics would be discussed.

2. Experimental

2.1. *Raw Materials*

The red mud and coal gangue were supplied by industrial enterprises in Shanxi, CHINA. The chemical composition of the major raw materials used here is shown in Table 1. Aluminum powders and NaOH solution were added as the foaming agents.

2.2. *Preparation of porous materials*

All the raw materials were ball-milled in a wet condition for 15 min, their mass percentage for each ingredient was shown in Table 2. After drying the slurry in an electric oven at 110 ℃, the mixed materials was re-crushed in a dry condition and sieved pass 150 mesh. The mixture was then wet-mixed with aluminum powders and NaOH solution in different proportions for 3 min.

Rectangular silicone molds were employed for molding the green specimens, the batch was dried at 110℃ for 1 h after foaming at 40 ℃ for 2 h. Finally, the dried specimens were calcined at 1060 ℃ at a heating rate of 6 ℃/min and naturally cooling.

2.3. *Characterization of porous ceramic samples*

The chemical composition of raw materials was determined by X-ray fluorescence analysis. The bulk density and apparent porosity of products were measured and calculated by the Archimedes method. The flexural strength of samples was directly measured with regular dimension. Appreciation of crystalline phases of the products was done on X-ray diffraction. The microstructure was analyzed through observing the cross sections of fractured samples via optical microscope.

Table 1. Chemical compositions of the raw material

Raw Materials	SiO_2	Al_2O_3	Fe_2O_3	CaO	MgO	K_2O	Na_2O	TiO_2	P_2O_5	ZrO_2
Red mud	27.22	25.76	5.72	21.78	0.90	1.32	11.81	4.85	0.42	0.16
Coal gangue	57.59	25.46	8.87	4.30	0.70	1.30	0.36	1.25	0.16	
Clay mineral	62.50	20.64	5.72	0.45	0.69	3.05		0.74	0.15	
Quartz	98.40	0.55	0.10	0.25	0.35					

3. Results and discussion

Porosity and flexural strength of the product and forming performance of the green body were found to depend on the water addition amount, concentration and volume of the NaOH solution, as well as the amount of the aluminum powders. In the present paper, a new moulding process was conducted by casting and foaming. To obtain a suitable slurry with good flow-ability and foaming behaviour, addition amount of water was ranged from 45~60 wt% according to the ingredient formula of the specimens. It was found that low water addition greatly increased the viscosity of the slurry, greatly impairing the foam formation and giving rise to a dense green body. With the increase of water addition, the viscosity became low, favouring the foam formation and the obtainment of suitable pore size. The appropriate water addition proportion should be controlled to 50~52 wt% for moulding and foaming of the green body.

Table 2. Specimen composition

Sample	Red mud (wt%)	Clay mineral (wt%)	Coal gangue (wt%)	Quartz (wt%)	NaOH (mol/L)	Al powders (wt%)
A1						0.03
A2	20-25	25-30	45-50	5-10	0.75	0.04
A3						0.05
B1						0.03
B2	20-25	25-30	45-50	5-10	0.5	0.04
B3						0.05
C1						0.03
C2	20-25	25-30	45-50	5-10	0.25	0.04
C3						0.05
D1						0.03
D2	20-25	25-30	45-50	5-10	0.10	0.04
D3						0.05

Figure1 shows the flexural strength and porosity of the ceramic specimens prepared with 68% red mud and coal gangue against the different sodium hydroxide solution concentration. With the increase of NaOH concentration, the flexural strength of porous ceramics increased firstly and then decreased, reaching a maximum value at 0.5mol/L NaOH. Under low concentration conditions, such as 0.1mol/L NaOH, since the amount of H_2 produced by the reaction of aluminum with water increased with its amount in the slurry, the internal pore size and pore content in the materials also increased, giving rise to a decrease in their flexural strength with the increase of the amount of aluminum powders. When the NaOH concentration was more than 0.5 mol/L, the reaction rate of aluminum with water was greatly accelerated, yielding more H_2 in a short time, which would lead to bubble coalescence easily.

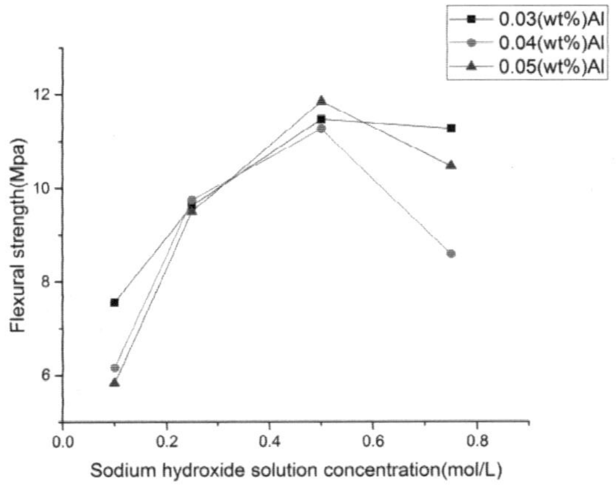

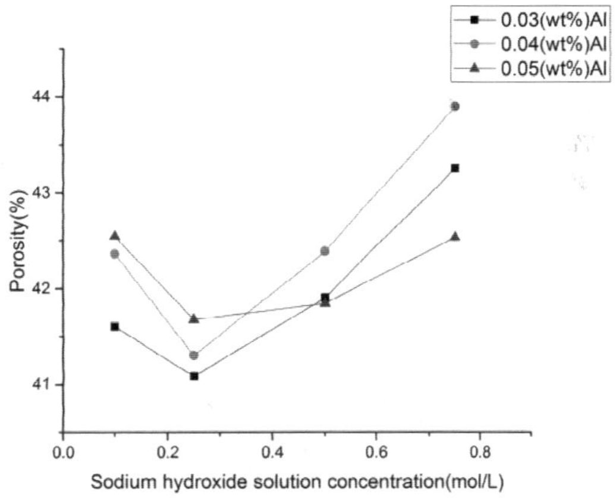

(a) Change in flexural strength (b) Change in porosity

Figure 1. Changes of flexural strength and porosity of specimens against NaOH solution concentration

In such a condition, further increasing the amount of aluminum powders would impair the foam formation. Thus, the foaming effect of aluminum powders was related to both the concentration of NaOH and the content of aluminum powders.

As shown in Figure 1b, the porosity of the samples increased with the increase of NaOH concentration, but the change was slightly when the NaOH concentration was in 0.1-0.5 mol/L, which was speculated to be related to the porosity and pore distribution of the materials. When the NaOH concentration was less than 0.25mol/L, the porosity increased with the amount of aluminum powders. However, when the NaOH concentration was > 0.5 mol/L, the porosity of the materials decreased due to the increase in the probability of occurrence of a large number of pores and bubbles.

Figure 2 shows the XRD pattern of specimens calcined at 1060℃ using 50% coal gangue and 20% red mud, which was formed using 0.03 wt% Aluminum powders and 0.5mol/L NaOH. Its major crystalline phases were mainly composed of quartz and $Ca(Al_2Si_2O_8)$ or $Na(AlSi_3O_8)$. High baseline also indicated the existence of the glass phase in the products. No obvious change in the phase composition was found for the samples by various amounts of Aluminum powders.

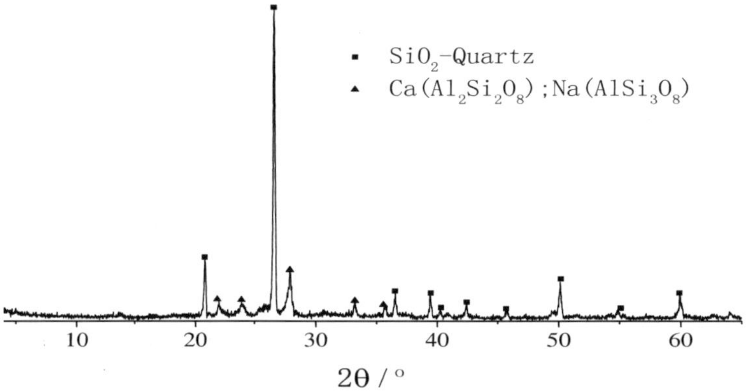

Figure 2. XRD patterns of porous ceramic prepared with 0.03 wt% aluminum powders

Figure 3 shows the surface morphology of the specimens prepared with 0.25mol/L, 0.5mol/L, 0.75mol/L NaOH and 0.03wt% aluminum powders. All the samples possessed high porosity. A large number of pores ranged from 0.5 to 2 mm could be seen on the fractured surface. With the increase of NaOH concentration, the pore size and porosity of the foam ceramics gradually increased, and their pore distribution was also improved gradually. The foaming performance of the materials was closely related with the content of aluminum powders and the reaction condition with water. With the increase of NaOH concentration, the more bubbles were generated by the reaction of Al powders and NaOH solution, the more pores were formed in the corresponding slurry, and the more pores are formed after forming, drying and sintering.

0.25mol/L NaOH 0.5mol/L NaOH 0.75mol/L NaOH

Figure 3. Surface morphology of specimens with 0.03wt% aluminum powders and different NaOH concentration 1000

4. Summary

Red mud and coal gangue was utilized to prepare the foam ceramic materials; its utilization amount could be up to 70 wt% when using aluminum powders as the foaming agents. The foaming

IOP Conf. Series: Materials Science and Engineering **281** (2017) 012033 doi:10.1088/1757-899X/281/1/012033

performance of the materials was closely related with the content of aluminum powders and the reaction condition with water solution. Utilization of red mud and coal gangue for preparing porous ceramics with good performances was expected to provide a feasible disposal of landfill and soil fail problems caused by deserted slag.

Acknowledgements
This research was financially supported by the Special Foundation for Science and Technology Innovation Service Platform of the Zhongshan Municipal Bureau of Science and Technology (No. 2014C1SQ0002), and supported by the Fundamental Research Funds for the Central Universities of Wuhan University of Technology (No. 2016-CL-B1-10).

References
[1] Xu H L, Song W J, W.B. Cao et al 2017 *J. Mater. Cycles Waste* **19** 1270–78
[2] Nan X L, Zhang T A and Liu Y, Dou Z H 2010 *J. Process Eng.* **10** 264-70.
[3] Liu D Y and Wu C S 2012 *J. Materials* **5** 1232-46
[4] Guo Y X, Zhang Y X and Chen F Q, 2014 *J. CIESC Journal* **65** 2443-53
[5] Singh M, Upadhayay S N and Prasad P M 1996 *J. Waste Management* **6** 665–70
[6] Zhang N, Sun H and Liu X 2009 J. Hazard. *Mater* **167** 927–32
[7] Kara M and Emrullahoglu F 2009 *J. Ceramic Industry* **7** 13–19
[8] Allaire C 1993 *J. Am. Ceram. Soc. Bull* **72** 59-64
[9] Liu S M, 2011 Comprehensive utilization of red mud, Mphil thesis, Northeastern University
[10] Wiecinska P and Bachonko M 2016 *J. Ceram. Int.* **42** 15057-64

Variation and Grey GM(1, 1) Prediction of Melting Peak Temperature of Polypropylene During Ultraviolet Radiation Aging

K Chen[1], T Y Zhang[2,*] F Zhang[2] and Z R Zhang[3]

[1]School of Bailie Mechanical Engineering, Lanzhou City University, Lanzhou, China
[2]Center of Information and Network, Lanzhou City University, Lanzhou, China
[3]School of Electronic and information Engineering, Lanzhou City University, Lanzhou, China

*Corresponding author e-mail: lzyunyun@126.com

Abstract. Grey system theory regards uncertain system in which information is known partly and unknown partly as research object, extracts useful information from part known, and thereby revealing the potential variation rule of the system. In order to research the applicability of data-driven modelling method in melting peak temperature (T_m) fitting and prediction of polypropylene (PP) during ultraviolet radiation aging, the T_m of homo-polypropylene after different ultraviolet radiation exposure time investigated by differential scanning calorimeter was fitted and predicted by grey GM(1, 1) model based on grey system theory. The results show that the T_m of PP declines with the prolong of aging time, and fitting and prediction equation obtained by grey GM(1, 1) model is T_m =166.567472exp(−0.00012t). Fitting effect of the above equation is excellent and the maximum relative error between prediction value and actual value of T_m is 0.32%. Grey system theory needs less original data, has high prediction accuracy, and can be used to predict aging behaviour of PP.

1. Introduction

Polypropylene (PP) has the excellences of abundant resources, low prices, high mechanical properties and easy forming, and has been applied widely in the industry of automobile, household appliance, machinery [1]. However, the existing of large amounts of unstable tertiary carbon atoms leads to the fact that PP is easily aged with the action of ultraviolet light and oxygen, which causes mechanical properties decline, discoloration and the changing of physicochemical properties, and makes PP one of the most ultraviolet aging unresistant polymers [2]. Therefore, it has great significance that researching the rule and mechanism of PP ultraviolet aging, modifying PP to improve its ultraviolet aging resistance and expanding the application range of PP. And these are also one of the major measures to promote the development of PP.

The service life of outdoor PP and its products is about 3 years under the action of ultraviolet light and oxygen [3]. But it can be extended to 5 years and even over 8 years when PP is modified through blend, copolymerization and filling, and so on [4]. Therefore, current PP and its modification research mainly adopt artificially accelerated aging which greatly shortens research time. But literature [5] indicates that artificial accelerated aging can't reflect the performance variation of materials under natural aging accurately.

Recently, according to experimental data related, forecasting properties of materials through mathematical model is one of the focuses of attention and study [6–8]. But simulation and prediction study about aging behaviour of PP was focused mainly on mechanical properties [9], and few about physical properties. Taking melting peak temperature as an example, Grey GM(1, 1) model, one of the typical data-driven modelling methods, is used in determining prediction equation of PP melting peak temperature changing along with ultraviolet radiation aging time.

2. Experiments

2.1. Ultraviolet radiation aging

PP specimens, provided by Gansu Langang Petrochemical Co. Ltd., was illuminated by 20W ultraviolet lamp with a distance of 20 cm under room temperature, and then taken out for Differential Scanning Calorimetry (DSC) characterization after regular time intervals.

2.2. DSC characterization

Differential scanning calorimeter (TA Company, q-10) was utilized to characterize melting peak temperature of different specimens. A specimen of about 3~5 mg was held in aluminum seal during each process at a heating or cooling rate of 10 K/min under nitrogen atmosphere. The specimen was melted at 200 °C for 5 min to eliminate thermal history and cooled to 100 °C, followed then by a reheating process which was recorded to 200 °C.

3. Results and discussions of DSC characterization

The effect of the ultraviolet aging time on the melting behavior of PP was evaluated by DSC. By the reheating DSC curves of PP after different aging time showing in Figure 1(a), the relationship between melting peak temperature and aging time was obtained in Figure 1(b). The melting peak temperature decreases with the extending of aging time. After 400 h ultraviolet aging, it decreases from 165.4 °C to 158.4 °C.

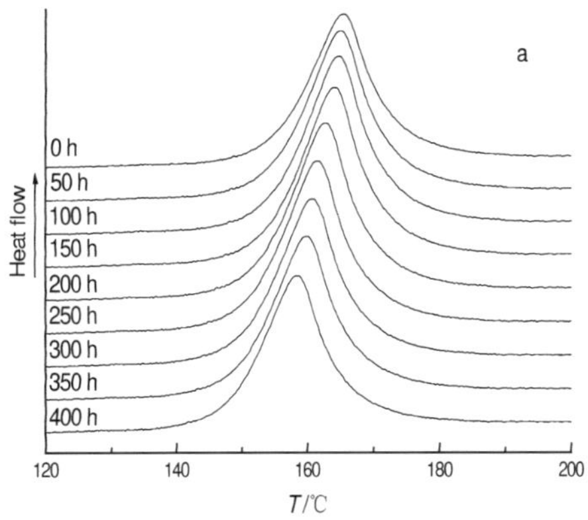

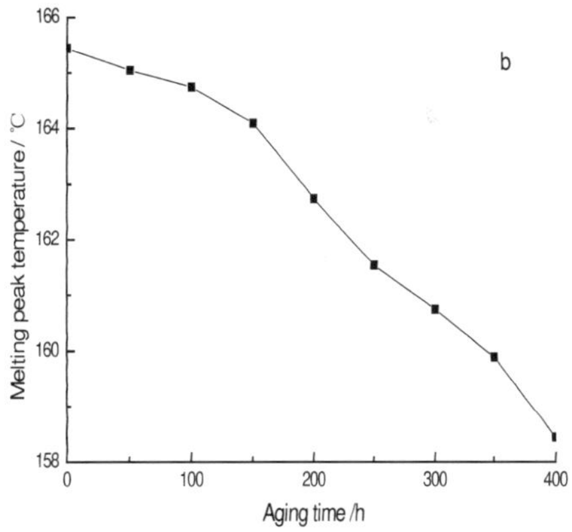

Figure 1. Reheating DSC curve (a) and melting peak temperature (b) of PP after different UV exposure time.

The decreasing of the melting peak temperature should be mainly due to the following three factors. Firstly, photo-oxidative degradation leads to decrease of crystal surface regularity of crystalline region till to the decomposition, which improves surface free energy of PP. Obviously; melting peak temperature of material declines with the increasing of surface free energy according to Hoffman-Weeks formula [10]

$$T_{\rm m} = T_{\rm m}^{0}\left(1 - \frac{2\sigma_{\rm e}}{l\Delta H}\right) \tag{1}$$

where $T_{\rm m}$ and $T_{\rm m}^{0}$ are melting peak temperature corresponding to thickness being l and ∞ respectively, ΔH is fusion heat of unit volume, and $\sigma_{\rm e}$ is surface free energy. Secondly, photo-oxidative degradation leads to molecular chains break of amorphous region, and causes molecular weight decline. Thus, the melting peak temperature of PP decreases accordingly. Moreover, molecular chains break in amorphous region can also destroy partial entanglement, which may result in lower temperature recrystallization happened at molecular chains of degradation with poor regularity, and brings about more crystallization defects. As a result, the melting peak temperature of PP moves in low temperature region [11, 12].

4. Fitting and prediction of grey GM(1,1) model

Grey system theory regards uncertain system in which information is known partly and unknown partly as research object, extracts useful information from part known, and thereby revealing the potential variation rule of the system [13]. Up to now, it has been widely applied in many fields. Taking time series as independent variable and melting peak temperature as dependent variable, melting peak temperatures of PP during ultraviolet aging are simulated and predicted with grey GM(1, 1) model. The specific process sees reference [13]. It is obtained that evolution parameter *a* and grey action quantity *b* of grey regression equation

$$\hat{x}^{(0)}(k+1) = \left(1 - e^{a}\right)\left(x^{(0)}(1) - \frac{b}{a}\right)e^{-ak}, \quad k=1, 2, \ldots, n \tag{2}$$

are 0.0060 and 167.0608 respectively. Thus, simulation and prediction equation can be expressed as

$$T_{\rm m} = 166.567472e^{-0.00012t} \tag{3}$$

where $T_{\rm m}$ is melting peak temperature, and *t* is aging time. Table 1 is simulation result obtained from Equation (3). Precision inspection shows that the accuracy grade of this model is first.

Table 1. Simulation result of melting peak temperature.

Aging time (h)	Actual value	Predicted value	Residual error	Relative error (%)
50	165.04	165.5710	-0.5310	0.3217
100	164.74	164.5805	0.1595	0.0968
150	164.09	163.5960	0.4940	0.3011
200	162.74	162.6173	0.1227	0.0754
250	161.54	161.6446	-0.1046	0.0647
300	160.74	160.6776	0.0624	0.0388
350	159.89	159.7164	0.1736	0.1086
400	158.44	158.7610	-0.3210	0.2026

It can be seen from Table 1 that, the maximum relative error of predicted value and actual value obtained from grey GM(1, 1) model is 0.32%, which is negligible. Therefore, grey GM(1, 1) model can be utilized in melting peak temperature prediction of PP during ultraviolet aging.

5. Conclusions

During the process of ultraviolet aging, the melting peak temperature of semi-crystalline PP declines with the increasing of surface free energy of crystalline region, and (or) molecular chains breaking and the formation of randomly ranged crystal of amorphous region.

The quantitative relationship between melting peak temperature and ultraviolet aging time of PP are obtained based on grey GM(1, 1) model. The maximum relative error between predicted value and actual value of melting peak temperature is 0.32% accordingly.

Owing to needing less experimental data and being easy to test, grey GM(1, 1) model is a simple and reliable way for melting peak temperature prediction during ultraviolet aging process since. But fitting and prediction of grey GM(1, 1) model can only be done via the time series. This insufficiency limits the further application of this method.

Acknowledgments

The article was financially supported by the National Natural Science Foundation of China (Grant No. 11464026) and the Scientific Research Foundation of Lanzhou City University (Grant No. LZCU-XZ2015-03).

References

[1] Wang Y Z, Yue C E, Ma D, Sun Z Y and Chen G L 2014 Study on anti-UV aging effect and reaction mechanism of PP-g-AN in PP/PP-g-AN blends *Eng. Plast. Appl.* **42** 82-87

[2] Kotek J, Kelnar I, Baldrian J and Raab M 2004 Structural transformations of isotactic polypropylene induced by heating and UV light *Eur. Polym. J.* **40** 2731-2738

[3] Leng L C, Zhang Y M, Han G T, Zhao J and Gao S W 2014 Study of different hindered amine light stabilizer modified anti light aging of polypropylene *Appl. Chem. Ind.* **43** 864-868

[4] Hu X J 2003 Study on weathering quality of Polypropylene *Synthetic Mater. Aging Appl.* **32** 1-3

[5] Azuma Y, Takeda H, Watanabe S, Nakatani H 2009 Outdoor and accelerated weathering tests for polypropylene and polypropylene/talc composites: A comparative study of their weathering behavior *Polym. Degrad. Stab.* **94** 2267-2274

[6] Rajaneesh A, Satrio W, Chai G B and Sridhar I 2016 Long-term life prediction of woven CFRP laminates under three point flexural fatigue *Compos. Part B-Eng.* **91** 539-547

[7] Ma X and Liu Z B 2016 Research on the novel recursive discrete multivariate grey prediction model and its applications *Appl. Math. Model.* **40** 4876–4890

[8] Celina Mathew C 2013 Review of polymer oxidation and its relationship with materials performance and lifetime prediction *Polym. Degrad. Stab.* **98** 2419-2429

[9] Chen K, Zhang T Y, Hu Y X and Zheng X P 2013 Application of residual error grey prediction model in aging behaviour prediction of plastic *Mod. Chem. Ind.* **33** 119-122

[10] Zoepfl F J, Markoviĉ V and Silverm Joseph 1984 Differential scanning calorimetry studies of irradiated polyethylene I. Melting temperatures and fusion endotherms *J. Polym. Sci. Pol. Chem.* **22** 2017-2032

[11] Rabello M S and White J R 1997 The role of physical structure and morphology in the photodegradation behaviour of polypropylene *Polym. Degrad. Stab.* **56** 55-73

[12] Xu L X and Zhong M Q 2006 Study on the effect of nano-TiO_2 on UV aging resistance of polypropylene *J. Zhejiang Univ. Techno.* **34** 302-305

[13] Chen K, Zhang T Y, Zheng X P and Hu Y X 2013 Application of grey theory for the prediction of plastic aging behavior *China Plastics Ind.* **41** 113-116

IOP Conf. Series: Materials Science and Engineering **281** (2017) 012035 doi:10.1088/1757-899X/281/1/012035

Mesoporous Akaganeite of Adjustable Pore Size Synthesized using Mixed Templates

Y Zhang, D L Ge, H P Ren, Y J Fan, L M Wu and Z X Sun*

School of Chemistry and Chemical Engineering, University of Jinan, 250022 Jinan, China

*Corresponding email: sunzx@ujn.edu.cn

Abstract. Mesoporous akaganeite with large and adjustable pore size was synthesized through a co-template method, which was achieved by the combined interaction between PEG2000 and alkyl amines with different lengths of the straight carbon chain. The characterized results indicate that the synthesized samples show comparatively narrow BJH pore size distributions and centered at 14.3 nm when PEG and HEPA was used, and it could be enlarged to 16.8 and 19.4 nm respectively through changing the alkyl amines to DDA and HDA. Meanwhile, all the synthesized akaganeite possess relativity high specific surface area ranging from 183 to 281 m^2/g and high total pore volume of 0.98 to 1.5 cm^3/g. A possible mechanism leading to the pore size changing was also proposed.

1. Introduction

Akaganeite (β-FeOOH) nanoparticles have extensive applications in the field of ion exchange [1], sorbents [2], and catalysts [3]. In addition, due to the fact that pores significantly increase the surface area/volume ratio of a material, mesoporous iron oxide films offer great potential as advanced materials for electrodes [4] and pharmaceutical drug delivery [5]. In order to synthesize nanocrystalline akaganeite, different methods such as precipitation, sol–gel as well as surfactant assisted methods have been employed [4, 6]. However, most of the synthesized akaganeite have either comparatively irregular pore size distributions [7] or have not reported on information about pores [6]. Up to now, synthesis of pore size adjustable mesoporous akaganeite material still represents a challenge. On the other hand, different templates have been successfully employed as the regulator to control the pore size and the pore channel architecture of silica and alumina [8, 9]. The idea in synthesizing size tunable silica and alumina mesoporous materials could also hopefully throw some lights in the synthesis of pore size adjustable mesoporous akaganeite materials.

To the best of our knowledge, however, there is no report on the synthesis of mesoporous akaganeite with large and adjustable pore size so far, although Bruce et al. [10] have synthesized two- and three-dimensional mesoporous iron oxides with micro-porous walls by using decylamine as the template. Furthermore, the mechanisms for the formation of mesoporous akaganeite using surfactant as a template are still unclear. Herein, we report, the synthesis of mesoporous akaganeite with large and adjustable pore size from 14 to 20 nm using PEG and alkyl amine with different chain length as mixed templates, and propose a possible mechanism leading to various pore sizes.

2. Experiments

2.1. Sample preparation

All the chemical reagents are of analytical grade and used as-received without any further purification, double distilled water was used throughout the experiment. In a typical synthesis, 12.0 mmol $FeCl_3 \cdot 6H_2O$ and 0.3 mmol PEG2000 were dissolved in 20 mL distilled water forming a yellow solution in a 100 mL beaker. 12.0 mmol n-Heptylamine (HEPA) (or other amine) was dissolved in 20 mL ethanol and added drop wise to the beaker forming a dark brown solution under magnetic stirring. The solution was then heated to 60 °C and maintained at this temperature for 3 h. After that, 16 mL 2.5% $NH_3 \cdot H_2O$ was added to the solution, which remained dark brown with some precipitation, and the system was then heated up to 80 °C. After washing three times with distilled water and ethanol respectively, the precipitates were dried for characterization.

2.2. Characterizations

Powder X-Ray diffraction (XRD) measurements of samples were carried out with a Bruker D8-Advance powder diffractometer in a scanning range of 20-70° at a scanning speed of 0.02° s^{-1} using a Ni-filtered Cu Kα radiation (wavelength 1.542 Å). The infrared spectra of samples were recorded on a Bruker VERTX-70 FT-IR spectrometer using KBr pellets containing about 1% weight sample in KBr. The surface area and pore parameter of prepared samples were determined by N_2 adsorption–desorption technique and relevant data were collected on a Nova 2000e Surface Area and Pore Size Analyzer at -196 °C. The surface area was calculated by the BET (Brunauer – Emmett – Teller) method and the pore size distributions were calculated from the desorption data of the isotherms by Barrett - Joyner - Halenda (BJH) model. Total pore volume was estimated from the amount adsorbed at $P/P_0 = 0.99$.

3. Results And Discussion

XRD measurements were carried out to investigate the crystal structure of the prepared samples and the result is shown in Figure 1. Only the results for samples synthesized with PEG2000 and HEPA are shown here because the others show basically the same XRD results.

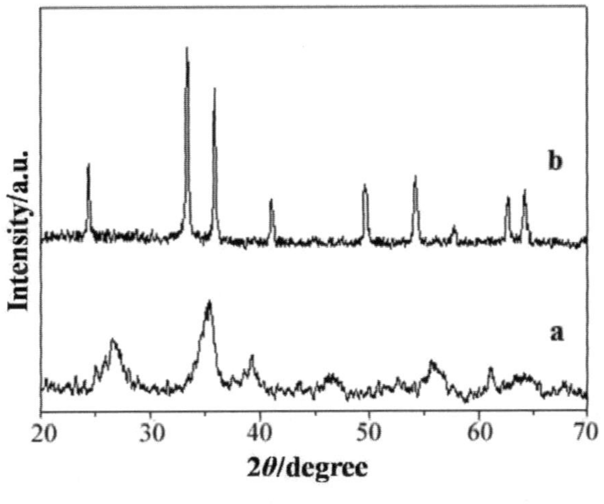

Figure 1. XRD patterns of samples synthesized with PEG2000 + HEPA: (a) precursor; (b) calcinated at 250°C.

It can be found from Figure 1 that the as synthesized material with broader peaks, which is typical for nanosized akaganeite [11, 12] (β-FeOOH, JCPDS No. 34-1266). According to the Scherrer's formula: $D = 0.89\lambda/(\beta \cos\theta)$, where D is the crystal size, λ is the X-ray wavelength, β is FWHM (full width at half maximum) of the diffraction peak and θ is the diffraction angle, the average crystallite size of as prepared akaganeite particle is about 6.6 nm. However, after being calcinated at temperature of 250 °C, it converted to hematite (α-Fe2O3, JCPDS No. 24-0072) with crystallite size of about 29.8 nm calculated using Scherrer's formula.

The surface area calculated from the Brunauer - Emmett - Teller (BET) method was determined to be more than 184 m^2g^{-1}. The diameter of a solid spherical particle may be calculated using the formula $S = 6/\rho d$ [13], where S, ρ and d denotes specific surface area, particle density and particle size, respectively. The average size calculated in this way is about 4 nm.

N_2 adsorption-desorption isotherms (Figure 2) for three template-extracted samples show type IV adsorption isotherm with an H1 hysteresis loop, which is a characteristic of mesoporous materials, as defined by IUPAC. The pore size distributions show a narrow range of 14.5 - 20.5 nm (centered at 16.8 nm), which is calculated from Barret - Joyner - Halenda (BJH) equation [14, 15] using the desorption data. The narrow pressure range for capillary condensation is also indicative of a narrow pore size distribution.

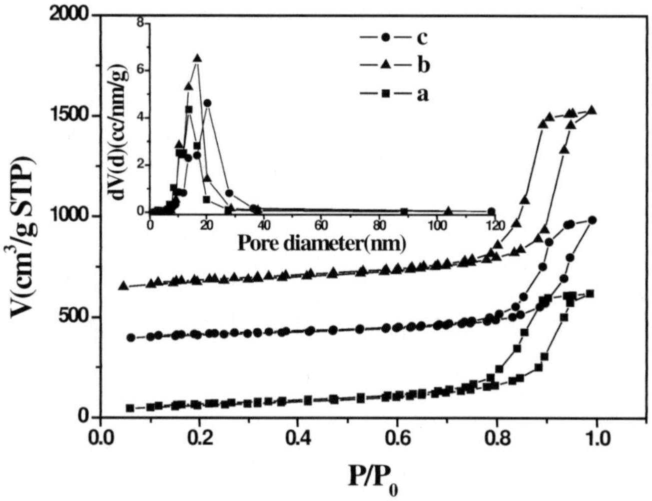

Figure 2. N_2 adsorption/desorption isotherms for three template-extracted samples (a) PEG + HEPA (b) PEG + DDA (c) PEG + HAD and the inset is the pore size distribution curves of the corresponding samples.

The BET specific surface area, average pore diameter and the pore volume of template-extracted samples are listed in Table 1.

Table1. BET surface area, pore diameter and pore volume data for mesoporous akaganeite samples prepared using different co-template

Composition of template	Surface area (m^2g^{-1})	BJH pore size (nm)		Pore volume (cm^3g^{-1})
		observed	calculated	
PEG	-	-	11.1	-
PEG + HEPA	218	14.3		0.99
PEG + DDA	281	16.8	16.9	1.50
PEG + HDA	184	19.4	19.0	0.98

Obviously, from table 1, the pore size increases as the alkyl amine chain length increases. However, the specific surface and pore volume have not the orderliness as the alkyl amine chain length increases, it is interesting to note that the specific surface area and pore volume of product prepared using PEG + DDA as co-template is the highest among them.

Restrained hydrolysis and condensation of the inorganic species appears to be important for forming mesophases of most of the non-silica oxides in order to slow down the hydrolysis and condensation rates [16]. According to our results, on the other hand, the PEG has a complexation tendency with Fe^{3+}, implying that the inorganic particles were more apt to combine with polar PEG than with the neutral alkylamine. On the basis of these considerations, a strategy was devised as presented in Scheme 1. The alkyl amines: n-Heptylamine (HEPA), Dodecylamine (DDA) and n-Hexadecylamine (HDA), used as the expanding agents, were supposed to occupy the interior of the micelle (aggregate formed by a physisorption).

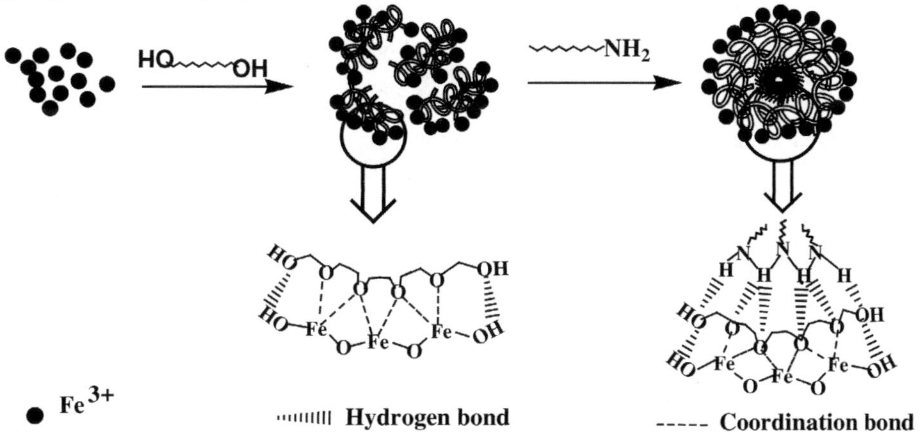

Scheme1. Schematic representation of the process synthesizing mesoporous iron oxide with various pore sizes using PEG2000 and alkyl amine as mixed templates

In order to explain these results, a possible mechanism leading to mesostructure and varying pore size is proposed, as illustrated in Scheme 1, which mainly comprises a two-step procedure: (i) the complexation interaction of the PEG and Fe^{3+} and (ii) the combination of alky lamines with PEG through weak hydrogen bonding. In fact, it is well-documented that alkyl amines are also good templates for mesoporous transition metal oxides [12]. Therefore, the alkyl amines are supposed to play the role of adjusting the pore size through changing its chain length in conjunction with PEG. The chain length of different alkyl amines have been calculated. Amazingly, the increases of pore size is nearly quantitatively agreeable with the increase of the carbon chain length (obsd. 16.8 nm and 19.4 nm, calcd. 16.9 and 19.0 for sample using template PEG + DDA and PEG + HDA respectively). For the product prepared from PEG + DDA systems, its highest specific surface area and pore volume may be due to DDA having the optimal chain length, with incorporation and organization properties different from the other two amines in the process of forming mesoporous iron oxide. Unlike the optimal swelling effects of decane which expands the pore diameter of mesoporous silica, reported by Ulagappan et al. [17], our results indicate the optimal effects of DDA, which fit in with the surface area and pore volume of iron oxide particles. The mechanisms involved are largely unexplored, it should be relevant with the physicochemical properties of the system, such as solubility, interaction among particles, PEG and the surfactants.

4. Conclusions

PEG and alkyl amines with different chain length as mixed templates can be used to produce pore size adjustable mesoporous akaganeite nanoparticles with high surface area and pore volume. We have proposed a mechanism with a two-step sorption effect, which comprises chemical interaction between ferric ions and PEG and hydrogen bonding between PEG and alkyl amines for the synthesis process.

Acknowledgement
Financial support from Chinese Natural Science Foundation (No.51274104; No.50874052; No. 20677022) is gratefully acknowledged.

References

[1] Cai J, Liu J, Gao Z, Navrotsky A, Suib S L, Synthesis and anion exchange of tunnel structure akaganeite. Chem. Mater. 13, 4595 (2001).
[2] Peak D, Regier T, Direct observation of tetrahedrally coordinated Fe(III) in ferrihydrite. *Environ. Sci. Technol.* 46, 3163 (2012).
[3] Mazeina L, Deore S, Navrotsky A, Energetics of bulk and nano-akaganeite, β-FeOOH: Enthalpy of formation, surface enthalpy, and enthalpy of water adsorption. *Chem. Mater.* 18, 1830 (2006).
[4] Poizot P, Laruelle S, Grugeon S, Dupont L, Tarascon J M, Nano-sized transition-metal oxides as negative-electrode materials for lithium-ion batteries. *Nature* 407, 496 (2000).
[5] Wu P C, Wang W S, Huang Y T. et al, Porous iron oxide based nanorods developed as delivery nanocapsules. *Chem. Eur. J.* 13, 3878 (2007).
[6] Xiong H, Liao Y, Zhou L, Xu Y, Wang S, Biosynthesis of nanocrystal akaganéite from $FeCl_2$ solution oxidized by acidithiobacillus ferrooxidans cells. *Environ. Sci. Technol.* 42, 4165 (2008).
[7] Gash A E, Tillotson T M., Satcher J H. et al, Use of epoxides in the sol-gel synthesis of porous iron (III) oxide monoliths from Fe (III) salts. *Chem. Mater.* 15, 3268 (2003).
[8] Zhao D, Feng J, Huo Q, Melosh N. et al, Triblock copolymer syntheses of mesoporous silica with periodic 50 to 300 angstrom pores. *Science.* 279, 548 (1998).
[9] Bagshaw S A, Pinnavaia T. J, Mesoporous alumina molecular sieves. *Angew. Chem. Int. Ed..* 35, 1102 (1996).
[10] Jiao F, Bruce P G, Two- and three-dimensional mesoporous iron oxides with microporous walls. *Angew. Chem. Int. Ed.* 43, 5958 (2004).
[11] Chitrakar R, Tezuka S, Sonoda A. et al, Phosphate adsorption on synthetic goethite and akaganeite. *J. Colloid Interf. Sci.* 298, 602 (2006).
[12] Yusan S, Erenturk S A, Adsorption equilibrium and kinetics of U (VI) on beta type of akaganeite. *Desalination* 263, 233 (2010).
[13] Sun Z X, Su F W, Forsling W, Samskog P O, Surface characteristics of magnetite in aqueous suspension. *J. Colloid Interf. Sci.* 197, 151 (1998).
[14] Barrett E P, Joyner L G, Halenda P P, The determination of pore volume and area distributions in porous substances. I. Computations from nitrogen isotherms. *J. Am. Chem. Soc.* 73, 373 (1951).
[15] Kruk M, Jaroniec M, Sayari A, Application of large pore MCM-41 molecular sieves to improve pore size analysis using nitrogen adsorption measurements. *Langmuir* 13, 6267 (1997).
[16] Antonelli D M, Ying J Y, Synthesis of hexagonally packed mesoporous TiO2 by a modified sol-gel method. *Angew. Chem, Int. Ed.* 34, 2014 (1995).
[17] Ulagappan N, Rao C N R, Evidence for supramolecular organization of alkane and surfactant molecules in the process of forming mesoporous silica. *Chem. Commun.* 24, 2759 (1996).

IOP Conf. Series: Materials Science and Engineering **281** (2017) 012036 doi:10.1088/1757-899X/281/1/012036

The Film Forming Mechanism of Chromium-free Passivation in the Tannic Acid System for Electrogalvanized Coating

K Meng[1,2], J Rong[2], X H Yu[2] and Y Wang*[1]

[1] College of Mechanical and Manufacturing Engineering, Southwest Forestry University, No. 300 Bailong temple, Panlong District, Kunming, Yunnan, China
[2] Faculty of Materials Science and Engineering, Kunming University of Science and Technology, No. 253, Xuefu Road, Wuhua District, Kunming, Yunnan, China

*Corresponding email: wyuan88@126.com

Abstract: In this paper, tannic acid, fluorine titanate and silane coupling agent passivation solution were used to prepare hot dip 55% Al-Zn steel sheet passivation coating. And then the mechanism of film formation was studied. The microstructure, surface composition, layer by layer element distribution and molecular structure of 55% Al-Zn steel sheet passivation coating were revealed in scanning electron microscopy, X ray photoelectron spectroscopy, glow discharge atomic emission spectrometry and infrared spectrometry, respectively. The results show that: in one hand, using the silane coupling agent would generate many sol-gel particles in the passivation solution. On the other hand, tannic acid would produce spatial mesh structure. Therefore, passivation solution would generate a smooth, uniform and compact 55% Al-Zn steel sheet passivation coating or membrane.

1. Introduction

The traditional chromate passivation film has been widely used in industrial production because of its high corrosion resistance, high cost performance and simple process[1,2]. However, Cr^{6+} in the passivation film is carcinogenic and has great harm to the human body and environment. Therefore[3-5], the production containing Cr^{6+} has been gradually banned and eliminated by other countries. In order to find the corresponding substitute products, a great deal of research work has been done on the formulation of environment-friendly passivation liquid for hot-dip galvanized sheet at home and abroad[6-10]. Among them, hot galvanized plate and environment-friendly passivation liquid formulation of the field of [6], Japan, Germany Henkel Parker is a leading manufacturing field [7] environmental protection passivating hot galvanized plate, the Japanese iron and steel enterprises JFE, Nippon Steel in 2003 which has formed a series of products. In addition, South Korea, Pohang and the East have also successfully developed environment-friendly passivation hot-dip galvanizing products in 2000~2005, and have been widely used in home appliance manufacturing industry [8].

The related research and development work started late in our country. Although a lot of experimental studies have been carried out, there are still no industrial application achievements due to the problems of corrosion resistance, process adaptability and cost. Therefore, the domestic appliance manufacturers with galvanized steel are imported, the Chinese higher Parker passivation agent to produce environmentally friendly hot galvanized products [9,10]. In fact, using imported passivation liquid prices high, the production cost of the domestic iron and steel enterprises increased significantly, and the import of passivation solution is all foreign to provide raw materials, according to the rules in the domestic fixed simple preparation, unclear mechanism of film and passivation liquid into the name of the drug formulation and their respective functions, making the domestic steel enterprise product quality stability. In summary, according to the current common trivalent chromium, molybdenum (tungsten) acid salt, titanium salt, zirconium salt and silicate, rare earth, tannic acid and silicon based organic coating related results of [11-14], this paper intends to study the tannic acid based chromium free passivation liquid film forming mechanism, provide a reference for the development of passivation solution.

2. Experimental materials and methods

D K7716 type wire cutting machine is used to heat the 55% Al-Zn alloy steel plate into 50 * 50 * 2 mm size, which then underwent accurate grinding by 240#, 360#, 400#, 600#, 800#, 1000#, 1200#, 1500#, 1800#, 2000# sandpaper, respectively, and wash with deionized water. Finally, rinse with acetone and deionized water and dry after removing the surface grease.

The tannic acid solution with mass fraction of 2%~7% is prepared at room temperature, and then the fluorine titanate and sol silane coupling agent with mass fraction of 2% is dripped in sequence, and the mixture is quickly stirred and evenly mixed. Among them, the main effect of tannic acid solution is the formation of the three-dimensional network structure, the effect of silane coupling agent to improve the corrosion resistance of the coating, promote the crosslinking between components in solution, another major effect of silane coupling agent is to improve coating film-forming properties, reduce curing temperature and increase the oxidation resistance of the coating [12, 13]. Finally, the heat treated 55% Al-Zn alloy steel plate is dipped in the mixed solution for about 1 s, and the coating [14] is cured at about 80 DEG C after taking out.

Japanese Ricoh JSM1600-LV scanning electron microscope (SEM) was used to observe the coating surface microstructure and composition by INCA, EDS analysis, the film was scanned by layer element GDS-850A glow discharge atomic emission. Molecular structure of Nicolet Magna coating type was detected by ir-560 infrared spectrometer.

3. Results and discussion

3.1. Microscopic morphology analysis

Figure 1 shows the microscopic morphology of the sample treated with chromate free passivation solution and without any passivation solution. As can be seen from figure 1 (a), the surface of the sample is uniform, dense and smooth without using passivation liquid. figure 1 (b) using tannic acid without surface chromium passivation liquid samples despite many obvious uneven Al-Zn structure, but the passivation membrane reaction passivation solution and Al-Zn layer after the formation of a good adhesion, no passivation powder, loss and other defects. From the figure 1 (a) and figure 1 (b) we can found that the use of tannic acid system chromium free passivation solution after the heat treatment, 55% Al-Zn alloy passivation coating has better coverage and cohesiveness.

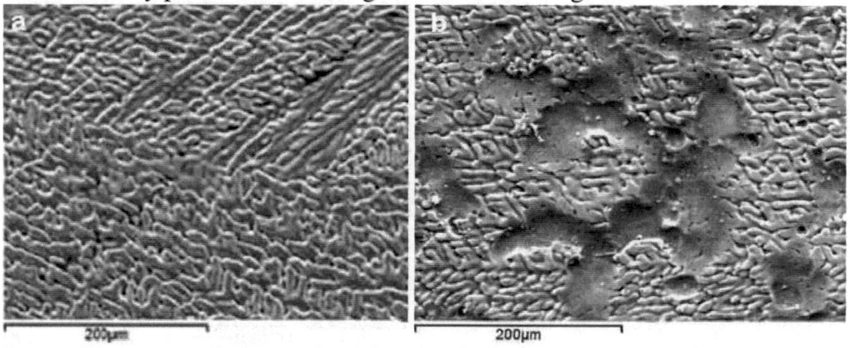

Figure 1 Microstructure morphology of coating (a) 55% Al-Zn alloy coating (b) Compared with a galvanized coating.

IOP Conf. Series: Materials Science and Engineering **281** (2017) 012036 doi:10.1088/1757-899X/281/1/012036

3.2. Coating elements and surface analysis

In order to further investigate the surface morphology and element distribution of the coating, the microstructure and composition of the sample surface were analyzed by scanning electron microscope (SEM) and energy dispersive spectrometer (EDS). The results were shown in figure 2.

As can be seen from the diagram, there is a network structure on the surface of the sample when treated with chromate chrome free passivation solution. The film is uniformly adhered to the surface of Al-Zn coating and has a corrosion pit. In addition, INCA spectrometer points out that except for Al and Zn, the surface coating elements are mainly C, O and Si.

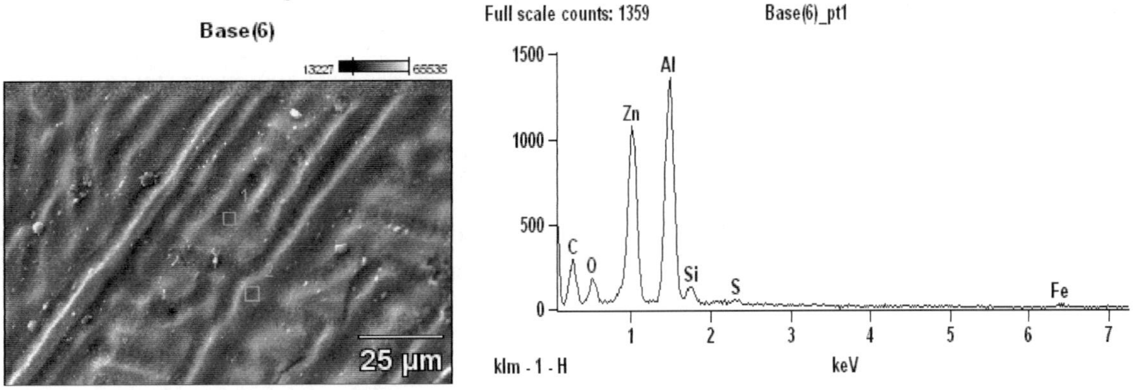

Figure 2 SEM and INCA analysis results.

The study of the element distribution of coating layer by layer is of great reference value to the existing form and distribution state of the coating. GDS-850A glow discharge atomic emission spectrometry was used to scan the film by layer by layer element. The scanning results were shown in figure 3. From the distribution of elements, it is found that the carbon decreases with the depth of the film. This shows that organic film forming material mainly exists on the surface of film. In addition, there is a cross between the passivation layer and the aluminized layer, which indicates that the passivation film and the aluminized layer are combined with a chemical bond.

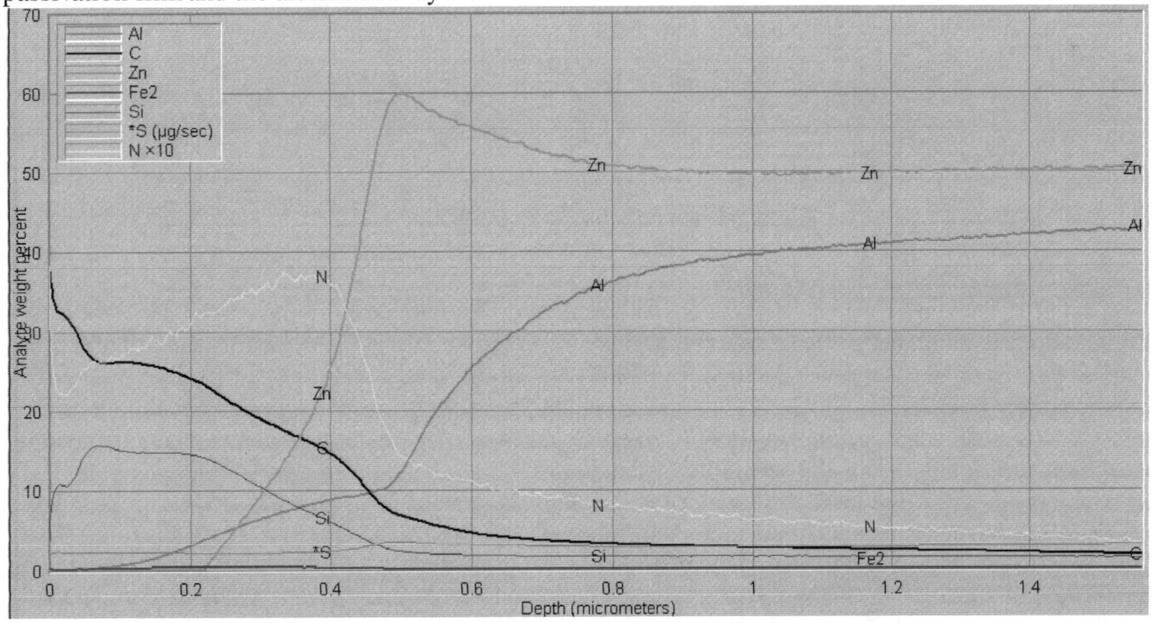

Figure 3 Analysis results of film glow spectra.

IOP Conf. Series: Materials Science and Engineering **281** (2017) 012036 doi:10.1088/1757-899X/281/1/012036

The infrared spectrum of the film can characterize the change of molecular structure during hydrolysis. As shown in figure 4, with the increase of hydrolysis time, the symmetric distortion absorption peak of Si- (CH3) at 453 cm-1, the symmetrical stretching peak of Si-O-C at 702 cm-1 and the intensity of asymmetric stretching peak at 1107 cm-1 gradually decrease. In the range of 1000-1176 cm-1, the absorption peaks of Si-O-Si, Si-O-C and Si-O-A1 overlap each other. The absorption peak at 1031 cm-1 is continuously enhanced, which is the A1-O-Si bond formed by the effect of the silanol hydroxyl group on the surface oxide of the aluminum alloy. The peak at 1450~1608 cm-1 is the result of bending vibration of the C-C bond on the benzene ring. The peaks at 1735 cm-1 are produced by stretching vibrations of the C=O group. The strongest peak at 3000-3500 cm-1 is the stretching vibration peak produced by - OH. In addition, at 760~800 cm-1, the absorption peak strength of the hydrolysis group decreases with the increase of hydrolysis time. The absorption strength of silanol group (Si-OH) at 2870~3050 cm-1 at the silane film layer of aluminum alloy decreases with the increase of aging time. It is shown that with the prolonging of curing time, the condensation of the silanol groups, silicon alcohol at 3033 cm-1 (Si-OH) absorption peak gradually weakened, and the formation of siloxane (Si-O-Si) and aluminum (Si-O-A1) siloxane bond absorption peak (1000 ~ L176 cm) continuously variable width reinforced, and showed that silane the surface of Al-Zn alloy in the passivation solution of the basic completion of the bonding reaction.

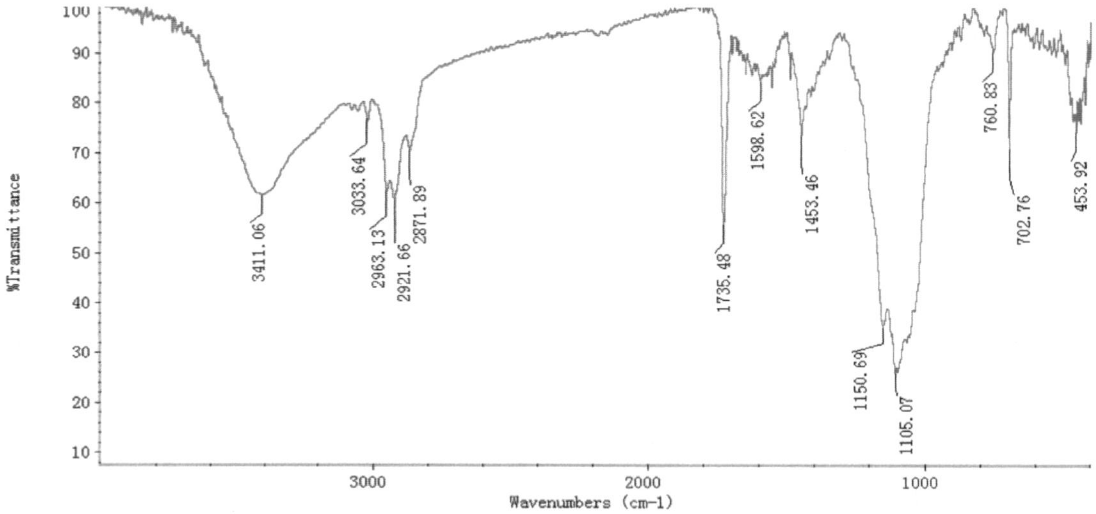

Figure 4 IR spectrum analysis of the film.

3.3. Experimental results are discussed

According to the above analysis results, it can be inferred that the film forming mechanism of the hot dip aluminized chromate free passivation liquid is as follows:

The chromate free chromate passivation solution in turn occurs in figure 5 (a), (b), (c), and (d) reactions. With the development of hydrolysis, the concentration of SiOH increases continuously. The concentration of hydrogen ion in solution increases, and the condensation of SiOH or the binding of oxygen to macromolecule will accelerate.

Subsequently, in the range of 1000-1176 cm^{-1} band, the absorption peaks of different groups have changed significantly. For example, the absorption peak of the Si-O-Si bond at 1110 cm-1 moves gradually to the high frequency band, which results in the formation of linear siloxane chains during cross-linking and curing. Moreover, in the range of 1000 ~1105 cm-1 band, the absorption peak is continuously enhanced. At this time, the aluminum siloxane (Al-O-Si) and siloxane (Si-O-Si) bonds form a complete spatial network structure at the aluminum alloy interface and sub interface. Because has not reached a steady state, so as to gradually form the film in silicone and aluminum siloxane network structure, crosslinking density increasing, the impedance of fingerprint resistant film is also increasing, the corrosion resistance of continuous improvement.

On the other hand, due to the increase of crosslinking density in the films, and nanometer styrene acrylic resin and nano colloidal SiO_2 particles filled into the void space mesh of these cross-linked, the coating porosity, permeability and swelling degree were significantly decreased, the film began to increase the uniformity. With the condensation reaction between the film without hydrolysis of the ester irreversible hydrolysis reaction and the silanol groups of the time, the film crosslink density increasing, the hydrophilic group (Si-OH) also decreased, resulting in film density and hydrophobic coating which can prevent the increasing infiltration of electrolyte effectively impedance, has been further improved, and the corrosion resistance increased again.

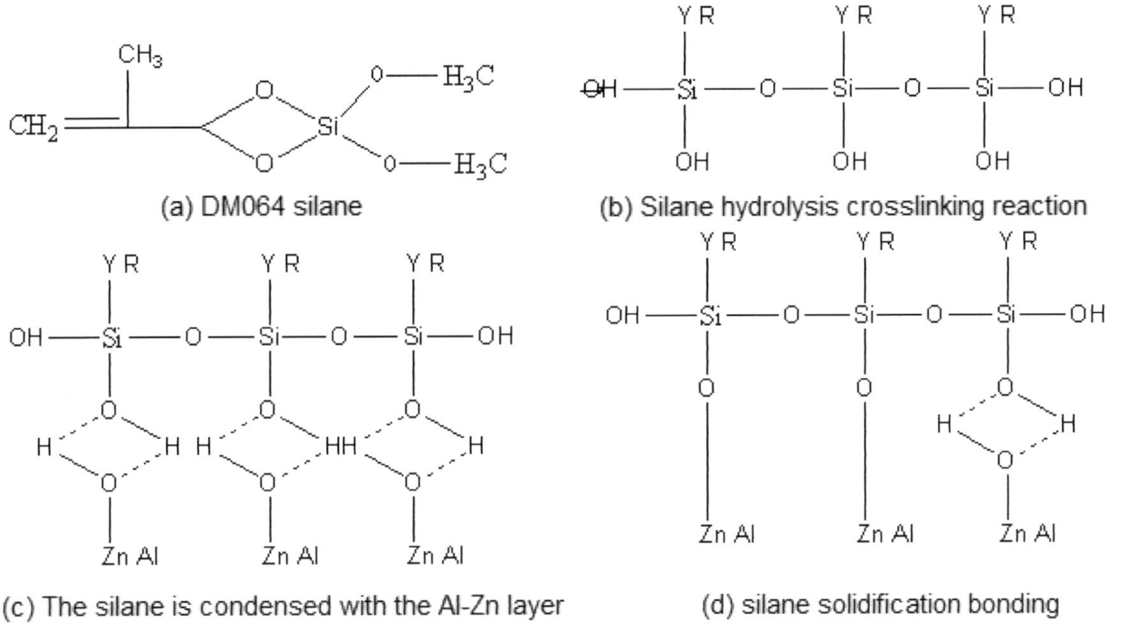

Figure 5 Mechanism of formation of chromium free passivation solution.

4. Conclusions

1) The main effect of tannic acid zinc coating system components of chromium in the passivation solution: tannic acid solution is the role of the formation of the three-dimensional network structure, 2tif6 role is to improve the corrosion resistance of the coating, promote the crosslinking of each element between the solution and the effect of silane coupling agent is the film-forming properties of the coatings to improve, reduce the curing temperature and increase antioxidation.

2) Tannic acid based hot aluminium zinc chromium free passivation liquid film forming mechanism is as follows: first to form three-dimensional network structure in the tannic acid solution, then sol type silane coupling agent reaction sol gel particles and filled in the three-dimensional network structure, finally the formation of smooth, uniform and compact combination of passivation coatings.

3) By using aluminium zinc fluoride solution and tannic acid, titanate sol silane coupling agent prepared by non chromium passivation solution has better passivation effect, surface passivation coating is uniform, dense and smooth, for industrial manufacturing environmentally friendly hot aluminium zinc plate provides a reference (has been applied in actual production).

Acknowledgements

This project have been supported by the Chinese National nature science foundation (Grant. 51301144), Yunnan eight key industrial brand professional construction project foundation (Grant.

51600631) and Yunnan Provincial Education Department Science Research Foundation (Grant. 2017YJS090)

References

[1] Silva C G D, Correia A N, Lima-Neto P D, Margarit I C P and Mattos O R 2005 *Corros. Sci.* **47** 709-22
[2] Hinton B R W and Wilson L 1989 *Corros. Sci.* **29** 967-85
[3] Wippermann K, Schuitze J W, Kessel R and Penninger J 1991 *Corros. Sci.* **32** 205-07
[4] Trabelsia T, Dhouibia L, Trikia E, Ferreira M G S and Montemor M F 2005 *Surf. Coat. Tech.* **192** 284–90
[5] Mochizuki K 1999 *Kawasaki Tech. Rep.* **41** 36-41
[6] Yong Z, Zhu J, Qiu C and Liu Y 2008 *Appl. Surf. Sci.* **255** 1672–80
[7] Chang Y T, Wen N T, Chen W K, Ger M D, Pan G T and Yang C K 2008 *Corros. Sci.* **50** 3494–99
[8] Zou Z L, Li N, Li D Y, Liu H and Mu S 2011 *J. Alloy. Compd.* **509** 503–07
[9] Szczygieł B, Winiarski J and Tylus W 2011 *Mater. Chem. Phys.* **129** 1126-31
[10] Tsaia Y T, Houb K H, Baic C Y, Lee J L and Ger M D 2010 *Thin Solid Films* **518** 7541-44
[11] Tsai C Y, Liu J S, Chen P L and Lin C S 2011 *Surf. Coat. Tech.* **205** 5124-29
[12] Figueira R B, Silva C J R and Pereira E V 2015 *J. Coat. Technol. Res.* **12** 1-35
[13] Padilla V and Alfantazi A 2013 *Corros. Sci.* **69** 174-85
[14] Célia R T, Cecilia I E and Alejandro R D S 2012 *J. Mater. Sci. Eng.* A **2** 693-11.

IOP Conf. Series: Materials Science and Engineering **281** (2017) 012037 doi:10.1088/1757-899X/281/1/012037

Enhanced Stability of Retained Austenite by Quenching and Double Partitioning Process

Z. R. Hou and X. M. Zhao*

State Key Laboratory of Rolling & Automation, Northeastern University, Shenyang 110819, People's Republic of China.
*Corresponding author. Tel. /fax: +86 24 83688544, +86 24 23906472
Postal address: Box 105, No.11, Lane 3, Wenhua Road, Heping District, Shenyang 110819, People's Republic of China

E-mail address: zhaoxm@ral.neu.edu.cn

Abstract: In the present study, a novel quenching and double partitioning (Q&DP) process is proposed. The quenching and partitioning (Q&P) process was accompanied by a second partitioning process, which led to the enhanced stability of retained austenite. The chemical composition of the investigated steel was 0.24C-1.9Mn-1.85Si, without an excess addition of alloying elements. This process aims to enhance the stability of retained austenite by controlling martensitic transformation and carbon partitioning. By applying this process, the uniform and total elongation is increased, without obvious reduction in ultimate tensile strength. An optimum combination of strength and ductility (ultimate tensile strength: 1304MPa; total elongation: 23%) was achieved by quenching to 65°C and subsequent second partitioning treatment, after the first quenching to 300°C and partitioning process. The enhanced mechanical properties were attributed to the increased amount of stabilized film-like austenite.

1. Introduction

The quenching and partitioning(Q&P) process, a promising method to produce materials with high strength and excellent ductility [1], has attracted increasing attention in the past decade. The Q&P process starts from either a full austenization state or an intercritical annealing state. Then, the steel is quenched to a temperature(QT) between martensite starting transformation temperature(MS) and martensite finishing transformation temperature(Mf), followed by a partitioning process either at or above the quenching temperature, and finally quenched to room temperature [2]. The austenite is enriched with carbon during the partitioning treatment and stabilized to room temperature [3].

Based on the CCE model, there is an optimum quenching temperature(OQT) at which the maximum austenite stabilization and retention can be obtained [4]. Fresh martensite, containing a higher concentration of carbon than tempered martensite, is detrimental for ductility [5]. It has been reported [6] that the presence of fresh martensite readily triggers microcrack and subsequently deteriorates toughness; and it also has a negative influence on the stability of austenite transformation. A second martensitic transformation will apply if the austenite is still unstable after the partitioning process. Meanwhile, second martensitic transformation temperature (M_S') is reduced to a lower temperature [7]. When the unstable austenite is quenched to a temperature between the M_S' and the M_f' again, and followed by a second partitioning process, the fresh martensite is reduced to a lower fraction and the retained austenite is enriched with higher carbon content. Gao's study [8] on the tempering behavior of Q&P process focused on the different tempering temperature. However, there are less detailed studies about the mechanism of the subsequent process.

In this paper, the microstructural evolution before and after the second partitioning process of a 0.24C-1.9Mn-1.85Si (wt.%) steel was analyzed. The effect of the second partitioning process was investigated based on experimental investigation, and the relationship between the retained austenite and fresh martensite was discussed in detail.

2. Experimental procedure

A traditional Q&P steel with chemical composition of Fe–0.24C–1.9Mn–1.85Si (wt.%) was used in this study. An ingot with thickness of 30 mm was firstly cast after melting in vacuum induction furnace, then homogenized at 1200 °C for 2h, and finally hot rolled to 3 mm sheets followed by air-cooling to room temperature. The sheet was further cold rolled to the final thickness of 1.8 mm after pickling in 10% hydrochloric acid.

The dilatometry study was carried out on a Bähr D805L quenching device equipped with quartz push-rods. The critical temperatures of A_{C1}, A_{C3} and M_S of the investigated steel are 725°C, 875°C and 360°C respectively. Specimens with dimensions of $200\times60\times1.8$ mm^3 were machined for heat treatments. The specimens were austenitized at 910°C for 5min in a tube furnace, and then quenched to 260°C, 280°C, 300°C and 320°C for 10s in a salt bath, which was marked as Q260, Q280, Q300 and Q320 respectively. After that, they were transferred to another salt bath with a temperature of 400°C for 100s. Finally, the specimens were water quenched to room temperature. Heat treatments were applied to investigate the influence of the second partitioning. After quenched to 300°C and partitioned at 400°C, the specimens were quenched to 18°C (room temperature) and 65°C again, which was marked as Q300-18 and Q300-65 respectively. After that, the samples were partitioned at 400°C for 100s again, followed by quenching to room temperature.

Tensile specimens with 25mm in gauge length and 12.5mm in gauge width were prepared along the rolling direction, and the tensile tests were performed at room temperature with the strain rate of approximately 1.33×10^{-3} s^{-1}. Metallographic specimens were ground, polished and etched with 4% nital, and observed by field emission scanning electron microscopy (FE-SEM). The specimens for TEM observation were firstly ground to a thickness of 50 μm and then twin-jet electro-polished at -30°C in an electrolyte consisting of 8% perchloric acid 92% acetic acid. The volume fraction of retained austenite was measured via X-ray diffraction (XRD) instrument equipped with a Cu tube. Samples were scanned from 40° to 100° at a scanning rate of 2°/min. The amount of retained austenite was obtained using the following equation [9]

$$V_\gamma=1.4I_\gamma/(I_\gamma+1.4\,I_\alpha) \tag{1}$$

Where V_γ, I_γ and I_α are the volume fraction of retained austenite, the average integral intensity of the (200), (220) and (311) austenite peaks and the average integral intensity of the (200) and (211) ferrite peaks. The carbon content was calculated by the following equation [10]

$$\alpha\gamma\ (nm)=0.3556+0.00453XC+0.000095XMn \tag{2}$$

Where α_γ is the austenite lattice parameter in nm; X_C and X_{Mn} are the concentrations of carbon and manganese in austenite in wt.%, respectively.

3. Results and discussion

3.1. Mechanical properties and work hardening behavior

As is shown in Fig. 1a, the ultimate tensile strength of the samples ranges from 1300MPa to 1320MPa. The Q280 sample exhibits higher elongation than the other samples. The increase in elongation is due to the higher fraction of retained austenite which provides more effective Transformation-Induced Plasticity (TRIP) effect than the other samples. Similarly, the ultimate tensile strength of the samples changes little in Fig. 1b, whereas the Q300-65 shows better elongation, even better than Q280 which contains more retained austenite. The enhanced amount of retained austenite in the partitioning process contributes to the TRIP effect during the deformation and also favors strength and elongation [11].

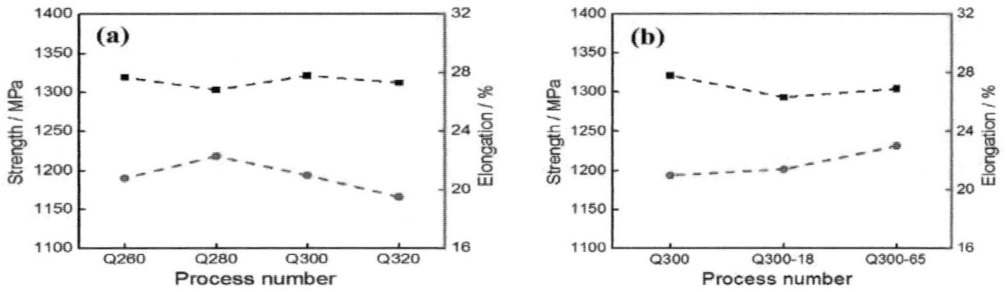

Fig. 1 Mechanical properties of the samples: (a) Q260-Q320, (b) Q300, Q300-18 and Q300-65

The engineering stress-strain curves of Q300 and Q300-65 are illustrated in Fig. 2. It is concluded in Fig. 2a that the elongation of Q300-65 increased by 2% without apparent sacrifice in strength. The increase in elongation is caused by the improved amount of film-like austenite which provides a continuous TRIP effect and effectively enhances elongation. The film-like austenite is more stable for high strains than blocky retained austenite [12]. The fresh martensite formed in the quenching process is tempered and the rest austenite is carbon-rich. In the end, there is a slight decrease in ultimate tensile strength and a slight increase in elongation.

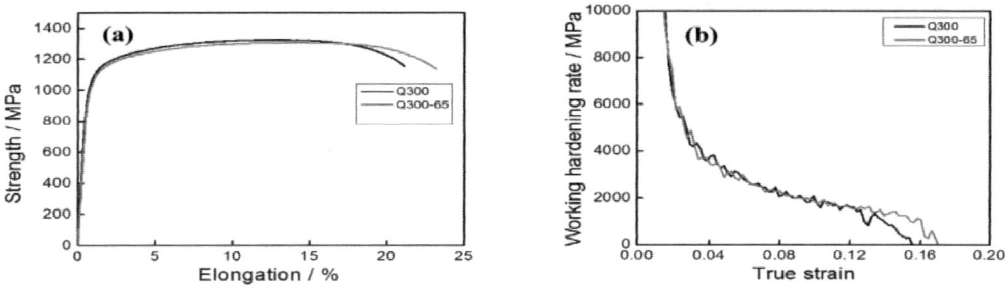

Fig. 2 (a) nominal stresss-strain curve and (b) work hardening of Q300 and Q300-65

The working hardening rate and true stress curves of Q300 and Q300-65 are presented in Fig. 2b. The work hardening rate descends quickly as the strain lower than 0.04, and decreases slowly at a strain higher than 0.04. The Q300-65 sample shows a steady and prolonged decline in working hardening rate than Q300 one. The transformation of the retained austenite to martensite can cause a local work hardening effect to delay the start of necking, which finally contributes to the higher uniform elongation. The TRIP effect of retained austenite improves the work hardening rate at high strain level and thus the uniform elongation as mentioned above.

3.2 Volume fractions and carbon content of retained austenite

The volume fraction and carbon content of retained austenite determined on the basis of XRD results are presented in Fig. 3. The volume fraction of retained austenite increases when the quenching temperature rises from 260°C to 280°C, and decreases gradually with further increase in quenching temperature (Fig. 3a). It has been mentioned above that there exists an optimum temperature which yields a maximum amount of retained austenite. It is evaluated by the lever rule on the base of the dilatometer data that the quenching temperature of 280°C gives rise to about 18% retained austenite, which is close to the optimum temperature[2,5]. However, XRD results of the retained austenite are below the calculated values. There are some reasons that affect the ultimate amount of retained austenite. First of all, for the calculated results, it is assumed that the carbon is all concentrated into austenite after the partitioning process. Nevertheless, the distribution of carbon is inhomogeneous, which may increase the time needed for homogeneous state. Secondly, the formation of bainite during the partitioning process competes with carbon partitioning and leads to the decrease in the volume fraction of retained austenite. Finally, though a substantial fraction of Si was added to the steel, the formation of cementite or transition carbide in martensite is observed during the partitioning step [13], and the remaining amount of carbon in martensite is reduced.

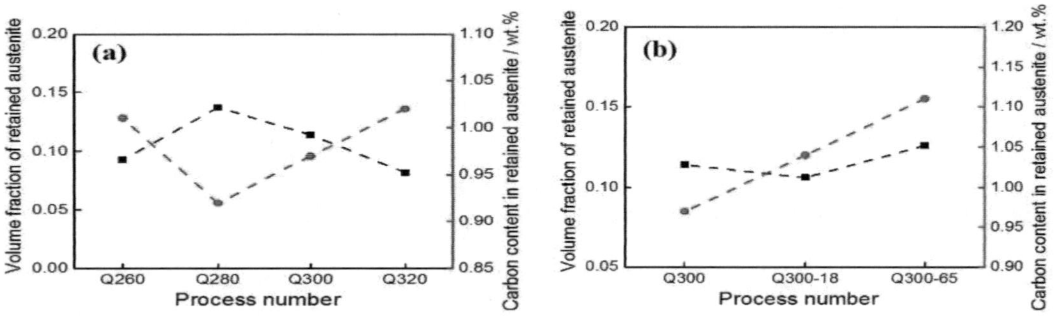

Fig. 3 Volume fraction and carbon concentration of retained austenite in the samples

However, as shown in Fig. 3b, when the second partitioning process is added to the Q300 process, the carbon content increases obviously, especially when the quenching temperature before second partitioning is 65°C. A comparison between Q300 and Q300-65 indicates that both the volume fraction and the carbon content increase in the latter process. The second partitioning progress enriches the austenite with carbon and increases the amount of the film-like austenite.

3.3 Microstructure properties relationship

Fig. 4 shows the TEM micrographs of samples treated by Q300 and Q300-65 process. Fig.4a shows the grain boundary of Q300. Martensite laths are parallel, and no retained austenite is identified. Fig. 4b~g shows the original austenite grain boundary of Q300-65. There is film-like austenite around the grain boundary, which is not found in Q300 (Fig.4b and e). Fig.4c and f are the dark field images of Fig.4b and e, in which the retained austenite is white. Fig.4d and g are the corresponding selected area electron diffraction (SAED) images of Fig.4b and e. It demonstrates that the second partitioning process increases the fraction of film-like austenite by controlling the martensitic transformation, especially the microstructure around the grain boundary, which is supposed to be blocky martensite by traditional process. The partially transformed austenite offers carbon for the untransformed austenite, and the stable austenite with higher carbon content is finally conserved to room temperature. Fig. 4c and f show that the film-like retained austenite around the grain boundary is accompanied by the martensite laths. It is considered that the film retained austenite is more stable than the blocky retained one. Additionally, the hard martensite phase surrounding film-like retained austenite provides high three-dimensional hydrostatic pressure which constrains the volume expansion and shear deformation [14], sustaining work hardening behavior at higher strain level.

In summary, the Q&DP process increases the fraction of stable film-like austenite compared with the Q&P process. The higher carbon content of the retained austenite of the Q&DP process improves the work hardening rate and the elongation by the TRIP effect. These characteristics provide the superior combination of strength and ductility (the ultimate tensile strength: 1304MPa; total elongation: 23%) by Q300-65 process.

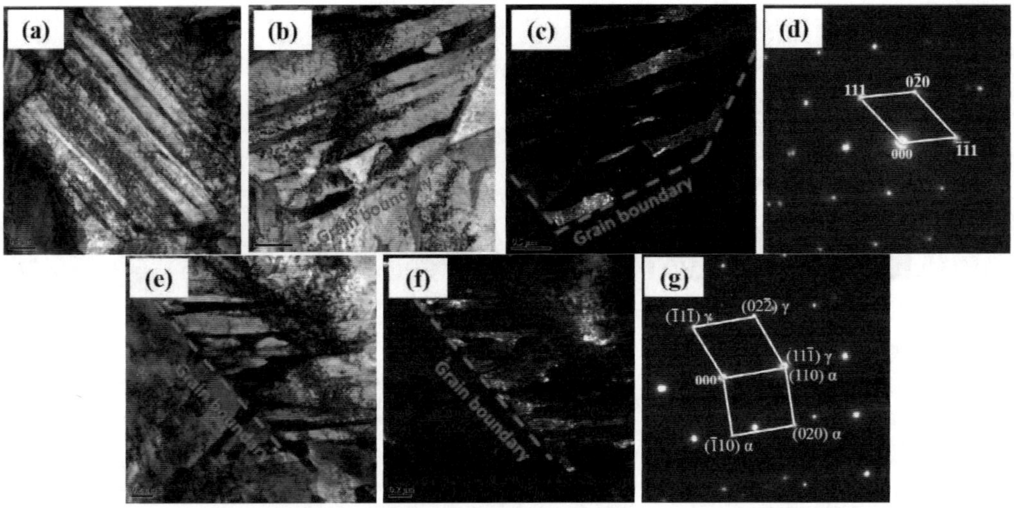

Fig. 4 TEM images of (a) Q300 and (b) ~ (g) Q300-65

4. Conclusions

The microstructure and mechanical properties of a traditional Q&P steel treated by Q&DP process have been studied. The main conclusions are as follows:

(1) The final microstructure of the steel is mainly composed of the initial martensite, retained austenite, fresh martensite and a part of bainite formed during the partitioning process. The Q280 process which is close to the optimum temperature yields a maximum amount of retained austenite.

(2) When the carbon is not enough to stabilize all the austenite in the first partitioning process, fresh martensite will form during the subsequent cooling process. By controlling the martensitic transformation, the second partitioning process can reduce the fraction of fresh martensite and enhance the amount of film-like retained austenite.

(3) The sample Q300-65 treated by the Q&DP process contains more film-like retained austenite, which offers a better combination of strength and ductility than the microstructure from the traditional Q&P process.

References

[1] Speer J G, Matlock D K, De Cooma B C n, Schroth J G, *Acta Mater.* 51 (2003) 2611-2622.
[2] Gao G, Zhang H, Tan Z, Liu W, Bai B, *Mater. Sci. Eng.* A 559 (2013) 165–169.
[3] Speer J G, Edmonds D V, Rizzo F C, Matlock D K, Curr. *Opin. Solid State Mater. Sci.* (8) 2004 219-237.
[4] Clarke A J, Speer J G, Matlock D K, Rizzo F C, Edmonds D V, Santofimia M.J., *Scripta Mater.* 61 (2009) 149–152.
[5] Akbary F H, Santofimia M J, Sietsma J, *Adv. Mater. Res.* 829 (2014) 100-104.
[6] Knijf D D, Petrov R, Föjer C, Kestens L A I, *Mater. Sci. Eng.* A 615 (2014) 107–115.
[7] Gao G, Zhang H, Gui X, Luo P, Tan Z, Bai B, *Acta Mater.* 76 (2014) 425-433.
[8] Gao G, Zhang H, Gui X, Luo P, Tan Z, Bai B, *J Mater Sci. Technol.,* 31 (2015), 199-204
[9] Sugimoto K I, Usui N, Kobayashi M, Hashimoto S I, *ISIJ Int.* 32 (1992) 1311–1318.
[10] Santofimia M J, T N Minh, Zhao L, Petrov R, Sabirov I, Sietsma J, *Mater. Sci. Eng.* A 527 (2010) 6429–6439.
[11] Santofimia M J, Zhao L, Petrov R, Kwakernaak C, Sloof W G, Sietsma J, *Acta Mater.* 59 (2011) 6059–6068.
[12] Xiong X C Xiong, Chen B, Huang M X, Wang J F, Wang L, *Scripta Mater.* 68 (2013) 321-324.
[13] Toji Y, Matsuda H, Herbig M, Choi P P, Raabe D, *Acta Mater.* 65 (2014) 215-228.
[14] Timokhina, P Hodgson, E Pereloma, *Metall. Mater. Trans.* A 35 (2004) 2331–2341.

IOP Conf. Series: Materials Science and Engineering **281** (2017) 012038 doi:10.1088/1757-899X/281/1/012038

Working Performance Analysis of Rolling Bearings Used in Mining Electric Excavator Crowd Reducer

Y H Zhang[1], G Hou[1*], G Chen[2], J F Liang[2] and Y M Zheng[2]

[1] School of Mechanical Engineering, Dalian Jiaotong University, Dalian 116028, Liaoning, China
[2] Branch of Gear Transmission, Taiyuan Heavy Industry Co., Ltd, Taiyuan 030024, Shanxi, China

Email: geng.hou@qq.com

Abstract. Refer to the statistical load data of digging process, on the basis of simulation analysis of crowd reducer system dynamics, the working performance simulation analysis of rolling bearings used in crowd reducer of large mining electric excavator is completed. The contents of simulation analysis include analysis of internal load distribution, rolling elements contact stresses and rolling bearing fatigue life. The internal load characteristics of rolling elements in cylindrical roller bearings are obtained. The results of this study identified that all rolling bearings satisfy the requirements of contact strength and fatigue life. The rationality of bearings selection and arrangement is also verified.

1. Introduction

Large mining electric excavator with standard bucket capacity of $55m^3$ is an important equipment used for intermittent mining opencast pit which annual output is more than 20 million tons [1,2]. The digging process of mining electric excavator is completed by coordinating of crowd mechanism and hoist mechanism [3], as shown in figure 1.The crowd reducer is an important component of crowd mechanism of electric excavator used for transfer crowd force. In digging process, dipper driving bucker to press material is completed due to crowd force (as F_c shown in figure 1). And bearings used in crowd reducer play the role of supporting reducer shaft system. The bearings installed on the input shaft and the medium shaft are rolling bearings (paper studying objects), and which installed on the output shaft (crowd shaft) are sliding bearings.

Based on the design of crowd reducer of large mining electric excavator, the computer aided analysis software of *Romax Designer* is used for simulation analysis of rolling bearings internal load distribution and contact stresses [4], in order to provide reference for related design and research.

2. Basic information and working conditions of rolling bearings

Kinematic sketch of electric excavator crowd mechanism (as shown in figure 2) shows that excavator crowd reducer contain 3 shafts that are input shaft, medium shaft and output shaft respectively. The rack is fixedly connected with excavator dipper. Both ends of each shaft are supported by a bearing. Thereinto, the bearings supporting the input shaft and the medium shaft are rolling bearings, and which supporting the output shaft are sliding bearings. Because crowd reducer adopts two stages involute straight spur gear transmission, theoretically, there is no axial force to act on shafts. But it is inevitable that slight axial force acts on shafts in practice. Therefore, cylindrical roller bearings are chosen to support input shaft and medium shaft. Bearing on the one side of shaft has no capacity of axial fixing, and the bearing on the other side of shaft has capability of axial bidirectional fixing. Through this bearing combination, radial and axial supporting for input shaft and medium shaft of crowd reducer are realized.

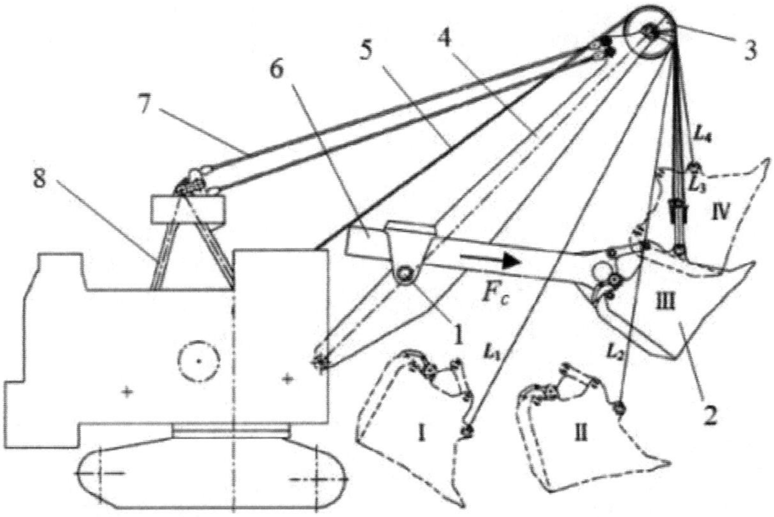

1-crowd shaft; 2-bucket; 3-head sheave; 4-jib; 5-hoist wirerope;
6-dipper; 7-luffing wirerope; 8- A type bracket

Figure 1. Diagram of excavator digging process

Bearings arrangement of excavator crowd reducer as shown in figure 3, bearing 1 to bearing 4 are rolling bearings, bearing 5 and bearing 6 are sliding bearings. The model and description of each rolling bearing are shown in table 1. (Bearing selection and combination of designed excavator crowd reducer in paper are different from existing product. Compared to tapered roller bearings, this design can almost ignore the negative influence of axial force on drive system)

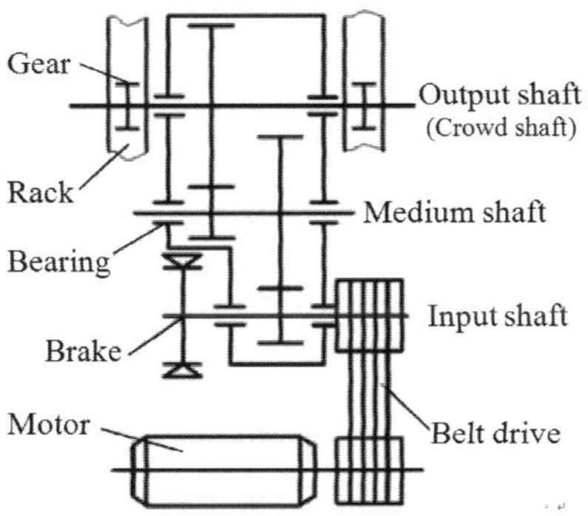

Figure 2. Kinematic sketch of crowd mechanism.

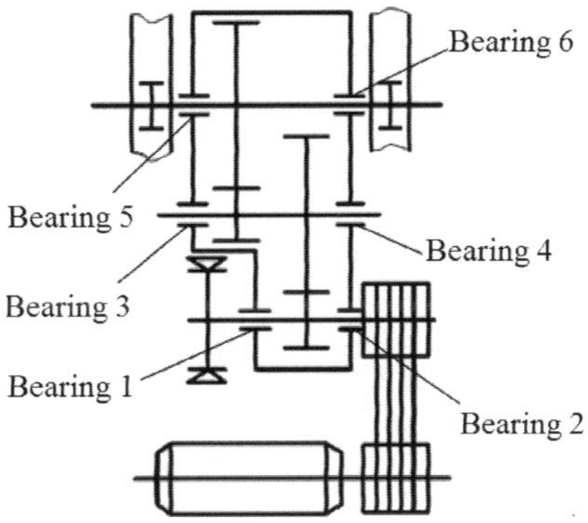

Figure 3. Diagram of bearings arrangement of crowd reducer.

IOP Conf. Series: Materials Science and Engineering **281** (2017) 012038 doi:10.1088/1757-899X/281/1/012038

Table 1. Model and description of rolling bearings of crowd reducer

Bearing number	Bearing model	Description
Bearing 1	NH332R	Both inner ring and outer ring have bilateral ribs
Bearing 2	NU2332R	Outer ring has bilateral ribs, inner ring has no rib
Bearing 3	NH2336	Both inner ring and outer ring have bilateral ribs
Bearing 4	NU336	Outer ring has bilateral ribs, inner ring has no rib

In addition, the lube of designed crowd reducer is VG 220 synthetic oil, and its working temperature is 70 °C.Bearing and shaft are connected by shrink fit with basic bore system; bearing and housing are connected by transition fit with basic shaft system [5]. Radial clearance of each bearing conforms to its standard radial clearance [6].

The three-dimensional model created by engineering software *Romax Designer* is shown in figure 4. Since the difference of statistical maximum crowd force of electric excavator at different digging position is very small, and the maximum crowd force within one digging process occurs at the digging position III in usual, the boundary conditions applied to the bearing analysis all conform to the statistical maximum crowd force condition at digging position III[7]. Under this digging condition, loads on the crowd reducer are shown in figure 5. In this diagram, crowd force is F_c=1272.9kN; The force belt acting on pulley is F_B=65kN; Output torque is M=222.76N▪m;Input revs of reducer is 463r/min; Input power is 528kW; α=45° ,θ=90° .

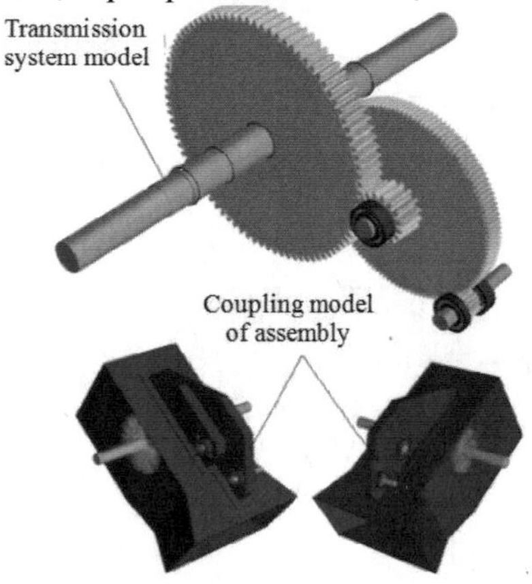

Figure 4. Model of excavator crowd reducer

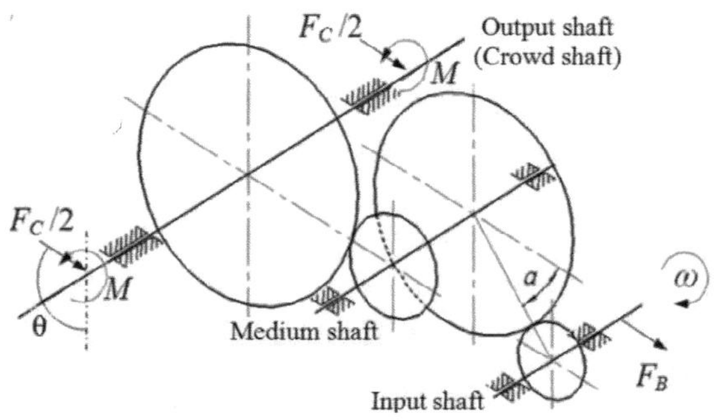

Figure 5. Force diagram of crowd reducer at digging location III

3. Analysis of rolling bearings working characteristics

3.1. Analysis of internal load distribution in rolling bearings

The revs of each bearing of crowd reducer is low, therefore, the impact of rolling elements inertia force is neglected in analysis. Under radial load, due to radial clearance and contact deformation of cylindrical roller bearings, the load magnitudes of each rolling element within angle range of load area are not exactly the same [8] (as shown in figure 6). For cylindrical roller bearings, the loads on each roller conform to [9]

$$Q_{\psi} = Q_{\max}\left[1-\frac{1}{2\varepsilon}(1-\cos\psi)\right]^{\frac{10}{9}} \tag{1}$$

Among this equation, the ε is written as

$$\varepsilon = \frac{1}{2}\left(1 - \frac{P_d}{2\delta}\right) \tag{2}$$

In equation (1) and equation (2), Q_ψ is normal load of rolling element within load range from different angles (ψ is location angle); Q_{max} is the normal load value of rolling element subjected to the maximum normal load (i.e., F_0 in figure 6); ε is load distribution factor; P_d is bearing radial clearance; δ is radial movement of bearing ring in location of ψ=0° .

Based on above theory, by simulation of software *Romax Designer*, the internal load distributions of cylindrical roller bearings used in excavator crowd reducer are obtained. The polar diagrams of load distribution at the contact point between rolling elements and raceway (including inner raceway and outer raceway) are shown in figure 7. (In this figure, the polar angle is location angle, its unit is °; and the pole diameter is normal load of the rolling element, its unit is N.) The number of loaded rolling elements and the load value of rolling element under maximum normal load are shown in table 2.

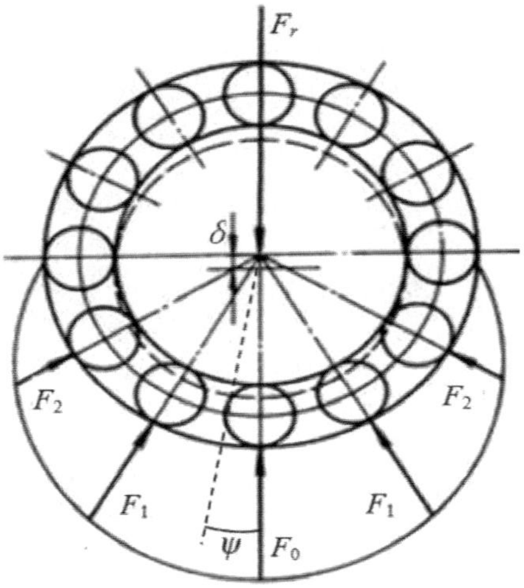

Figure 6. Diagram of rolling bearing internal load distribution

Figure 7 shows that the loads applied to each rolling element by inner raceway and outer raceway are almost equal. It is attributed to the radial balance of rolling elements of the rolling bearings. And table 2 shows that the number of rolling elements under load is more than 1/3 of total number of rolling element.

Table 2. The status of each bearing internal load distribution of crowd reducer

Bearing number	Total number of rolling element	The number of rolling element under load	Maximum load of rolling bearing Q_{max}/N
Bearing 1	14	5	23040
Bearing 2	14	5	57035
Bearing 3	17	7	82776
Bearing 4	17	7	59345

3.2. Analysis of internal contact stresses in rolling bearings

The contact stress analysis of rolling bearing includes the contact stress between the rolling elements and inner raceway, and the contact stress between the rolling elements and outer raceway. For cylindrical roller bearings, the contact region between rolling element and the raceway is line before loading, and then it will change into small surface contact after loading. Based on the analysis of rolling bearing internal loads of excavator crowd reducer, using software *Romax Designer*, the contact

IOP Conf. Series: Materials Science and Engineering **281** (2017) 012038 doi:10.1088/1757-899X/281/1/012038

stresses between rolling elements and raceways (inner raceway and outer raceway) are solved. The simulation results are shown in figure 8. (In this figure, the polar angle is location angle, its unit is ° ; and the polar diameter is the normal contact stress of rolling element, its unit is MPa.) The contact stress values of rolling element under the maximum load are shown in table 3.

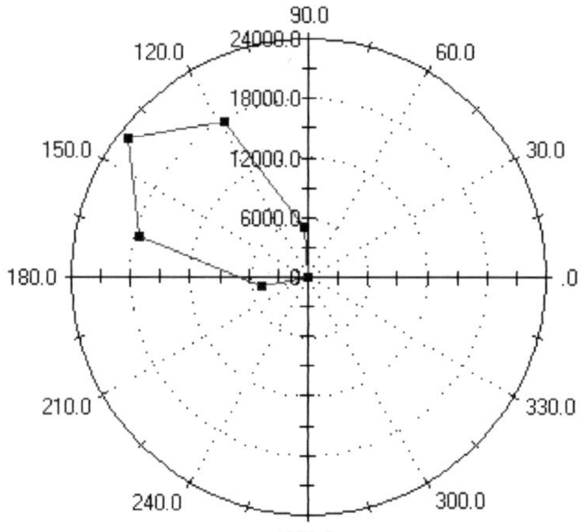

(a) Internal load distribution of bearing 1

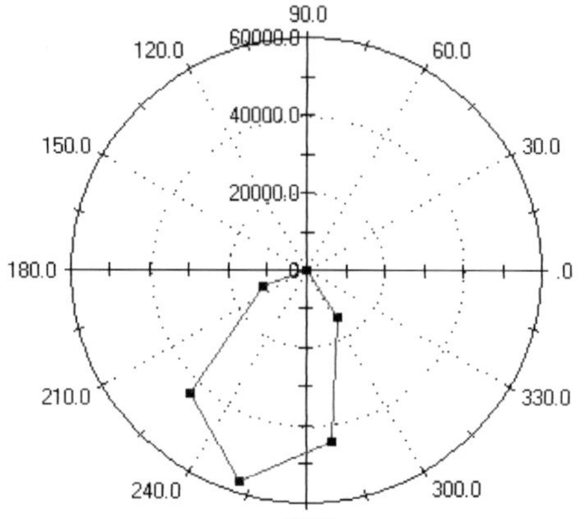

(b) Internal load distribution of bearing 2

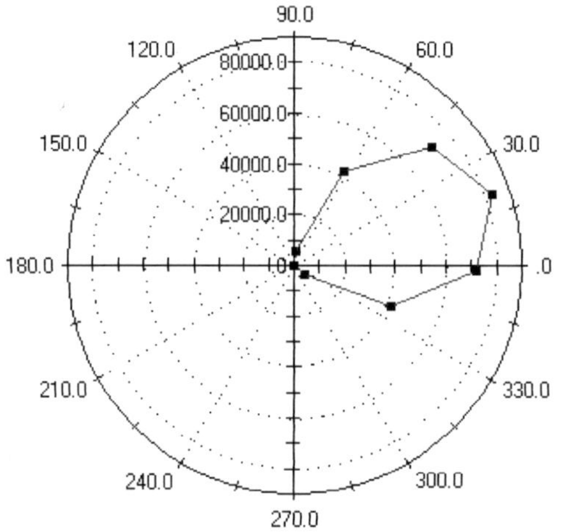

(c) Internal load distribution of bearing 3

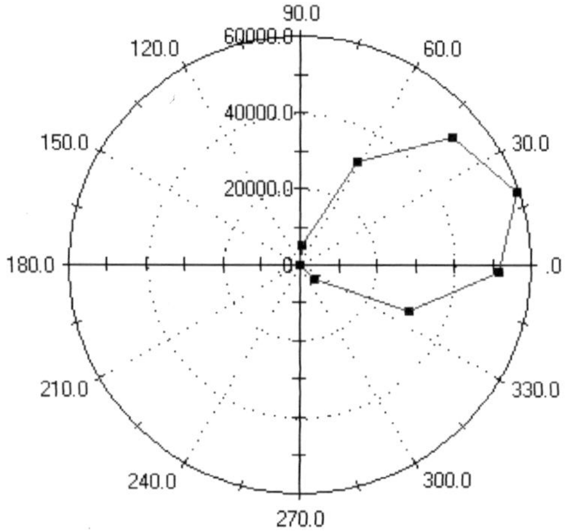

(d) Internal load distribution of bearing 4

Inner raceway Outer raceway

Figure 7. Polar diagram of each rolling bearing internal load distribution of crowd reducer

Table 3. The status of each bearing internal contact stress distribution of crowd reducer

Bearing number	Contact value between rolling element and inner raceway/MPa	Contact value between rolling element and outer raceway /MPa
Bearing 1	1276	1059
Bearing 2	1426	1173
Bearing 3	1716	1450
Bearing 4	1867	1578

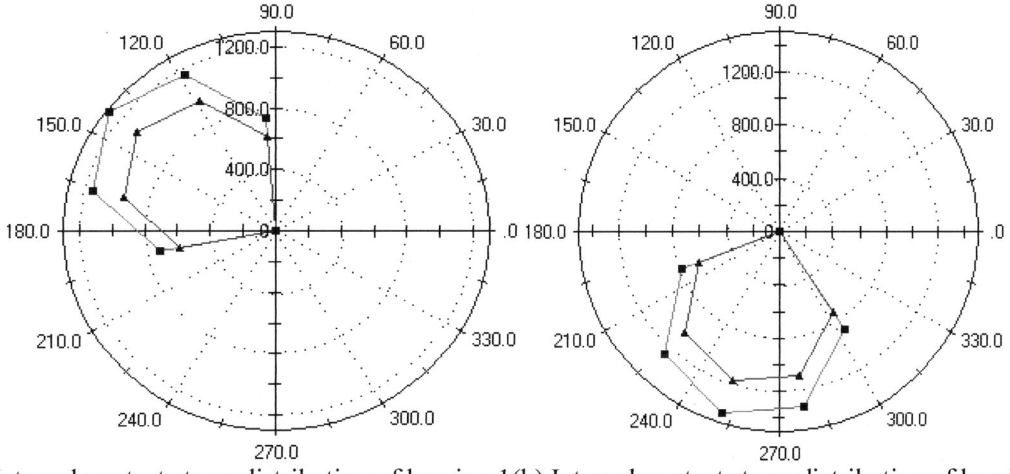

(a) Internal contact stress distribution of bearing 1(b) Internal contact stress distribution of bearing 2

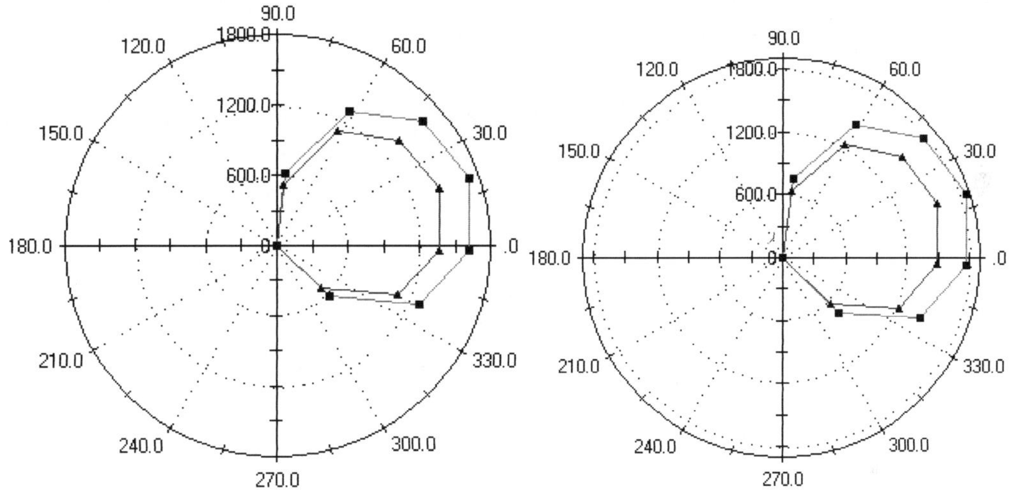

(c) Internal contact stress distribution of bearing 3(d) Internal contact stress distribution of bearing 4

—■—Inner raceway —▲— Outer raceway

Figure 8. Polar diagram of each rolling bearing internal stress distribution of crowd reducer

The allowable contact stress of cylindrical roller bearings is 4000MPa [10], therefore, all bearings used in excavator crowd reducer satisfy the requirements of contact strength. Furthermore, it can be known from figure 8 and table 3 that the contact stresses between rolling elements and inner raceway are greater than contact stresses between rolling elements and inner raceway. This is because the synthetic curvature of contact between the rolling element and inner raceway is greater than synthetic curvature of contact between rolling element and the outer raceway. According to Hertz contact theory [11], the greater synthetic curvature is, the greater contact stress is.

3.3. Analysis of fatigue life of rolling bearings

According to international standard of ISO/TS 16281, the fatigue life of rolling bearings used in excavator crowd reducer is obtained [12]. The simulation analysis results are shown in table 4.

Table 4. The fatigue life of each roller bearing

Bearing number	Fatigue life/h
Bearing 1	10194000
Bearing 2	1161100
Bearing 3	138240
Bearing 4	92148

The simulation results show that the fatigue life of each bearing satisfy the design requirements of 40000 hours at least.

4. Conclusions

Through the working performance simulation analysis of rolling bearings used in excavator crowd reducer in the process of operation, the following conclusions can be drawn. For the same rolling element, the force acted from inner raceway is equal to the force acted from outer raceway, but the contact stress with inner raceway is greater than contact stress with outer raceway. Besides, all rolling bearings satisfy the requirements of contact strength and fatigue life. Therefore, the rationality of bearings selection and arrangement is also verified.

Acknowledgments
This work is supported by the 2011 Collaborative Innovation Center Project of Ministry of Education of China, and the Ph.D. Initial Scientific Research Fund of Liaoning Province, China (NO. 201501183).

References

[1] Yan S W 1991 *Design of Mechanical Excavator* (Beijing, China: China Machine Press) pp. 1-2.
[2] Zhao H Z and Shi M X. The Machining of Hoist Drum for WK-55 Electric Rope Shovel J. *Mechanical Management and Development* Vol.5, No.10, (2012) pp.118-119.
[3] Di L W, Kong D W and Bai X. Parameterization design of mechanical type mine excavator equipment based on simulative prototype technology J. *Construction Machinery* No.1, (2013) pp.78-82.
[4] Zhang L H and Yi J. Analysis on contact status of cylindrical roller bearings for railway passenger car axle box based on Romax J. *Journal of Harbin Bearing* Vol.38,No.1, (2017) pp.3-7.
[5] Zhang X J 2013 *Interchangeability and Basic of Measuring Technology* (Beijing, China: Tsinghua University Press) pp.238-250.
[6] International Organization for Standardization. Rolling bearings clearance — Part 1: Radial internal clearance for radial bearings (ISO 5753-1:2009) (2009) pp.5-6.
[7] Yuan H, Load spectrum and statistical dynamics analysis of large mining shovel. Unpulished M.Sc. dissertation, Northeastern University, Shenyang, China, 2014 pp.167-192.
[8] Chen J Q, Zhou H and Wu S Q. Theoretical Study on Load Distribution of Cylindrical Roller Bearing J. *Bearing* No.6, (2001) pp.8-11.
[9] Harris T A and Kotzalas M N 2010 Rolling Bearing Analysis Vol.1— *Essential Concepts of Bearing Technology* (Beijing, China: China Machine Press) pp.135-138.
[10] International Organization for Standardization. Rolling bearings — Static load ratings (ISO 76:2006) (2006) p.v.
[11] Johnson K L 1992 *Contact Mechnics* (Beijing, China: Higher Education Press) pp 96-122.
[12] International Organization for Standardization. Rolling bearings—Methods for calculating the modified reference rating life for universally loaded bearings (ISO/TS 16281) (2008) pp.1-19.

IOP Conf. Series: Materials Science and Engineering **281** (2017) 012039 doi:10.1088/1757-899X/281/1/012039

Research on Treatment Technology and Device of Oily Sludge

J Q Wang[1], F S Shui[1*] and Q F Li[1,2]

1 School of Petroleum Engineering, Xi'an Shiyou University,No.18 Dianzi 2nd Road,Xi'an 710065, China
2 Yanchang Petroleum Group Exploration Company, Zaoyuan Road,Yan'an 716099, China

E-mail: wjq_xasy@163.com; 279653937@qq.com; lf646332814@qq.com

Abstract. Oily sludge is a solid oily waste, which is produced during the process of oil exploitation, transportation, refining and treatment of oily sewage. It contains a great number of hazardous substance, and is difficult to handle with. To solve the problem of waste resources of oil sludge with high oil content and usually not easy to aggregate during the preparation of profile control agent, a new oily sludge treatment device was developed. This device consists of heat supply unit, flush and filter unit, oil removal unit and dehydration unit. It can effectively clean and filter out the waste from oily sludge, recycle the oil resources and reduce the water content of the residue. In the process of operation, the water and chemical agent are recycled in the device, eventually producing little sewage. The device is small, easy to move and has high degree of automation control. The experimental application shows that the oil removal rate of the oily sludge is up to 70%, and the higher the oil content rate the better the treatment.

1. Introduction

In the process of development and transportation, oil field will inevitably produce a large amount of oily sludge, which mostly contains petroleum hydrocarbons, if directly emitted without treatment or below the standard, it will cause serious harm [1]to the surrounding. Various ways have been used to combat oil sludge both at home and abroad. These methods include: thermal desorption technology [2], conditioning-mechanical separation technology [3], incineration technology [4] and microbial technology [5, 6] and other technologies. The matching device of oily sludge treatment is mostly used for refinement. The Chemical Equipment Research Institute of China University of Petroleum has developed a skid mounted oily sludge treatment device, which can realize the oil content of oily sludge less than 1%, and then through the deeply pyrolysis can make the mineral oil content less than 0.3% [7]; Qing Li et al developed the thermochemical separation skid mounted oily sludge treatment device, it can reduce the irreducible oil of oily sludge to 0. 21% ~ 0. 65% [8]; The residues and sewage treated by the oily sludge treatment mobile device developed by Wenya Wang can meet the requirements of the national environmental protection, but the whole process should be carried out under the steam protection [9]. These above oily sludge treatment technologies and devices are inadequate in complex treatment process, low economic efficiency and resources waste.

In recent years, the research on the oily sludge profile control technology has been carried out in major oil fields, and the different formula of profile control agent system for oily sludge has been developed according to the different reservoir characteristics [10,11,12]. Oily sludge as the main

IOP Conf. Series: Materials Science and Engineering **281** (2017) 012039 doi:10.1088/1757-899X/281/1/012039

ingredients of profile control agent, to a certain degree, reduce the treatment difficulty and costs of oily sludge. However, most of the current research is to directly add a certain proportion of chemical additives in the oily sludge to make the profile control agent, which will cause serious waste of resources for oily sludge with high oil content. In addition, the polymerization effect of oily sludge with high oil content is not good, and after the oil removal agent, the sludge polymerization is complete, strong and have a good granulation performance [13].

In view of this situation, the focus of this paper is to design a set of oily sludge treatment device, cleaning and filtering the waste and reducing the oil content from oily sludge; Moreover, the water content of the treated oily sludge is low, which is convenient for transportation and storage. So it can be used as the main material for preparation of profile control agent.

2. Technological process

The technological process adopts the procedure of "hot water flushing, medicine demulsification, gravity oil removal and centrifugal dewatering", which is combined with the waste utilization of the oily sludge profile control. The main process is shown in Figure 1.

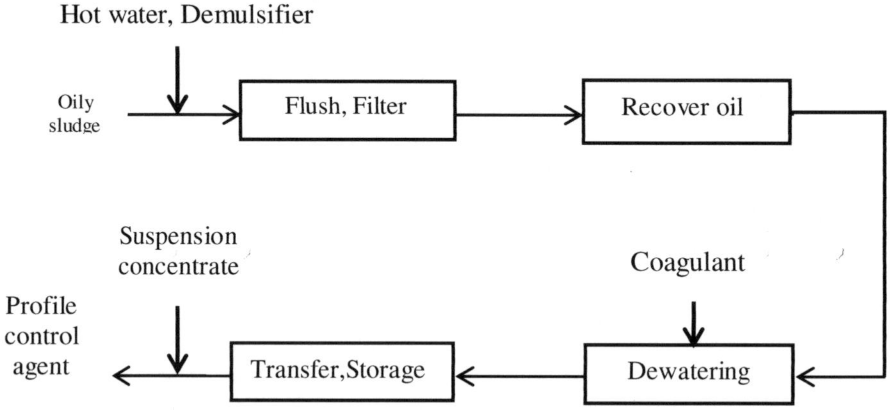

Figure. 1 Technological process of oily sludge

Oily sludge treatment can be used for profile control agent of injection well, it divides the oily sludge treatment system into the four section. In the process, according to the sources and the oil content of oily sludge, under the control of PLC, changing the process conditions and controlling the proportion of chemical agents, to realize the function of the same set of devices to deal with various types of oily sludge.

Impurity removal section: In this section, the method is combined with chemical demulsification, high temperature and high pressure water flushing, using filter drum and filter screen. It is effective for oily sludge, especially for mixture consist of oil and soil. The working principle is: The oily sludge enters into the device through the inlet point, and under high pressure, using more than 80°C water mixed with chemicals to flush oily sludge, while the trommelsieve rotating and then the waste in the oily sludge is filtered.

Gravity separation-Oil removal section: The function of this section depends on gravity, making oil and water stratification, and then recycle oil in oily sludge as much as possible. The working principle is: After flushing and filtering, the oily sludge enters the separation tank, stirred for a certain time to make water and oily sludge mixed evenly and fully reacted. After stirring, holding 50mins to 60mins, and taking the upper layer of oil, the lower water and the silt, the oily sludge through the sludge pump, and then enters the centrifuge for dehydration treatment, in this section, the temperature is controlled at about 60°C.

Chemical flocculation-Centrifugal dehydration section: The main function of this section is to separate the water from the silt in the centrifuge, to reduce the water content of the silt, so as to facilitate the transportation and storage of the oily sludge. The working principle is: When the high water content of the oily sludge enters the centrifuge, the centrifugal force generated by the high-speed rotation of the centrifuge rotor, speeding up the settling velocity of particles in the liquid, to achieve the purpose of separating water and silt. The separated silt can be transported or stored, and the separated water can be directly reused after heating.

Physical dispersion-Chemical suspension: After the oily sludge treatment,, the new condensed "muddy structure" is difficult to disperse evenly under the mixing action of the liquid preparation pool during the transport or storage process. For this, special means are needed for decentralized treatment. It's working principle is similar to the flush and filter section: The injected water disperse mud under the high pressure, and the not up to standard of the solid particles were filtered out. With the filtrate enters the liquid preparation pool, based on a certain ratio, we can add suspension concentrate, dispersant and other additives to make up the profile control agent of oily sludge.

3. Design of device

According to the process sequence of oily sludge pretreatment, the oily sludge treatment device design is shown in Figure 2, based on the sectional functions, it can be divided into flush and filter section, oil section, sludge dehydration section and boiler section. In the working process, the water can be recycled, and the last of a small amounts of residual sewage can be treated in the sewage treatment station; In the whole process, the oily sludge can closed flow in tank and pipeline, which can effectively ensure the cleaning of the working site; In addition, the separated waste oil can be sent back to the joint station for treatment, and the dehydration mud cake can be stored in the destination, which can be used for profile control of water well.

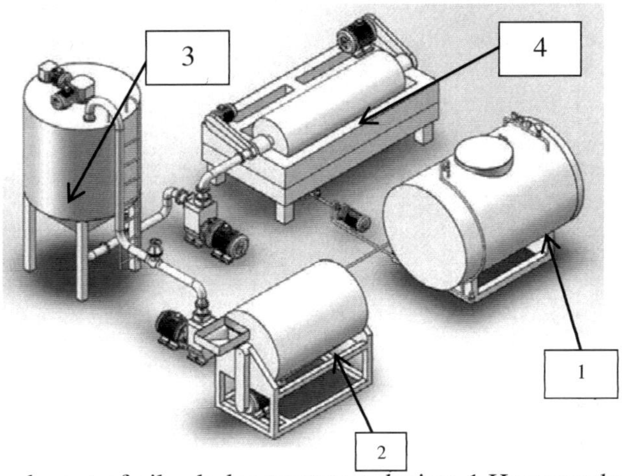

Figure. 2 Main unit space layout of oily sludge treatment device: 1 Heat supply unit 2 Flush and filter unit 3 Oil removal unit 4 Dehydration unit

The whole device can be divided into four units according to the space layout:

Heat supply unit: Electric heating boiler is the heat supply unit. Three of electric heating pipe, with triangular connection, connected into a group. It is covered with a connecting tube, connected with the external electrical equipment and installed on the flange plate. Each group of electric heating tube is covered with an electric heating shield, and the heating process of oily sludge is completed in the cylinder without arranging the pipeline. Electric heating boiler adopts metal tubular electric heater, so that the electric energy is directly converted into the heat energy. this process does not need air and fuel for combustion, and will not discharge harmful gases and ash in line with environmental requirements.

Flush and filter unit: The flush and filter unit mainly consist of cylindrical filter screen, impeller, high-pressure water nozzle, window, motor, hollow shaft and insulating layer, etc. The high-pressure water nozzle is distributed on the hollow shaft, after material enters the unit, the cylindrical filter screen rotates around the hollow shaft. Hot water flush material by nozzle, and the filtrate produced by flushing and filtering enters the settling tank. It is shown in Figure 3.

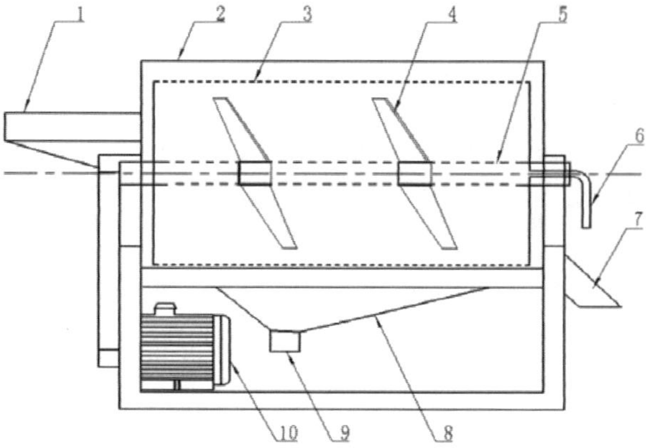

Figure. 3 Flush and filter unit layout: 1 Inlet point 2 Envelope 3 Cylindrical filter screen 4 Impeller 5 Water nozzle 6 Water inlet 7 Residue outlet 8 Liquid hopper 9 Liquid outlet 10 Motor

Oil removal unit: The oil removal unit mainly includes settling tank, oil collector, agitator, cat ladder, window, insulating layer, trace band, the liquid level and the temperature sensor, etc. The upper part of the settling tank is a cylinder, the lower part is a conical funnel and the outer layer is sequentially fitted with an insulating layer and a trace band. The blade of the agitator has a certain angle, which allows the material inside the tank to roll up and down. The effective volume of the tank is $2m^3$. It is shown in Figure 4.

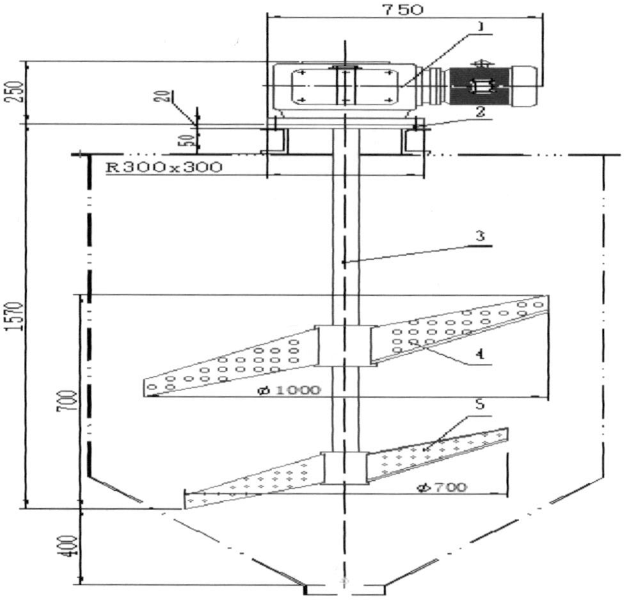

Figure. 4 Oil removal unit layout: 1 Decelerator 2 Base plate 3 Agitator shaft 4 Impeller 5 Tank

Dehydration unit: The core of the dehydration unit is the centrifuge, which includes the pulley, transmission, feeding screw, rotary drum and bearing pedestal, etc. Rotary drum and screw with a certain velocity contrast rotating rapidly, while the material continuously enters the feeding screw inner from feeding tube and then enters the rotary drum after acceleration. Under the action of centrifugal force, the heavier solid components are deposited on the wall of rotary drum, and then the sediment layer is formed. The solid component of the deposit is continuously pushed to the end of the rotary drum by the feeding screw, and is discharged out from the slag-drip outlets. The lighter liquid component forms an inner liquid ring, which is continuously overflowed from the main overflow port of the rotary drum and discharged out through the fluid outlets. It is shown in Figure 5.

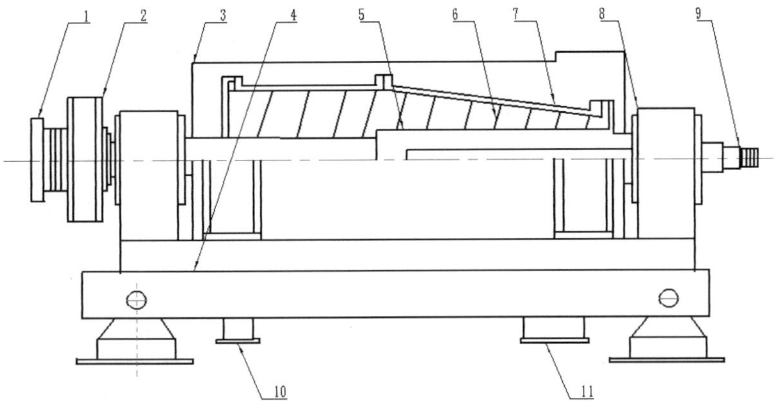

Figure. 5 Dehydration unit layout: 1 Pulley 2 Transmission 3 Envelope 4 Machine base 5 Outlet 6 Feeding screw 7 Rotary drum 8 Bearing pedestal 9 Feeding tube 10 Clarified liquid outlet 11 Slag-drip outlet

At present, the oily sludge in oil field is mostly stored in woven bags. Because of the high viscosity of oily sludge, it is very laborious and time-consuming to dump and feed. In order to solve this problem, this paper designs unloading device of sacked oily sludge [14] (Figure 6). The working principle of the device is as follows: The sacked oily sludge is inverted on the transmission belt, and the sludge is discharged from the oily sludge bag under the extrusion of the progressive gyro wheel. Finally, the extrusion of two large pressure gyro wheels complete the dumping of the oily sludge. The whole process is accompanied by washing with the high pressure water to prevent the oily sludge clogging the crack between the gyro wheels. The design can effectively solve the problem of discharging sacked oily sludge and improving the treatment efficiency of oily sludge, which has been granted a Chinese patent certificate.

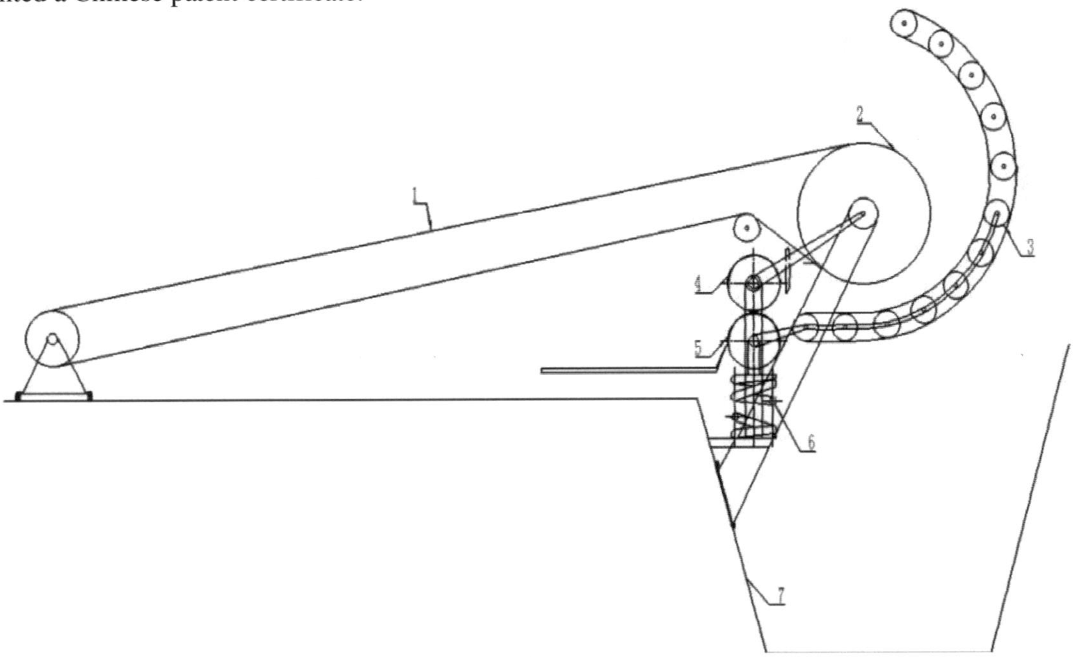

Figure. 6 An unloading device of sacked oily sludge Oily sludge treatment test: 1 Transmission belt 2 Large diameter gyro wheel 3 Tactic roller 4 Stationary pressure wheel 5 Movable pressure wheel 6 Spring 7 Sludge sump

IOP Conf. Series: Materials Science and Engineering **281** (2017) 012039 doi:10.1088/1757-899X/281/1/012039

4. Oily sludge treatment test

During the experiment the manual mode of the oily sludge added to the feed hopper of the filter. Oily sludge into the filter after a certain proportion of chemicals with high pressure hot water washing, the waste was cleaned and removed under the separation of the cylindrical filter screen. From the sludge pump into the settling tank,the filtrate through the stirring and sedimentation for the oil and water separation. After separation, the upper oily sludge was recovered, and the lower mud is dehydrated under the action of the centrifuge. The sludge piled up to the designated place, sewage back to the boiler, and enter the next circulation

4.1. Experimental conditions

According to the technological conditions of oil removal from oily sludge, the experimental parameters of oily sludge treatment device are set up: The water temperature in the boiler is 100°C; the flushing and filtering time is 15min; the stirring time is 20min; the minimum temperature of the separation tank is 60°C; the holding time is 40min; the solid-liquid ratio is 1:4.

4.2. Sample properties

According to the sources of oily sludge, the natures of the crude oil, the particle size of the silt and the proportion of each component all are not all the same. In view of this situation, the mud samples from different sources are selected to experiment. The specific parameters of oily sludge are shown in Table 1.

Table 1 Sources of oily sludge

Sample number	Resource	Oil content (%)	Water content (%)	Solid content (%)
1	United station	32. 5	25. 8	41. 7
2	Mixture consist of oil and soil	10. 3	21. 8	67. 9
3	Workover job	20. 3	29. 5	50. 2

4.3. Treatment effect

The above three different sources of oily sludge were treated by using the oily sludge treatment device. The results showed: that the oil content of the treated oily sludge decreased with the increase of the concentration of the drug. When the concentration of the drug was more than 1.2%, the rangeability in oil content of the treated oily sludge is decreased; When the concentration of the drug is more than 1. 6%, the oil content of the treated oily sludge tends to be constant (Figure 7). Consider the water is recycled in the device during the oily sludge treatment process, so the amount of additives can be controlled between 1% and 1.6%.

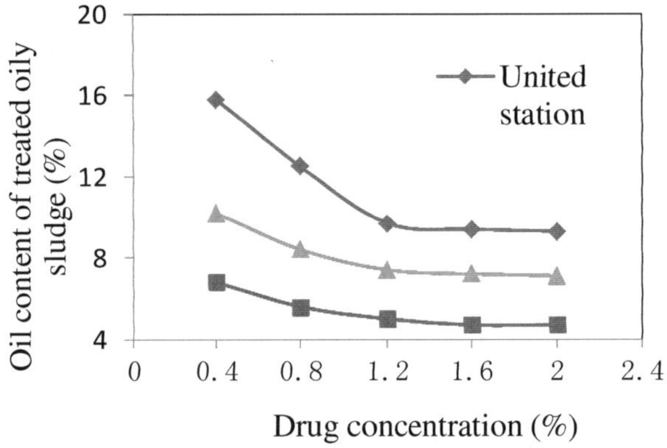

Figure. 7 Effect of drug concentration on oily sludge treatment

Figure. 8 The untreated sacked oily sludge

Figure. 9 The treated oily sludge

5. Conclusions

The oily sludge treatment device designed in this paper is small and easy to move; The parameter setting is simple; High degree of automation; The chemical agents and water are added in the operation process, it will be recycled in device, and produced less residual sewage; In addition, the greater the final oil removal rate, the relative treatment effect is better. There is little oil residue on the surface of the filtered residue, it is better to treat oily sludge with high oil content. The successful development of this device not only realized the small shape and the mobility, but also can be used in combination with the profile control agent. Therefore, it has wide application prospects in oil field.

References

[1] Hu G, Li J and Zeng G 2013 Recent development in the treatment of oily sludge from petroleum industry: A review *J Hazard Mater* 261 470-490.

[2] Yang X X, Li X N, Cheng G 2012, 27 (5) Study on Co-pyrolysis Characteristics of Oily Sludge and Coal *J Journal of Xi'an Shiyou University (Natural&Science Edition)* 82-85.

[3] Yu L L, Wang D, Ji J 2011, 38 (4) Treatment of oily sludge in oil field by quenching and high temperature tempering-mechanical separation technology *J Chemical engineering and machinery* 413-416.

[4] Liu L, Luo Y, Liu Q Y 2014, 43 (2) Study on incineration technology of oily sludge in Jianghan oil field *J Chemical Engineering of Oil and Gas* 200-203.

[5] Bao M T, Wang B, Li X M, 2007, 22 (6) Study on biological treatment technology of oily sludge *J Journal of Natural Resources* 865-870.

[6] Vasudevan N, Rajaram P 2001, 26 (6) Bioremediation of oily sludge-contaminated soil *J Environmental International* 409-411.

[7] Jiang N, Liu G R, Guo F T, and so on 2015, 25 (3) Structural design and development of skid-mounted oily sludge treatment device *J Journal of Filtration&Separation*4-6.

[8] Li Q, Liu X Y, Wang W B 2015, 25 (1) Thermochemical separation of skid-mounted treatment device of oily sludge *J Environmental Protection of Oil and Gas Fields* 40-43.

[9] Wang W Y 2014, 35(5) Development and application of oily sludge treatment mobile device *J Chemical Equipment Technology* 9-13.

[10] Wu Y, Yang T, Chen S J 2016, 31 (4) Development of compound profile control system for oily sludge and weak gel *J Journal of Xi'an Shiyou University (Natural &Science Edition)* 92-97.

[11] Chang H, Yu C T, Baikun Wang 2014, 34 (2)Preparation of oil field profile control agent with oily sludge *J Environmental Protection of Chemical Industry* 160-164.

[12] Zhao Z X, Liu G L, Li R L 2006, 37 (7) Study on Oil Removal Technology of Oily Sludge for Preparation of expanded Particulate profile control agent *J China Petroleum Processing Petrochemical Technology* 67-70.

[13] Tang J L, Du X Y, Hao Z Y 2004, 27 (3) Research and application of profile control technology of oily sludge J *Drilling and Production Technology* 86-87.

[14] Li Q F 2016 An unloading device for sacked oily sludge *P* China: CN2015208591090.

IOP Conf. Series: Materials Science and Engineering **281** (2017) 012040 doi:10.1088/1757-899X/281/1/012040

Study on Elastoplastic Crack Propagation Behavior of Laser-welded 6061 Aluminum Alloy Using Digital Image Correlation Method

Y Chen[1], Z K Lei[*1], R X Bai[1], Y P Wei[1] and W Tao[2]

[1] State Key Laboratory of Structural Analysis for Industrial Equipment, Dalian University of Technology, Dalian 116024, China
[2] State Key Laboratory of Advanced Welding and Joining, Harbin Institute of Technology, Harbin 150001, China.

E-mail: leizk@163.com

Abstract. Laser welding is commonly used in the aluminum alloy welding of aerospace field as an efficient and precision welding method. Local high temperature and electrode material change the microstructure of aluminum alloy in the laser welding process. The nonuniform and weakened material properties in the weld zone will result in localized stress concentration, which induced the plastic deformation and the crack initiation of the weld. In this paper, the digital image correlation method was used to study the elastoplastic crack propagation of laser welding 6061 aluminum alloy. According to the method that determines the nominal crack-tip of the weld, we obtained the CTOA and CTOD. The experimental results showed that CTOA, CTOD and J-integral increased nonlinearly with the crack propagation in the heat affected zone.

1. Introduction

Aluminum alloy is widely used in the aerospace field as common component materials. And laser welding is an effective welding method for joining components. The fracture parameters that characterize the crack propagation behavior of crack-tip include crack-tip opening angle (CTOA), crack-tip opening displacement (CTOD), energy release rate and J-integral [1]. CTOA and CTOD can explain the stability of the crack propagation. Laser welding will result in uncoordinated stress, crack tunneling and crack deflection at the crack-tip due to the different strength and strain hardening index between welded metal and base metal [2-3].

Aluminum alloy laser welded joint is nonuniform materials. It is imprecise to analyze the elastoplastic crack propagation behavior by linear elastic damage mechanics and previous measuring method [4-5]. Digital image correlation (DIC) is an optical method for measuring displacement contour, deformation, vibration and strain on the specimen surface. The method can calculate the strain and stress distribution on material surface, observe fluid flow and analyze material's fracture [6]. Digital image correlation method has the advantages of full-field and non-contact. In this paper, it was applied to measure the elastoplastic deformation of mode-I weld crack in laser welded joint of 6061 aluminum alloy. In the process of crack propagation, we analyzed the strain distribution and plastic zone size of the welded joint, and studied the propagation behavior of the crack in laser seam by changed CTOA, CTOD, J-integral.

2. Measurement principle of crack parameters

2.1. Elastic crack parameters

For homogeneous brittle materials, the plastic zone at crack-tip can be neglected because of its tiny size, and the mode-I crack normally propagates along the precrack of these materials. A linear fitting method can be used to determine the CTOA at the crack-tip [7]. Two reference lines in elastic zone intersect at a nominal crack-tip (NCT) position and the CTOA is the angle of determined from the angle of the lines, as shown in figure 1(a). The CTOD is the vertical distance at 0.5 mm behind the NCT.

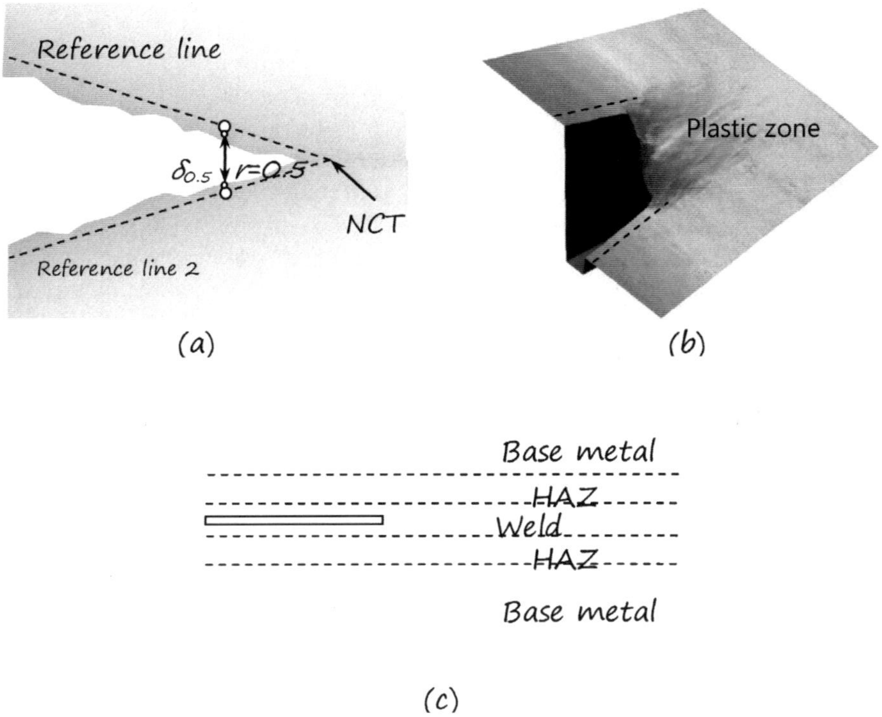

Figure 1. (a) NCT determined by the linear fitting method, **(b)** Effect of plastic zone on NCT and **(c)** deflection trajectory of weld crack propagation

2.2. Elastic-plastic crack parameters

For laser welded joint of 6061 aluminum alloy, there is a larger plastic zone near the crack-tip. The linear fitting method can not obtain the fracture parameters for the following reasons.

Firstly, the mode-I crack in laser welded joint was blunted and propagated along the precrack. The nonlinear changed y-displacement in the blunted region is shown in figure 1(b). It is unreasonable to define the NCT by the linear fitting method.

Secondly, the propagation process of mode-I weld crack in 6061 aluminum alloy can be divided into three stages, as shown in figure 1(c). In the first stage, the weld crack was blunted and initially propagated in the fusion zone (FZ) along the precrack. In the second stage, the crack deflected rapidly from the FZ to the heat affected zone (HAZ), as shown in figure 2(a). Finally, the crack propagated along the interface between the HAZ and the base metal (BM) because of the weak material performance in the HAZ. Therefore, a multi-stage linear fitting method is used to determine the NCT by considering the influence of the plastic zone, as shown in figure 2(b).

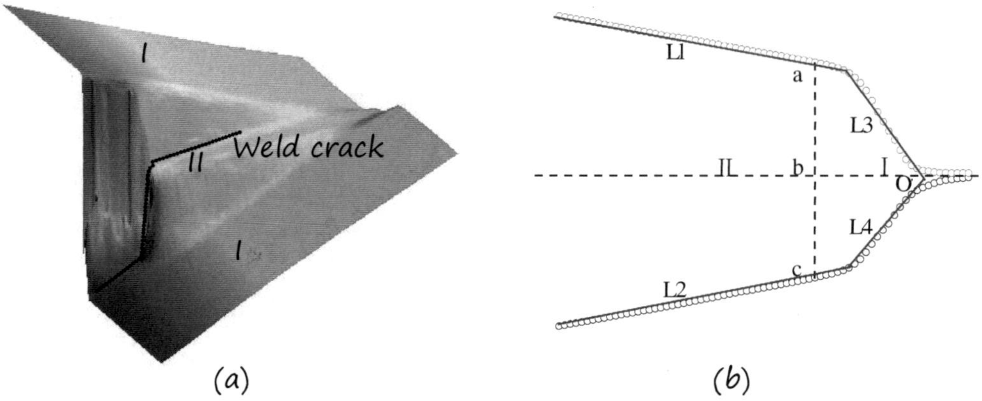

Figure 2. (a) Y-displacement near the weld crack, which deflects to the HAZ and **(b)** a multi-stage linear fitting method

The elastic zone data (I zones in figure 2a) was used to fit the straight line by Lei's method [7]. The CTOA can be obtained from the plastic data of II zone, as shown in figure 2(b). So the area near the crack-tip can be seemed as a combination of elastic zone and plastic zone. L1 and L2 are fitted in the elastic zone. L3 and L4 are fitted in the plastic zone. The NCT is determined by the intersecting fitting lines (L3 and L4). The CTOA is the angle between L3 and L4, as shown in figure 2(b). The CTOD is the vertical distance at 0.5 mm behind the NCT.

2.3. J-integral calculation

J-integral that characterize the fracture properties is expressed by a contour integral,

$$J=\int_{\Gamma}\left(Wn_x-T_i\frac{\partial u_i}{\partial x}\right)ds \quad (i=x,y) \tag{1}$$

where Γ is an arbitrary contour around the crack-tip, $T_i=\sigma_{ij}n_j$ is the force acting on the contour unit, σ_{ij} is the stress tensor and n_j is the normal vector of the contour. u_i is the displacement field and $W=\int\sigma_{ij}d\varepsilon_{ij}$ is the strain energy density.

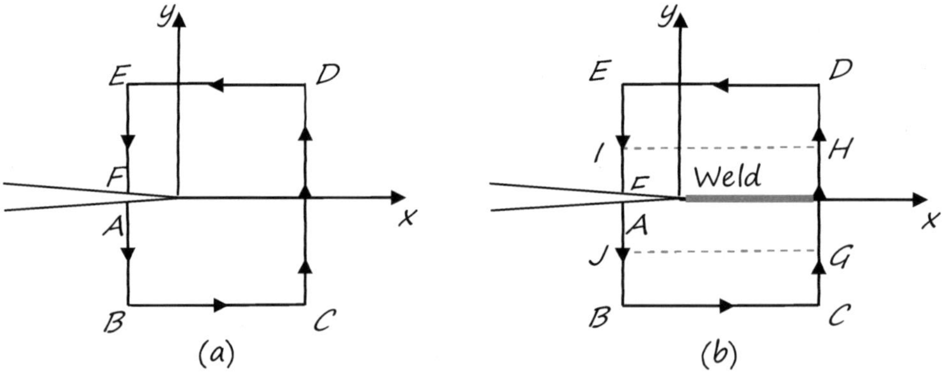

Figure 3. The rectangular integral path of J-Integral

For the mode-I crack in 6061 aluminum alloy, the J-integral can be expressed by a rectangular path (figure 3a) as

$$J=J_{ABCDEF} \tag{2}$$

For the mode-I weld crack in 6061 aluminum alloy, we considered the different properties between weld material and base metal. The J-integral is calculated by three rectangular paths (figure 3b),

$$J = J_{BCGJB} + J_{AJGHIF} + J_{IHDEI} \tag{3}$$

Above J-integral calculations are based on the Ramberg-Osgood constitutive model,

$$\varepsilon = \frac{\sigma}{E} + \frac{\sigma}{E}\alpha\left(\frac{\sigma}{\sigma_0}\right)^{n-1} \tag{4}$$

where σ and ε are the uniaxial stress and strain, respectively. α and n are the hardening coefficient and hardening index, respectively. E and σ_0 are the elastic modulus and yield stress, respectively. The first term of the formula indicates the linear elastic behavior, and the second term indicates the stage of plasticity.

3. Experiments

The experimental material is 2 mm thick 6061 aluminum alloy specimen. ER4047 aluminum wire with 1.2 mm diameter was used as welding filler. The welding system consists of YLS-10000 laser and KR16-2 industrial robots. The welding rate is 3m/min, and the welding power is 1900W. A precrack with 0.17 mm width and 7 mm length was made by wire cutting. The specimen was painted with speckles and fixed in an Instron 3345 tensile tester with a loading rate of 0.5 mm/min. A CCD camera (Guppy F080b) was used to capture the deformed images with a size is 1024×768 pixels, as shown in figure 4. The load increased linearly and then entered the yield stage. The maximum load is 2.0 kN, as shown in figure 5.

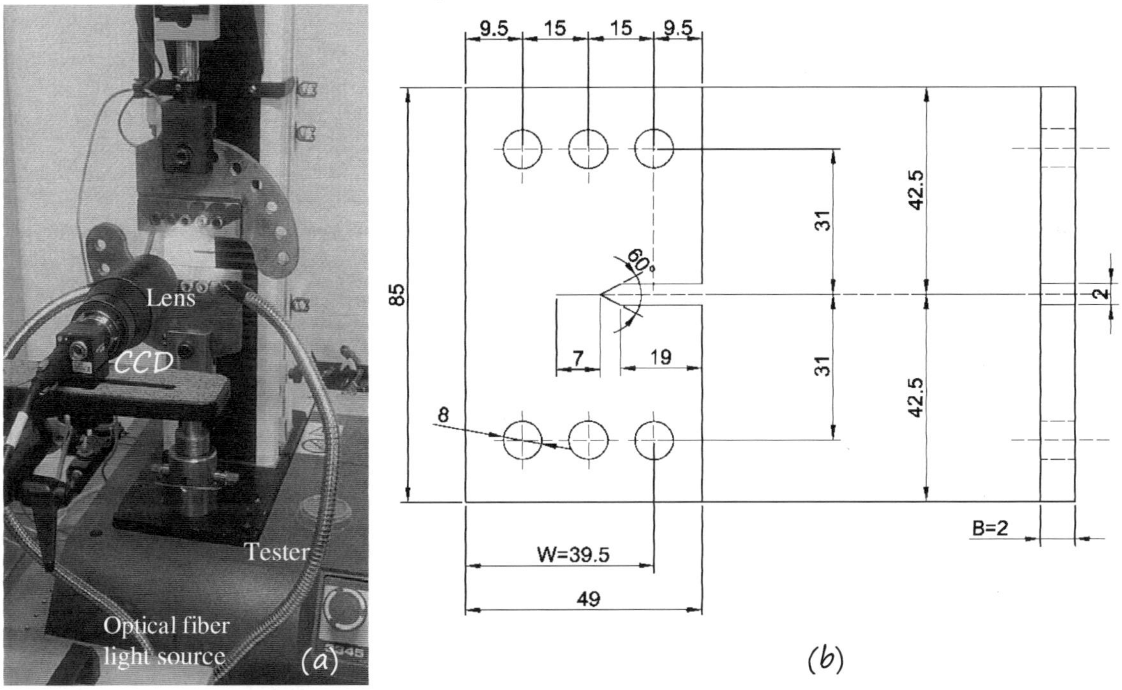

Figure 4. (a) Experimental setup for DIC and (b) specimen with a precrack (in mm)

IOP Conf. Series: Materials Science and Engineering **281** (2017) 012040 doi:10.1088/1757-899X/281/1/012040

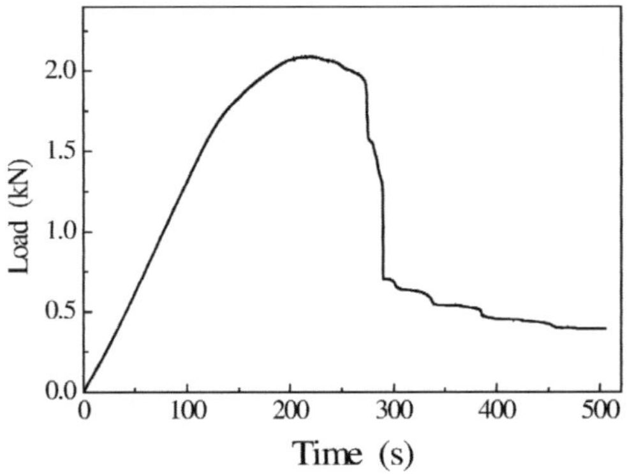

Figure 5. The Loading history

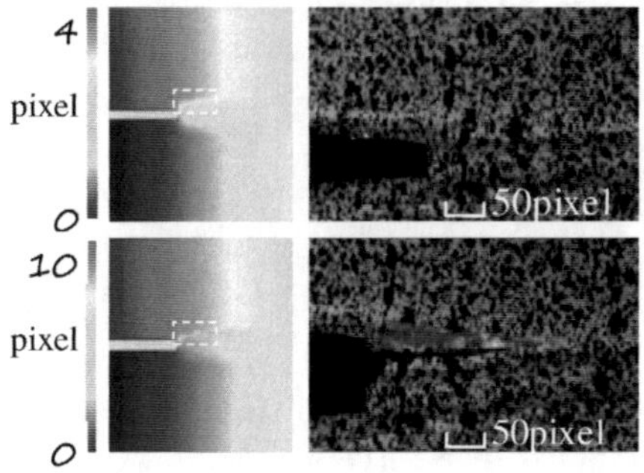

Figure 6. Y-displacement and crack propagation

4. Results and discussion

4.1. NCT-time curve

The cracked expansion can be seen from the y-displacement obtained by the DIC method. The size of each pixel is 0.0441mm, as shown in figure 6. Four horizontal lines, y coordinates are 86 pixel, 87 pixel, 88 pixel and 89 pixel, were selected near the crack tip. The four y-displacement curves related to four horizontal lines intersect at 104.54 pixel, which is consistent with the NCT determined by the multi-stage linear fitting method, as shown in figure 6(a). Finally, the relationship between NCT position and time was given, as shown in figure 6(b).

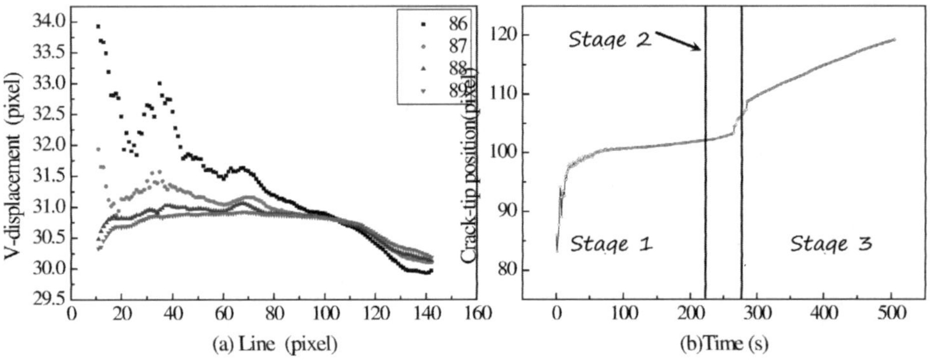

Figure 7. (a) Four y-displacement curves intersected at the crack tip and (b) a NCT-time curve

The NCT position is at 86th pixel before the propagation. The crack propagation process can be divided into three stages, as shown in figure 7(b). In the first stage (<275s), the initial NCT position was in the FZ and fluctuated due to the blunting phenomenon. After the crack trapping initiation, the NCT moved to 102nd pixel. In the second stage (from 275 s to 289 s), the crack expanded from the FZ to the HAZ in a short period. And the crack-tip moved to 106th pixel at the end of this stage. In the third stage, the crack had a stable expansion in the FAZ.

4.2. CTOA and CTOD

The CTOA and CTOD curves can be given by the NCT, as shown in figure 8(a) and figure 8(b). In the first stage, the CTOA had a slow increase, and then grew rapidly to 53.77° at 220 s because of the blunting and trapping effect of crack-tip in the FZ. The second stage, the weld crack propagated from the FZ to the FAZ. The CTOA and CTOD is almost zero. Finally, the crack spreaded at the interface between the HZA and the BM, because the material properties of the HAZ are lower than those of the FZ and BM.

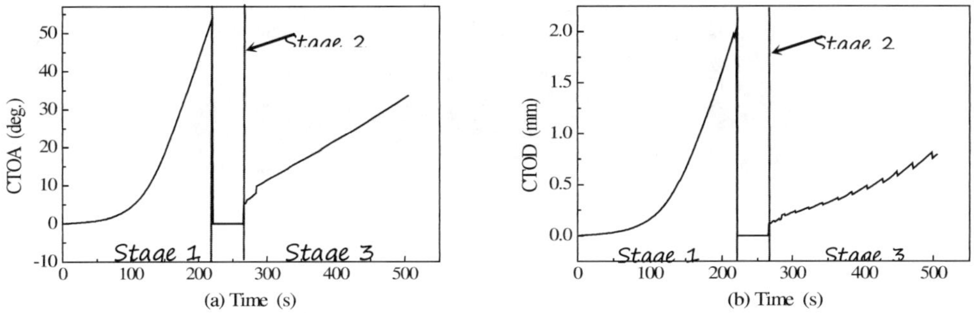

Figure 8. (a) CTOA-time curve and (b) CTOD-time curve

4.3. J-integral and plastic zone

The fracture toughness of weld crack in 6061 aluminum alloy was calculated by equation (3). The constitutive parameters of Ramberg-Osgood model are shown in Table 1.

Table 1. The properties of laser-welded 6061 aluminum alloy

	Elastic Modulus (MPa)	Poisson's ratio	Strain hardening index	Hardening coefficient	Yield stress (MPa)	Scale factor (Pixel/mm)
Base metal	43420	0.33	17.811	0.257	121.5	0.0441
Laser weld	13898	0.38	4.29	4.33	70.80	0.0441

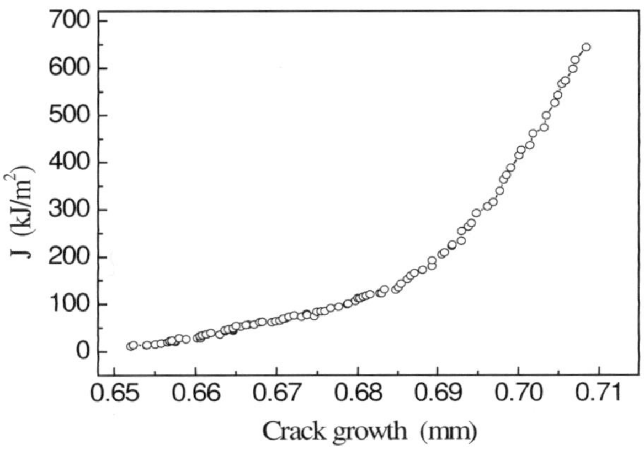

Figure 9. J-integral and time curve in the first stage

In the first stage of weld crack propagation, the J-integral increased slowly, as shown in figure 9. At the crack initiation time, the crack growth determined by the grabbed images is 0.6515mm. The corresponding J-integral value is 7.07 kJ/m^2. It is lower than 23 kJ/m^2, the fracture toughness of 6061 aluminum alloy, due to the lower performance of weld than that of the base material [8].

During the propagation process of weld crack, the strain ε_y distribution is shown in figure 10(a). The high strain shows the lowest performance of the HAZ. So the weld crack propagated to the HAZ. The strain ε_y distribution of 6061 aluminum alloy without laser welding is also given in figure 10(b). The crack propagated in the high strain zone near the crack tip. Comparing the two plastic zone shape shown in figure 10(a) and figure 10(b), we know that the laser weld affected the plastic zone shape.

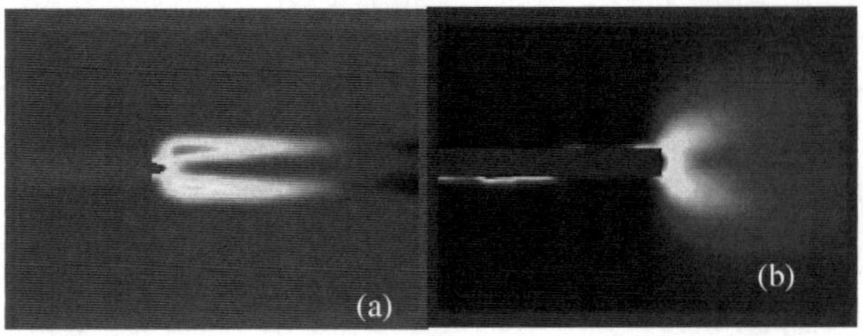

Figure 10. Plastic zone shape of **(a)** weld crack and **(b)** non-weld crack in the aluminum alloy

5. Conclusions

About the full-field displacement measured by the DIC, a new method to determine the NCT and J-integral of the laser weld crack in 6061 aluminum alloy was proposed. And we studied the CTOA, CTOD and J-integral by the method. The experimental results show that the multi-stage linear fitting method is feasible. The crack propagation was divided into three stages: the crack blunting and initiating process, the crack propagation from the HZ to the HAZ and the crack propagation in the HAZ. The conclusions obtained in this study are as follows:

(1) The CTOD and CTOA remained growing during the weld crack blunting and initiating process in the FZ, and then dropped to zero in the stage of the crack propagation from the HZ to the HAZ. Finally, the CTOD and CTOA stably increased in the HAZ.

(2) *The* critical J-integral value is 7.07 kJ/m^2. It is lower than 23 kJ/m^2, the fracture toughness of 6061 aluminum alloy, due to the lower performance of weld than that of the base material.

(3) The weld affected the plastic zone shape near crack tip because of the highest ε_y strain of the HAZ.

Acknowledgments

The authors thank the National Natural Science Foundation of China (Nos. 11472070, 11772081, 11572070), the Natural Science Foundation of Liaoning Province of China (No. 2015020145).

References

[1] Newman J C, James M A and Zerbst U 2003 A review of the CTOA/CTOD fracture criterion Engineering Fracture Mechanics **70**(3) pp 371-385.

[2] Sutton M A, Helm J D and Boone M L, 2001 Experimental study of crack growth in thin sheet 2024-T3 aluminum under tension-torsion loading International journal of fracture **109**(3) pp 285-301.

[3] Heerens J and Schoedel M, 2009 Characterization of stable crack extension in aluminium sheet material using the crack tip opening angle determined optically and by the δ_5 clip gauge technique Engineering Fracture Mechanics **76**(1) pp 101-113.

[4] Zhang H and Senkara J, 2011 Resistance welding: fundamentals and applications. CRC press.

[5] Lacroix R, Lens A, Kermouche G, Bergheau J M and Klocker H, 2012 Determination of CTOA in the molten material of spot welds using the Digital Image Correlation technique Engineering Fracture Mechanics **86** pp 48-55.

[6] Sahlabadi M, Valiollahi A, Konh B and Soltani N, 2017 Evaluating J-integral and Q parameter in high - density polyethylene using a combined experimental finite element method. Fatigue & Fracture of Engineering Materials & Structures, 40(6) pp 924-938.

[7] Zhenkun L, Ruixiang B, Libo D and Wei Q, 2012 Noncontact optical measurement of CTOA and CTOD for interface crack in DCB test Optics and Lasers in Engineering **50**(7) pp 964-970.

[8] Budzakoska E, Carr D G, Stathers P A, Li H, Harrison R P, Hellier A K and Yeung W Y, 2007 Predicting the J integral fracture toughness of Al 6061 using the small punch test Fatigue & Fracture of Engineering Materials & Structures **30**(9) pp 796-807.

IOP Conf. Series: Materials Science and Engineering **281** (2017) 012041 doi:10.1088/1757-899X/281/1/012041

Effect of different filler wires on weld formation for fiber laser welding 6A02 Aluminum alloy

F Xu*, L Chen, W Lu and E G He

Science and Technology on Power Beam Processes Laboratory, Beijing Aeronautical Manufacturing Technology Research Institute, Beijing 100024, China

*Corresponding email:17963418@qq.com

Abstract. 6A02 aluminum alloy was welded by fibre laser welding with two different filler wires (ER4043 and ER5356). The weld apperance, microstructure and mechanical properties were analysed. The results show the welding course with ER4043 is more stable than that with ER5356, and the welding spatters of the former are smaller than that of the latter. The microsturtrue of the weld zone, including columnar-grains near the fusion zone and mixed microstructures (columnar grains and equiaxed grains) in the weld center zone, is finer with ER5356 than that with ER4043. So the average microhardness value of the former is higher than the latter. A great number of low melting point eutectic phases disperse in grains boundary. Due to the eutectic phases distributing more in two zones (overheat zone near the fusion zone and the weld center zone) than other zones, the welded joints have these two low hardness and weak strength zones. The ultimate strength and the elongations after fracture of the welded joints with ER4043 are lower than that with ER5356 slihgtly. However, the former are improved obviously and higher than the latter after heat treatment. The tensile properties of all joints can reach to the base material level. And the tensile fractures always occur near the fusion zone.

1. Introduction

The filler wire is used during the laser welding course, which could reduce assembly accuracy before welding, improve the weld surface forming and mechanical properties of welded joints [1-3]. So the possibilities and the application scope of the laser welding are expanded significantly [4]. Recently, the fiber laser is developed rapidly, which has better light beam quality and brightness [5] than the other types of laser. Compared with the conventional YAG laser or CO_2 laser, the fiber laser has very high power density. Therefore, the fiber laser is easier to realize the full penetration welding of the thin-walled structures of aluminum alloy, and the good quality of the welded joints in the condition of high welding speed is easier to achieve [6]. Now the weldability, structures and mechanical properties of the aluminum alloy by using YAG laser or CO_2 laser have been widely studied [1, 4]. However, the studies about fiber laser welding are relatively limited, especially in the condition of high welding speed. For these reasons, the experimental of 6A02 aluminum alloy was carried out by using the system of the fiber laser welding with filler wire. Two different filler wires were used during the welding course, and some full penetration welds were obtained. The different chemical compositions of filler wires may have some influence on the microstructure and properties of the welded joints, which will be systematically analysed. The results will provide data reference for accelerating the application of aluminum alloys in the aviation field.

2. Experimental

The 6A02-T6 aluminum alloy sheet with 1.0mm thickness and two types of filler wires (ER4043 and ER5356) are used for butt-welding. Their chemical compositions are shown in table 1. The 6A02 alloy is Al-Mg-Si aluminum alloy, which can be strengthened by solid solution and aging heat treatment. It not only has good plasticity and moderate strength, but also is a kind of structural material with good resistance to corrosion. So it is often used in aircraft and engine parts with high plasticity and corrosion resistance [7]. The filler wire ER4043 is rich in Si element, and has good crack resistance. The filler wire ER5356 is rich in Mg element, which has good thermal cracking resistance and high strength. The size of workpieces is 200mm×100mm. And the surfaces of the workpieces were cleaned by chemical method before welding. The diameters of two types of filler wires are Ø1.0mm.

Table 1. Chemical compositions of 6A02 aluminum alloy and filler wires (wt %)

Material	Mg	Cu	Ti	Fe	Si	Mn	Zn	Cr	Al
6A02-T6	0.45-0.9	0.2-0.6	0.15	0.5	0.5-1.2	0.15-0.35	0.2	-	Bal
ER4043	≤0.05	≤0.3	≤0.2	≤0.8	4.5-6.0	≤0.05	≤0.1	-	Bal
ER5356	4.5-5.5	0.1	0.15	0.4	0.25	0.05-0.2	0.25	0.05-0.2	Bal

The experiments have been carried out by the welding system which is based on high power fiber laser and push-pull wire feeding system. The laser beam wavelength is 1.06μm. The normal output power of the laser is 5kW. The laser beam is transferred through fiber to the workpiece after being reflected and focused by lens. The focal distance of the focus lens is 160mm, and the focused spot diameter is Ø0.28mm. The axle direction of the laser beam leans 10° to the normal direction of the workpiece. The workpiece is static, and the laser head is driven by a six-axis robot to realize the welding. The high temperature zone near the molten pool is protected by argon during the welding. The optimized welding parameters, including the welding speed of 6m/min, the laser power of 2kW, and the wire feed rate of 2m/min are chosen in the fiber laser welding with filler wire for the butt-joints.

Metallographic specimens and standard tensile samples are intercepted from the workpieces after welding. The alkali solution with 2～3% NaOH is used to corrode the metallographic specimen. The microstructure of the joint is observed by optical microscope. And microhardness of the joint are measured on hardness tester (HXD-1000) by holding a test load of 0.1N for 10s. The tensile samples are shown in Fig. 1 and the tensile tests were carried out at room temperature by using Z100kN universal test system. The tensile properties of joints after reasonable heat treatment were also tested. The heat treatment includes solution and aging. The former is carried out at 530°C for 50 minutes and the latter is carried out at 165°C for 8 hours.

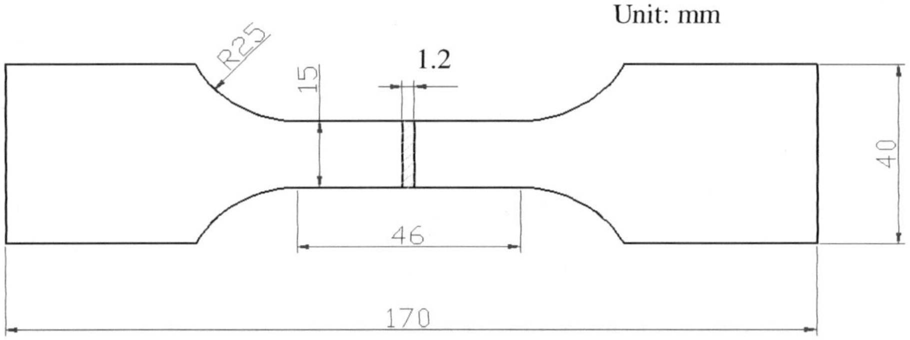

Fig. 1 Dimension of the tensile sample

3. Results and discussion

3.1. Macroscopic morphology analysis of the welded joints

The distance between laser and filler wire is important geometric parameter during the laser welding with filler wire. It should be controlled in the range of 1.0mm to 1.5mm. The wire is heated by plasma/metal vapor cloud, direct laser irradiation and the radiation of the molten pool rapidly. So the wire is transformed into small droplets and filled into the molten pool with the mode of surface tension transition, which is similar to the short circuit transition of argon arc welding, as shown in Fig. 2. The high power density of the fiber laser is good to maintain the dynamic balance of the keyhole. The molten pool with small volume could be obtained easily under the condition of high welding speed. And the filler wire could not only make up the evaporation loss of some alloy elements (such as Mg), but also be helpful to backfill of liquid metal [8]. Therefore, the weld bead is very full, and the surface appearance of the weld is very good. The spatter is closely related to the stability of welding course and joint quality [9]. When the welding spatters are less, the welding course is more stable and the joint with high quality is easier to be obtained. The results show that the spatters are produced very little by filling either ER4043 or ER5356 wire. And no excessive defects are found in the welded joints with the help of X-ray inspection. The Si element could increase the fluidity of the molten pool, which cause that spatters is rarely produced in the welding process. Therefore, compared with the spatters by filling ER5356 wire, the spatters by filling ER4043 wire is less. And the welding course of the latter is more stable. Meanwhile, element Si of the filler wire could be helpful to increase the volume of molten pool under the same welding condition. It is calculated that the weld area of the cross section is about 4.02mm^2 with ER4043 wire, while the weld area is about only 2.37mm^2 with ER5356 wire, as shown in Fig. 3.

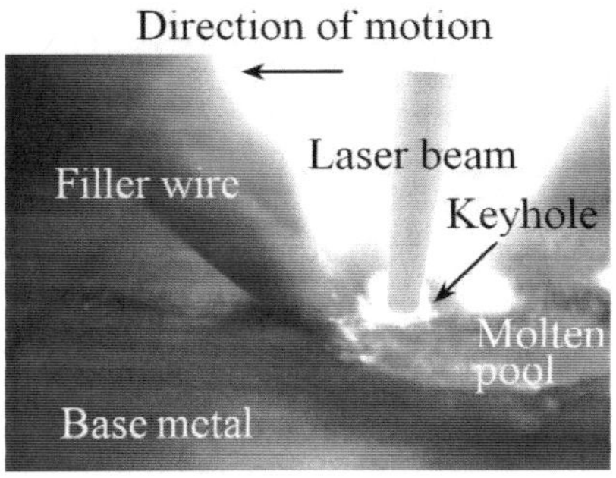

Fig. 2 Droplet transition of filler wire

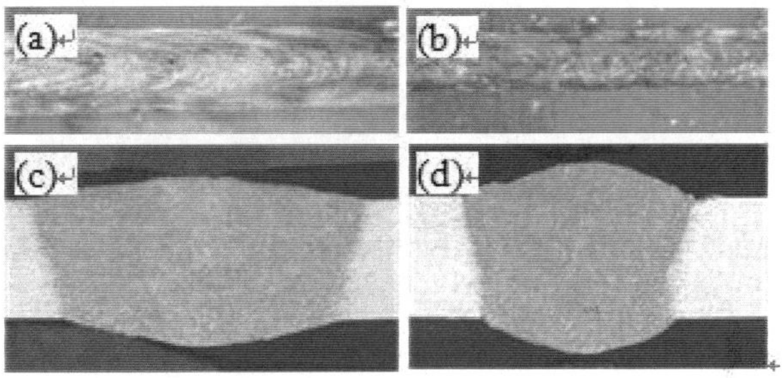

Fig. 3 Cross-section of welded joints of 6A02 aluminum alloy by fiber laser welding with filler wire ER4043 (a) , (c) and filler wire ER5356 (b) , (d)

(a) and (b) the front surface of weld bead; (b) and (d) cross-section of joints

3.2. Microstructure of welded joint by fiber laser welding with filler wire

The microstructures of the welded joints by using filler wire ER5356 and ER4043 are shown in Fig. 4. The fusion zone and the heat-affected zone (HAZ) are very narrow. And their width is about 3 to 6 grains of the base metal. Due to the liquid metal of the molten pool is always in an overheat state, spontaneous nucleation is very difficult, and heterogeneous nucleation is the chief solidification mode of liquid metal [10-11]. Columnar crystals microstructure depends on the sectional melted grains of the fusion zone, and rapidly grows along the maximal temperature gradient direction. In general, the maximal temperature gradient direction is always vertical to the fusion line. So the columnar crystals microstructure is formed perpendicular to fusion line. The high welding speed means the heat input is very low, which is only 20J/mm by using present welding parameters. This causes that the temperature gradient of solid-liquid interface of molten pool is very large and the linear velocity of the grain growth is very fast. So the columnar microstructure is very small, and its width is only 1/4 to 1/2 grains of the base metal. The columnar microstructure is distributed from the fusion line to the weld center. Mass of solute elements is pushed to the center of the molten pool by the columnar crystals' growth. When the composition overcooling reach to a certain degree, unmelted suspending particles in the liquid pool are suitable for nucleation and growth of equiaxed grains. Therefore, the mixed microstructure characteristics of columnar grains and equiaxed grains are formed in the weld center zone. The microstructure of the weld bead presents hypoeutectic characteristic, including primary α-Al and fine eutectic structure (α-Al + Si binary phases, α-Al + Mg_2Si binary phases, and α-Al + Mg_2Si + Si ternary phases).

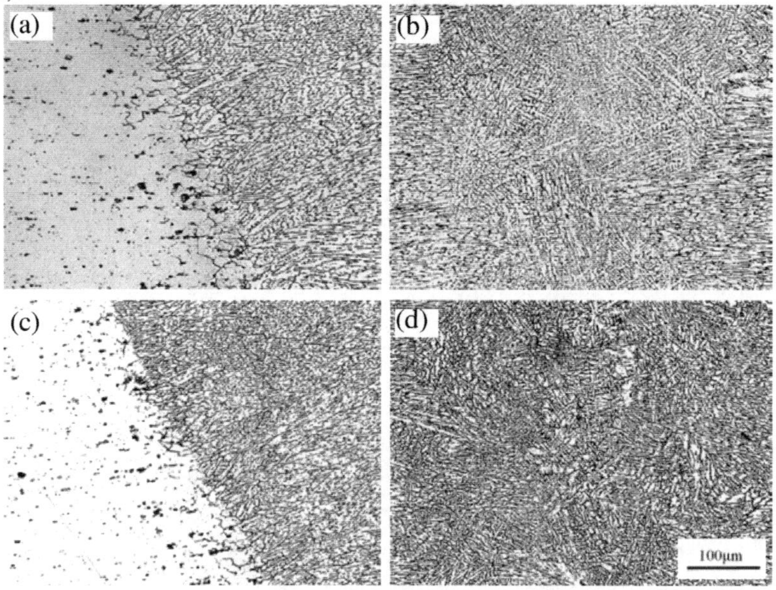

Fig. 4 Microstructure of joint with filler wire ER4043 (a), (b) and ER5356 (c), (d)(a) and (c) the fusion zone in cross-section of joints; (b) and (d) weld center in cross-section of joints

The weld area of cross section by using wire ER4043 is bigger than that by using wire ER5356, which means the volume of molten pool of the former is larger than that of the latter during the welding course. The larger volume of molten pool means the longer high temperature residence time of the liquid metal. It could result in the more obvious growth of the grains. Meanwhile, element Si could promote formation of low melting point eutectic structures in the molten pool, which could increase the composition overcooling in front of solid-liquid interface. And the growth of the grains in weld zone is accelerated. Therefore, the microstructure in the weld zone by using wire ER4043 are coarser than that by using wire ER5356. The grains in the fusion zone are sectional melted and some low melting point eutectic structures gather at triangle grain boundary near the HAZ. In addition, the microstructure near the fusion zone of the welded joint is different obviously, so the position near the fusion zone is likely to become the weak part of the welded joint.

3.3. Microhardness of welded joint by fiber laser welding with filler wire

The 6A02 aluminum alloy is a typical aging strengthening material. The joint softening phenomenon will appear after rapid melting and solidification of the base metal. The softening degree of the joint by filling wire ER4043 is more serious than that by filling wire ER5356, as shown in Fig. 5. This is not only closely related to the content of element Si and Mg, but also affected by the microstructure of the welded joints. The wire ER4043 is rich in element Si, which could promote the formation of low melting point eutectic structures. So the softening degree of the joint with wire ER4043 is increased.

The wire ER5356 is rich in element Mg, which could be conducive to generate strengthening phases Mg2Si. And the microstructure of the joint with wire ER5356 is relatively refined. So the microhardness of the weld zone with wire ER5356 is higher than that with wire ER4043.

Meanwhile, two low hardness zones are located near the fusion zone and the weld center. A large number of low melting point eutectic structure are dispersed in the grain boundaries of the weld center zone, which cause the microhardness of this zone is lower than other zones. The grains of the fusion zone are sectional melted, and some low melting point eutectic phases always gather near the fusion zone. So the softening degree near the fusion zone is more serious than adjacent zones slightly. The microstructure of the base metal includes mass of fine strengthening phases Mg_2Si, and those phases present characteristic of dispersion distribution. So the microhardness value of base metal is higher than that of the other zones. When the tensile properties of the welded joints are tested, the fracture of the samples will be possible to occur near the two low hardness zones.

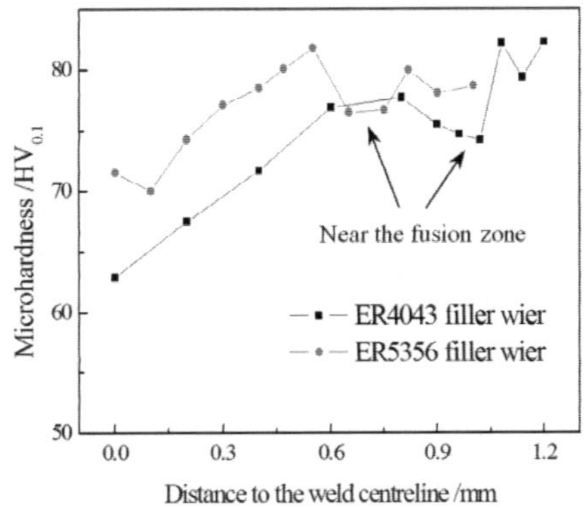

Fig. 5 Microhardness distribution of welded joints in transverse direction

3.4. Tensile properties of welded joint by fiber laser welding with filler wire

The welded joints in the welding state and heat treatment state were made into standard tensile specimens respectively, and all weld reinforcement were removed. Then the tensile test of those joints was carried out, and the results are shown in stable 2. It is found that the tensile properties of joints with filler wire ER5356 is better than that with filler wire ER4043 in welding state. This is not only related to grain size of the weld zone, but also closely related to the content of element Mg. According to the Hal-Petch formula, the finer grain size means the higher tensile properties. When the wire ER5356 is used, the grain structure is relatively fine and the strengthening element Mg is added into the molten pool, which is conducive to the formation of more strengthening phase Mg_2Si. The fracture occurs near the fusion line, which means that the zone near the fusion line is the weakest zone of the joints. The differences of the grain morphology near the fusion zone are larger than that in weld centre zone. So the position near the fusion zone is easy to cause non-uniform deformation and formation of stress concentration in the process of tensile deformation. And the deformation extent of the fusion zone is very small during the tensile testing, which causes this position is easily to be broken. However, in the weld center zone, the strain difference between the inner of grain and grain boundary is tiny and the deformation of the weld zone is uniform under the same external force. So the crack is not easy to initiate and propagate in the fine grains.

Excitedly, the heat treatment is carried out after welding, and the joints with better properties are obtained easily. The strengthening phase Mg_2Si would maximize precipitation during solution and aging course of heat treatment. The tensile strength and yield strength of 0.2% reach the base material level. The elongation is also improved significantly, which could reach 20.99%. Meanwhile, it is also found that the tensile properties of welded joint with filler wire ER4043 are better than that with filler wire ER5356.

Table 2 Tensile test average results of welding joints of 6A02 aluminum alloy

State of the tensile specimens	Ultimate strength /MPa	Yield strength of 0.2%/MPa	Elongation after fracture/%
With filler wire ER4043 in welding state	236.92	186.86	2.93
With filler wire ER5356 in welding state	267.78	193.64	3.28
With filler wire ER4043 in heat treatment state	356.78	318.12	20.99
With filler wire ER5356 in heat treatment state	336.52	310.77	20.58
The base material level	295	225	11

4. Conclusions

1) The weld bead by fiber laser welding with filler wire is very full. It can be inferred that the welding course with filler wire ER4043 is more stable than that with filler wire ER5356 from the welding spatters amount. And the volume of the molten pool of the former is larger than that of the latter.
2) The microstructure characteristics of 6A02 aluminum alloy welded by fiber laser welding with filler wire are the columnar microstructure near the fusion line and the mixed microstructures (including columnar grains and equiaxed grains) in the weld center zone. And the microstructure presents hypoeutectic characteristic. Moreover, the microstructure in the weld zone with filler wire ER4043 are coarser than that with filler wire ER5356, and the microhardness of the former is lower than that of the latter.
3) Two low hardness zones are located near the fusion zone and the weld center of the welded joints. And the fracture of the joints usually occurs near the fusion zone either with or without the solution and aging treatment. The tensile properties with filler wire ER5356 is better than that with filler wire ER4043 in welding state. However, the latter is better than the former after the heat treatment. The tensile strength and yield strength of 0.2% could reach the base material level. And the elongation is also improved significantly, which could reach 20.99%.

References

[1] Zuo T C, Xiao R S, Chen K, et al. *Laser materials processing of high strength aluminum alloys*. Beijing: National Defence Industry Press, 2002.
[2] Beijing Aeronautical Manufacturing Technology Research Institute. *Aeronautical manufacturing technology*. Beijing: Aviation Industry Press, 2013.
[3] Xu F. *Aluminum alloy's welding by laser with filling wire and laser-arc hybrid*. Beijing: Beijing University of Technology, 2009.
[4] Chen W Z. *Quality control of laser welding and cutting*. Beijing: China Machine Press, 2010.
[5] Chen L, Gong S L. *Application and development of laser welding technology for aluminum alloy*. Aeronautical Manufacturing Technology, 2011(11), 46-49.
[6] Xu F, Gong S L, Chen L, et al. *Characteristics of titanium alloy by fiber laser welding*. Aeronautical Manufacturing Technology, 2013(16): 90-92.
[7] China Aviation Materials Handbook Editional Board. *China Aviation materials handbook (Vol. 3) Aluminum alloy, magnesium alloy series*. Beijing: Standards Press of China, 2002.
[8] Ţălu Ş. *Micro and nanoscale characterization of three dimensional surfaces. Basics and applications*. Napoca Star Publishing House, Cluj-Napoca, Romania, 2015.
[9] Gao X D, Long G F, Wang R L, et al. *Analysis of characteristics of spatters during high-power disk laser welding*. Acta Physica Sinica, 2012, 61(9): 098103.
[10] Chen B L. *Principles of Welding Metallurgy*. Beijing, Tsinghua University Press. 1991.
[11] Zhou Z F. *Welding Metallurgy (Metallic weldability)*. Beijing: China Machine Press, 1995.

IOP Conf. Series: Materials Science and Engineering **281** (2017) 012042 doi:10.1088/1757-899X/281/1/012042

Deposition of Cubic AlN Films on MgO (100) Substrates by Laser Molecular Beam Epitaxy

Z K Mo, W J Yang, Y Weng, Y C Fu*, H He and X M Shen

Center of Ecological Collaborative Innovation for Aluminum Industry in Guangxi, Guangxi Key Laboratory of Processing for Non-ferrous Metal and Featured Materials, College of Materials Science and Engineering, Guangxi University, Nanning, Guangxi 530004, China

* Corresponding email: fuyuechun@126.com

Abstract. Cubic AlN (c-AlN) films were deposited on MgO (100) substrates by laser molecular beam epitaxy (LMBE) technique. The crystal structure and surface morphology of deposited films with various laser pulse energy and substrate temperature were investigated. The results indicate that c-AlN films exhibit the (200) preferred orientation, showing a good epitaxial relationship with the substrate. The surface roughness of c-AlN films increases when the laser pulse energy and substrate temperature increase. The film grown at laser pulse energy of 150 mJ and substrate temperature of 700 °C shows the best crystalline quality and relatively smooth surface.

1. Introduction

The III-V semiconductor cubic AlN (c-AlN) is becoming one of the promising optoelectronic and microelectronic materials for its excellent properties. For example, decreased phonon scattering and easier doping are expected due to its higher crystallographic symmetry [1,2]; there does not exist polarization electrical fields in the (001) growth direction [3], etc. However, high quality c-AlN film is difficult to fabricate for it is metastable. As the development of laser molecular beam epitaxy (LMBE) technique, the precise controlling and highly nonequilibrium characteristics of LMBE process are suitable to deposit metastable c-AlN films.

According to reported literatures [4,5], high oriented c-AlN films have been obtained on lattice-matched substrates. MgO has a cubic NaCl structure, and shows a small lattice-mismatch with c-AlN. Thus, c-AlN films are expected to show high quality on MgO substrates. In this work, c-AlN films were deposited on MgO (100) substrates by LMBE, and the dependences of microstructure on laser pulse energy and substrate temperature were investigated.

2. Experimental procedure

A LMBE system was used to deposit AlN films on MgO (100) substrates. The growth chamber was first evacuated to 10^{-5} Pa, and then MgO (100) substrates were placed opposite to the AlN (99.9% purity) target with a distance of 5 cm. Subsequently, the target was irradiated by a KrF excimer laser. During the depositions, the laser pulse energy ranged from 80 to 200 mJ and the substrate temperature ranged from 650 to 750 °C for 40 min respectively. Meanwhile, N_2 gas (99.999% purity) was introduced into the chamber, and the pressure was maintained as 10^{-1} Pa.

The crystal structures of films were investigated by X-ray diffraction (XRD, Rigaku D/MAX-RB) and Fourier transform infrared (FTIR) spectroscopy (Nexus 470). The surface morphologies were examined by atomic force microscopy (AFM, Agilent 5500).

3. Results and discussion

3.1. Crystal structure

Figure 1 displays the XRD patterns of deposited AlN films with different laser pulse energy at the substrate temperature of 750 °C. At laser pulse energies of 80, 110 and 150 mJ, deposited films exhibit a single AlN peak at around $2\theta = 45.5^{\circ}$, which is attributed to the c-AlN (200) reflex. Obviously, c-AlN films are successfully deposited on lattice-matched MgO (100) substrates. However, no AlN peaks are detected for the film deposited at 200 mJ, indicating an amorphous structure. The influence of substrate temperature on the XRD patterns of films at the laser pulse energy of 150 mJ is shown in figure 2. (200)-oriented growth is dominant for all the c-AlN films, and the intensity of (200) peak changes apparently with increasing substrate temperature. A maximum value of (200) peak is obtained at 700 °C, suggesting the good crystalline quality at this temperature.

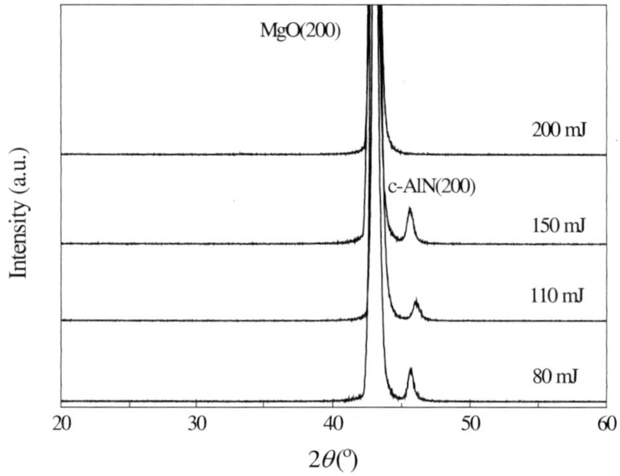

Figure 1. XRD patterns of deposited AlN films as a function of laser pulse energy at the substrate temperature of 750 oC.

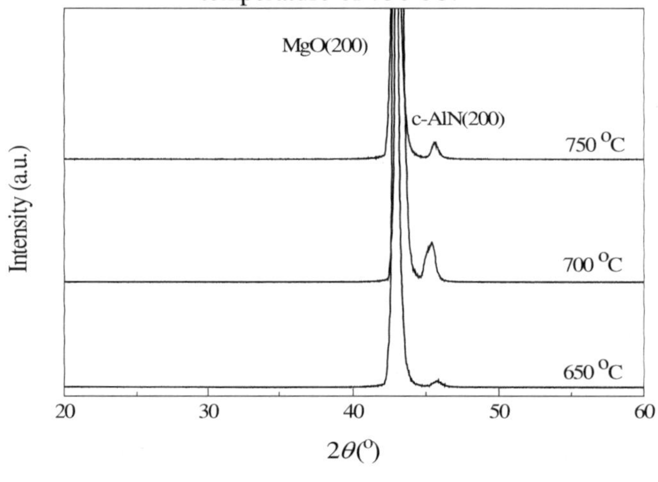

Figure 2. XRD patterns of deposited AlN films at different substrate temperature at the laser pulse energy of 150 mJ.

Based on the further analysis of XRD patterns, the grain size and lattice constant of c-AlN films are shown in table 1. The grain size changes little when the laser pulse energy increases, but the substrate temperature has more effect on it. Compared with reported results [6,7], the films in this work have a smaller grain size, indicating a good surface quality. While, the lattice constants of all films are smaller than the constant of bulk c-AlN (a = 0.4045 nm), which proposes that the lattice mismatch cannot be fully released during the depositions. It can be noted that the film grown at 700 °C and 150 mJ shows the minimum grain size of 14.7 nm, and its lattice constant (a = 0.39888 nm) is closest to the standard data.

Table 1 Grain size and lattice constant of c-AlN films with different laser pulse energy and substrate temperature.

	Laser pulse energy at 750 °C (mJ)			Substrate temperature at 150 mJ (°C)		
	80	110	150	650	700	750
Grain size (nm)	23	20.8	21.2	18.7	14.7	21.2
Lattice constant (nm)	0.39705	0.39347	0.39722	0.39547	0.39888	0.39722

Metastable c-AlN films are difficult to deposit on Si (100) [2] or Al_2O_3 (0001) [8] substrates for the large lattice-mismatch between them. In the present study, the lattice-matched MgO substrate acts as a template for the growth of c-AlN. And this phenomenon was also observed on the TiN (NaCl structure) buffered Si substrates [9] and 3C-SiC substrates [4]. So it is deduced that the epitaxial growth of c-AlN is significantly controlled by the crystalline plane of the substrate. As for the processing conditions investigated here, the substrate temperature plays a more important role in the crystallinity of c-AlN films compared with the laser pulse energy. In general, the mobility of atoms on the surfaces can be greatly enhanced at higher substrate temperature, which will improve the c-AlN phase formation [10].

3.2. FTIR analysis

Deposited AlN films are further characterized by FTIR absorbance spectra as shown in figure 3 and figure 4, in which the influence of substrates is deducted. In figure 3, two obvious absorption peaks at about 810 cm^{-1} and 980 cm^{-1} are observed for the films deposited with different laser pulse energy. From the reported vibration modes of c-AlN [11-12], the two peaks are assigned to the A_1 transverse optical (TO) and A_1 longitudinal optical (LO) phonon vibration modes respectively, but they both shift to the higher wavenumber. This variation may be due to the residual stress induced during the depositions, which is consistent with the XRD results. For the film deposited at 200 mJ, a weak peak at 667 cm^{-1} is assigned to the TO phonon vibration mode of h-AlN [13], showing the poor crystalline quality of this film. The absorbance spectra of AlN films deposited at different substrate temperature show the similar absorption peaks, see figure 4. An extra peak at 636 cm^{-1} for the film deposited at 700 °C is also detected, which corresponds to the E_1 TO phonon mode of c-AlN. This observation confirms the good crystalline quality of the film under this condition.

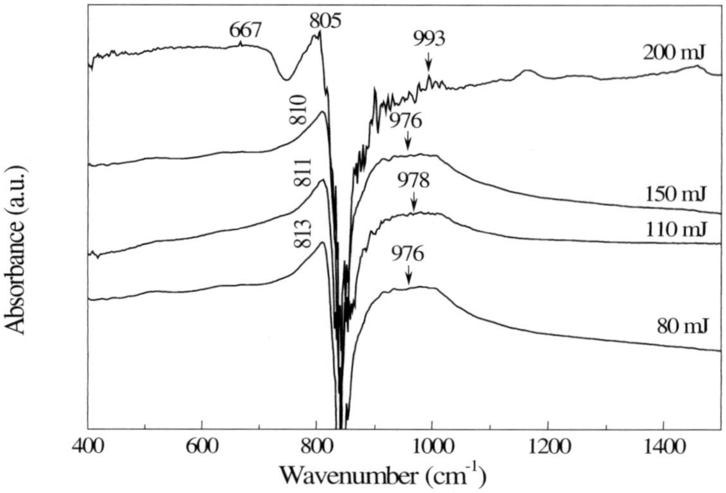

Figure 3. FTIR patterns of AlN films deposited at various laser pulse energy.

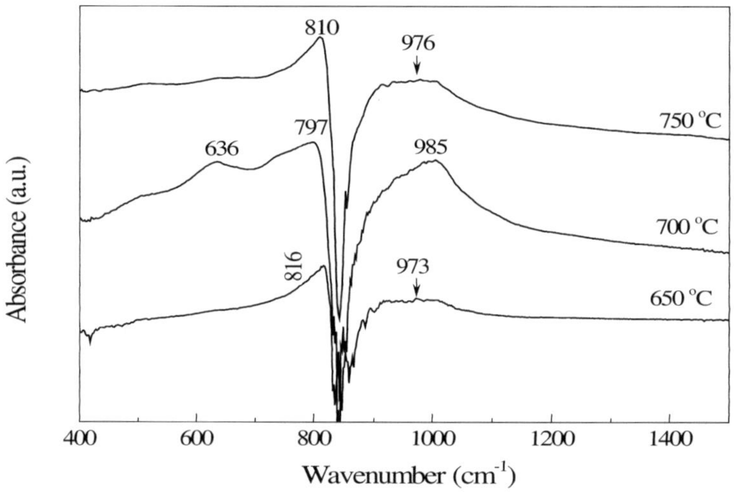

Figure 4. FTIR patterns of AlN films deposited at different substrate temperature.

3.3. Surface morphology

Figure 5 and figure 6 show the AFM surface morphologies of AlN films at different laser pulse energy and substrate temperature, respectively. It is clear that all the films are uniform, dense and relatively smooth. Both islands and terraces are observed on these surfaces owing to the Stranski-Krastanov growth mode of AlN films. The surface roughness of AlN films increases slowly when the laser pulse energy and substrate temperature are below 150 mJ and 700 °C, respectively. While a rough surface is obtained at higher laser pulse energy and substrate temperature as shown in figure 5 (d) and figure 6 (c).

The surface roughness of AlN films deposited by LMBE technique is affected mainly by two aspects. One is the atom mobility on substrate surfaces. When the deposition rate is higher than atom mobility, the surface roughness will increase. It is known that high substrate temperature leads to high atom mobility, but atoms would be too active to settle along the low energy sites at higher temperature, resulting in large surface undulation. The other is the deposition rate on substrate surfaces. Though

IOP Conf. Series: Materials Science and Engineering **281** (2017) 012042 doi:10.1088/1757-899X/281/1/012042

higher laser pulse energy can increase the deposition rate, the sputtering damage and insufficient migration of atoms would increase the surface roughness. Thus, the proper laser pulse energy and substrate temperature for depositing c-AlN films in this study are 150 mJ and 700 °C.

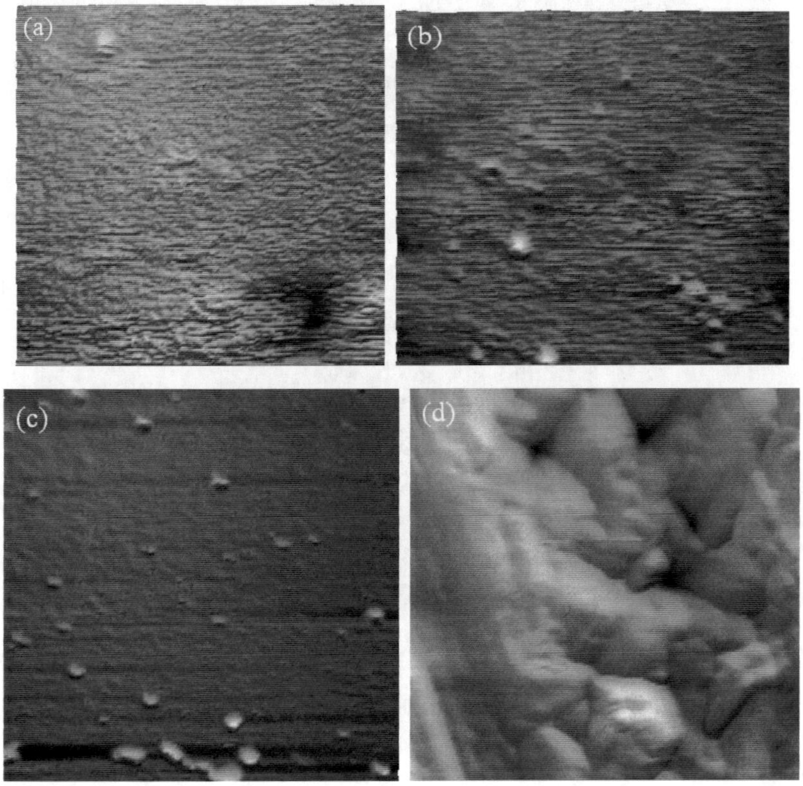

Figure 5. $2\times2\mu m^2$ AFM surface morphology of AlN films deposited at various laser pulse energy: (a) 80 mJ, (b) 110 mJ, (c) 150 mJ, (d) 200 mJ.

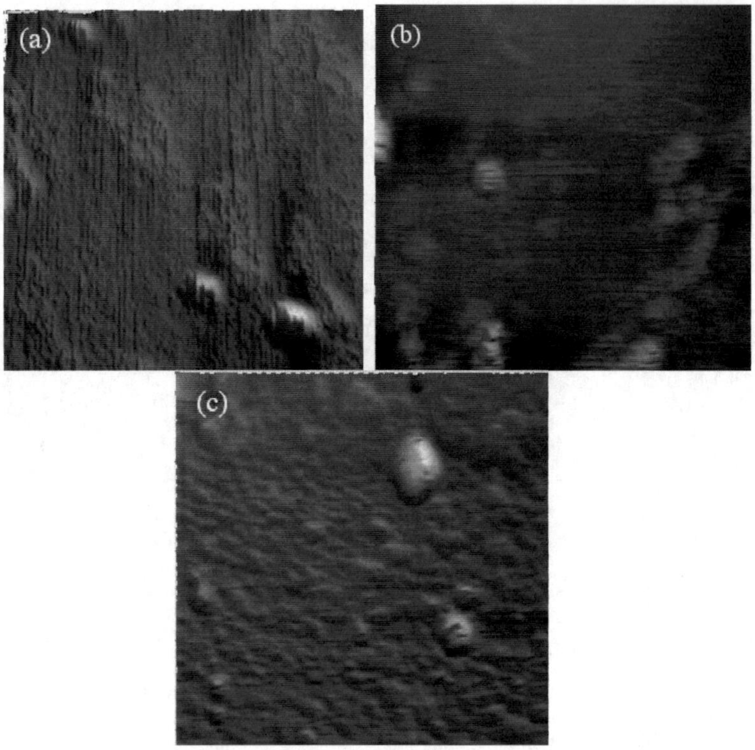

Figure 6. 2×2μm^2 AFM surface morphology of AlN films deposited at various substrate temperature: (a) 650 °C, (b) 700 °C, (c) 750 °C.

4. Conclusions

Cubic-AlN films with high quality were grown on MgO (100) substrates using LMBE technique. (200)-oriented growth of c-AlN films is dominant, showing the orientation relationship of AlN (200) // MgO (100). The surface morphology of AlN films is featured as islands and terraces, and the surface roughness increases when the laser pulse energy and substrate temperature increases. At the laser pulse energy of 150 mJ and substrate temperature of 700 °C, c-AlN film shows the best crystalline quality and relatively smooth surface.

Acknowledgments

This work was supported by the Guangxi Natural Science Foundation (Grant number 2015GXNSFAA139265) and National Natural Science Foundation of China (Grant number 61474030).

References

[1] Petron I, Mojab E, Powell R C, Greene J E, Hultman L and Sundgren J E 1992 *Appl. Phys. Lett.* 60 2491-3
[2] Lei T, Fanciulli M, Molnar R J, Moustakas T D, Graham R J and Scanlon J 1991 *Appl. Phys. Lett.* 59 944-6
[3] As D J 2009 Microelectron. J. 40 204-9
[4] Felice R Di, Bertoni C M and Catellani A 1999 *Appl. Phys. lett.* 74 2137-9
[5] Röppischer M, Goldhahn R, Rossbach G, Schley P, Cobet C, Esser N, Schupp T, Lischka K and As D J 2009 J. *Appl. Phys.* 106 076104
[6] Zhu J, Zhao D, Luo W B, Zhang Y and Li Y R 2008 J. *Cryst. Growth* 310 731-7
[7] Cibert C, Chatras M, Champeaux C, Cros D and Catherinot A 2007 *Appl. Surf. Sci.* 253 8151-4

IOP Conf. Series: Materials Science and Engineering **281** (2017) 012042 doi:10.1088/1757-899X/281/1/012042

[8] Mohri S, Yoshitake T, Hara T and Nagayama K 2008 *Diamond Relat. Mater.* 17 1796-9
[9] Fu Y C, Meng X H, Yang W J, He H and Shen X M 2011 *Vacuum* 85 1037-41
[10] Lv L, Li Q S, Li L, Zhang L C, Wang C F and Qi H X 2007 *Chin. Phys. Lett.* 24 552-4
[11] Kuang J C, Zhang C R, Zhou X G, Liu Q C and Ye C 2005 *Mater. Lett.* 59 2006–10
[12] Li L, Hao X P, Yu N S, Cui D L, Xu X G and Jiang M H 2003 J. *Cryst. Growth* 258 268–71
[13] Ristoscu C, Ducu C, Socol G, Craciunoiu F and Mihailescu I N 2005 *Appl. Surf. Sci.* 248 411-5

IOP Conf. Series: Materials Science and Engineering **281** (2017) 012043 doi:10.1088/1757-899X/281/1/012043

Coefficients of Thermal Expansion for Composites with Agglomerated Carbon Nanotubes

J Pan* and L C Bian

Key Laboratory of Mechanical Reliability for Heavy Equipments and Large Structuresof Hebei Province, Yanshan University, Qinhuangdao 066004, PR China

E-mail: panjing1234@qq.com

Abstract. In this paper, a micromechanics model is presented, which can predict the thermal expansion coefficients (CTE) of composites with agglomerated carbon nanotubes (CNTs). In the present model, two aggregation parameters are introduced. The Mori-Tanaka method and rule of mixture model are adopted to estimate the effect of CNT agglomeration on the thermal expansion coefficients of composites. The aim of the present paper is to investigate the CTEs of CNTs reinforced composites, with an emphasis on the influence of two agglomeration parameters on the CTEs. It is noted that the carbon nanotube agglomeration has a strong influence on the thermal expansion coefficients of composites.

1. Introduction

Carbon nanotubes are known to possess exceptional mechanical stiffness and strength. Since the discovery of carbon nanotubes, most of the research on CNTs composites focused on predicting the mechanical properties [1]. The CNT dispersion has been assumed to be uniform in the matrix and the analytical studies have neglected the agglomeration effects of CNTs. Inherent characteristics of CNTs, such as low bending stiffness (due to a small elastic modulus in the radial direction) and high aspect ratio, make CNTs easy to agglomerate in a polymer matrix [2-3]. Analyses of thermo elastic properties of composite materials have drawn considerable interest due to a very broad range of their current and projected applications [4-5]. Many micro mechanics models and bound relations have been suggested to predict effective coefficients of thermal expansion of various composites.

To date, many mechanics models have been developed to predict the effect of CNT aggregation and the thermal property on the effective elastic modulus. Barai and Weng [6] developed a two-scale micromechanical model to analyze the effect of CNT agglomeration and interface condition on the plastic strength of CNT/metal composites. It is shown that CNTs are indeed a very effective strengthening agent, but CNT agglomeration and imperfect interface condition can seriously reduce the effective stiffness and elastoplastic strength. Karevan et al. [7] researched the effect of filler content and the presence of interphase on the effective Young's modulus of polypropylene based nanocomposites with exfoliated graphite nanoplatelets and carbon nanotubes. Jarali et al. [8] derived the effective elastic properties of multifunctional carbon nanotube composites with agglomeration of straight circular carbon nanotubes dispersed in soft polymer matrices. Hassanzadeh-Aghdam et al. [9] investigated the coefficients of thermal expansion (CTEs) of unidirectional glass fiber-reinforced polyimide composites containing silica nanoparticles. Lee and Song [10] investigated the microstructure anisotropy of multiphase polymer composites with talcs and glass fibers and characterized the anisotropic thermomechanical properties of composites.

2. Basic theory

Based on the experimental results and micrograph images, the CNTs are easy to agglomerate in a polymer matrix. In order to study the influence of agglomeration of CNTs on the effective elastic property of composites, we adopt the morphology of Fig.1 in our model. The regions with concentrated CNTs are assumed to be spherical in shape and are considered as inclusions. The randomly oriented CNTs embedded in the original matrix to form a new matrix. The total volume V of representative volume element (RVE) can be divided into the following two parts:

$$V = V_r + V_m \tag{1}$$

where V_r and V_m denote the volume of CNTs and original matrix, respectively. The average volume fraction of CNTs in the composite is c_r

$$c_r = \frac{V_r}{V} \tag{2}$$

The total volume V_r of CNTs can be denoted as follows:

$$V_r = V_r^{in} + V_r^{m} \tag{3}$$

where V_r^{in} and V_r^{m} denote the volumes of CNTs dispersed in the inclusions (concentrated regions) and in the original matrix, respectively.

To reflect such an agglomerated state, we introduce two aggregation parameters ξ and λ in the following analysis.

$$\xi = \frac{V_{in}}{V}, \lambda = \frac{V_r^{in}}{V_r} \tag{4}$$

where V_{in} is the volume of spherical inclusions in the representative volume element (RVE). The parameters ξ and λ denote the volume fraction of inclusions with respect to the total volume of the RVE and the volume ratio of CNTs that are dispersed in inclusions with respect to the total volume of nanotubes, respectively.

The effective elastic modulus of inclusions and new matrix can be calculated by Mori-Tanaka method provided by Qiu and Weng [11]. The bulk and shear modulus of the inclusions (K_{in}, G_{in}) and new matrix (K_{out}, G_{out}) can be written as follow

$$K_{in} = \frac{(\xi - \lambda c_r)K_m + \lambda c_r \xi^{LA}}{\xi - \lambda c_r + 3\lambda c_r \xi^{A}} \tag{5}$$

$$G_{in} = \frac{(\xi - \lambda c_r)G_m + \lambda c_r \eta^{LA}}{\xi - \lambda c_r + 3\lambda c_r \eta^{A}} \tag{6}$$

$$K_{out} = \frac{[1 - \xi - c_r(1-\lambda)]K_m + c_r(1-\lambda)\xi^{LA}}{[1 - \xi - c_r(1-\lambda)] + 3c_r(1-\lambda)\xi^{A}} \tag{7}$$

$$G_{out} = \frac{\left[1-\xi-c_r(1-\lambda)\right]G_m + c_r(1-\lambda)\eta^{LA}}{\left[1-\xi-c_r(1-\lambda)\right]+3c_r(1-\lambda)\eta^{A}} \tag{8}$$

where the detailed express of parameters ξ^{A}、η^{A}、ξ^{LA} and η^{LA} can be found in the study [11].

Based on the rule of mixture model, the thermal expansion coefficients of inclusions α_{in} and new matrix α_{out} can be obtained:

$$\alpha_{in} = \frac{\lambda c_r}{\xi}\alpha_r + \frac{\xi-\lambda c_r}{\xi}\alpha_m \tag{9}$$

$$\alpha_{out} = \frac{c_r(1-\lambda)}{1-\xi}\alpha_r + \frac{1-\xi-c_r(1-\lambda)}{1-\xi}\alpha_m \tag{10}$$

where α_r and α_m denote the thermal expansion coefficients of CNTs and original matrix, respectively. The effective thermal expansion coefficients of composites can be obtained by Schneider model.

$$\alpha_{11} = \frac{\xi\alpha_{in}E_{in} + (1-\xi)\alpha_{out}E_{out}}{\xi E_{in} + (1-\xi)E_{out}} \tag{11}$$

$$\alpha_{22} = \xi\alpha_{in}(1+v_{in}) + (1-\xi)\alpha_{out}(1+v_{out}) - \alpha_{11}\left(\xi v_{in} + (1-\xi)v_{out}\right) \tag{12}$$

where α_{11} and α_{22} are longitudinal CTE and transverse CTE of composites. E_{in} and E_{out} denote the Young's modulus of inclusions and new matrix. v_{in} and v_{out} denote the Poisson's ratio of inclusions and new matrix. The detailed express can be found as follow:

$$v_{in} = \frac{(3K_{in}-2G_{in})}{2(3K_{in}+G_{in})}, \quad v_{out} = \frac{(3K_{out}-2G_{out})}{2(3K_{out}+G_{out})} \tag{13}$$

$$E_{in} = \frac{9K_{in}G_{in}}{3K_{in}+G_{in}}. \quad E_{out} = \frac{9K_{out}G_{out}}{3K_{out}+G_{out}} \tag{14}$$

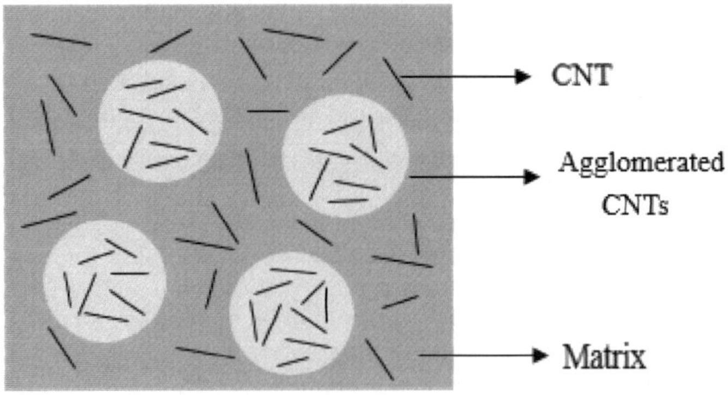

Fig 1. Micromechanical model of agglomerated CNTs

3. Result and discussion

In the following, numerical calculations are depicted in Figs. 2-5. The elastic constants of CNTs and matrix are $E_m = 1.9\text{GPa}$, $E_r = 450\text{GPa}$, $\nu_m = \nu_r = 0.3$ The thermal expansion coefficients of materials are $\alpha_m = 52.5\times10^{-6} / ^\circ C, \alpha_r = 5\times10^{-6} / ^\circ C$.

Fig. 2 and Fig. 3 present the variation of longitudinal CTE and transverse CTE with CNTs volume fractions at different agglomeration parameters ξ. The value of λ=0.5 is used in Fig. 2 and Fig. 3. It can be found that the longitudinal CTE increases with the increase of agglomeration parameters ξ. However, the transverse CTE decreases with the increase of agglomeration parameters ξ. The reason could be that with the increase of parameter ξ, the agglomerated region, i.e., spherical inclusions applies to a reinforced effect on the longitudinal CTE.

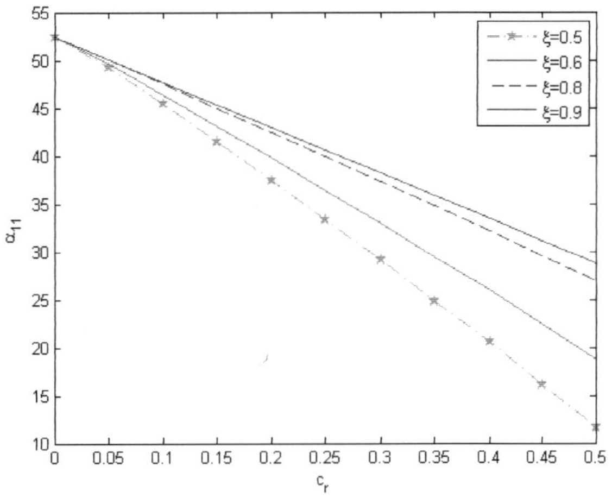

Fig 2. The variation of longitudinal CTE with the agglomeration parameter ξ

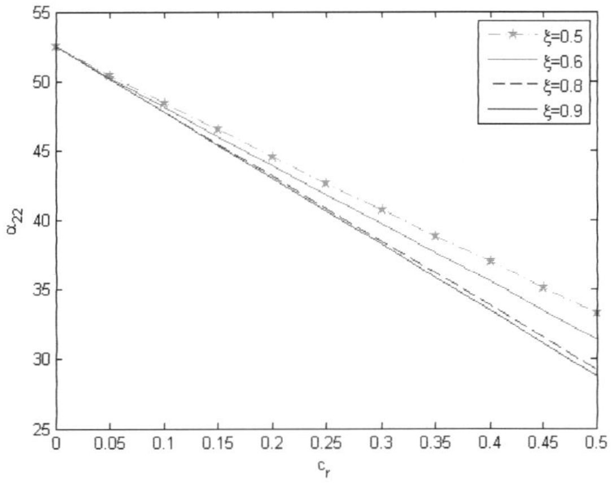

Fig 3. The variation of transverse CTE with the agglomeration parameter ξ

The longitudinal CTE and transverse CTE changed with the agglomeration parameter λ are given in Fig. 4 and Fig. 5. The value of ξ=0.9 is used in Fig. 4 and Fig. 5. As shown in Fig. 4 and Fig. 5, the longitudinal CTE decreases with increasing an agglomeration parameter λ, whereas the transverse CTE increases with it. With increasing parameter λ, the spherical inclusions region contains more agglomerated CNTs and the new matrix encloses more of the original matrix. So, the parameter λ is characterized by weakening effect in the longitudinal CTE with increasing it.

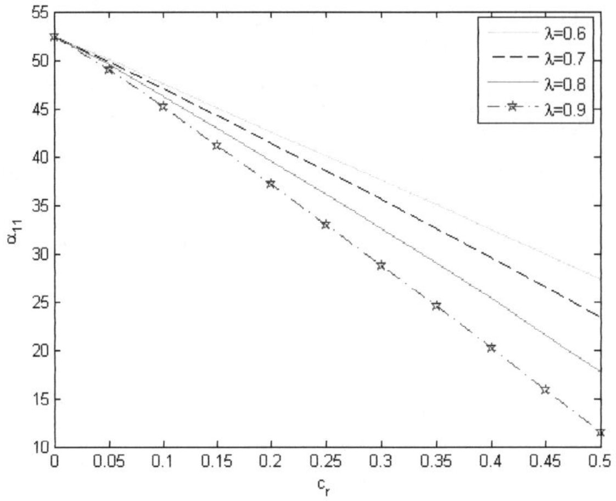

Fig 4. Effect of agglomeration parameter λ on the longitudinal CTE

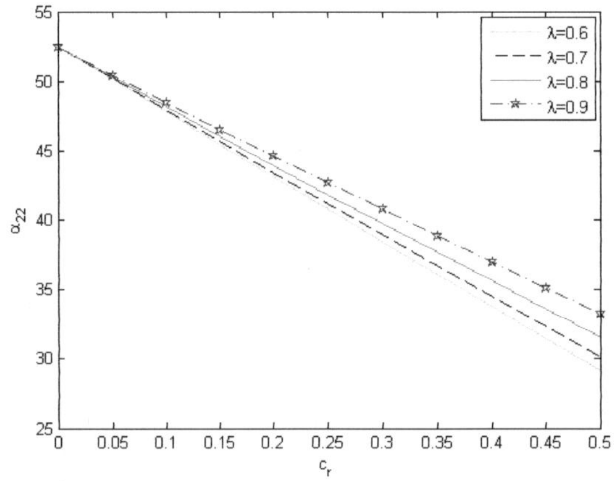

Fig 5. Effect of agglomeration parameter λ on the transverse CTE

4. Conclusions

In the present analysis, the mechanical and thermal properties of CNTs reinforced composites are studied. A new micromechanics method, which involves two agglomeration parameters, is presented and the thermal expansion coefficients of composites are derived. In the process of derivation, the Mori-Tanaka method and rule of mixture model are adopted to predict the mechanical and thermal properties of composites. The results show that the CNTs agglomeration has a critical effect on the longitudinal CTE and transverse CTE of composites. Besides, it is found that the two agglomeration parameters have a different effect on the longitudinal CTE and transverse CTE. The agglomeration parameter ξ results a reinforced effect on the longitudinal CTE and the agglomeration parameter λ has a weakening effect on the longitudinal CTE with its increase.

Acknowledgments
This research was funded by the Science Research Foundation of Hebei Advanced Institutes (ZD2017075) and Graduate Innovation Foundation of Hebei Province in China (CXZZBS2017053).

IOP Conf. Series: Materials Science and Engineering **281** (2017) 012043 doi:10.1088/1757-899X/281/1/012043

References

[1] Jarali C S, Patil S F and Pilli S C 2015 *Mechanics of Advanced Materials & Structures* 22(6) 428-439.

[2] Pan J, Bian L C, Zhao H C and Zhao yang 2016 *Computational Materials Science* 113 21-26.

[3] Pan J, Bian L C 2017 *Acta Mechanica* 228(6) 2207-2217.

[4] Lu P 2013*Polymer* 54(6) 1691-9.

[5] Jarali C S, Basavaraddi S R, Pilli S C, Raja S and Karjinni V V 2015 *International Journal for Multiscale Computational Engineering* 13(3) 231-248.

[6] Barai P and Weng G J 2011 *International Journal of Plasticity* 27(4) 539-559.

[7] Karevan M, Pucha R V, Bhuiyan M A and Kalaitzidou K 2010 *Carbon Lett* 11(4) 325-331.

[8] Jarali C S, Basavaraddi S R, Kiefer B, Pilli S C and Lu Y C 2013 *Journal of Applied Mechanics* 81 (2) 1-11.

[9] Hassanzadeh-Aghdam M K, Ansari R and Darvizeh A 2017 *Applied Science & Manufacturing* 96 110-121.

[10] Lee D J and Song Y S 2017 *Composite Structures* 174 329-337.

[11] Qiu, Y P and Weng, G J 1990 *Int. J. Eng. Sci* 28 1121–37.

IOP Conf. Series: Materials Science and Engineering **281** (2017) 012044 doi:10.1088/1757-899X/281/1/012044

Investigation of the Transverse Spread of Neodymium Ions Implanted in SOI

X F Qin*, L Chen and F Y Xu

College of Science, Shandong Jianzhu University, Jinan 250101, China

Email:xfqin@sdjzu.edu.cn

Abstract. It is very important to consider the range distribution and transverse distribution of ions implanted into semiconductor materials in design and fabrication of semiconductor integration devices by ion implantation. The Nd (neodymium) ions with energy of 400 keV and dose of 2×10^{15} ions/cm^2 were implanted into SOI (Silicon-on-insulator) samples at room temperature under the angles of 0°, 30° and 45°, respectively. The transverse distribution of 400 keV Nd ions implanted in SOI samples were measured by Rutherford backscattering technique. The measured results are compared with Monte Carlo code SRIM2012 predictions. It can be found that the experiment values were in good agreement with the prediction of SRIM2012 code.

1. Introduction

SOI planar waveguide is promising for future low-cost photonic integrated circuits because of its excellent electronic and optical properties and compatibility with mature Si CMOS integrated circuit technology. The rare-earth (RE) particles doping of Si-related materials in the last years has been extensively investigated [1-7] for the development of light-emitting devices in the Si-based optoelectronics. As for RE doped materials, because Er's characteristic emission at 1.54 μm is occurring in the minimum absorption of optical fibers, people have paid much attention to the Er doped Si-related semiconductors [8-13]. Few reports can be found on Nd doped Si-related materials. Nevertheless, it is significant meaningful to investigate the luminescence in Nd^{3+} doped Si-related semiconductors owing to the excitation of strongly quenched high energy levels[14, 15], because not only infrared (IR) emission is produced from the $^4F_{3/2}$ level, but also UV and visible luminescence from upper levels, such as $^4D_{3/2}$. And Nd ion is the popular RE element used for solid state lasers as active medium. It is Essential to study the transverse spread of Nd ions implanted in semiconductors, especially for today's deep-submicrometer devices. For all we know, there has been no report on the research of the transverse spreads of implanted keV Nd ions in SOI.

In this work, the 400 keV, 2×10^{15} ions/cm^2 Nd ions were implanted into SOI samples at the angles of 45°, 30° and 0°, respectively. The depth distributions and the transverse spreads of implanted Nd ions at the energy of 400 keV in SOI were obtained by Rutherford backscattering (RBS) technique. The measured transverse spreads were compared with those obtained by SRIM2012 (Stopping and Range of Ions in Matter, the predictions from Monte Carlo code) [16].

IOP Conf. Series: Materials Science and Engineering **281** (2017) 012044 doi:10.1088/1757-899X/281/1/012044

2. Experiment

The implantation was performed on the 500 keV ion implanter at the Institute of Semiconductors, Chinese Academy of Sciences. The SOI Smart Cut® UNI-BOND wafers used in the work consisted of a 300 nm thick crystalline Si cap layer, on a 500 nm thick buried SiO_2 layer, and a 720 μm thick Si substrate. The SOI samples were cleaned by the standard procedure before the implantation. All the SOI samples were implanted at room temperature (RT) with energy of 400 keV Nd ions at the dose of 2×10^{15} ions/cm^2. Sample I was implanted with Nd ion beams vertically (In fact, ion implantation has been performed at titled 7° off the normal plane of the sample surface to avoid the channeling effect, which would result in an error less than 1%).To measure the transverse distributions, tilted angle implantations have been Carried out at 30° and 45° for samples II and III. The voltage (it decided the energy of the implanted ions) of the implanter was calibrated by high-tension voltmeter. To avoid excessive heating of the samples, the incident current density was less than 1 μA/cm^2.

The RBS experiments were performed at 1.7 MV tandem accelerator of the Shandong University. The depth range of Nd ions implanted in SOI was measured by the RBS technique. The RBS spectrometry was carried out using 2.1 MeV He^{2+} ion beam produced by a 1.7 MV tandem accelerator with a ion beam current about 10 nA. The SOI samples were fixed on a three axis goniometer actuated by pulse motor in a vacuum chamber. All the measurements were carried out in a vacuum of 2.66×10^{-4} Pa. The backscattering spectra were performed by a multi-channel analyzer.

3. Results and discussion

The depth distribution of Nd ions implanted in SOI is nearly Gaussian, so the deepness profile could be depicted by the range straggling and the projected range, the standard deviation of the Gaussian distribution in deepness. The range straggling can be calculated from the surveyed full width at half maximum (FWHM)[17].

In term of the composition by Furukawa and Matsumura[18], the transverse spread of an deposited ion in SOI can be put into the depth range straggling of a target under canted implantation. If the spatiality probability distribution of the implanted ions is a three-dimensional Gaussian, the deepness distribution for a target tilted at an angle to the incident ion beam also becomes a Gaussian with standard straggling ΔD. The transverse spread and range straggling (longitudinal straggling) are denoted by ΔX_{L} and ΔR_{p}, respectively. Then the transverse distribution could be obtained by using Furukawa's formula[18]:

$$(\Delta D)^2 = (\Delta R_{\mathrm{p}})^2 \cos^2\theta + (\Delta X_{\mathrm{L}})^2 \sin^2\theta . \quad (1)$$

The transverse spread ΔX_{L} can be calculated from two measurements of ΔD for two different tilted implantation angles θ. Generally speaking, the deepness distributions of the implanted ions are not Gaussian absolutely; only in a small incident energy range of the implanted ion is distributed close to a Gaussian. Hence, Furukawa and Matsumura's method can only be approximately correct.

The backscattering spectra of Nd ions implanted in SOI show a nearly Gaussian distribution. To get the numerical values of the peak position and FWHM of longitudinal distribution of implanted Nd ions, these spectra were fitted by a computer program . These are relevant to the range straggling and projected range of Nd ions, respectively.

Figure 1 shows a typical RBS spectrum of the 400 keV, 2×10^{15} ions/cm^2 Nd ions perpendicularly implanted in SOI sample I at RT. The average projected range R_{p} and the range straggling ΔR_{p} of the implanted Nd ions in SOI can be calculated from the spectrum by the surface energy approximation[17].

In figure 2 the deepness distribution of 400 keV Nd ions perpendicularly implanted in SOI sample I by normal RBS spectrum of 2.1 MeV He ions is presented. The distribution shape denoted by triangles. The Gaussian fit curve is represented by the solid line. It is evident that the distribution shape shows nearly Gaussian for the incident ions. It is interesting that there is no obvious channeling tail in the spectrum of Nd ions perpendicularly implanted in SOI sample in Fig. 2, which manifests that the channeling effect in the surfase of SOI sample was nearly avoided by ion implantation performed at an tilted of 7°. The value of the range straggling ΔR_p can be obtained from the measured FWHM after thinking about the energy resolution of the measuring system and the energy straggling of He ions in the target. The key step in data analysis is the transform of the RBS energy spectra to deepness profiles. The work was done by the theory of Chu, Mayer, and Nicolet [17]. The ΔR_p value is 44.1 nm.

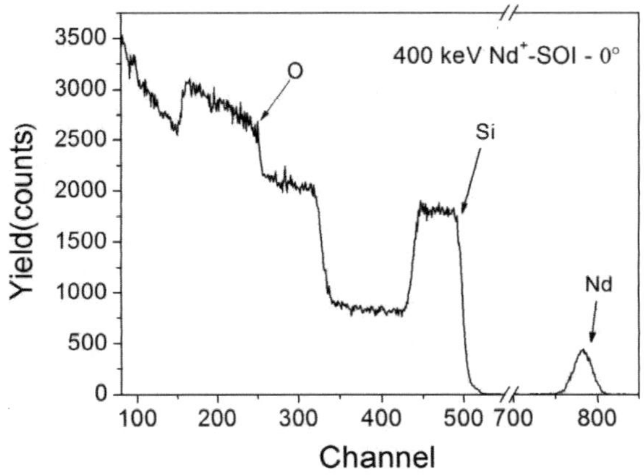

Figure 1. Typical RBS spectrum of the 400 keV, 2×10^{15} ions /cm^2 Nd ions vertically implanted in SOI sample I

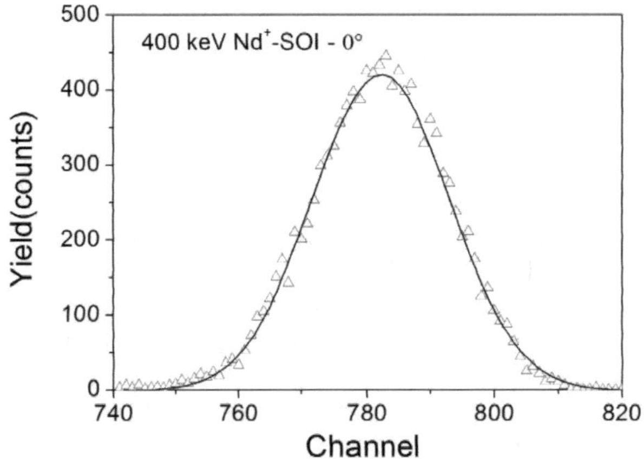

Figure 2. Normal RBS spectra of the deepness distribution of 400 keV Nd ions vertically implanted in SOI sample I . The data of the ions implanted at 0° and the Gaussian fit is represented by the triangles and solid line , respectively.

Figure 3 represents the RBS spectrum of the deepness distribution of 400 keV , 2×10^{15} ions/cm^2 Nd ions implanted in SOI at angle 30° (sample II). The distribution shape denoted by squares. The Gaussian fit curvehe is represented by solid line. It can be seen the distribution shape of the implanted ions shows nearly Gaussian profile. The standard deviation ΔD_1 of Nd ions implanted in SOI sample

IOP Conf. Series: Materials Science and Engineering **281** (2017) 012044 doi:10.1088/1757-899X/281/1/012044

II at 30° is obtained by the surface energy approximation[17]. The ΔD_1 value of is 40.6 nm. The lateral spread ΔX_L of 400 keV Nd ions implanted in SOI sample II at angle 30° was calculated by using Furukawa's formula (1) , ΔX_{L1}=27.6 nm.

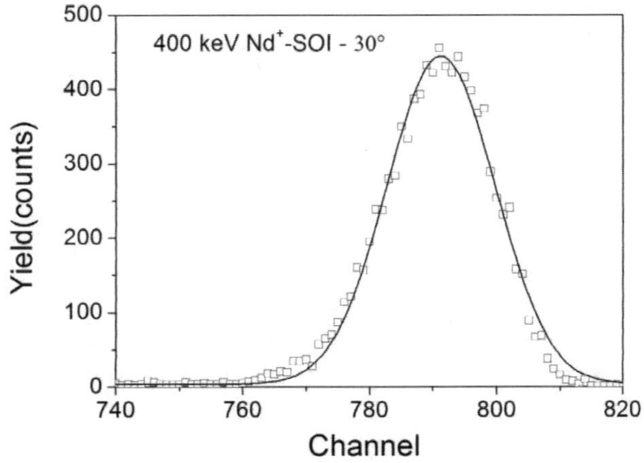

Figure 3. Normal RBS spectra of the deepness distribution of 400 keV Nd ions implanted in SOI sample II at 30°. The data of the ions implanted at 30° and the Gaussian fit are represented by the squares and solid line, respectively.

Figure 4 gives normal spectrum RBS of the deepness distribution (circles) of 400 keV, 2×10^{15} ions/cm^2 Nd ions implanted in SOI at 45° (sample III). The Gaussian fit curves (solid lines) was given too. It is obvious that the distribution shape shows nearly Gaussian profile for the 45° implanted Nd ions. Same, using the surface energy approximation [17], the range straggling ΔD of the 45° implanted Nd ions is calculated, ΔD_2=37.2 nm. Then another value of the transverse spread ΔX_L of 400 keV Nd ions implanted in SOI sample III was calculated by using formula (1), ΔX_{L2}=28.7 nm.

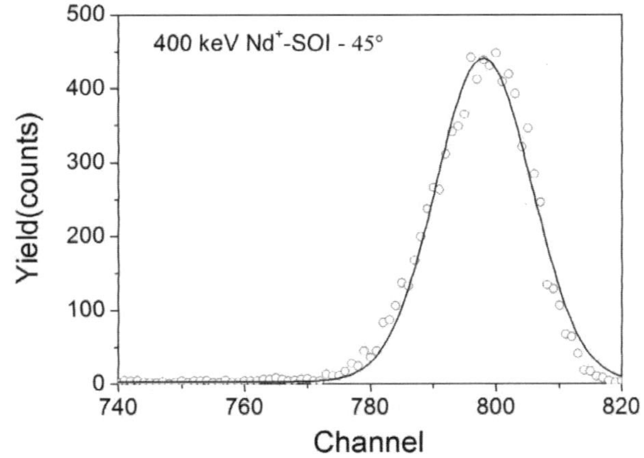

Figure 4. Normal RBS spectra of the deepness distribution of 400 keV Nd ions implanted in SOI sample III at 45°. The data of the ions implanted at 45° and the Gaussian fit are represented by the circles and solid line, respectively.

The mean experimental value of the transverse spread ΔX_L of 400 keV, 2×10^{15} ions/cm^2 Nd ions implanted in SOI samples was calculated from the data of ΔX_{L1} and ΔX_{L2}, ΔX_L=28.15 nm.

One of the primary aim of this work is to compare the experimental transverse spread with its theoretic value. The lateral spread ΔX_L had been calculated by using SRIM2012 code. The value of the lateral straggling of 400 keV Nd ions implanted in SOI calculated by SRIM2012 is 26.6 nm. Compared the theoretical simulation value with the experimental data, it can be seen that the experimental value of the transverse spread of 400 keV Nd ions implanted in SOI is a bit of larger than the theoretical simulation data. The difference between the theoretical value of SRIM2012 and the experimental result is about 5.5%, which is a good agreement.

4. Summary

400 keV, 2×10^{15} ions/cm^2 Nd ions were implanted in different SOI samples at room temperature at the angles of 0°, 30° and 45°, respectively. The range distribution and the transverse distribution of Nd ions implanted in SOI were measured by RBS technique. All deepness distributions of the 400 keV, 2×10^{15} ions/cm^2 Nd ions implanted in SOI samples at 0°, 30° and 45° are approximately Gaussian. The transverse spread ΔX_L of 400 keV Nd ions implanted in SOI sample was calculated using Furukawa's formula (1) and the data of Chu, Mayer and Nicolet. The experimental value of the transverse spreads the 400 keV Nd ions implanted in SOI sample were compared with its data of SRIM2012 prediction. It is shown that the experimental transverse spread agrees well with the theoretical value.

Acknowledgments

This work was supported by the National Natural Science Foundation of China (Grant No. 11205096), the Natural Science Foundation of Shandong Province, China (Grant No. ZR2015AM021), the Shandong Jianzhu University Doctoral Foundation, China (Grant No. XNBS1341), the Science and Technology Plan Project of Jinan City, China (Grant No. 201202092).

References

[1] Chen C Y, Chen W D, Wang Y Q, Song S F and Xu Z J 2003 *Acta Phys. Sin.* **52** 736
[2] Wang J Z, Shi Z Q, Lou H N, Zhang X L, Zuo Z W, Pu L, Ma E, Zhang R, Zheng Y L, Lu F and Shi Y 2009 *Acta Phys. Sin.* **58** 4243
[3] Ennen H, Schneider J, Pomrenke G and Axmann A 1983 *Appl. Phys. Lett.* **43**(10) 943
[4] Liang J J, Chen W D, Wang Y Q, Chang Y and Wang Z G 2000 *Chinese Phys.* **9** 783
[5] Ding W C, Liu Y, Zhang Y, Guo J C, Zuo Y H, Cheng B W, Yu J Z and Wang Q M 2009 *Chinese Phys. B* **18** 3044
[6] Hansson G V, Du W X, Elfving A and Duteil F 2001 *Appl. Phys. Lett.* **78** 2104
[7] Lei H B, Yang Q Q and Wang Q M 1998 *Acta Phys. Sin.* **47** 1201
[8] Fu Y C, Li B S, Yang Y T, Zhang C H, Zhang H H and Zhou L H 2009 *Acta Phys. Sin.* **58** 3302
[9] Qin X F, Chen M, Wang X L, Liang Y and Zhang S M 2010 *Chinese Phys. B* **19** 113403
[10] Wang K M, Shi B R, Guo H Y, Wang W and Ding P J 1996 *Mater. Sci. Eng., B* **39** 133
[11] Biersack J P 1982 *Z. Phys. A* **305** 95
[12] Qin X F, Wang F X, Liang Y, Fu G, Zhao Y M 2010 *Acta Phys. Sin.* **59** 6382
[13] Tan N and Zhang Q Y 2006 *Chin. Phys.* **15** 2165
[14] Lenth W and Macfarlane R M 1990 *J. Lumin.* **45** 346
[15] Macfarlane R M, Tong F, Silversmith A J and Lenth W 1988 *Appl. Phys. Lett.* **52** 1130
[16] Ziegler J F *http://www.srim.org*
[17] Chu W K, Mayer J W, Nicolet M A 1978 *Backscattering Spectrometry* (New York: Academic) chap 5, p137
[18] Furukawa S and Matsumyra H 1973 *Appl. Phys.* **22** 97

IOP Conf. Series: Materials Science and Engineering **281** (2017) 012045 doi:10.1088/1757-899X/281/1/012045

Research on the EDM Technology for Micro-holes at Complex Spatial Locations

J Y Liu [1,2*], J M Guo [1,2], D J Sun [1,2], Y H Cai [1,2], L T Ding [1,2] and H Jiang[1,2]

1 Beijing Electric Machining Institute, Beijing 100191;
2 Beijing Key Laboratory of EDM Technology, Beijing 100191

* Corresponding author: (J Y Liu) Jianyong Liu (1983-), associate research fellow, E_mail:Liu_18778@163.com

Abstract. For the demands on machining micro-holes at complex spatial location, several key technical problems are conquered such as micro-Electron Discharge Machining (micro-EDM) power supply system's development, the host structure's design and machining process technical. Through developing low-voltage power supply circuit, high-voltage circuit, micro and precision machining circuit and clearance detection system, the narrow pulse and high frequency six-axis EDM machining power supply system is developed to meet the demands on micro-hole discharging machining. With the method of combining the CAD structure design, CAE simulation analysis, modal test, ODS (Operational Deflection Shapes) test and theoretical analysis, the host construction and key axes of the machine tool are optimized to meet the position demands of the micro-holes. Through developing the special deionized water filtration system to make sure that the machining process is stable enough. To verify the machining equipment and processing technical developed in this paper through developing the micro-hole's processing flow and test on the real machine tool. As shown in the final test results: the efficient micro-EDM machining pulse power supply system, machine tool host system, deionized filtration system and processing method developed in this paper meet the demands on machining micro-holes at complex spatial locations.

1. Introduction

As the development of aviation, aerospace, automobile and biomedical engineering fields, a lot of components are developed with complex spatial location micro-hole structure [1]. Because many such kinds of components have small micro-hole aperture, low structure rigidity and special material, it is difficult to meet the machining requirements through common machining. Due to the advantages of micro-EDM machining such as high machining precision without macro cutting force during machining, it plays the irreplaceable and key role for machining such kind of components [2,3]. For the EDM forming machining of complex spatial location, five-axis combined machining can meet the machining requirements commonly. However, to complex spatial location micro-hole EDM machining, one micro-feed axis shall be designed for meeting the feeding and guiding demands of electrode wire of micro diameter [4-7]. Therefore, for such kind of components (for example: oil spray nozzle of automobile engine high-pressure common rail system, micro spray hole on aerospace divert and attitude control engine, aviation engine blade micro-hole), the six-axis EDM machining equipment with four linear axes and two rotation axes shall be developed for meeting the requirements on technique and technology. For the machining of such kind of components, this paper conducts systematic researches from the aspects of power supply system development, machine tool host design, deionized water system development, technique test and other aspects, to conquer the technical problems of EDM machining. Development of power supply system

For the six-axis discharging machining demands on micro-EDM machining narrow pulse and high frequency, the precise micro-EDM machining power supply system developed in this paper composes of numerical control system and EDM machining control system, power amplifier unit, main circuit unit, automatic voltage regulation unit and kinematic axis drive unit. EDM machining control system mainly includes: clock control and discharging pulse signal generation part; parameters such as current and voltage setting part; special machining waveform control part (equal pulse width); micro servo clearance and discharging state (voltage and current) detection part; abnormal discharging treatment part (increasing the pulse spacing, cutting the pulse width) etc.. The power amplifier unit can amplifier the power of the machining pulse voltage and current, and composes of Vertical Metal Oxide Semiconductor (VMOS) power switching tube and prepositional drive amplification circuit. The main circuit unit includes: low-voltage machining circuit; high-voltage machining circuit; high and low voltage compound circuit; low wear circuit; super precision machining circuit etc.. The automatic voltage regulation unit provides stable machining voltage with the Pulse-Width Modulation (PWM) automatic voltage stabilizing circuit, and is capable of achieving numerical control and varied protection functions. Please see Figure 1.

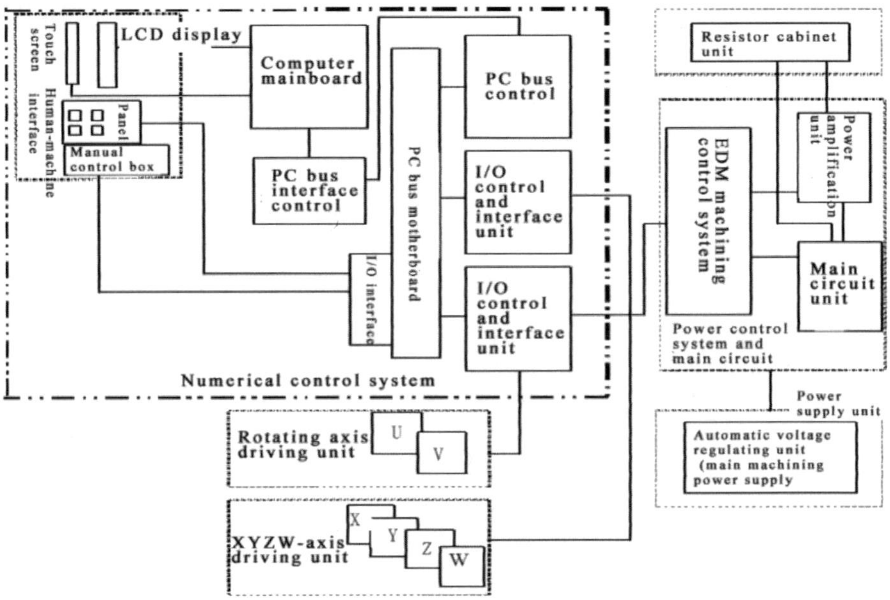

Figure 1. Overall framework of the power supply system.

The machining energy of micro-EDM is provided by the pulse power supply; therefore, each circuit design is closely related with the micro-hole machining quality.

In this paper, a low-voltage automatic voltage regulation unit is designed to provide stable machining voltage for low-voltage power supply circuit. It adopts high speed PWM automatic voltage stabilizing circuit to achieve full closed-loop numerical control and varied protection functions of power output, as well as ensure the ripple factor is lower than 1%, voltage stabilizing precision reaches 0.5%, thus precisely controlling the discharging energy and improving the machining stability.

The high-voltage circuit with the principle of high speed PWM modulation is adopted to achieve high-voltage (300V) and high power voltage stabilizing power supply, and ensure the ripple factor is lower than 1% and voltage stabilizing precision reaching 1%. During the period of micro-EDM machining, it mainly works for auxiliary breakdown, which can greatly improve the breakdown rate and utilization rate of pulses, reduce the breakdown time delay, and benefit to increase the machining speed.

With the combined advantages of relaxation type pulse power supply and transistor pulse power supply, the micro and precision machining circuit, which is capable of providing small enough current and narrow enough Tr-RC pulses, is developed; thus, small discharging etch pits are produced on the machining surface, and then machining surfaces with small roughness are got, achieving the micro machining of small holes.

During micro high frequency discharging machining, extremely high requirement is presented to the real-time monitoring under machining state, the precision and stability during feedback has direct influence on the discharging energy and machining effect. In the clearance detection system, the 12bit high speed analog to digital converter is adopted to monitor the abnormal discharging state, high speed converting time and high precision sampling result can quickly show the state of the discharging machining clearance. Super treatment capacity (30MIPS) is achieved via the most advanced programmable circuit, which is capable of detecting the pulse status within 30ns, effectively eliminating risks of harmful discharging, such as electric arc and short circuit, as well as improving the machining efficiency and surface quality.

2. Development of the host structure

The mechanical system structure design rationality of the complex spatial location micro-hole EDM machining machine tool is the key factor to achieve precise micro machining. With the method of combining CAD structure design, CAE simulation analysis (shown in Figure 2-3), modal test (shown in Figure 4), ODS (Operational Deflection Shapes) test (shown in Figure 5) and theoretical analysis [8,9], optimization design is conducted for the components and key axes of the machine tool.

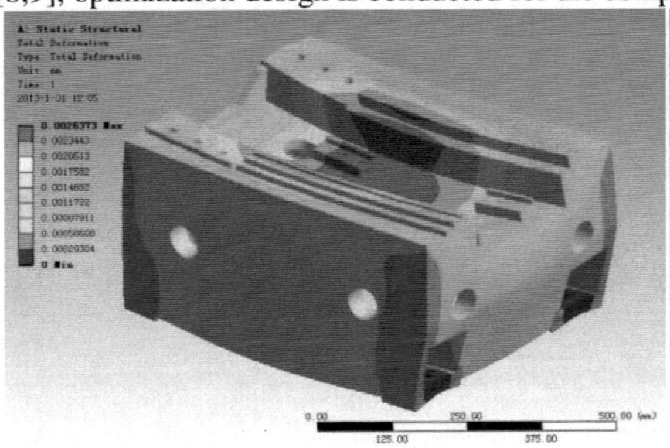

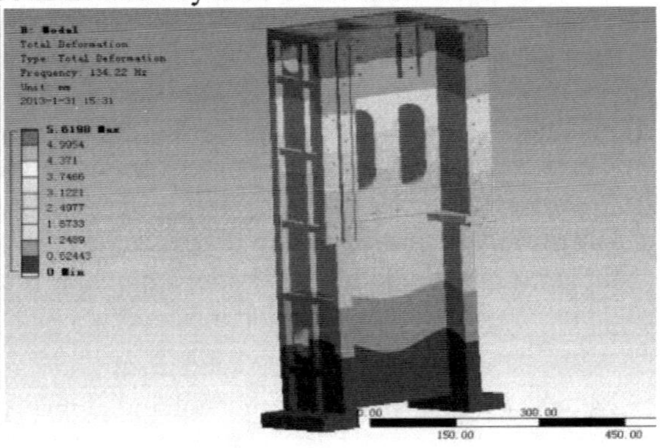

Figure 2. Machine body deformation nephogram. **Figure 3.** First-**order** modal vibration nephogram.

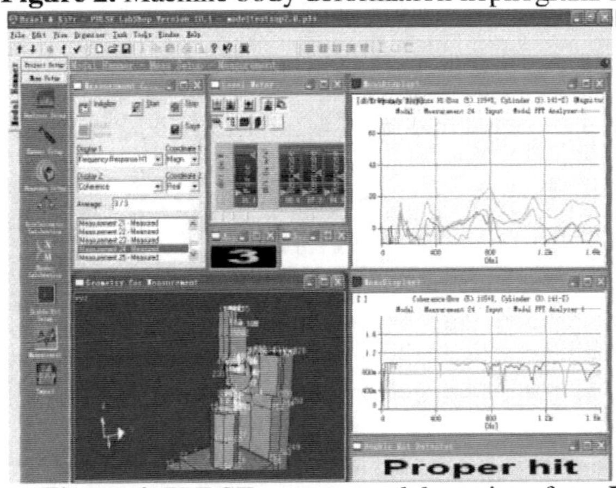

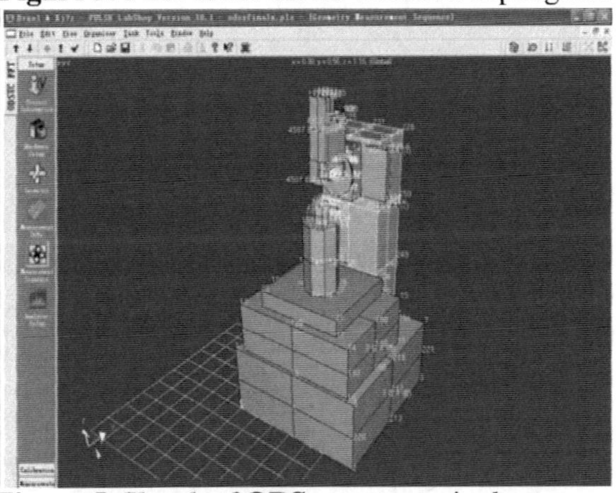

Figure 4. PULSE system modal test interface. **Figure 5.** Sketch of **ODS** measure point layout.

Through the above analysis and test, casting part structures of the main components of the machine tool are optimized, and the construction of the machine tool has been improved; and the host structure is capable of meeting the EDM discharging machining requirements on complex spatial location micro-hole with aperture dispersion of ±0.003mm. The body of the machine tool is shown in Figure 6.

IOP Conf. Series: Materials Science and Engineering **281** (2017) 012045 doi:10.1088/1757-899X/281/1/012045

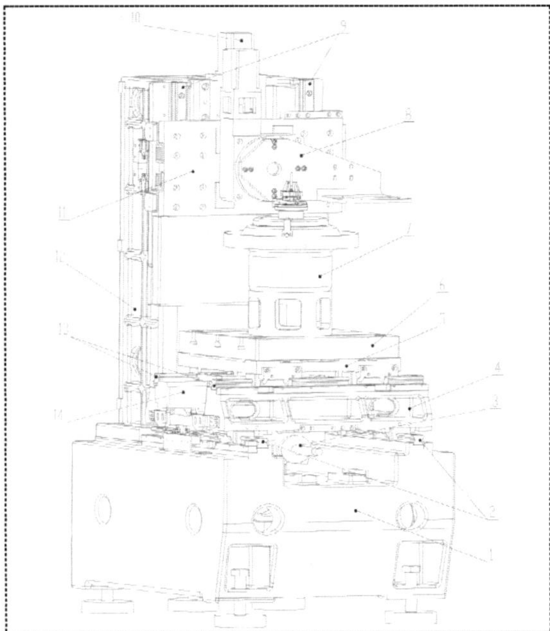

Figure 6. the host structure of EDM machine tool

1-machine body; 2-Y axis guide rail; 3-Y axis servo drive mechanism; 4-middle slide plate; 5-upper slide plate; 6-workbench; 7-C axis; 8-B axis; 9-Z axis guide rail; 10-Z axis servo drive mechanism; 11-B axis support plate; 12-upright column; 13-X axis guide rail; 14-X axis servo drive mechanism.

3. Development of deionized water filtration system

In order to achieve the precise, reliable and stable machining of micro-holes, the water circulation system shall meet the following conditions:

(1) The electrical conductivity of deionized water must be lower than 0.8μS/cm;

(2) The flow rate of deionized water ejected from the water spray nozzle shall be adjustable in the certain range;

(3) During machining, the water temperature shall be controlled in a certain range.

According to the demands on working condition, the schematic diagram of the planned deionized water circulation system is shown in Figure 7.

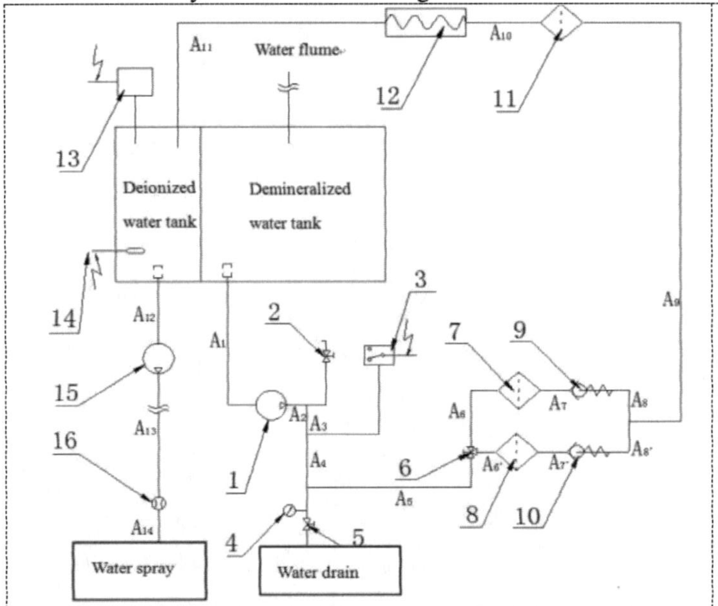

Figure 7. Schematic Diagram diagram of deionized water circulation system

1-filtration pump; 2-overflow valve; 3-electronic pressure switch; 4-pressure meter; 5-drain valve; 6-filter selector valve; 7-filter I; 8-filter II; 9, 10-one-way valve; 11-filter III; 12-panel radiator; 13-electrical conductivity sensor; 14-water level sensor; 15-water spray pump; 16-flow rate meter.

The work principle of this water circulation system is: the demineralized water is sucked from demineralized water tank with the filtration pump 1, flows from the filtration system with the filter I7 or filter II8 and filter III11, and the demineralized water is filtered into the deionized water, after cooling with panel radiator 12, the water flows into the deionized water tank; deionized water is pumped from the deionized water tank with the water spray pump 15 for precisely machining micro inverted cone holes; observing with the overflow valve 2 if the demineralized water added into the demineralized water tank has reached the requirements; the pressure meter 4 is adopted for observing the water pressure of filtration pump; the drain valve 5 is adopted for waste water drainage when replacing demineralized water; the filter selector valve 6 is adopted to control water path to cross filter I7 or filter II8; one-way valves 9 and 10 are adopted to ensure the single flow of water path and avoid interference between filter I7 and filter II8; the electrical conductivity meter 13 is adopted for detecting if the deionized water in the closed deionized water tank meets the requirements; the water level sensor 14 is adopted for detecting the deionized water level; the flow rate meter 16 is adopted for detecting the ejection flow rate of deionized water.

In order to get deionized water of high electrical conductivity, 1μm polypropylene meltblown PP filter is adopted for primary filtration; the filter structure is coarse fiber for external layer fine fiber for internal layer, loose external layer and tight internal layer. This filter is capable of removing 1μm or bigger impurity particles, including the metal particles formed during EDM machining. Polished resin is adopted for secondary filtration to replace Na+, Cl- from water with H+ and OH- radical group in the filtration material, H+ and OH- in the water are combined to product water, thus reducing the electrical conductivity.

4. Development and test machining of process technology

The complex spatial location micro-hole machining subprogram technique flow is shown in Figure 8. Machining with reversed polarity of certain time is achieved with the electrode parameters correction, thus correcting the profile of the electrode tip, and then preparing for detection, workpiece clamping, cooling pump opening and electrode electrical contact positioning. The purpose of electrode contact positioning is to achieve automatic feed compensation of electrode with the automatic feed function and electrical contact electrode positioning function, and position the electrode and workpiece to the best place of servo hole machining through the voltage change of the electrode electrical contacting workpiece. Then, with the servo hole machining function, it is identified as open circuit, short circuit and normal machining state with the real-time acquired discharging state current, controlling the W-axis to do quick servo movement to achieve hole machining. Meanwhile, the control system will identify if it has achieved the machining depth through real-time detection, which means if it has achieved the set machining lower limit, and the machining is completed after reaching lower limit positions of the precise micro-hole machining.

In this paper, This paper takes the machining of precise reverse taper micro-hole of high-pressure common rail system and injector head in aerospace field as example to verify if the micro-hole EDM machining equipment and process technology developed in this paper can meet the micro-hole machining demands of complex spatial location, the micro-hole EDM machining equipment adopted in this paper is shown in Figure 9, the oil spray nozzle of high-pressure common rail system is shown in Figure 10,the injector head in aerospace field is shown in Figure 11. Records of the inside and outside diameters of the finished oil spray nozzle's micro-holes are shown in Table 1, and records of the outside diameters of the finished injector head's micro-holes are shown in Table 2.

As shown in Table 1, the max outside aperture of reverse taper micro-holes in oil spray nozzle of the high-pressure common rail system is 0.224mm, minimum is 0.220mm; the max inside diameter is 0.241mm, minimum is 0.245mm. The dimensional tolerance range of micro-holes is in the range of ±0.003mm. As shown in Table 2, the max diameter of micro-holes in injector head in aerospace field is 0.173mm, minimum diameter is 0.168mm. The dimensional tolerance range of micro-holes is in the range of ±0.003mm.

IOP Conf. Series: Materials Science and Engineering **281** (2017) 012045 doi:10.1088/1757-899X/281/1/012045

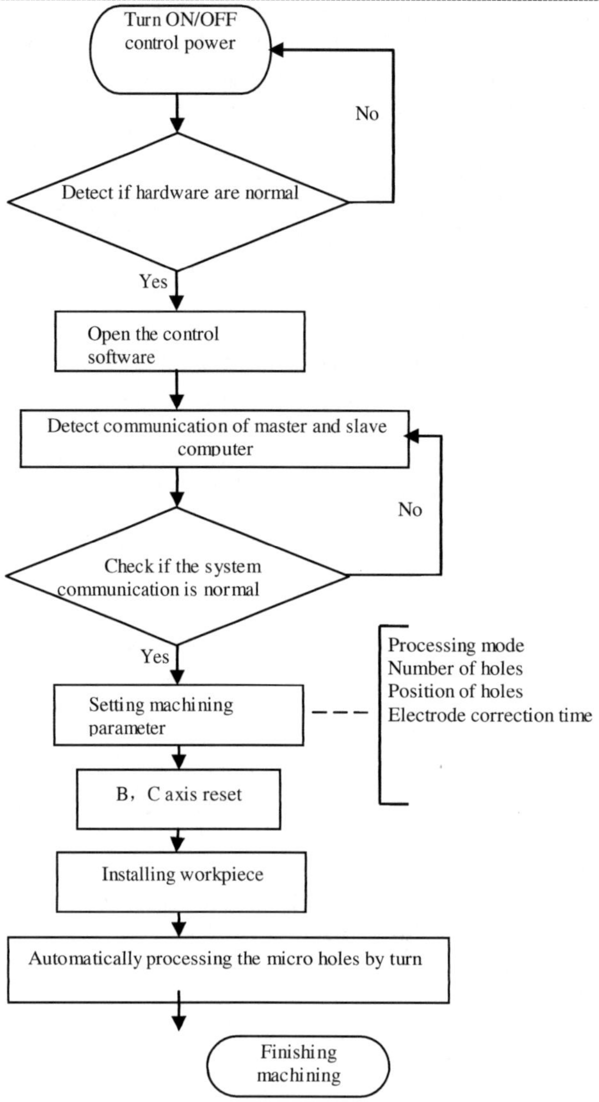

Figure 8. Technique flow of fully automatic machining control.

IOP Conf. Series: Materials Science and Engineering **281** (2017) 012045 doi:10.1088/1757-899X/281/1/012045

Figure 9. Micro-hole EDM equipment.

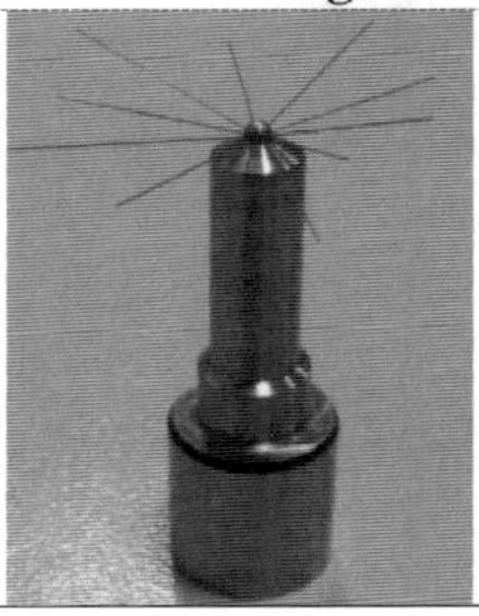

Figure 10. Finished oil Spray Nozzle.

Figure 11. Injector head.

Table 1. Test records of oil spray nozzle

Hole No.	Out-Dia (mm)	In-Dia (mm)	Taper (mm)	Vert Ang Degree	Dev. Degree	Hor Ang Degree	Dev. Degree	A-dim (mm)	Dev. (mm)	D-dim (mm)	E-dim (mm)	Tan Dev (mm)
1	0.223	**0.244**	**-0.021**	77.2	-0.3	35.1	-0.1	1.07	0.01	0.73	1.16	0.00
2	**0.223**	0.242	**-0.019**	77.3	-0.2	71.2	0.0	1.07	0.01	0.73	1.16	-0.03
3	**0.223**	0.242	**-0.019**	77.1	-0.4	143.0	-0.4	1.07	0.01	0.73	1.15	-0.02
4	0.221	**0.245**	**-0.024**	**76.8**	**-0.7**	179.2	-0.4	1.07	0.01	0.73	1.15	0.01
5	0.220	**0.245**	**-0.025**	77.2	-0.3	215.2	-0.4	1.07	0.01	0.73	1.16	0.05
6	0.220	0.243	**-0.023**	77.3	-0.2	287.1	**-0.6**	1.08	0.02	0.73	1.16	0.06
7	**0.223**	0.241	**-0.018**	77.4	-0.1	323.0	**-0.8**	1.08	0.02	0.73	1.16	0.03
8	**0.224**	0.242	**-0.019**	77.0	-0.5	359.7	-0.2	1.07	0.01	0.73	1.16	0.00
Mean	**0.222**	0.243	**-0.021**	77.2	-0.3		-0.4	1.07	0.01			0.01
Nom□	0.000											
Range	**0.004**	**0.004**	**0.008**	**0.5**	0.5		0.7	**0.01**	**0.01**			**0.09**
NomR	0	0	0		0		0		0			0
Sigma	2.289	1.261	3.227	0.2	0.2	117.1	0.2	0.00	0.00	0.00	0.00	0.03

Table 2. Test records of injection holes

No.	1	2	3	4	5	6	7	8
diameter (mm)	0.17	0.169	0.171	0.17	0.173	0.168	0.171	0.17

5. Summary

Based on the technical characteristics of EDM technology for micro-holes at complex spatial locations, in this paper, the micro-hole EDM power supply system is developed. And the host construction is developed with the method of combining theoretical analysis, CAD optimization design, CAE simulation analysis, modal test and ODS test. And the deionized water filtration system is developed according to the demands on micro-hole reliable machining. Based on these studies, the fully automatic machining control technique flow of the complex spatial location small hole is developed. Finally, the tests are conducted with the typical components (Oil spray nozzle of the high-pressure common rail system, injector head in aerospace field) of complex spatial location micro-hole. The test results show: the efficient micro-EDM pulse power supply system, host constrcution of the machine tool, deionized filtration system and machining process method developed in this paper can meet the demands on machining micro-holes at complex spatial locations.

Acknowledgement

This project is supported by Beijing Natural Science Foundation(L150006)

References

[1] Zhai L J, Fu J J, Liu J Y et al..2016. *Metal Machining (Cold Machining)*.21:1
[2] Jahan M P, Wong Y S, Rahman M. 2009.*Journal of Materials Machining Technology*. 209(4):1706
[3] Jiang Y, Zhao W, Xi X. 2012.*Journal of Materials Machining Technology*. 212(7):1463
[4] ELECTRICAL DISCHARGE MACHINE APPRATUS FOR REVERSE TAPER BORES [P]. World Intellectual Property Organization. A. WO 2008/144141 A1.2008.
[5] Floating Cover Electrode Guide System For Electric Dischage Maching [P].US.A. US005951883A,1999.
[6] Zhao W S, Li L, Li Z Y. 2004. *Computer Integrated Manufacturing System*. 10(10):1263
[7] Cao F G 2005 *EDM Technology* (Beijing. Chemical Industry Press)
[8] Liao P, Deng F P. 2011. *Journal of Zhengzhou University (Engineering Science)*.32(3): 76
[9] Tan W J, Yang L, Wu X R et al.. 2011. *Journal of Vibration Engineering*. 24(5): 498

Large Efficient Intelligent Heating Relay Station System

C Z Wu*[1], X G Wei[1] and M Q Wu[2]

[1]School of Mechanical Engineering, University of Jinan, Jinan250022, China
[2]Shandong vocational college of transportation and technology,Weifang261206, China

Email:changzhongwu@163.com,2530149460@qq.com,324240336@qq.com

Abstract. The design of large efficient intelligent heating relay station system aims at the improvement of the existing heating system in our country, such as low heating efficiency, waste of energy and serious pollution, and the control still depends on the artificial problem. In this design, we first improve the existing plate heat exchanger. Secondly, the ATM89C51 is used to control the whole system and realize the intelligent control. The detection part is using the PT100 temperature sensor, pressure sensor, turbine flowmeter, heating temperature, detection of user end liquid flow, hydraulic, and real-time feedback, feedback signal to the microcontroller through the heating for users to adjust, realize the whole system more efficient, intelligent and energy-saving.

1. Introduction

The design of large and efficient intelligent heating relay station system by improving the heat exchanger, research and development of intelligent control system, to solve the traditional central heating system of low efficiency, high energy consumption, serious pollution problems[1], and gradually extended to the whole country, realize the efficient nationwide heating system, energy-saving and intelligent.

2. Design of intelligent heating system

2.1. The design of the control part.

ATM89C51 is a powerful microcontroller, a wide range of uses, and has been applied in many fields [2]. Therefore, in the control part of this design, ATM89C51 series microcontroller is adopted. Compared with the existing PLC control system in foreign countries, the utility model has the advantages of simple structure, convenient operation, fast processing speed, less power consumption, strong environmental adaptability, etc.[3]. SCM schematic as follows(Figure 1):

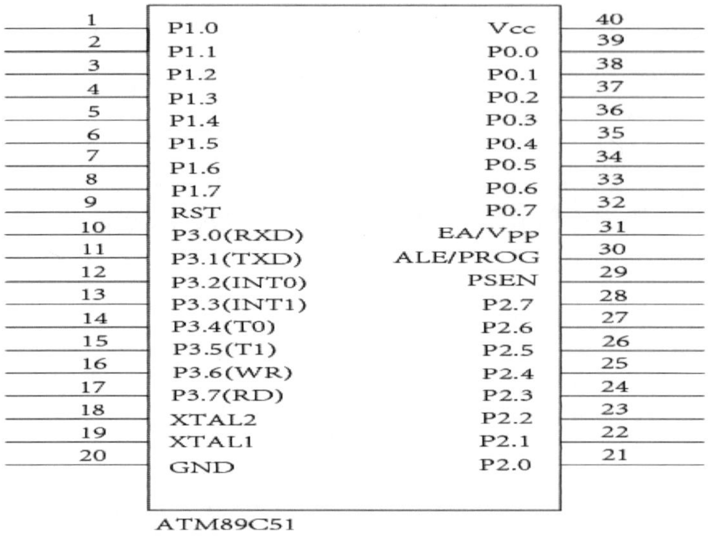

Figure 1. Single chip computer schematic diagram

The single chip microcomputer controls the electromagnetic pump directly through the chopper control circuit. Because the electromagnetic pump connects the plate heat exchanger, the single chip microcomputer will receive the feedback signal of the three sensors in real time (Figure 2):

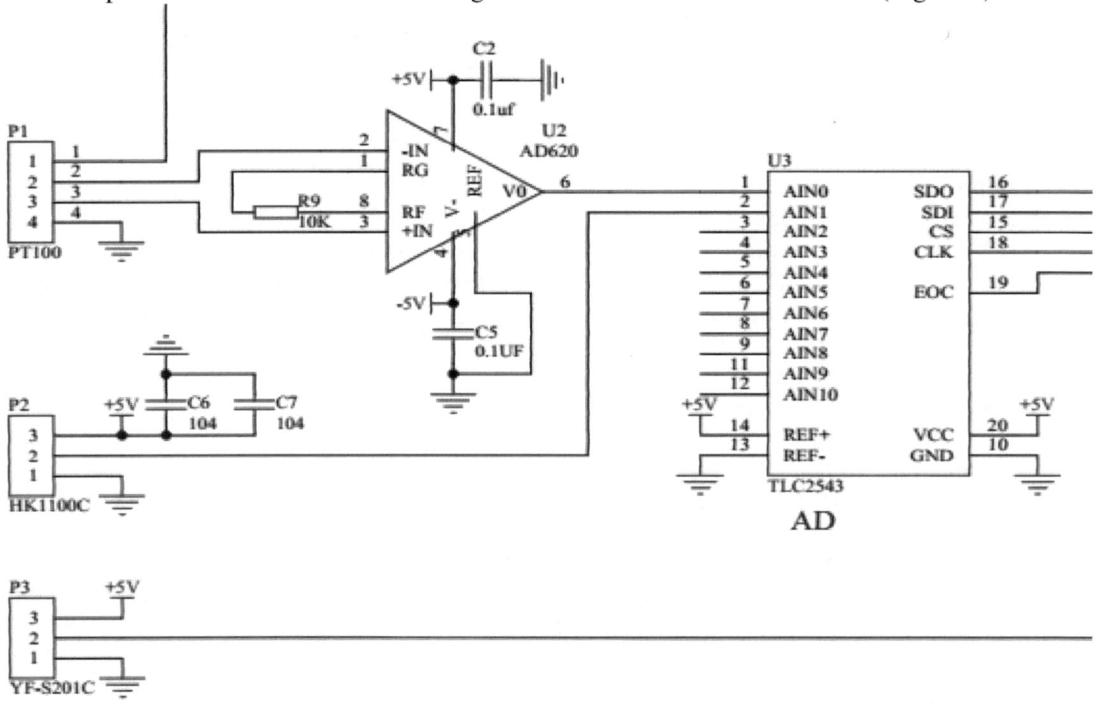

Figure 2. single chip microcomputer feedback signal of real time

The single chip microcomputer carries out real-time control of the electromagnetic pump through the signals returned by the three sensors, so as to control the heating of the plate heat exchanger and realize intelligent control. The three sensors convert analog signals into digital signals through the AD conversion circuit and feed them back to the microcontroller[4].

2.2. The design of the sensor temperature sensor

This design uses the PT100 temperature sensor. The PT100 temperature sensor is an instrument that transforms a temperature variable into a transmitted output signal. Usually consists of two parts: sensors and signal converters. The sensor is mainly a thermocouple or a thermal resistor, and the thermal resistance sensor is used in this design.

Works: The relationship between temperature and resistance change in PT100 temperature sensor .

Differential pressure sensor: The pressure acts directly on the diaphragm of the sensor, causing the diaphragm to produce a slight displacement, resulting in a change in the capacitance of the sensor. The electronic circuit in the sensor detects this change and converts it into a corresponding electrical signal that is fed back to the control unit. For this design, a change in the hydraulic pressure changes the capacitance in the sensor, which changes directly back to the control section.

Turbine flowmeter: Turbine flowmeter is a common sensor for testing liquid velocity in China. It belongs to the speed flow meter. Its main characteristics are compact structure, high reliability, intuitive reading, no interference from external power supply, and resistance to lightning stroke.

3. The design of plate heat exchanger components

3.1. Overall design of plate heat exchanger.

In the heat exchanger, with two water inlets and two outlets through the middle plate to the heat transfer effect, both ends of a set of inlet and outlet, can pass through the same medium, can also be through two different kinds of media, and about the heat transfer on both sides are not affected each other. Three in view of the plate heat exchanger as shown (Figure 3,4,5):

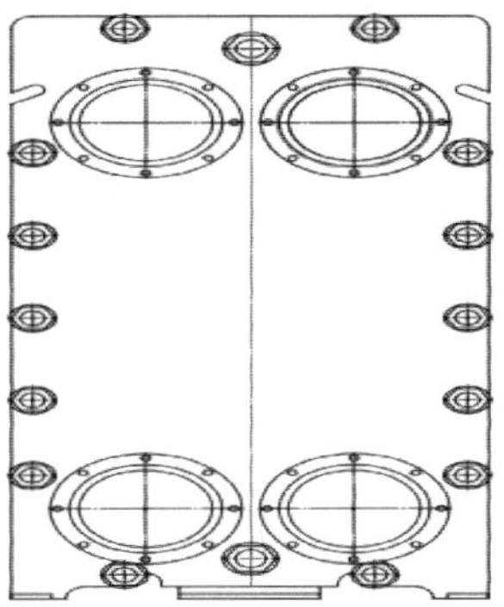

Figure 3. Main view of plate heat exchanger

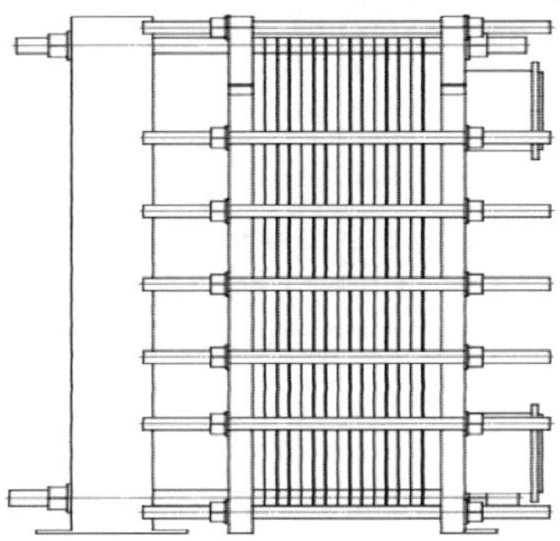

Figure 4. Top view of plate heat exchanger

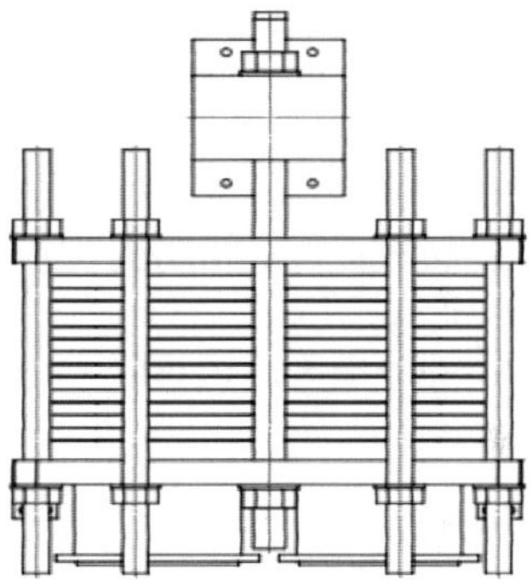

Figure 5. Left view of plate heat exchanger

Compared with shell and tube heat exchangers, plate heat exchangers have the following advantages: (1) The heat transfer coefficient of plate heat exchanger is higher; (2)Small temperature difference at the end; (3)Small footprint; (4) Overall light weight; (5) Small heat loss.

3.2. The component design of plate heat exchanger.

The structure of plate heat exchanger consists of metal plate, fixed compression plate (front end plate), movable compression plate (rear end plate), clamping bolt, upper guide rod, lower guide rod and strut.

3.2.1 Front and rear end plate

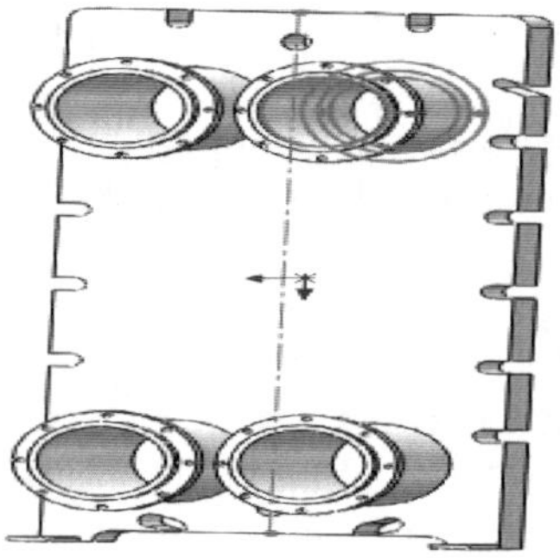

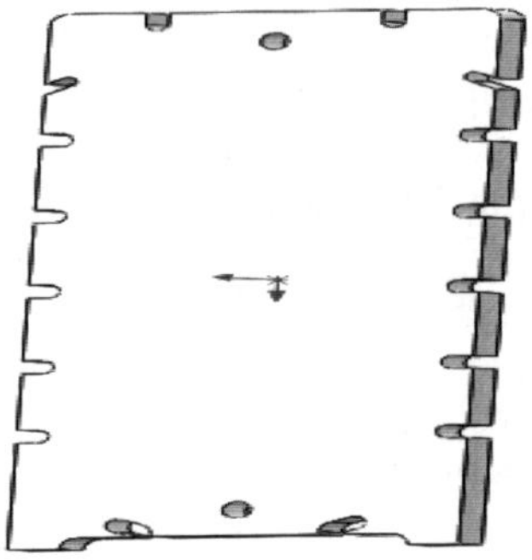

Figure 6. Front plate (fixed clamping plate) **Figure 7.** Back end plate (moving clamping plate)

After analysis, the end plate is not only in the middle of the clamping plate, the heat transfer process, the four corners of the front plate and the back end plate, each side will be in direct contact with the fluid, and some fluid is corrosive, so in the selection of materials, need to consider the material resistance corrosive, and need to have certain compressive strength. Considering the above factors, 0Cr18Ni14Ti stainless steel is selected. This material has sufficient compressive strength and corrosion resistance. For the front and rear end plates, corrosion resistance is the most important performance, so it is necessary to treat the front and rear end plates with solid solution. Figure 6 is a fixed clamping plate, and Figure 7 is a moving clamping plate.

3.2.2 Guide bar and screw rod.

Guide bar and screw rod need good wear resistance and good tensile strength, so 45 steel can be used. In the plate heat exchanger, the guide bar needs to bear the shear stress and the tensile stress in the horizontal direction, and the wear resistance of the threaded part is higher. Also, due to the plate screw with the nut on the middle of the clamping screw rod, so by the horizontal direction of the tensile stress is larger, the thread is more easy to wear, so for a guide rod and a screw rod, comprehensive mechanical properties need to be optimized, so the need for a guide rod and a screw rod for quenching and tempering.

3.2.3 Hexagon nut.

Because of the plate changing in use, the plate, the screw and the nut need to be removed and the loss is greater. Therefore, the 45 steel material with higher cost performance is chosen. When the nut matches with the guide bar and the screw rod, the thread is worn more seriously, and the nut is not affected by other external force except the thread force. Therefore, the nut can be treated with low temperature tempering at the time of heat treatment.

3.2.4 Sheet bar.

When considering the sheet material, it is necessary to consider its corrosion resistance and service life. Compared with stainless steel, titanium has better corrosion resistance than stainless steel, and can adapt to more kinds of fluids, and its service life is longer. Moreover, the density of titanium is only 0.6 times of that of stainless steel, the quality is light, and the pressure on the guide bar is small, which can greatly reduce the total weight of plate heat exchangers. Therefore, in this design, the plate is made of titanium.

3.2.5 Pillar.

Pillars play a supporting and stabilizing role in plate heat exchangers and prevent them from tilting and moving during operation, so they need to have certain mechanical properties. Because of the poor mechanical properties of cast iron, Q345 structural steel is used here. As shown in figure 8. for the pillar.

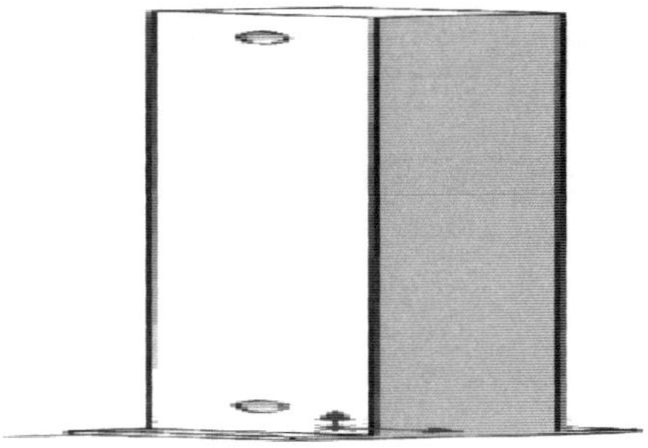

Figure 8. Stanchion

3.2.6 Materials and heat treatment.
Taking into account the functions and forces of the main components, select the appropriate materials to meet the work requirements, as shown in table 1.

Table 1 Materials and heat treatment details

Part name	Material	Heat treatment process
Front panel (moving clamping plate)	0Cr18Ni14Ti	Solid solution treatment
Back end plate (fixed clamping plate)	0Cr18Ni14Ti	Solid solution treatment
screw	45	Quench + high temperature tempering
Guide bar	45	Quench + high temperature tempering
nut	45	Harden and low temperature tempering
plate	Ti	No heat treatment
stanchion	Q345	No heat treatment

Conclusion
This paper designs a kind of energy saving heating relay station system. From the design process of intelligent heating system, first through the microcontroller and sensor to achieve the intelligent control of the relay station system, improve energy efficiency; secondly, we improved the existing heat exchanger, by replacing the improvement on the part of the guide rod and plate material, reduce the overall weight of plate heat exchangers the increase of the service life of the plate. This design makes up the deficiency of the existing heating system in China, and realizes the high efficiency, intelligence and energy saving of the heating system.

References
[1] Yang Y. *Application Research of intelligent control system for central heating* [D]. Hebei United University. 2013
[2] Yan L. *Research on intelligent control method of central heating network* [D]. Dalian University of Technology. 2003

[3] Xu D C. *Research on the Technology of Intelligent Control Terminal Personalized Heating Heating System*[J]. JiLin Normal University. 2014
[4] Wang W B, Mu D D, Wang X D, Wang S Y. *Intelligent Control of Heating Systems in Public Building[J].* Dalian Maritime University,IT College of Dalian Maritime University,Dalian Boiler and Pressure Vessel lnspection lnstitute. 2014

IOP Conf. Series: Materials Science and Engineering **281** (2017) 012047 doi:10.1088/1757-899X/281/1/012047

Thermodynamic Simulation of Viscosity of TiO_2-Ti_2O_3-CaO Ternary Slag

P Zhang[1], W H Ma[1*], S J Zhang[1,2], Y Lei[1] and J H Wen[1]

[1] School of Metallurgy and Energy Engineering, Kunming University of Science and Technology, Kunming 650000, China.
[2] Panzhihua University, Panzhihua 617000, China.

* E-mail: mwhsilicon@126.com

Abstract. The viscosity of high titanium slag at high temperature is one of the key factors of slag-iron separation. Based on the Einstein-Roscoe equation, thermodynamic simulation of viscosity of TiO_2-Ti_2O_3-CaO ternary slag is studied by using FactSage® software, and the effects of temperature, CaO content and solid-phase particles on the viscosity of slag were studied. The results show that the increase of CaO content has the effect of reducing melting temperature and viscosity of TiO_2-Ti_2O_3-CaO ternary slag. After the TiO_2-Ti_2O_3-CaO ternary slag is completely melted, the increase of temperature has little effect on viscosity of slag, and the viscosity is about 110~125mPa·s. When the temperature is lower than melting temperature, TiO_2-Ti_2O_3-CaO ternary slag will precipitate solid-phase particles, and the precipitation process is carried out in stages, and with the decrease of temperature, the precipitation will increase and the viscosity will sharply increase. TiO_2-Ti_2O_3-CaO ternary titanium slag has obvious characteristics of short slag.

1. Introduction

There is a wealth of titanium resources in China, but it is mainly for the ilmenite with the low grade and high impurity content, which will limit the production of titanium dioxide and sponge titanium. With the rapid development of titanium dioxide and sponge titanium industry, the demand for high-grade titanium concentrates is increasing. Therefore, the processing of rich ilmenite resources into titanium-rich material is the urgent task for the development of titanium dioxide and titanium materials industry in China [1].

High titanium slag is a high quality raw material in the production process of titanium dioxide and sponge titanium. Electric furnace smelting is one of the main methods to prepare high titanium slag at present. The good effect of slag-iron separation in the smelting process is the key to ensuring the smooth production and the quality of the product. The viscous flow characteristics of high titanium slag in the furnace are especially important [2].

Due to the high melting point and complex composition of high titanium slag, the related research on the viscosity of high titanium slag at home and abroad is rare. In 1957, Researchers at the Soviet Academy of Sciences pioneered a study of the impact of the composition on the viscosity of titanium slag [3]. In the 1960s, a group of foreign researchers for viscosity of titanium slag was related to explore. Until today, foreign reports on the viscosity of titanium slag is still less [4,5]. In the last century 80's, the domestic began to study the viscosity of titanium slag [6].

There is large inequality on high titanium slag with its high demand and research results of viscosity. Solving the blank problem of viscosity of some slag system has become a task that researchers need to solve urgently.

In the past, the researchers mainly got through the experimental method to study the change rules of viscosity of high titanium slag. Due to high melting point and other properties of high titanium slag, it leads to difficult test and high cost. In this study, thermodynamic simulation of viscosity of TiO_2-Ti_2O_3-CaO ternary slag was studied by using FactSage® software. The effects of temperature, CaO content and solid-phase particles on the viscosity of slag were studied. Fill the gaps in the domestic viscosity data of TiO_2-Ti_2O_3-CaO ternary slag.

2. Research methods

In 1906, Einstein [7] proposed a method for describing the viscous behavior of melting containing solid-phase particles, as shown in equation (1). Einstein pointed out that the equation is applicable regardless of the size distribution of the sphere.

$$\eta = \eta_0(1 + 2.5c) \tag{1}$$

Where η is the viscosity of melting containing the solid-phase particles; η_0 is the viscosity of melting in the fully molten state; and c is the volume concentration.

With the advancement of the theory, the researchers have improved the equation (1), where the typical model is Einstein-Roscoe equation [8], as shown in equation (2).

$$\eta = \eta_0(1 - af)^{-n} \tag{2}$$

Where f is the volume fraction of solid-phase particles; a and n are constants. Roscoe assumes that the solid-phase particles are spherical, and he gives a and n values. According to Seong-Ho S [9], it is assumed that the density values of the solid-phase particles and the melting are the same, that is, the solid volume fraction is equal to the solid mass fraction.

FactSage is one of the representative thermodynamic software, developed by the FACT thermodynamics database developed by Canada, and integrated with the Germany developed ChemSage database to become a large thermal chemical database running on Windows. It is widely used in pyrometallurgy, materials science, hydrometallurgy and other fields [10].

This paper bases on the Einstein-Roscoe equation, uses the Equilib and Viscosity modules of FactSage® thermodynamics software and selects the FToxid database to study the viscosity of TiO_2-Ti_2O_3-CaO ternary slag. According to the phase diagram of slag, the temperature range of the study is 1620 ~ 1750 °C (step 5 °C). The pressure is a standard atmospheric pressure. The phase diagram of TiO_2-Ti_2O_3-CaO ternary slag is shown in Figure 1.

Simulation scheme of TiO_2-Ti_2O_3-CaO ternary slag was determined by the composition of high titanium slag in Wuding Company of Yunnan Metallurgical New Titanium Industry: reduction degree was 0.35 ~ 0.39 and the CaO content was 0.4% ~ 2%. The chemical composition of high titanium slag of the factory is shown in Table 1, simulation scheme of TiO_2-Ti_2O_3-CaO ternary slag system is shown in Table 2.

The reduction degree (R) is calculated as:

$$R = \frac{M_{Ti_2O_3}}{1.111 \times M_{Ti_2O_3} + M_{TiO_2}} \times 100\% \tag{3}$$

Where M_{TiO_2} and $M_{Ti_2O_3}$ respectively represent the mass of TiO_2 and Ti_2O_3.

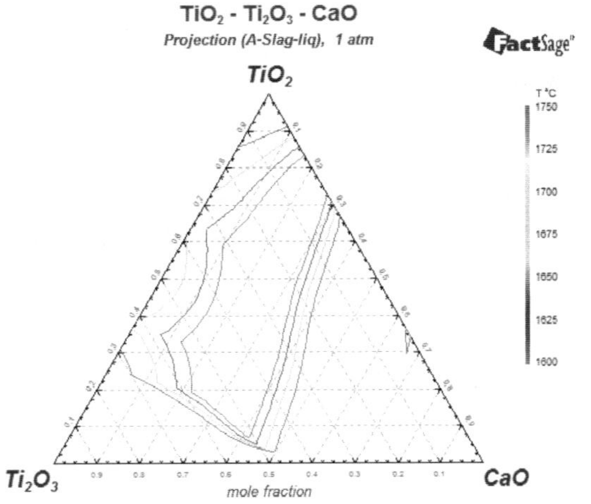

Figure 1. The phase diagram of TiO_2-Ti_2O_3-CaO ternary slag

Table 1. The chemical composition of high titanium slag (*wt%*).

TiO_2	Ti_2O_3	FeO	MnO	CaO	MgO	SiO_2	Al_2O_3	Cr_2O_3	ZrO_2
55.441	35.172	2.417	1.245	0.177	0.861	2.104	1.824	0.058	0.186

Table 2. Simulation Scheme of TiO_2-Ti_2O_3-CaO Ternary Slag System (*wt%*).

	TiO_2	Ti_2O_3	CaO
	63.33	36.27	0.40
	63.08	36.12	0.80
R=0.35	62.82	35.98	1.20
	62.57	35.83	1.60
	62.31	35.69	2.00
	61.17	38.43	0.40
	60.92	38.28	0.80
R=0.37	60.68	38.12	1.20
	60.43	37.97	1.60
	60.19	37.81	2.00
	59.00	40.60	0.40
	58.76	40.44	0.80
R=0.39	58.52	40.28	1.20
	58.29	40.11	1.60
	58.05	39.95	2.00

3. Results and discussion

3.1. Effect of temperature on the viscosity of TiO_2-Ti_2O_3-CaO ternary slag

Temperature is one of the main factors on affecting viscosity. The melting property of the high titanium slag is expressed by melting temperature and melt-property temperature. Melting temperature is the temperature at which the solid-phase disappears completely during the heating of titanium slag. Melt-property temperature indicates the lowest temperature at which the slag is free to flow. Melting temperature (A) and melt-property temperature (B) of TiO_2-Ti_2O_3-CaO ternary slag is simulated by

FactSage® software as shown in Figure 2. It can be seen from Figure 2 that the reduction degree and the CaO content can affect melting temperature and melt-property temperature of TiO_2-Ti_2O_3-CaO ternary slag system. When the CaO content is the same, melting temperature and melt-property temperature of high titanium slag decreases with the increase of reduction degree.When the reduction degree is the same, melting temperature and melt-property temperature of high titanium slag decreases with the increase of CaO content. In this study, melting temperature of TiO_2-Ti_2O_3-CaO ternary slag is 1675 °C ~ 1705 °C, melt-property temperature of TiO_2-Ti_2O_3-CaO ternary slag is 1655 °C ~ 1700 °C. Under the same conditions, melt-property temperature is lower than melt temperature of 5 °C ~ 20 °C.

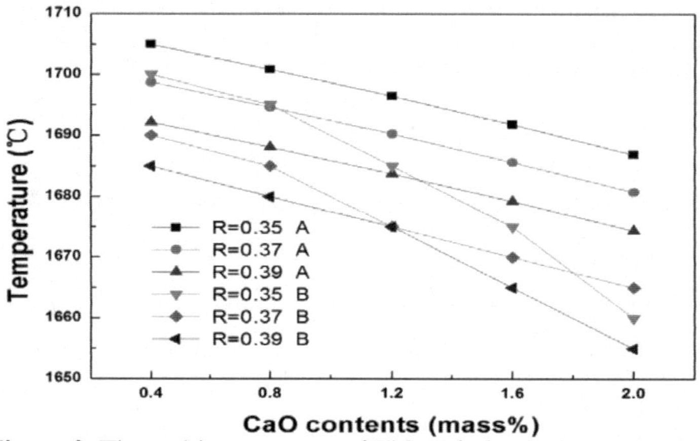

Figure 2. The melting property of TiO_2-Ti_2O_3-CaO ternary slag

Effect of temperature on the viscosity of TiO_2-Ti_2O_3-CaO ternary slag was calculated. The results are shown in Figure 3, for the given composition of slag, when the temperature is higher than melting temperature, with the temperature increases, the viscosity value is almost unchanged, and the viscosity value is low, about 110 ~ 125mPa·s; when the temperature is lower than melting temperature, with the temperature decreases, the viscosity value increases rapidly. There is a distinct turning-point in the viscosity-temperature curve, TiO_2-Ti_2O_3-CaO ternary slag has obvious characteristics of short slag.

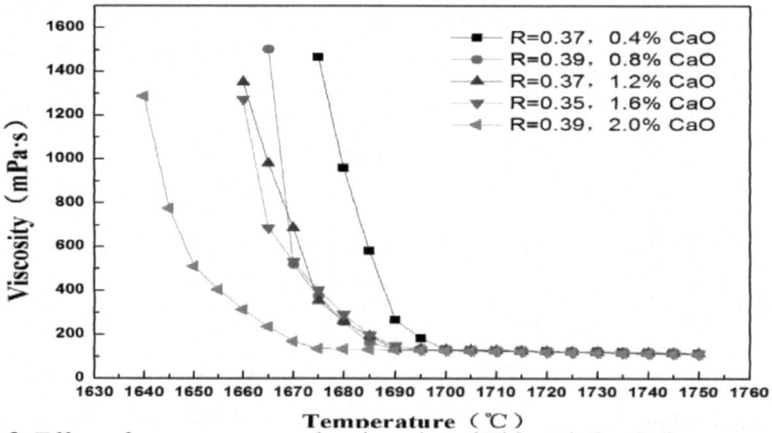

Figure 3. Effect of temperature on the viscosity of TiO_2-Ti_2O_3-CaO ternary slag

3.2. *Effect of CaO content on the viscosity of TiO_2-Ti_2O_3-CaO ternary slag*

The effect of CaO content on the viscosity of TiO_2-Ti_2O_3-CaO ternary slag is shown in Figure 4. The slag temperature of both groups (1) and (2) is higher than their melting temperature, and high titanium slag is completely melted. With the increase of CaO content, the viscosity value decreases slowly. (3) and (4), with the decrease of CaO content, the slag temperature is lower than melting temperature, and the solid-phase particles precipitate from molten slag, and the viscosity value increases. The increase

IOP Conf. Series: Materials Science and Engineering **281** (2017) 012047 doi:10.1088/1757-899X/281/1/012047

of CaO content can reduce the viscosity of slag, and the flowability becomes better, that is to say, CaO is a good cosolvent.

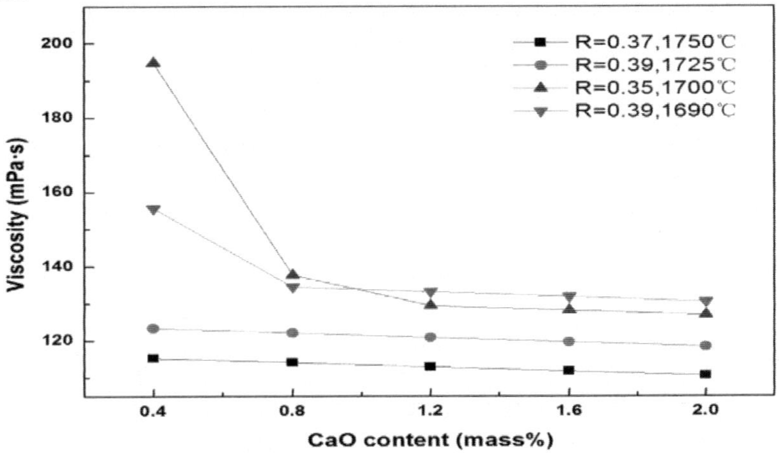

Figure 4. Effect of CaO content on the viscosity of TiO_2-Ti_2O_3-CaO ternary slag

3.3. Effect of solid-phase particles on the viscosity of TiO_2-Ti_2O_3-CaO ternary slag

The phase and viscosity of TiO_2-Ti_2O_3-1.2%CaO (R=0.37) ternary slag are shown in Figure 5. It can be seen from Figure 5 that melting temperature of TiO_2-Ti_2O_3-1.2%CaO (R=0.37) ternary slag is 1690.24 °C, when the temperature is higher than melting temperature, titanium slag is in the completely molten state, the viscosity value is almost constant; when the temperature is lower than melting temperature, solid-phase particles in the molten titanium slag are formed, which is not conducive to the movement. With the decrease of the temperature, the precipitation of solid-phase particles increases abruptly, the viscosity value increases rapidly, and the flowability becomes poor. When the temperature drops to 1660 °C, the mass fraction of solid-phase particles in the slag is 43.731%, the viscosity value is about 1351 mPa·s, and the flowability was poor. When the temperature drops to about 1375 °C, the liquid phase completely disappears, leaving only Ti_xO_y and Perovskite ($CaTiO_3$).

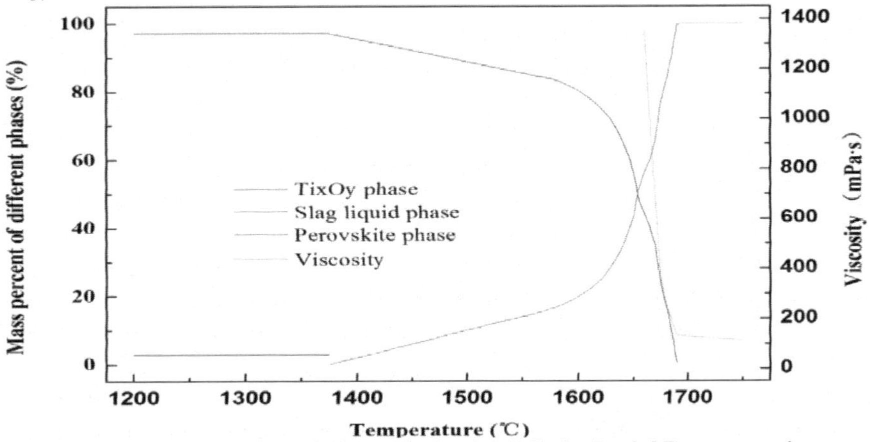

Figure 5. Phase and viscosity of TiO_2-Ti_2O_3-1.2%CaO (R=0.37) ternary slag.

The effect of solid mass fraction on the relative viscosity of TiO_2-Ti_2O_3-CaO ternary slag is shown in Figure 6. It can be seen from Figure 6 that the relative viscosity increases with the increase of the mass fraction of solid, and the curve have obvious turning-point when the mass fraction of solid is about 53%.

IOP Conf. Series: Materials Science and Engineering **281** (2017) 012047 doi:10.1088/1757-899X/281/1/012047

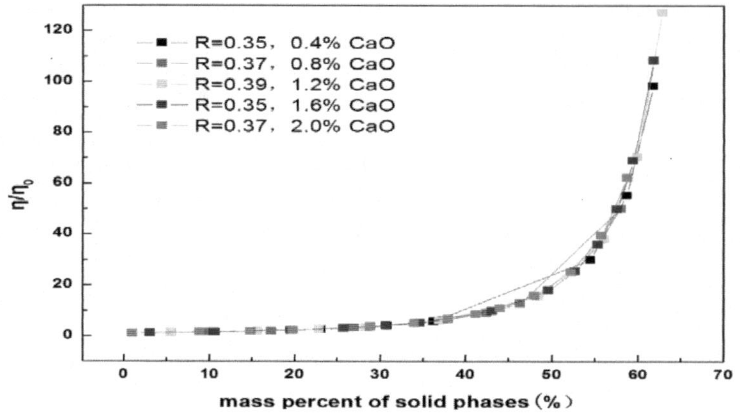

Figure 6. Effect of solid mass fraction on the relative viscosity of TiO_2-Ti_2O_3-CaO ternary slag

4. Conclusions

1) When the TiO_2-Ti_2O_3-CaO ternary slag is completely melted, the temperature has little effect on the viscosity of slag, and the viscosity value is low (about 110 ~ 125mPa·s). When the temperature is lower than melting temperature, the viscosity value increases sharply with the decrease of the temperature; the slag system has obvious characteristics of short slag.

2) CaO content can reduce the melting temperature and viscosity of TiO_2-Ti_2O_3-CaO ternary slag.

3) When the temperature is lower than melting temperature, TiO_2-Ti_2O_3-CaO ternary slag will precipitate solid-phase particles, the viscosity value will increases rapidly.

References

[1] Zhang W, Zhu Z, Cheng C Y. A literature review of titanium metallurgical processes [J]. *Hydrometallurgy*, 2011, **108**(3):177-188.

[2] Huang R, Lv X W, Bai C G, et al. Solid state and smelting reduction of Panzhihua ilmenite concentrate with coke [J]. *Canadian Metallurgical Quarterly*, 2013, **51**(4):434-439.

[3] Zhou L, Lei T. Research Actuality and Developing Trend of Titanium Slag in the World [J]. *Titanium Industry Progress*, 2009.

[4] Pesl J, Eriç R H. High-temperature phase relations and thermodynamics in the iron-titanium-oxygen system [J]. *Metallurgical & Materials Transactions B*, 1999, **30**(4):695-705.

[5] Handfield G, Charette G G. Viscosity and structure of industrial high TiO_2 slags [J]. *Canadian Metallurgical Quarterly*, 1971, **10**(3):235-243.

[6] Li S P, Lv X W, Song B, Miao H J, Han K X. Rheological property of TiO_2-CaO-(SiO_2, CaO, MgO) ternary slag [J]. *Transactions of Nonferrous Metals Society of China*, 2016, (09): 2015-2022.

[7] Einstein, A. Ann. *Phys.*, *Lpz.*, **19**, p. 289 (1906); **34**, p. 591 (1911).

[8] Roscoe R. The viscosity of suspensions of rigid spheres [J]. *British Journal of Applied Physics*, 2002, **3**(8):267.

[9] Seong-Ho S, Sung-Mo J, Young-Seok L, et al. Viscosity of Highly Basic Slags [J]. *Isij International*, 2007, **47**(8):1090-1096.

[10] Cao Z, Song X, Qiao Z. Thermodynamic Modeling Software FactSage and Its Application [J]. *Chinese Journal of Rare Metals*, 2008.

IOP Conf. Series: Materials Science and Engineering **281** (2017) 012048 doi:10.1088/1757-899X/281/1/012048

Research on Thermal Topology Design Method of Spindle Based on HCAM

S B Weng[1], X L Deng[2,3*], H Lin[2], J C Wang[2,3] and C X Xie[2]

[1]College of Entrepreneurship & Innovation, Quzhou Universtiy, Quzhou 324000, China
[2]College of Mechanical Engineering, Quzhou Universtiy, Quzhou 324000, China
[3]ZhejiangYonglida CNC Technology Co., Ltd., Quzhou 324000, China

Email: dxl@zju.edu.cn

Abstract. Spindle thermal error is one of the key factors in the machining accuracy of CNC machine tool. In order to reduce the thermal error of the spindle and improve the machining precision of the CNC machine tool, the idea of the thermal topology design based on the hybrid cellular automaton method (HCAM) is proposed. In this paper, the existing research results of hybrid cellular automaton are summarized, and the main achievements of domestic and foreign research work are analysed and reviewed from the aspects of structural topology optimization. Finally, the realization and prospect of HCAM self-optimizing thermal topology design method for spindle are presented.

1. Introduction

A large number of researches have shown that in the factors that affect the machining accuracy, the thermal error of the spindle is the biggest error source of precision machining machines such as CNC machine tool [1]. Spindle is an important component of the machine tool, which is directly involved in processing. The thermal error control problem of spindle is very important for the improvement and maintenance of machining precision for machine tool. It has been a hot topic for domestic and foreign scholars [2-4].

Spindle thermal design problems mainly come from the engineering field; meanwhile, the self-optimization phenomenon exists in nature, such as the distribution of different wood in the trunk and the distribution of different material in the animal bones, which are all worth studying in thermal design. Hybrid Cellular Automaton Method (HCAM) is a biomarker topology optimization method proposed by Tovar [5] bone remodelling biology process (shown in Figure 1 and Figure 2). According to the theory of bone reconstruction, simulating the bone remodelling process and reshaping the internal structure of the skeleton to have a uniform strain energy density. According to the theory of bone reconstruction, CA (Cellular Automaton) method and FEM (Finite Element Method) are combined. The CA cells record the material distribution in the design space. The FEM calculates the material density. The results of the changes which are caused by the CA cells state change through a one-to-one correspondence mapping to the finite element model. The method is simple and easy to understand, the convergence is good, the calculation efficiency is high, the numerical instability is reduced, and the structure topology and even the solution to the multi-objective problem can be solved automatically.

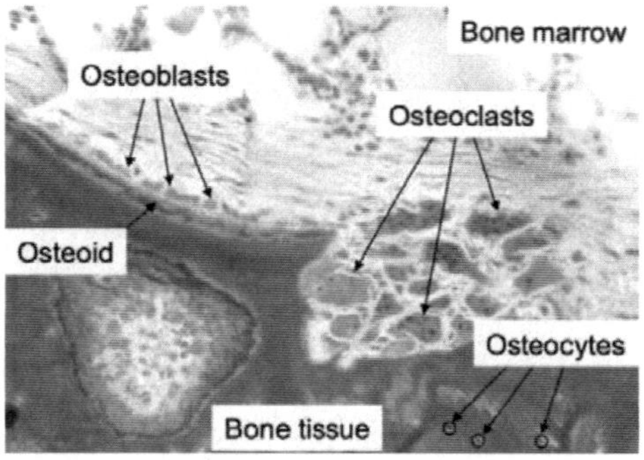

Figure 1. Bone cells [5]

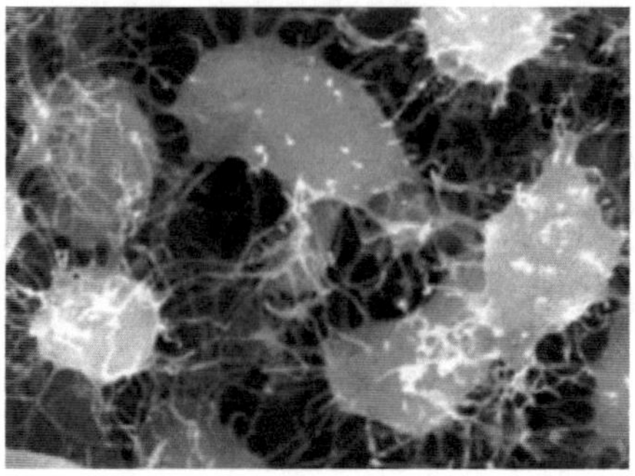

Figure 2. Cellular network of osteocytes [5]

2. Research status of topology optimization design based on HCAM

The structural design and optimization of the continuum structure of the mechanical structure has been a major problem in the automatic design of the structure. Nowadays, the frequently-used continuum topology optimization algorithms are Homogenization Method, Variable Density Method, Progressive Structure Optimization Method and CA method and so on. HCAM [5] is based on Tovar's theory of bone reconstruction, coupling CA method and FEM. A topological optimization bionic design method is proposed to solve the problem that CA method does not solve the global analysis of field state. HCAM uses the FEM analysis to evaluate the field state (strain energy), that is, using FEM to calculate the strain energy as the state information of CA. This method is easy to understand, and has good convergence, high computational efficiency and reduces numerical instability. Those multi-objective problem solving could provide new analytical ideas and technical means [6,7] for the effective structure of the topology optimization design. In recent years, Tovar and other scholars have also combined the energy absorption control method [8], weighted multi-objective method [9], harmonic vibration load method [10] with HCAM application successfully in the vehicle anti-collision design, thin-walled tube structure optimal design, the energy collector shell and the outdoor air-conditioning cover [11] are optimized even on military projects such as multi-material explosion-proof armour plate optimization design [12].

At home, Professor Chen [13] studied the structural topology optimization based on HCAM and the method of solving the topology of stress constraint and displacement constraint. By employing the penalty function method, the strain energy density target value, to achieve the organic combination with the CA local rules. Professor Guo [14] et al. proposed a method for topology optimization of multi-space continuum structures based on HCAM for large-deformation nonlinear structural topology optimization problems. In 2011, a hybrid cellular automaton framework optimization method was proposed. The method was based on the application of strain and multi-domain topology optimization algorithm to HCAM. As shown in Figure 2, the research on vehicle crashworthiness in the case of large deformation is studied, including transient, non-linear finite element structure analysis and local control rules on the cell [15]. Wang [16] et al. implemented an optimization design on the aluminium alloy front impact beam structure by HCAM to optimize calculation model based on explicit dynamic finite element analysis software. Dr. Cui [17] applied HCAM to the flexible mechanism design with interval parameters. The local rules were used to control the distribution of materials in the design domain and converge with less iteration and finally got the structure topology optimization design results. Ni[18] et al. focused on the target of minimization of strain energy and the maximization of mutual strain energy, constrained by the structure volume, proportional controlled by local control law and applied HCAM to simple two-dimensional compliant mechanism (displacement inverter) topology optimization, effectively eliminating the different nature of the objective function in the order of magnitude difference.

IOP Conf. Series: Materials Science and Engineering **281** (2017) 012048 doi:10.1088/1757-899X/281/1/012048

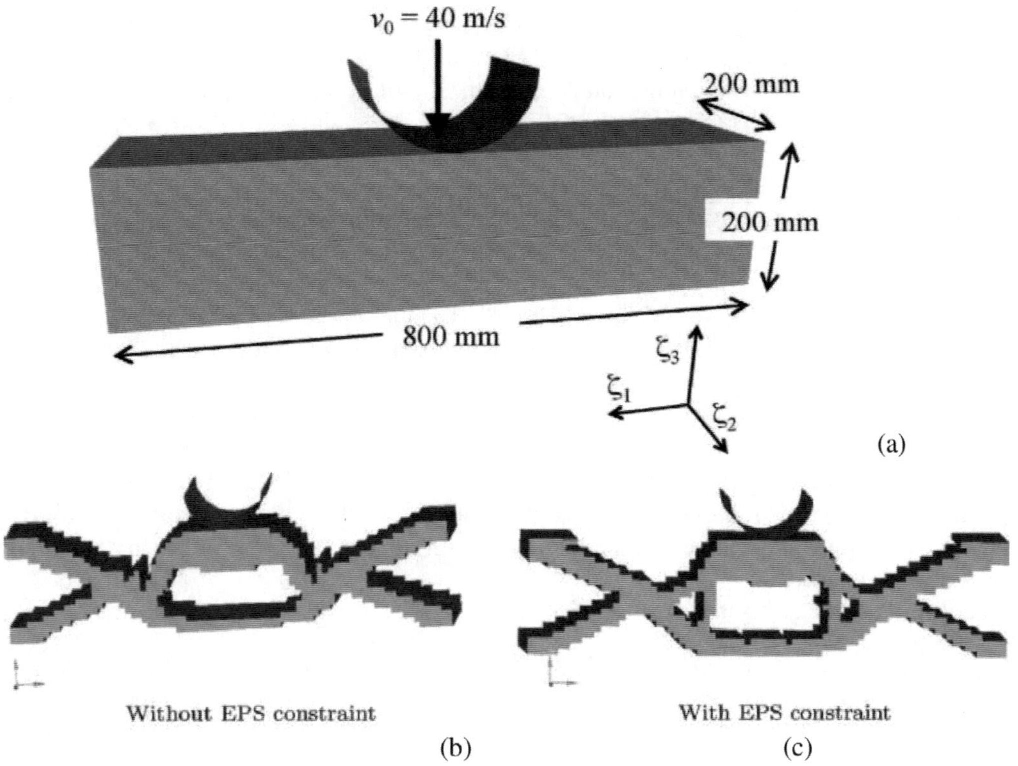

Figure 3. With the EPS and multi-domain HCAM collision avoidance structure optimization [15]

The literatures' research show that HCAM has been applied in the optimization design of different product topology under stress and displacement constraints. It has high convergence, high efficiency, stable numerical calculation and other obvious advantages. Its development is rapid and has strong vitality and application prospect. Therefore, the application of HCAM based on the theory of bone reconstruction will provide a new bionic method for thermo-design and optimization of the spindle.

3. Realization of HCAM self-optimizing thermal topology design method for spindle

The difference between bone and general engineering materials is that it is constantly changing as a biological material. From the perspective of mechanics, bones are a functional adaptive structure, and its growth and reconstruction are related to the mechanical environment. Through the precipitation and absorption in different parts, the optimal structural form of the mechanical environment is formed [19], that is, self-optimizing function (skeletal function adaptive rule). Based on the theory of bone reconstruction and the theory of structural optimization, the self-optimizing design model is established based on volume ratio, heat flux density and local temperature. The local evolution rule is established. The CA cells and the finite element model's grids have one-to-one correspondence. It firstly constructs a single component, a single optimization space optimization design method, and then extended to multi-component integration system and multi-space domain optimization. And the optimization results are compared with the spindle prototype. Finally, the optimization model and algorithm are validated on the thermal equilibrium test platform.

The main process of using HCAM spindle self-optimizing bionic thermal topology design algorithm is as follows:

1. Based on the fluid-temperature-structure field coupling mathematical model, the parameters such as design area, material property and thermal load are defined, and the design value $x^{(0)}$ is initialized.

2. The solid isotropic penalty microstructure model based on variable density method is employed to design the domain material parameterization. The penalty factor is used to punish the intermediate density value. And the heat conduction coefficient $E^{(t)}$ of each element in the model is expressed by the exponential function of the unit relative density P, which is

$$E^{(t)} = x^{(t)p} E_0 \qquad (p = 1) \tag{1}$$

where, E_0 is the relative thermal conductivity of the material which value is 1. So that the intermediate density values converge to both ends of 0/1, and the topological optimization model of continuous variables can approach the optimization model of 0-1 discrete variables.

3. Calculate the field variable $S(x^{(t)})$ with FEM, and then assign it.
4. The material part of the design field is determined according to the local PID feedback control rule of HCAM. The change of the set value S of the state field variable S_{avg} and the set value S of the state field variable is minimized to calculate the variation Δx of the design variable. To avoid numerical instability, set the upper and lower limits for each iteration design process.
5. In the process of iterative self-optimizing thermal design, the local control rule is applied to update the design variable $x^{(t+1)}$, and the change of the design variable is used as the convergence criterion to judge whether the minimum difference is set for the given difference or not. Please see Figure 4.
6. Check convergence, when the convergence criteria is met, the final topology structure will be got, or return to step 3 to continue to run.
7. Through the comparison with the original design of the program, you can get a better design. Figure 3 is a flow.

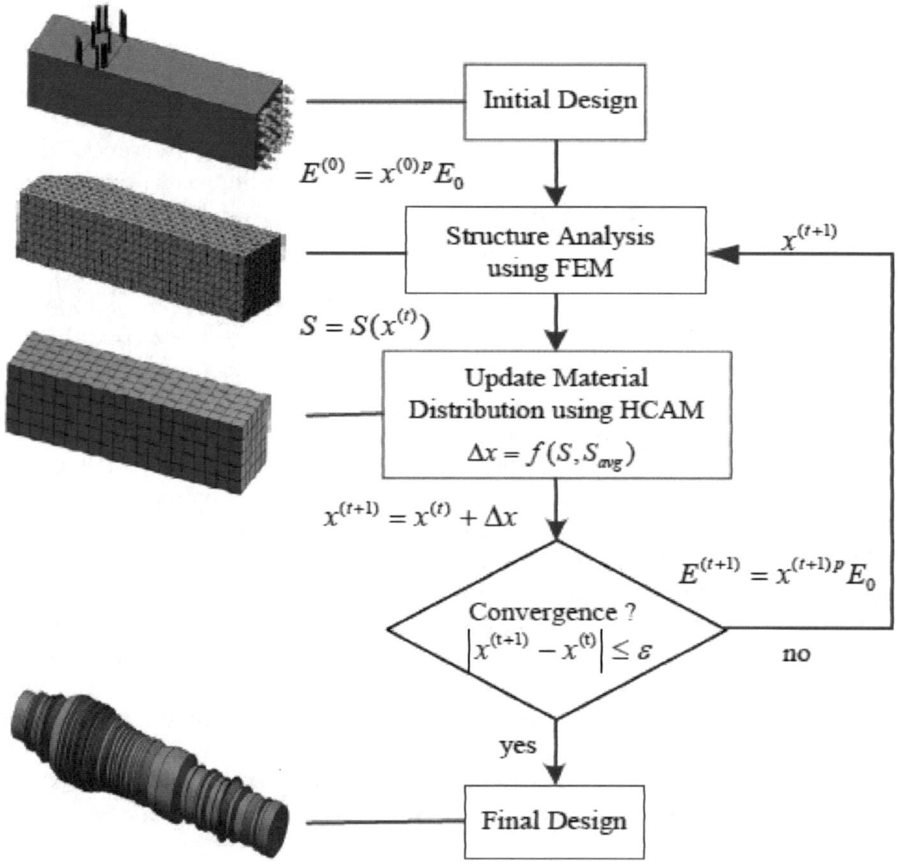

Figure 4. Structure Thermal Topology Self-Optimizing Algorithm Process of Spindle

4. Conclusions

Spindle, as one of the key component of CNC machine tools, its thermal error control has a very important role in improving the machining precision for CNC machine tool. HCAM is a method can be used to solve the topology design and optimization problem of spindle structure thermo-design in the future, and decrease the thermal error of spindle. It is useful for precision machine tool designer in research, analysis, decision-making, and any other related activities.

Acknowledgments

This research was financially supported by National Natural Science Foundation of China (No. 51605253), Zhejiang Provincial Natural Science Foundation of China (No. LY16E050011) and Quzhou Science and Technology Planning Project (2016Y003 & 2015Y007).

References

[1] Mayr J, Jedrzejewski J, Uhlmann E, et al. Thermal Issues in Machine Tools [J]. *CIRP Annals-Manufacturing Technology*, 2012, 61(2): 771-791.
[2] Xiaolei Deng, Jianzhong Fu and Yuwen Zhang. A Predictive Model for Temperature Rise of Spindle-Bearing Integrated System [J]. *Journal of Manufacturing Science and Engineering-Transactions of the ASME.* 2015, 137.
[3] Abele E, Altintas Y and Brecher C. Machine Tool Spindle Units [J]. *Annals of the CIRP,* 2010, 59(2): 781-802.
[4] Xiaolei Deng, Jianzhong Fu,Chenhui Xia, et al. Multi-objective Correction Method for Thermal Model Parameters of CNC Machine Tool Spindle System [J]. *Journal of Mechanical Engineering,* 2014, 50 (15) :119. (In Chinese)
[5] Andr´es T, Wilson I Q, Neal M P, et al. *Hybrid Cellular Automata with Local Control Rules: A New Approach to Topology Optimization Inspired by Bone Functional Adaptation*[C]// 6th World Congresses of Structural and Multidisciplinary Optimization Rio de Janeiro, 30 May - 03 June 2005, Brazil: 1-11.
[6] Tovar A, Neal M. Patel, Amit K. Kaushik, et al. Optimality Conditions of the Hybrid Cellular Automata for Structural Optimization [J]. *AIAA Journal,* 2007, 45 (3): 673-683.
[7] Andres T, Neal M P, Glen L N, et al. Topology Optimization Using a Hybrid Cellular Automaton Method with Local Control Rules [J]. *Journal of Mechanical Design,* 2006, 128: 1205-1216.
[8] Punit B, James P S and Andre´s T. Design of Crashworthy Structures with Controlled Energy Absorption in the Hybrid Cellular Automaton Framework [J]. *Journal of Mechanical Design,* 2013, 135: 1-11.
[9] Satyajeet S, Duane D and Andre´s T. S*tructural Optimization of Thin-walled Tubular Structures using Weighted Multi-objective Approach* [C]// 10th World Congress on Structural and Multidisciplinary Optimization, USA, 2013: 1-12.
[10] Soobum L and Andre´s T. Topology Optimization of Piezoelectric Energy Harvesting Skin Using Hybrid Cellular Automata [J]. *Journal of Mechanical Design,* 2013, 135: 1-11.
[11] Zeng D and Duddeck F . Improved hybrid cellular automata for crashworthiness optimization of thin-walled structures [J]. *Structural&Multidisciplinary Optimization,* 2017: 1-15.
[12] Goetz J, Tan H, Renaud J and Andres T. Two-material Optimization of Plate Armor for Blast Mitigation Using Hybrid Cellular Automata [J]. *Engineering Optimization,* 2012, 44(8): 985-1005.
[13] Chen J Q and Zhou Z Z. Structural Topology Optimization Based on the Hybrid Cellular Automaton Method [J]. *Journal of Wuhan University of Technology,* 2014, 36(1): 140-144. (In Chinese)
[14] Guo L S, Punit B and John E R. Method of multi-domain topology optimization for continuum structures [J]. *Journal of Beijing University of Aeronautics and Astronautics,* 2009, 35(2): 227-230. (In Chinese)

IOP Conf. Series: Materials Science and Engineering **281** (2017) 012048 doi:10.1088/1757-899X/281/1/012048

[15] Guo L S, Andr´es T, Charles L P, et al. Strain-based Topology Optimisation for Crashworthiness Using Hybrid Cellular Automata [J]. *International Journal of Crashworthiness,* 2011, 13(3): 239-252.
[16] Wang G, Zhou J, liu Z W, et al. Lightweight design and crash performance analysis of automotive aluminum bumper [J]. *The Chinese Journal of Nonferrous Metals,* 2012, 22(1): 90-98. (In Chinese)
[17] Cui M T, Chen J J, Tuo Y F, et al. Compliant mechanism design with interval parameters using the hybrid cellular automaton method [J].*Chinese Journal of Mechanical Engineering,* 2007, 43(7): 39-43. (In Chinese)
[18] Ni Y, Zhan J Q and Zhu D C. Multi-objective Topological Optimum Design of Compliant Mechanisms Using Hybrid Cellular Automata Method [J]. *China Mechanical Engineering,* 2012, 7: 860-863. (In Chinese)
[19] Chen B Z, Zhao W Z and Sun Y B. Bone Remodeling Numerical Simulation on the Basis of Bone Adaptive Theory [J]. *Journal of Biomedical Engineering,* 2008, 25(2): 363-367. (In Chinese)

IOP Conf. Series: Materials Science and Engineering **281** (2017) 012049 doi:10.1088/1757-899X/281/1/012049

Finite Element Analysis of the Integral Hoisting of 49m Flange Connection Flare

J B Liu*

Daqing Oilfield Engineering Construction Co., Ltd,Chengfeng Street No. 295,RangHu Lu,Daqing, Heilongjiang Province, 163411, China

Jingbo Liu(J B Liu), *email:280081407@qq.com

Abstract. Installation of flange connection structure in flare stack commonly adopt parts assembly or segmental hoisting. In this paper,due to the harsh conditions of construction,an innovation of flange connection type flare stack's one-time integral hoisting technology and method is proposed.This paper use CATIA finite element analysis simulate force of flare stack under two extreme conditions during the integral hoisting: horizontal and vertical conditions. The strength and stability of the hoisting stack are checked and calculated.Then we find out the dangerous point and propose the reinforcement solutions for weak structure,providing the basis for hoisting process design.

1. Preface

In construction of petroleum chemical station, the flare are usually required to be installed, due to the large volume of the component assembly, equipment height, installation position limitation and space limitation, which cause difficulties for installation[1]. Conventional flare stack using aerial parts assembly or segment hoisting method to install. Laerqianling Station flare system maintenance station needs to construct a new 49.00m flange connection type flare system. The flange connection structure of flare stack generally adopts a sectional hoisting, but during construction the Daqing area is in the winter. The temperature is lower than -25°C, it is windy and snowy. There is high risk in climbing construction and high labor intensity. By combination of ideas from each sides, we determine the reasonable reinforcements and the integral hoisting of the stack, this paper conduct check calculation and analysis for the whole flare stack hoisting process.

2. **Flare stack structure parameters**[2]

The flare stack is equilateral triangle in cross-section. Corner section is 9m×9m, the top section of the stack is 3×3m,and total height is 45.8m. The lower end of the flare cylinder is Φ630×14 steel pipe, height is 7.31m. Upper end is Φ508×12 steel pipe,its height is 39.19m, and the height of flare cylinder is 49m. There flare tip is on the top platform of the stack, the height of flare head is 3.5m. Total height of flare cylinder and head is 50m.Total lifting weight of flare cylinder, stack and accessories is 29.5 tons,and height is 45.8 meters. Flare head weighs 1 ton,its height is 3.5 meters.Flare base weighs 2.8 tons,its height is 7.3 m.Reinforced attachment weighs 2 tons. Total weight of flare stack Q=29.5+2.8+1+2=35.3 tons. Height from top of the flare stack foundation to the ground is 0.3m.

3. Model establishment and mechanical properties query of the flare stack

3.1. Model building[3]

In CATIA 3D design and computer aided analysis integration platform, we establish a 3D model of flare stack(Figure 1), applying CATIA's tetrahedral adaptive nonlinear meshing technique to divide the grid of the overall model. And we refine the mesh division at position where stress deformation increase in the stack to improve the accuracy of calculations.

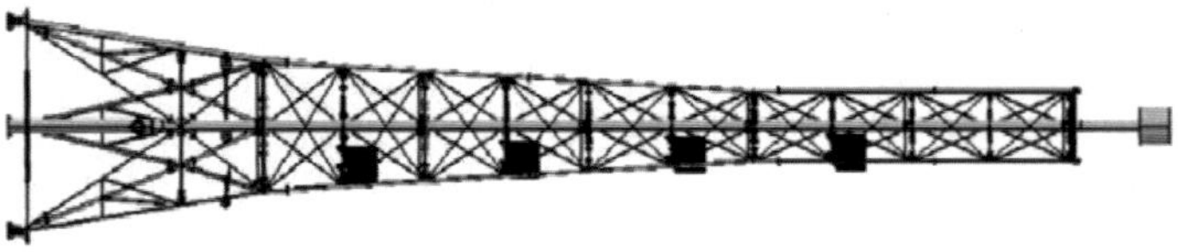

Figure 1. 3D model of flare stack

3.2. Flare weight, center of gravity and moment of inertia check results

After the calculation model was established, we check the mechanical properties of the three-dimensional model of the flare stack in the CATIA software. Then we evaluate, analyze and calculate the total weight, coordinates of gravity center location and volume mass. Parameter data are shown in Figure 2. Total weight of flare stack is 27.4 tons, the overall height is 50m.The centre of gravity of the integral is 20.496m from the end by calculation.

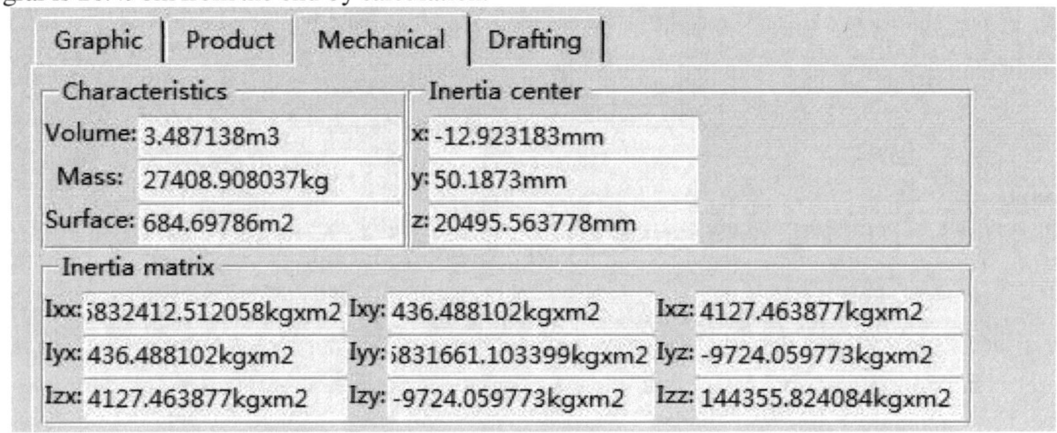

Figure 2. Mechanical properties of flare stack model

4. Flare stack reinforcement

(1) Reinforcement of the bottom of the flare stack

Bottom of the Flare stack adopt Φ159x6 seamless steel tube stack with welding, and the C12.6 U-steel and flare stack cylinder body is connected to the reinforcement (Figure 3).

(2) Sliding tail reinforcement

As shown in Figure 4, tailing lug is 9.3 meters from the stack bottom plane. We use a Φ159x6 seamless steel pipe connecting lug and oblique beam to reinforce support.

(3) The reinforcement of the flare tube

The barrel adopts C12.6, C10.0 U-steel connected with the beam reinforcement (Figure 5), there are 4 sets of interlayer reinforcement in total.

IOP Conf. Series: Materials Science and Engineering **281** (2017) 012049 doi:10.1088/1757-899X/281/1/012049

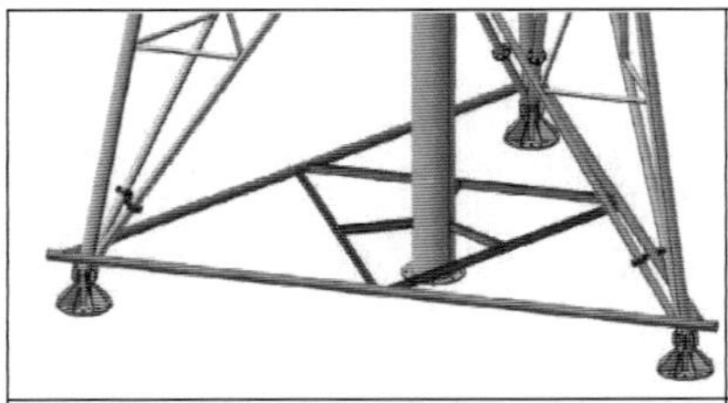	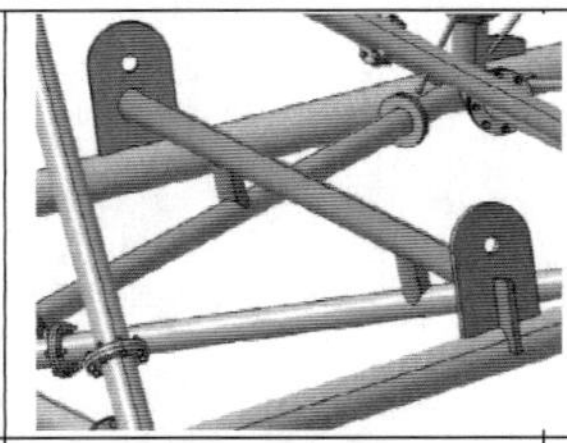	
Figure 3. Reinforcement at the bottom	**Figure 4.** Reinforcement of hanging ear of sliding tail	**Figure 5.** Flare tube reinforcement

5. CATIA finite element analysis and check of flare stack

In lifting process, we analyze the force situations of flare stack in the two extreme positions. The strength and stability of the flare stack will be checked. First, when the flare stack is in horizontal state, truck crane and crawler crane wire rope jointly bear the weight of the flare stack.This is in accordance with the law that center of gravity is inversely proportional to the distance from the flare stack. At this time, load of the lifting lug and steel wire rope is not high.But under the force of gravity, the moment in the middle of the flare stack is too large. The second is when the flare stack is in upright position, total weight of the flare stack is hold by the upper portion of the main lug, so lug bear the full weight of the main flare stack. At this time, the stress of lifting lug and hanging ear near the flare stack is higher. This need to be checked. The specific strength and stability check analysis are as follows:

5.1. Strength, stability check of the flare stack in horizontal state

When flare stack frame is hoisted in the horizontal state, main and auxiliary hanging ear wire rope are vertically upward and the upper end of the steel wire rope, are set by fixed constraint. Calculation of the Von Mises stress is shown in Figure 6. The stress level of the structure as a whole is very low, which is in the order of 10~40 MPa. This ia far less than the material yield strength value, which is 240Mpa. So stress of overall structure in horizontal position is low and the strength satisfy the safety requirements.

When the structure is in the horizontal state, the stack structure is under the effect of gravity. This will induce a downward displacement. If the displacement is too large, structure in the lifting process may produce unstable damage. We can check whether the horizontal lifting point is arranged reasonably through the horizontal displacement .

Figure 7 shows the displacement of flare stack frame in horizontal position, maximum displacement of the middle part is 27.8mm, distance between two hanging point is 36.5m, maximum deflection [F](m) =L/300=0.1217m>0.0278m,stability satisfy safety requirements.

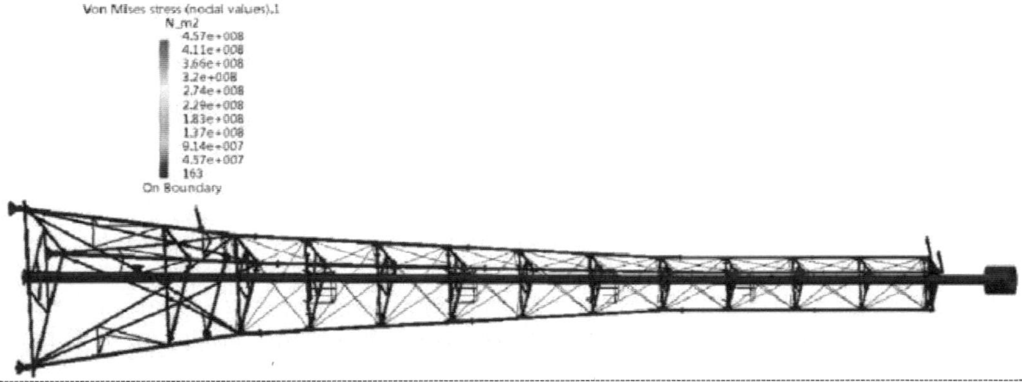

Figure 6. Calculation results of overall stress

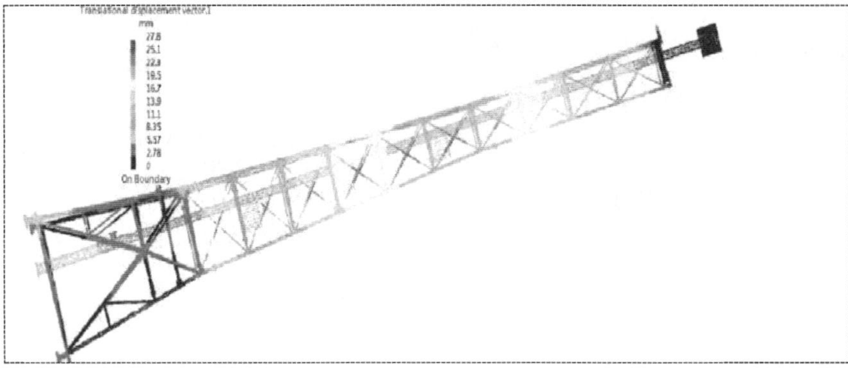

Figure 7. Overall displacement results in horizontal position

5.2. Strength, stability check of flare stack in vertical state

When whole flare stack is lifted in a vertical state, we calculate the Von Mises stress cloud of overall flare stack shown in Figure 8. The stress of overall structure is low, in the order of 10~50MPa, far below the yield strength of the material value, which is 240 MPa. The strength satisfy the safety range, because the structure is only subjected to gravity load. There is no other external load.

Calculation of overall displacement in flare stack shows (as shown in Figure 9), in the vertical position, flare stack structure does not bear the external force.It only bear the gravity. We calculate the displacement size, and the maximum displacement of the structure is 15.9mm, which is located in the bottom of the flare stack. Results are conform to the requirements of the maximum deformation and satisfy the safety conditions.

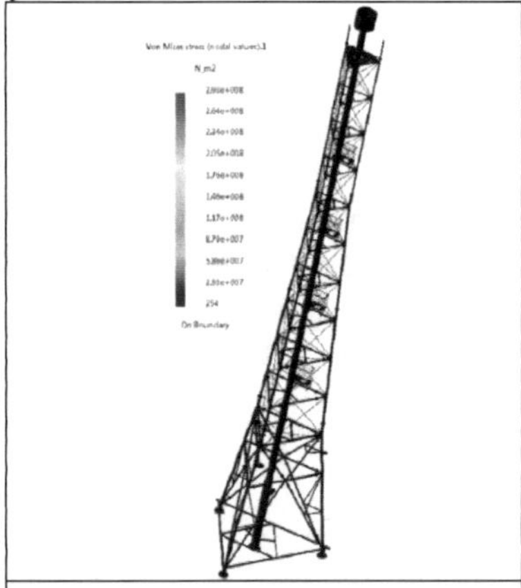

Figure 8. Integral stress contour in vertical position

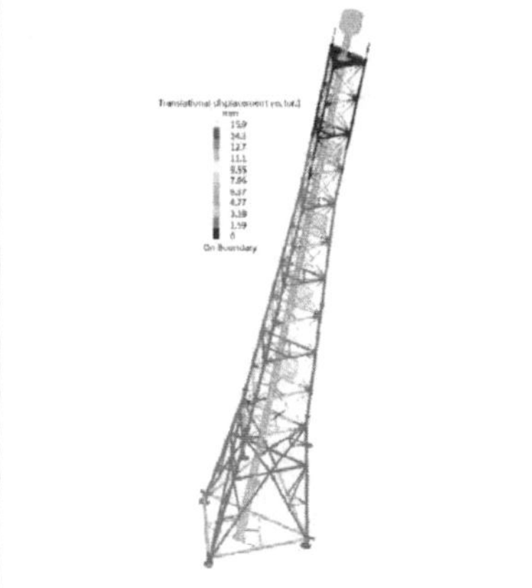

Figure 9. Overall stress contour in vertical position

5.3. Partial dangerous point inspection and treatment measures

The partial structure flare stack will be checked one by one (as shown in Figure 10) in horizontal state. The stress monitoring of main column flange bolts shows that, at horizontal state the 2,3,4-level flange bolt stress in the upper part of the main column is large, which is between 190-260 MPa. The stress of third flange bolts is over 400MPa.The slices are shown in Figure 11. The reason is that the third flange locates near the junction of the oblique column and main straight column, so the bolt has high stress there. Therefore, at site we must reinforce flange in these positions. The maximum stress of main, vice hanging lugs is around 100 MPa, and it is less than the yield strength of the material. So the design of the lifting lug is safe and can satisfy the safety requirements when the lifting lug is in horizontal position.

IOP Conf. Series: Materials Science and Engineering **281** (2017) 012049 doi:10.1088/1757-899X/281/1/012049

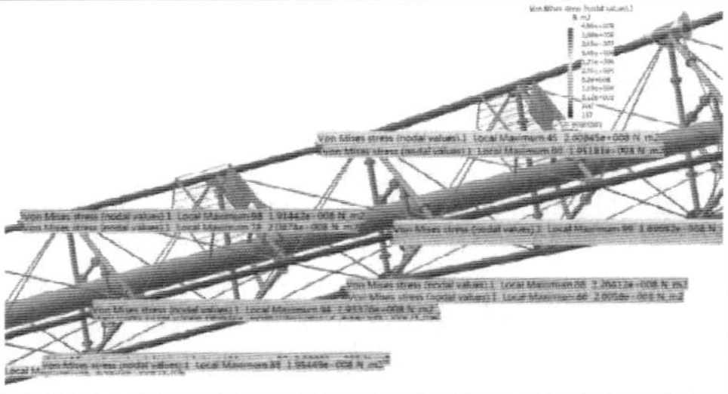

Figure 10. Inspection results of dangerous point

Figure 11. Slice show of the maximum stress

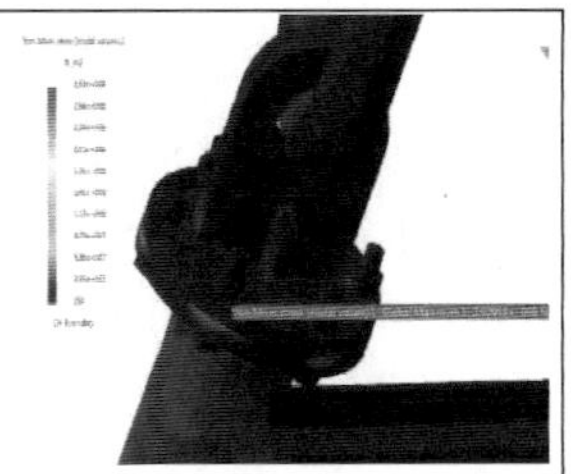

Figure 12. Partial maximum stress contour

At upright state, we conduct dangerous points detection for partial flare stack structure, without outer force the whole stack can be freely deformed, it is suffered less stress. Due to the weight of a fully loaded stack on the main lifting lugs flange bolts, the structure is exposed to the maximum equivalent stress which is located at the upper part of the main flare stack lug flange bolts (in Figure 12). The maximum stress of the overall stack is the tensile stress suffered by main lug bolt flange (value is 293MPa), so we need to use a bolt with higher intensity level (level 8.8 M20×100)to connect the main bolt lugs.

6. On-site lifting implementation

Refer to the CATIA finite element checking calculation and results analysis, and according to the reinforcement guidance program, on-site engineering completion of the flare stack, main lug, side lug and column flange reinforcement work are shown in Figure 13, 14, 15. On 2016 January 14, after 1 hour of lifting operation, the integral hoisting work of flare stack frame is achieved successfully (as shown in Figure 16).

Figure 13. Slide tail hanging ear reinforcement

Figure 14. Main lifting lug reinforcement

Figure 15. Column flange reinforcement

Figure16. Integral hoisting in place

7. Conclusions

By applying the accurate, reliable, mature computer aided engineering(CAE) calculation, for 49.00m flange connection type flare stack hoisting process, flange bolt stress, overall structure deformation and lug partial stress are possibly too larger. Focus on these, we conduct 100% solid modeling, then we check and analyze overall strength and stability of flare stack under extreme horizontal and vertical condition, which satisfy the safety requirements. By reasonable setting of the main and side hanging ear, column flange reinforcement measures and the selection of high strength grade of the main lifting lug flange bolt, we ensure intensity and stability of the flare stack in integral hoisting better satisfy the safety requirements, and ensure flare stack integral hoisting could be completed successfully. The integral hoisting process of flare stack flange connection structure solve many difficulties of aerial parts assembly and segmental hoisting operation in winter construction, and lay a solid foundation for promotion of this hoisting method.

References

[1] Wang Y L. Hoisting of large equipment management of [J]. *China High-Tech Enterprises*, 2010, Volume 21: 131-132.
[2] SH/T 3029-2014. *Petrochemical exhaust tube and the flare stack design specification* [M], Beijing: Ministry of Industry and Information Technology of the People's Republic of China, 2014.
[3] Sheng XY, Tang S Q. Finite-element-analysis of the CATIA: commands and the examples[M], Beijing: *Mechanical Industry Press*, 2005

IOP Conf. Series: Materials Science and Engineering **281** (2017) 012050 doi:10.1088/1757-899X/281/1/012050

Research on Modeling Technology of Virtual Robot Based on LabVIEW

Z Wang , J L Huo , L Y Sun * and X Y Hao
School of Mechanical Engineering, Hebei University of Technology, NO.8 Dingzigu one road ,street ,Tianjin,China

Email : sunlyu78@126.com

Abstract. Because of the dangerous working environment, the underwater operation robot for nuclear power station needs manual teleoperation. In the process of operation, it is necessary to guide the position and orientation of the robot in real time. In this paper, the geometric modeling of the virtual robot and the working environment is accomplished by using SolidWorks software, and the accurate modeling and assembly of the robot are realized. Using LabVIEW software to read the model, and established the manipulator forward kinematics and inverse kinematics model, and realized the hierarchical modeling of virtual robot and computer graphics modeling. Experimental results show that the method studied in this paper can be successfully applied to robot control system.

1. Introduction

With the progress and development of science and technology, human activity space is no longer limited to the field of human survival. It has penetrated into the deep sea, outer space, nuclear radiation, toxic and harmful environment, and other dangerous environment. At this point, dangerous operating robots are being used. Instead of human beings, they enter the job site, operate under human control or do their work autonomously, [1]. The robot in this paper is a robot which is used in the underwater environment of nuclear power station to identify and salvage foreign bodies. Due to the particularity of the environment, the operators are not able to work close observation of the robot, with the help of robot carrying a camera, can realize the identification and tracking of foreign body, but because of the camera installation position and angle limit, still unable to observe the full range of the position and pose of the robot, the robot control becomes very difficult. In order to solve this problem, we join the virtual robot model is highly consistent with the real robot, motion driven real-time virtual robot with real robot pose information, through the observation of the movement of virtual robot, obtain the real robot working position. The operation difficulty of the operator is greatly reduced, and the operability of the whole system is improved.

There are many modelling methods for virtual robots, and various modelling software are springing up. Different methods and software have different characteristics:

(1) using OpenGL graphics library combined with VC, VC++ or VB tools.

This method is relatively mature, there have been many successful examples of [2,3], but the procedure is too complicated, need to draw a graphical programming language, when the graphics is complex, or the accuracy is relatively high, the workload alarming, and graphics effect is not realistic, the effect is not very ideal.

(2) using virtual reality commercial software.

IOP Conf. Series: Materials Science and Engineering **281** (2017) 012050 doi:10.1088/1757-899X/281/1/012050

There are many mature software, such as VRP in China, Virtools in France, EON Studio in USA, visual simulation driver software, WTK, /VTK, /STK, Quamtum3D, OpenGVS and so on. These software has made great progress in modeling virtual scene. But when it is applied to the virtual simulation of industrial robot, it is not accurate, convenient and quick to construct and modify the geometric characteristics of mechanical products. And it can not exactly reflect the position information of real manipulator. In teleoperation robot, the maximum efficiency of virtual reality can not be realized[4,5].

2. Robot structure and kinematics analysis

2.1. Robot structure analysis

The robots studied in this paper are applied to the underwater environment of nuclear power plants, and the identification and salvage of foreign bodies in the high radiation environment of nuclear power plants are realized. The main body includes the following parts, such as car body, large camera, small camera, waist, big arm, small arm, wrist and gripper, as shown in figure 1.

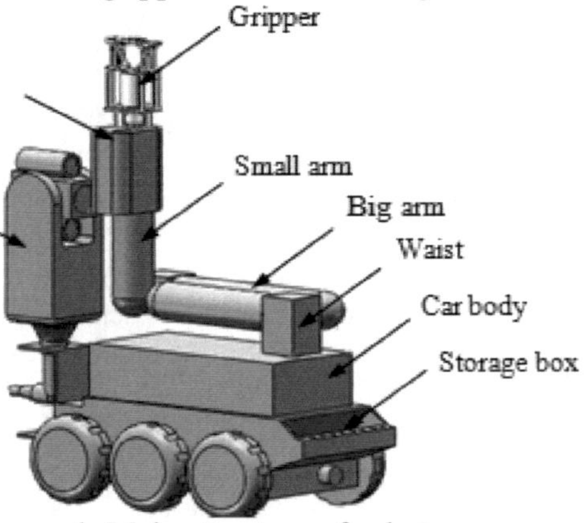

Figure 1. Main structure of robot

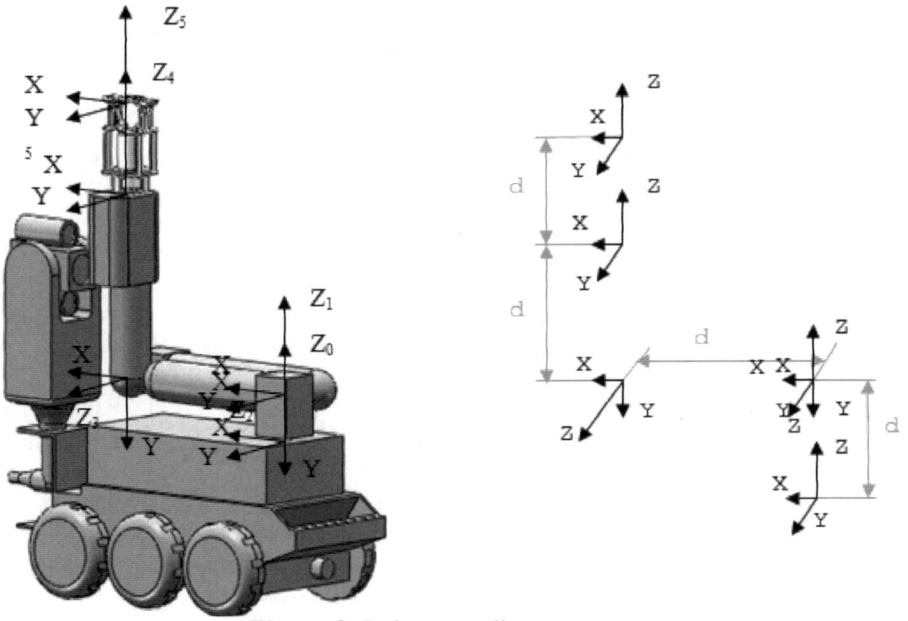

Figure 2. Robot coordinate system

IOP Conf. Series: Materials Science and Engineering **281** (2017) 012050 doi:10.1088/1757-899X/281/1/012050

2.2. kinematics analysis of robot

According to the law of motion of robot, coordinate system is set up, as shown in figure 2. When a robot does a job, he usually needs to do the following. With the movement of the robot in the working environment, the vehicle mounted large camera visits around, when the foreign body is found, the recognition and tracking of the vehicle are realized. At the same time, the information transmitted by various sensors on the vehicle body is detected. When the foreign body enters the manipulator's grab range, the crawl mode is started and the virtual robot model is enabled. It is only necessary to analyze the kinematics of the robot in the grasping process. As shown in Figure 2, the waist rotates around the Z_1 axis, the big arm revolves around the Z_2 axis, the small arm revolves around the Z_3 axis, the wrist rotates around the Z_4 axis, and the gripper movement is realized by a special mechanism, and the manipulator end point coordinate system is O_5-$X_5Y_5Z_5$. The connecting rod size information is shown in table 1.

Table 1. Every connecting rod size of robot

connecting rod i	twist angle α_{i-1}	Length of connecting rod a_{i-1}	Twist angle θ_i	distance d_i
1	0°	0	θ_1	d_1=0.065m
2	-90°	0	θ_2	0
3	0°	a_2=0.300m	θ_3	0
4	90°	0	θ_4	d_4=0.270m
5	0	0	0	d_5=0.141m

Because, ${}^{i-1}_{i}T=R_x(\alpha_{i-1})D_x(a_{i-1})R_z(\theta_i)D_z(d_i) = \begin{bmatrix} c\theta_i & -s\theta_i & 0 & a_{i-1} \\ s\theta_i c\alpha_{i-1} & c\theta_i c\alpha_{i-1} & -s\alpha_{i-1} & -s\alpha_{i-1}d_i \\ s\theta_i s\alpha_{i-1} & c\theta_i s\alpha_{i-1} & c\alpha_{i-1} & c\alpha_{i-1}d_i \\ 0 & 0 & 0 & 1 \end{bmatrix}$ [4] and

${}^0_5T={}^0_1T(\theta_1)\,{}^1_2T(\theta_2)\,{}^2_3T(\theta_3)\,{}^3_4T(\theta_4)\,{}^4_5T(\theta_5)$,so

$$
{}^0_5T=\begin{pmatrix} c_1c_4c_{2+3}-s_1s_4 & -s_4c_1c_{2+3}-s_1c_4 & c_1s_{2+3} & c_1s_{2+3}(d_4+d_5)+a_2c_1c_2 \\ c_4s_1c_{2+3}+c_1s_4 & -s_4s_1c_{2+3}+c_1c_4 & s_1s_{2+3} & s_1s_{2+3}(d_4+d_5)+a_2s_1c_2 \\ -s_{2+3}c_4 & s_{2+3}s_4 & c_{2+3} & c_{2+3}(d_4+d_5)-s_2a_2+d_1 \\ 0 & 0 & 0 & 1 \end{pmatrix}
$$

$$
\begin{aligned}
&n_x=c_1c_4c_{2+3}-s_1s_4 && o_x=-s_4c_1c_{2+3}-s_1c_4 && a_x=c_1s_{2+3} && p_x=c_1s_{2+3}(d_4+d_5)+a_2c_1c_2 \\
&n_y=c_4s_1c_{2+3}+c_1s_4 && o_y=-s_4s_1c_{2+3}+c_1c_4 && a_y=s_1s_{2+3} && p_y=s_1s_{2+3}(d_4+d_5)+a_2s_1c_2 \\
&n_z=-s_{2+3}c_4 && o_z=s_{2+3}s_4 && a_z=c_{2+3} && p_z=c_{2+3}(d_4+d_5)-s_2a_2+d_1
\end{aligned}
$$

3. The establishment of virtual model

3.1. Model building

In LabVIEW, there are two ways to build model. (1) draw the model (2) by using the geometric VI of LabVIEW, and import the model built by external CAD software. The first method is used to model simple geometry and complex assembly relations. Second methods, you can import geometry model format, ASE, STL, VRML, three kinds. ASE format used in 3D Max more, VRML and STL format is more common, in SolidWorks, Pro-E, AutoCAD and other software can be used[6].

The VRML (Virtual Reality Modeling Language) is a platform independent operation, based on 3DWeb, describing interactive 3D world and object language, it by creating a virtual scene to achieve real effects. Virtual reality scene created by VRML is a realistic simulation of the 3D product modeling, hierarchical, multi sensory object oriented characteristics of the virtual world, interactive,

IOP Conf. Series: Materials Science and Engineering **281** (2017) 012050 doi:10.1088/1757-899X/281/1/012050

real-time and support web browsing, and because the VRML file as HTML document upload / download, and start the VRML browser to watch the 3D model of the document described, a good solution to the problem of information exchange and sharing [7,8].

The shape and assembly relationship of robot in this research are complex, and second methods are adopted. Draw the model using SolidWorks, export VRML format file.

3.1.1. SolidWorks rendering model
A comprehensive analysis of the robot body during the crawling process not only exercise the waist, big arm, small arm, hand motion, simple procedures, the robot is divided into five parts, body, waist, big arm, small arm, hand, in the "clear outline structure" under the condition of more simple drawing better. The model, to fully consider the position and direction in the process of kinematics analysis of coordinate system, the coordinate system of each model is consistent with the real robot kinematics coordinate system. In the establishment of industrial robot 3D model with SolidWorks2011, using SolidWorks VRML conversion program will output VRML file: File - Save: save type VRML (3. WRL) - option.. - File Format: VRML version: VRML97, you can output VRML file. The use of SolidWorks VRML modeling, greatly reducing the modeling workload, improve work efficiency, more important is to realize the precise modeling of virtual industrial robot model accuracy is directly related to the effect of process control[9,10].

3.1.2. Virtual robot in LabVIEW
So far, the part model of the virtual robot has been constructed. It is necessary to display the assembly in the LabVIEW and simulate the motion of the real robot.
Analyze the relationships among the components, as shown in figure3

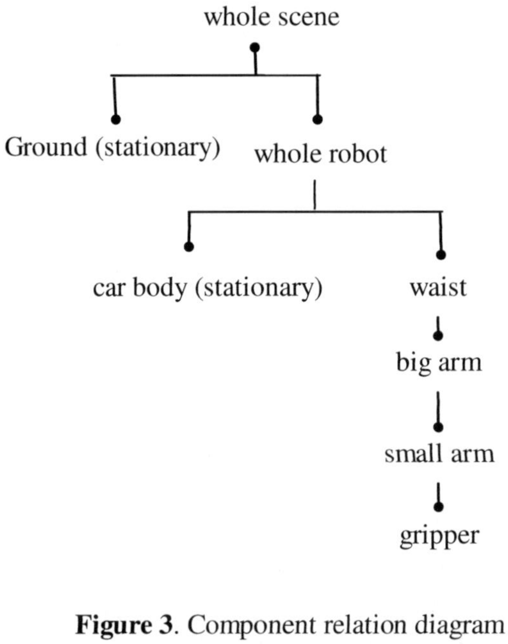

Figure 3. Component relation diagram

The whole scene is independent of an object, and the independent ground is a sub object. The robot, each degree of freedom requires the establishment of a relationship between two objects, an object is the relationship between father and son, son with the parent object object motion, shown in Figure 4 hierarchy, the next level is a component part of sub objects. The whole robot has four degrees of freedom relative to the scene. From the waist, each child object has only one degree of freedom relative to the parent object. Establish a hierarchical relationship between components in a program. As shown in figure 5.

According to the assembly relation of the real robot, the LabVIEW model is used to set up the spatial relation of each model in the program VI, and the virtual robot is assembled, as shown in figure 4.

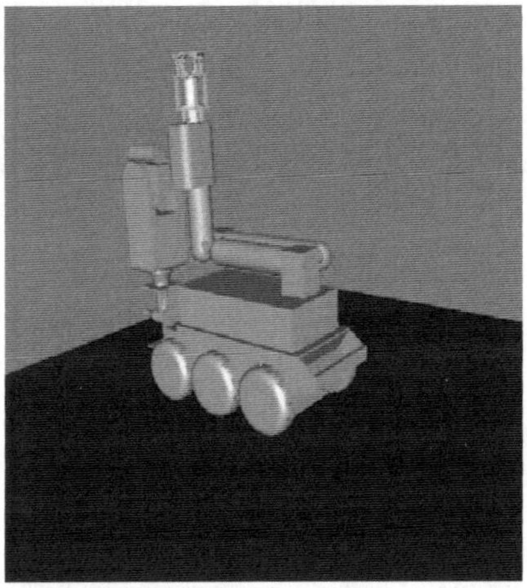

Figure 4. Virtual robot assembly model

3.2. Virtual robot data interface

So far, the virtual robot has been built. In order to realize the movement of the virtual model, it is necessary to set up the data interface. The motion of each joint has been analyzed in the 2.1 section. Figure 4 shows the initial position of the robot, and the interface data of the virtual model is the absolute position of each joint. No matter what data is passed, it needs to be transformed into the absolute position of each joint to realize the correct display of the virtual model.

Each joint rotation during the crawling process, using the LabVIEW VI "set rotation, rotation axis and angle of each joint position information set, will pass over the data transmitted to each joint, realize the model movement

4. Virtual model motion control

Introduced in the 1.2 section, the system has two major functions (1) real-time simulation movement (2) simulation movement. Each module is equipped with real-time data display window, with the movement of the virtual model, real-time display of the rotation angle of each joint, and the manipulator end position matrix, to provide detailed data support for the operator.

4.1. Real-time simulation motion model

The operator operates the real robot motion and collects the rotation angle of the motor at each joint. The collected data is directly communicated to the virtual model interface, and the virtual model is driven to follow the real robot motion.

4.2. Simulation movement model

When the operator is in real operation, the simulation operation can be carried out on the virtual system, and several exercises can be carried out to improve the operation proficiency. You need to start the simulation motion model. The robot can also solve the kinematics forward kinematics and inverse kinematics solutions, and realize a variety of control methods.

4.2.1. Real time manipulation simulation

Operators on the system simulation operation, training operation ability. The real robot uses speed control in the course of movement, and the operator sends a speed command to the robot through a desktop control system or a joystick. The operator uses the same method in the simulation operation. A speed interface is needed for the virtual model to capture the speed information that the operator passes over. As shown in figure 5 provides an interface for virtual velocity model, at the same time as the virtual mode of the movement, the real-time display of each joint and the turning angle of the manipulator end effector matrix, provides detailed data for the operator.

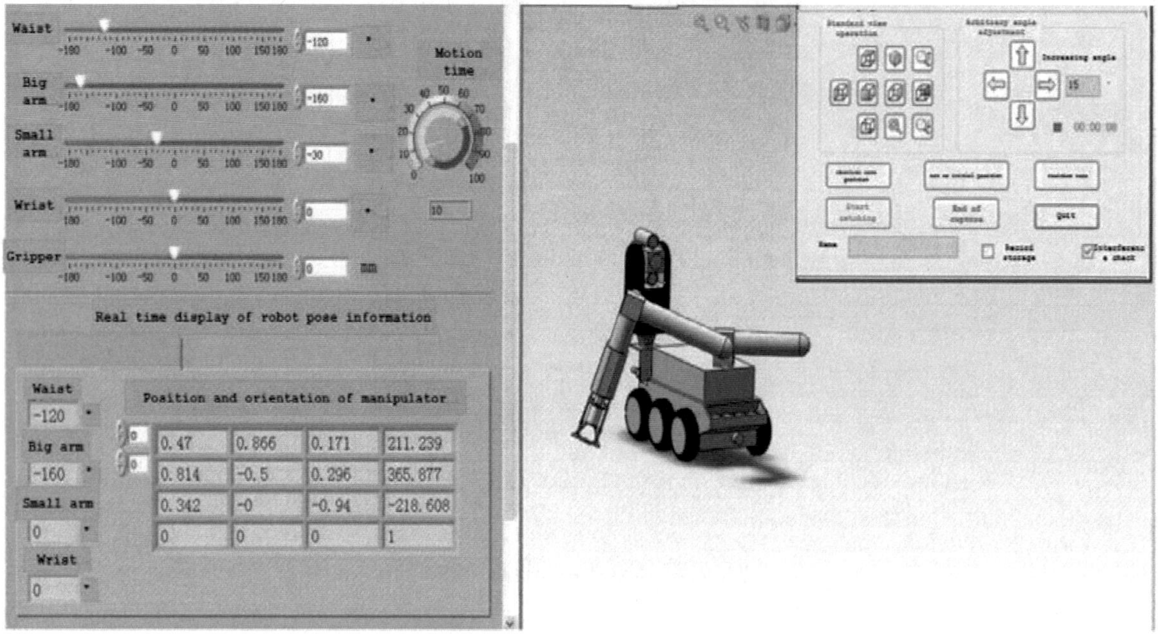

Figure 5. Interface for virtual velocity model

4.2.2. *The positive and inverse solutions*

Before the actual operation, the operator often on the robot kinematics analysis and trajectory planning, grasp the inverse kinematics analysis and so on a series of theoretical analysis, through the virtual model simulation, prediction model of real motion please condition, to plan the path with better attitude. This virtual system provides rich operation mode and powerful data support for it.

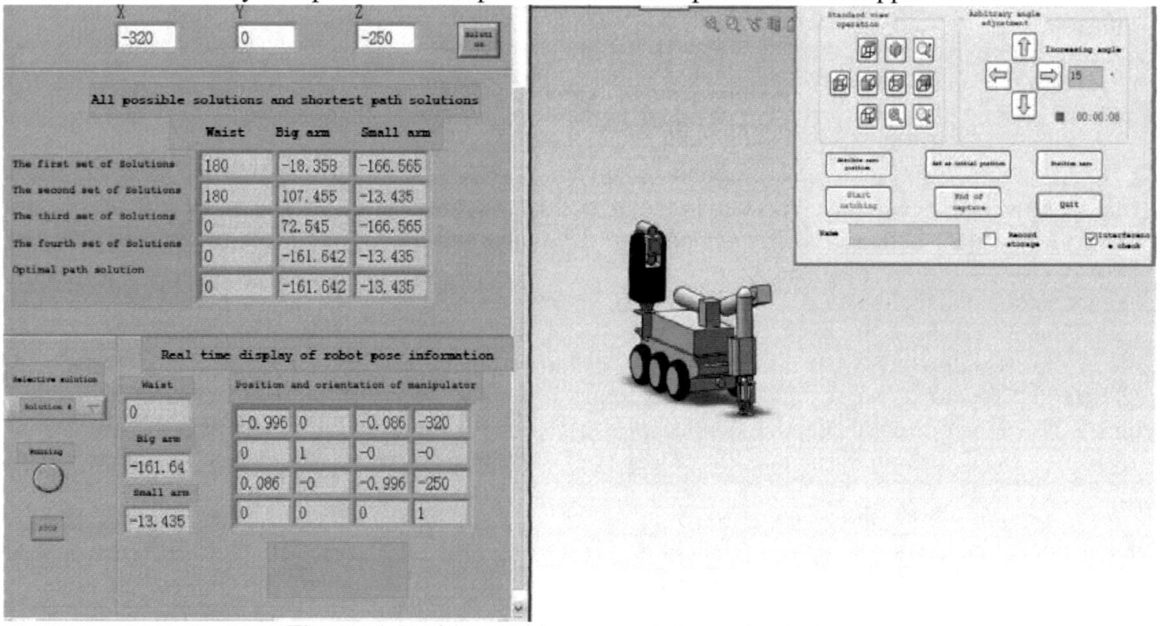

Figure 6. Positive and inverse solutions simulation

IOP Conf. Series: Materials Science and Engineering **281** (2017) 012050 doi:10.1088/1757-899X/281/1/012050

As shown in figure 6, the kinematics analysis interface for the robot is positive. The operator inputs each joint to the program, and sets the movement time, chooses the movement pattern: each joint moves step by step or the synchronized movement. The virtual robot moves according to the set data. By simulating the motion of virtual robot, the operator visually obtained position information of the robot, the robot can adjust the position of each joint, movement speed, movement mode, to plan the best path and attitude. At this point, the data are used to operate a real robot, to improve operational efficiency, and to reduce real robot casualties.

5. Conclusions

This study proposes the use of LabVIEW development of teleoperation virtual reality system, using SolidWorks model, modeling greatly simplifies the workload, using LabVIEW powerful programming functions and excellent interactivity and portability, developed a virtual reality system, provide various modes of operation and powerful data support for the operating personnel. Compact program, high portability. The system has been verified by experiments. The results show that the system is stable and reliable, and the real-time meets the requirements. It provides an important reference for future research.

In the next stage, this research will further improve the system, add robot scene and achieve accurate collision detection. Through many experiments, improve the performance of various aspects of the system.

References

[1] Shim H W, Jun B H, Lee P M, B Hyuk, Le J H.2010 J. Workspace control system of underwater tele-operated manipulatorson an ROV. *Ocean Engineering* .37: 1036–1047.

[2] Wang Z J. 2007.Research on 3D virtual scene modeling technology based on OpenGL, (Wuhan: Wuhan University).

[3] Zhang K M,2012. *Robot simulation research based on virtual reality*, (Tianjin: Tianjin University)

[4] John, Craig J. 2006,*Robotics introduction*, (Beijing: Mechanical industry press)

[5] Yuan F W, Li B W, Ho B.2007. *Accurate modeling of virtual reality based on SolidWorks 2VRML.* Mechatronics Engineering.10:104~105.

[6] Li L, Ye T, Tan M, Chen X J, "Present State and Future Development of Mobile Robot Technology Research ," *Robot*, 2002.7: 37~40. (In Chinese)

[7] Tirmant H, Baloh M, Vermeiren L,et al, "An Alternative Two Wheeled Vehicle for an Automated Urban Transportation System," *Proceedings of the 2002 IEEE International Conference on Intelligent Vehicle Symposium*, Yorozu, M. Hirano, 2002: 594~603.

[8] Yao K C. *Virtual Spectrum Analyzer Design and Implementation,* Shenyang, China , August 12-August 14 .

[9] Poor H., *An Introduction to Signal Detection and Estimation.* New York: Springer-Verlag, 1985, ch. 4. Sun Di-sheng et.al, Robot Control Technology, China Machine Press, 1997: 180~194. (In Chinese)

[10] Xu G H, Tan M, "Development state and tendency of Mobile Robot Fundamentals of erbium-doped fiber amplifiers arrays ," *Robot Technique and Application,* 2001.3: 7~15

IOP Conf. Series: Materials Science and Engineering **281** (2017) 012051 doi:10.1088/1757-899X/281/1/012051

Simulation of Ion Energy Distributions in ICP Reactor with Bias Voltages Using COMSOL

C K Han[1], Y Y Yang[1*], W F Liu[1], J Cheng[2], Y J Lu[2] and L H Ji[2]

[1] School of Engineering and Technology, China University of Geosciences (Beijing), Beijing 100083, China
[2] Department of Mechanical Engineering, Tsinghua University, Beijing 100084, China

E-mail: yangyy@cugb.edu.cn

Abstract. An inductively coupled plasma reactor driven by continuous wave (cw) or pulse source with dc bias is introduced to investigate the plasma parameters and ion energy distributions using COMSOL and compared with experimental results. The results show that with the increase of pressure, the electron density varies from $4.5\times10^{11}cm^{-3}$ to $16.7\times10^{11}cm^{-3}$, and electron temperature is in the range of 3.3eV to 3.9eV. For pulsed power, the change of electron temperature is corresponding to the duty cycle of pulse power. With the dc bias, the ion energy peaks shift from 8eV to 19.5eV, and the full width at half maximum (FWHM) of the IEDs varies in the range of 0.5eV to 2.7eV. Finally, application of a synchronous dc bias in the afterglow of pulsed plasma leads to a bi-modal IEDs with a sharper peaks at lower energy during the afterglow and a broader peak at higher energy when the power is on.

1. Introduction

The control of ion energy bombarding the wafer is important in anisotropic plasma etching and deposition process [1, 2]. With the decrease of line width and increase in the scale of wafer, precise and independent control of the ion energy distribution (IEDs) and fluxes which determine selectivity of plasma etching and etch rate becomes critically important [3, 4]. Many methods have been developed to control the plasma parameters and IEDs independently [5], such as dual-frequency capacitively coupled plasma source (DF-CCPs), inductively coupled plasma source (ICPs) with bias voltage and pulsed plasma source with synchronous bias voltage, et al. Due to the high density of plasma and the independent control of ion energy distributions, ICPs with bias voltages have been widely used in integrated circuit manufacture, especially deposition and etching process. In ICPs with bias voltage system, the coil power intends to generate and sustain plasma while bias power intends to control IEDs and ion fluxes.

To date, there are many the researches of investigating and controlling the structure of IEDs using analytical methods, experimental measurement and simulation models. The retarding field energy analyser (RFEA) has been introduced to investigated the IEDs for both rf-biased and moderating the shape of bias voltage waveform ICPs or CCPs in argon and helicon plasma [6-10]. The influence of discharge pressure, bias power and coil power on the IEDs and ion angle distributions (IADs) are investigated in the studies. An extended semi-analytic collisionless sheath model is used to obtained the IEDs and IADs for muti-frequency CCPs [11]. Hybrid fluid model and PIC/MCC model are introduced to analyse the plasma parameters and IEDs in both ICP and CCP discharge for Ar, CF_4 and

IOP Conf. Series: Materials Science and Engineering **281** (2017) 012051 doi:10.1088/1757-899X/281/1/012051

O_2 [12-16]. The influence of bias frequency, discharge pressure, voltage amplitude, gas species and bias power on the plasma parameters and IEDs are simulated. However, the pulsed ICP discharge with dc bias, which is used in industrial manufacture, has been rarely investigated by simulation method. In this paper, both cw and pulsed source power ICP argon discharge with dc bias is investigated by COMSOL under variety of discharge condition. The plasma parameters and IEDs are obtained under various discharge condition. The control of IEDs using inductively coupled pulsed plasma with dc bias is investigated.

2. Model description

A 2D axisymmetric schematic diagram of the ICP reactor used in this work is given in Figure1 [6]. Plasma is sustained in an 80 mm diameter dielectric tube with a three-turn spiral coil (10 mm sider length) around quartz window (the relative permittivity sets to 4.2). In this study, plasma source can be selected between cw powers at 13.56 MHz and pulsed at 10 KHz power. The input nozzle is placed at the top of chamber, which consists of three nested metal cylinders 30 mm tall and 13, 32 and 40 mm in radius (the electrical conductivity is set to 10.2×10^6 S/m). The walls of inner chamber except the symmetry axis and electrode are all grounded. The interfaces between plasma and quartz tube are dielectric (surface charge accumulation is used here). External boundaries of system except the symmetry axis are magnetically insulated ($n\times\mathbf{A}=0$, n is the surface normal and **A** is the magnetic vector potential). The air case surrounding the coils is considered to vacuum characteristics.

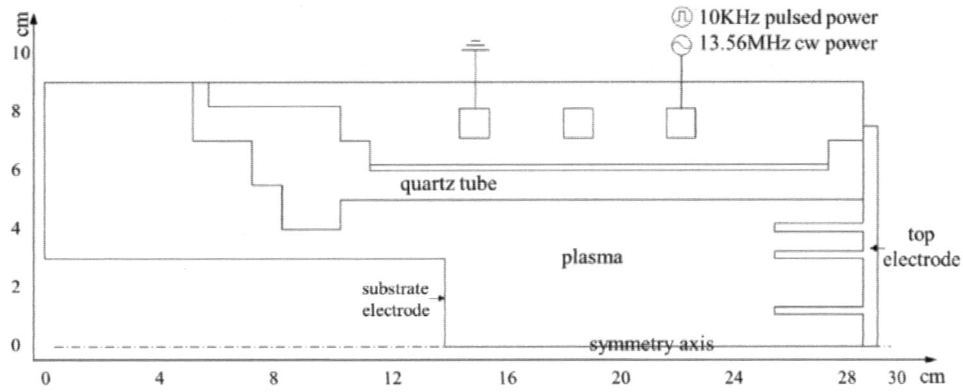

Figure 1. The schematic diagram of the ICP reactor

The main collision parameters of argon gas used in this simulation are listed in table 1:

Table 1. The main collision parameters in the argon discharge [17].

Formula	Collision	Energy loss (eV)
e+Ar=>e+Ar	Elastic	--
e+Ar=>e+Ar*	Excitation	11.5
e+Ar*=>e+Ar	Super elastic	-11.5
e+Ar=>2e+Ar+	Ionization	15.8
e+Ar*=>2e+Ar+	Ionization	4.24
Ar*+Ar*=>e+Ar+Ar+	Penning Ionization	--
Ar*+Ar=>Ar+Ar	Metastable quenching	--

In table 1 the symbols + and * denote positive ions and metastable atoms respectively.

3. The ORY description

The plasma module in COMSOL is based on fluid model [18-20]. In this study, the electron energy distribution function (EEDF) is assumed to Maxwellian. Since the discharges are operated at low pressure (up to 50 mTorr) and the electron energy is low (up to 6 eV), the assumption is appropriate in this study [21]. Monte Carlo (MC) model is applied to capture ions trajectories until they collide with the wafer surface [22].

3.1. Fluid module

The spatio-temporal distributions of particles and the electric field are obtained by solving the coupled Maxwell equations, continuity equations of electron, energy balance equation of electron, continuity equations of ions and Poisson's equation as follows [19],

$$\frac{\partial n_e}{\partial t}+\nabla\cdot\mathbf{\Gamma}_e=\mathrm{R}_e-(u\cdot\nabla)n_e;\mathbf{\Gamma}_e=-(\mu_e\cdot\mathbf{E})n_e-D_e\cdot\nabla n_e \tag{1}$$

$$\frac{\partial n_\varepsilon}{\partial t}+\nabla\cdot\mathbf{\Gamma}_\varepsilon+\mathbf{E}\times\mathbf{\Gamma}_e=R_\varepsilon-(\mathbf{u}\cdot\nabla)n_\varepsilon+\frac{Q}{q};\mathbf{\Gamma}_\varepsilon=-\frac{5}{3}((\mu_e\cdot\mathbf{E})n_e+D_e\cdot\nabla n_e)\varepsilon-\frac{5}{3}(D_e\cdot\nabla\varepsilon)n_e \tag{2}$$

$$\rho\frac{\partial w_k}{\partial t}+\rho(\mathbf{u}\cdot\nabla)w_k=\nabla\cdot\mathbf{j}_k+R_k \tag{3}$$

where parameters undefined are listed in table 2.

Table 2. The undefined parameters in the fluid model

n_e — the electron number density	$\mathbf{\Gamma}_e$ — the electron flux vector
R_e — source or sink of electrons	$\mathbf{u}$ — the averaged mass fluid velocity vector
n_ε — the electron energy density	$\mathbf{\Gamma}_\varepsilon$ — the electron energy flux vector
μ_e — the electron mobility	D_e — the electron diffusivity
R_ε — the energy loss by inelastic collision	Q — the external power
q — the charge of an electron	$\mathbf{E}$ — the electric field intensity
ρ — the mixture gas density	w_k — the mass fraction of the *k*th species
$\mathbf{u}$ — the averaged mass fluid velocity vector	$\mathbf{j}_k$ — the diffusive flux vector
R_k — velocity rate of the *k*th species	

3.2. Electromagnetic module

For a non-polarized plasma, a frequency domain Maxwell equations are applied to compute the electromagnetic distribution of the charge system. The magnetic vector potential **A** as the independent variable is computed in the frequency domain using the following Eq. (4) and (5),

$$(j\omega\sigma-\omega^2\varepsilon_0\varepsilon_r)\mathbf{A}+\nabla\times(\mu_0{}^{-1}\mu_r{}^{-1}\nabla\times\mathbf{A})=\mathbf{J}_e \tag{4}$$

$$-\varepsilon_0\nabla^2V=\rho \tag{5}$$

where parameters undefined are listed in table 3.

Table 3. The undefined parameters in the electromagnetic model

ω — the angular frequency of power source	$\mathbf{J}_e$ — the external current
σ — electrical conductivity of the system	ε_0 — the vacuum permittivity
ε_r — relative permittivity of the system	μ_0 — the vacuum permeability
μ_r — relative permeability of the system	V — the electric potential

3.3. MC module

The charge particle tracing (CPT) module based on MC model is applied in this work [22]. The ion speed reduces to zero after collision with wafer to ensure the ion energy is captured only once. The ion energy distribution on the wafer is calculated in this model.

The distribution equation is defined by Eq. (6),

$$f(v_i) = \sqrt{\frac{M}{2\pi kT}} \exp\left(-Mv_i^2/2kT\right) \tag{6}$$

where M is ion mass, k is the Boltzmann coefficient, T is the initial temperature of ions which is 300K in this work, v_i is the ion velocity.
The collision probability is defined by Eq. (7),

$$P = 1 - \exp(-\nu \Delta t) \tag{7}$$

where v is the collision frequency between ions and neutral particles, Δt is the time step .

4. Results and description

Compared with experimental results from [6], the plasma parameters are discussed in this paper. The discharge conditions are: an initial gas temperature of 300K, a coil power of 300W and initial electron energy of 3eV.

4.1. Discharge with pulsed source power

Figure 2 shows the electron number density along the chamber symmetrical axis under different pressures when the bias electrode is grounded. As shown in Figure 2, the electron number density (n_e) is higher in the central chamber than the electrode due to the diffusion and recombination of ions and electrons at the electrode and wall. The n_e increases from 4.5×10^{11} cm^{-3} to 16.7×10^{11} cm^{-3} with the increase of discharge pressure. The result shows that the variation of electron number density is in good agreement with the experimental results, especially for 14mTorr discharge. Because of the Maxwellian EEDF assumption and simplification of electron transport coefficient by Einstein's equation, the peak density position (215mm) in this work is smaller than experimental results (230mm). In addition, the electron density next to electrode is much higher than experimental results (the n_e is 11.2×10^{11} cm^{-3} in simulation and 2×10^{11} cm^{-3} in experiment at 170mm for 50mTorr). Because the fluid model used in this study does not consider the sheath region, the gradient of electron density adjacent boundary becomes smaller, resulting in a higher electron number density. Meanwhile, a quasi-neutral hypothesis is applied to the whole chamber.

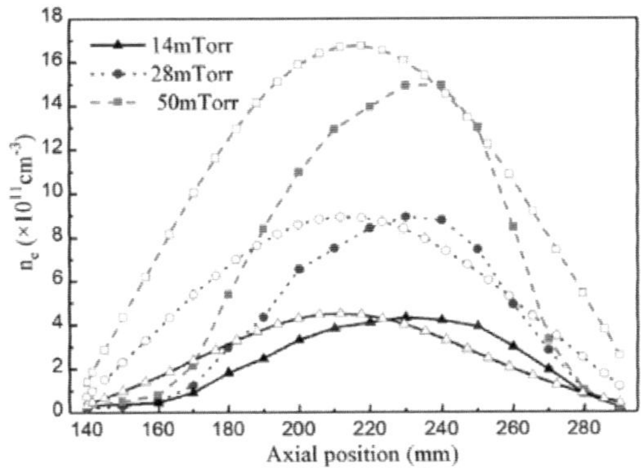

Figure 2. Electron number density plots as a function of symmetrical axis with cw power for different discharge pressures. The hollow icon is simulated results, the solid icon is experimental results from [6].

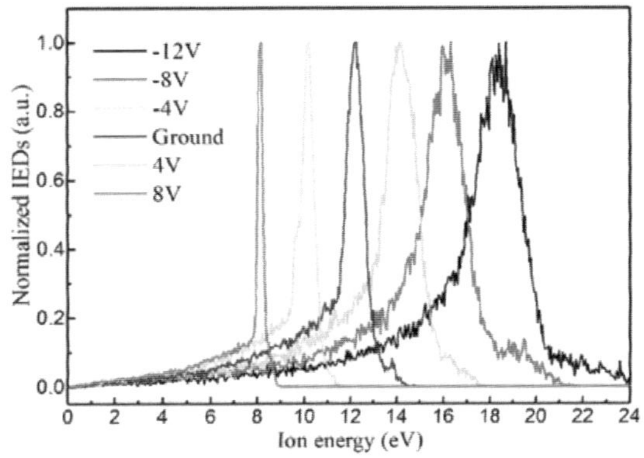

Figure 3. Normalized IEDs plots as a function of dc bias voltages applied to the bias electrode with cw source power.

Figure 3 demonstrates the normalized IEDs under different dc bias voltages applied to the substrate electrode. The bias voltage applied to the substrate electrode will shift the peak of ion energy. The dc bias voltages have influence on the FWHM of IEDs. When positive voltages are applied, ions may not reach the wafer due to the suppression of bias electric field. Since ion energy lower than the threshold energy can't be captured, the ion energy broadening will be narrower. The offset of IEDs peak is proportional to the increase of bias voltage. Bias voltages applied to the substrate electrode affect the damped potential which directly determine the movement of ion transits the plasma sheath region. Because of a more directly influence on the damped potential, the peak of ion energy has a larger offset and the energy broadening can become narrower when bias voltage is applied to the substrate electrode.

4.2. Discharge with pulsed source power

To obtain nearly menoenergetic IEDs, narrowing the ion energy broadening which can be achieved by lowering electron temperature can be efficient. Pulsed plasma power can reduce electron temperature and maintain a constant sheath damped voltage. With a dc bias voltage applied to the substrate electrode, a narrower broadening IEDs will be achieved.

Figure 4 and Figure 5 show the electron density and electron temperature under different pressures and power duty ratios. The results show that the ne is increase from 4.75×10^{10} cm^{-3} to 3.7×10^{11} cm^{-3} with the increase of discharge pressure, the peak density position decreases to 195mm. Electron temperature increases rapidly after the plasma source turns on, overshoots then decreases to a steady value. Since electrons obtain energy from power source, electron energy increases sharply and presents rapid increase in electron temperature. Because of the collision between electron and other particles, electron energy consuming in excitation and ionization process cause the decrease of electron temperature. As expected, electron temperature decreases with increasing pressures due to the increase of collision probability between electron and other particles. After the plasma source turns off, electron temperature decreases rapidly and then reaches steady in the afterglow. Meanwhile, electron temperature decreases faster in low pressure because of a faster diffusion rate which determines the cooling mechanism at low pressure.

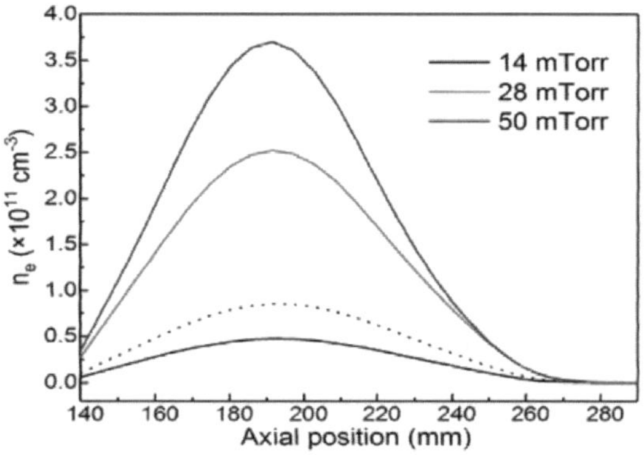

Figure 4. Electron number density plots as a function of symmetrical axis with pulsed power for different duty ratios and discharge pressure, with 120W time averaged power. Solid line is cycle of 20%, dot line is cycle of 50% duty ratio at 14 mTorr.

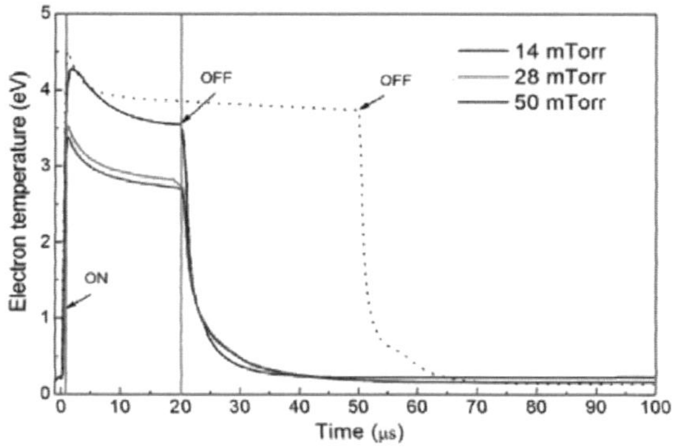

Figure 5. Electron temperature plots in a cycle for pulsed plasma in10KHz for different duty ratios and discharge pressure, with 120W time averaged power. Solid line is cycle of 20%, dot line is cycle of 50% duty ratio at 14 mTorr.

Figure 6 shows the normalized IEDs for pulsed source power coupled with dc bias voltage applied to bias electrode. With the decrease of dc voltage, the structure of IEDs transforms from uni-modal towards bi-modal with higher low energy peaks and broad high energy peaks. Because the dc bias directly affect sheath damped potential, the structure of IEDs will be changed in cycle of source power. When bimodal IEDs is obtained, the low energy peaks vary in the range of 4.2 to 12.2 eV, which is corresponding to the amplitude of bias voltage. The broader energy peak is corresponding to the condition when plasma source turns on. The narrower and higher energy peak is correspond to the afterglow of pulsed plasma condition. The sheath potential is equal to the bias voltage when plasma source turns off, ions accelerated by bias electric field results a sharper peak in low energy.

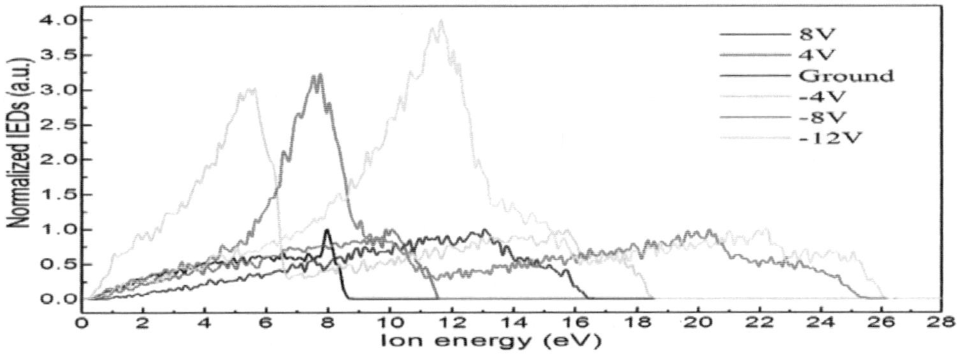

Figure 6. Normalized IEDs obtained under pulsed power with different dc bias voltage applied to the substrate electrode.

5. Conclusions

In this paper, dc bias ICP argon discharge driven by cw/pulsed source power is simulated and discussed by COMSOL. Both cw power and pulsed power source with dc bias voltage are investigated. The results indicate that the electron density is proportional to the discharge pressure. With the increase of discharge pressure, the electron density increases accordingly. Secondly, bias voltage has a strong influence on IEDs. Bias voltages can shift ion energy peak and affect the peak energy broadening. Finally, pulsed plasma discharge with dc bias voltage is investigated. A periodic change in plasma parameters and bimodal IEDs are found in a discharge cycle. Furthermore, a broad peak corresponds to ion energy during plasma on phase, while the sharper peak corresponds to the dc bias voltage during plasma off. Because of the Maxwellian EEDF assumption, there are some errors in the density peak position (it is at 215mm for cw power and 195mm for pulse power in simulation, which is 230 mm in experimental results). The electron transport coefficients calculated by Einstein equation cause error in electron potential which finally generates larger deviation in IEDs. In the future, EEDF obtained by experimental results or calculated by electron Monte-Carlo simulation model (EMCSM) will be coupled with fluid model.

Acknowledgement

This study is financially supported by the China Important National Science and Technology Specific Project No. 2 (2011ZX02403-004).

References

[1] Greene W M, Hartney M A, Oldham W G and Hess D W **1988** *J. Appl. Phys. 63* 1367
[2] Chen W C and Pu Y K **2014** *J. Phys. D:* Appl. *Phys. 47* 345201
[3] Economou D J **2013** *J. Vac. Sci. Technol., A. 31* 050823
[4] Agarwal A, Stout P J, Banna, S, Rauf S, Tokashiki K and Lee J Y **2009** *J. Appl. Phys. 106* 103305
[5] Lieberman, M A and Lichtenberg, A J **2005** *Principles of Plasma Discharges and Materials Processing,* 2nd ed. Wiley, New York p 351
[6] Shin H, Zhu W, Xu L, Donnelly V M and Economou D J **2011** *Plasma Sources Sci. Technol.20* 055001
[7] Wang S B and Wendt A E **2000** *J. Appl. Phys. 88* 643
[8] Woodworth J R, Riley M E, Miller P A, Hebner G A and Hamilton T W **1997** *J. Appl. Phys. 80* 1304
[9] Qin X V, Ting Y H and Wendt A E **2010** *Plasma Sources Sci. Technol. 19* 065014
[10] Ye C, He H, Huang F, Liu Y and Wang X **2014** *Phys. Plasmas. 21* 043509

IOP Conf. Series: Materials Science and Engineering **281** (2017) 012051 doi:10.1088/1757-899X/281/1/012051

[11] Kwon D C, Song M Y and Yoon J S **2013** *J. Phys. D: Appl. Phys.46* 025202
[12] Wang D, Ma T C and Deng X D **1973** *J. Appl. Phys. 74* 2986
[13] Zhang Y T, Kushner M J, Sriraman S, Marakhtanov A and Holland J **2015** *J. Vac. Sci. Technol., A. 33* 031302
[14] Dai Z L, Yue G and Wang Y N **2012** *Plasma Sci. Technol. 14* 240
[15] Kawamura E, Lieberman M A and Graves D B **2014** *Plasma Sources Sci. Technol. 23* 064003
[16] Zhang Y T, Kushner M J and Shannon S **2015** *J. Appl. Phys. 117* 233302
[17] https://fr.lxcat.net/home/
[18] COMSOL 4.4, Userbook
[19] Hagelaar G J and Pitchford L C **2005** *Plasma Sources Sci. Technol. 14* 722
[20] Key R J, Coltrin M E and Glarborg P **2003** *Chemically Reacting Flow Theory and Practice.* Wiley, New York
[21] Panagopoulos T, Kim D, Midha V and Economou D J **2002** *J. Appl. Phys. 91* 2687
[22] Boffard J B, Jung R O, Lin C, Aneskavich L E and Wendt A E **2011** *Plasma Sources Sci. Technol. 20* 055006

IOP Conf. Series: Materials Science and Engineering **281** (2017) 012052 doi:10.1088/1757-899X/281/1/012052

Dynamic Simulation Research on Chain Drive Mechanism of Corn Seeder Based on ADAMS

Y B Wang[1,2] **and H P Jia**[1,2*]

[1] School of Mechanical and Machinery Engineering, Southwest Forestry University, Yunnan, Kunming 650224, China

[2] These authors contributed equally to this work and should be considered co-first authors

Email:jhp_159@163.com

Abstract. In order to reduce the damage to the chain and improve the seeding quality of the seeding machine, the corn seeder has the characteristics of the seeding quality and some technical indexes in the work of the corn seeding machine. The dynamic analysis of the chain drive mechanism is carried out by using the dynamic virtual prototype. In this paper, the speed of the corn planter is 5km/h, and the speed of the simulated knuckle is 0.1~0.9s. The velocity is 0.12m/s, which is equal to the chain speed when the seeder is running normally. Of the dynamic simulation of the movement and the actual situation is basically consistent with the apparent speed of the drive wheel has changed the acceleration and additional dynamic load, the chain drive has a very serious damage, and the maximum load value of 47.28N, in order to reduce the damage to the chain , As far as possible so that the sowing machine in the work to maintain a reasonable uniform speed, to avoid a greater acceleration, the corn sowing machine drive the design of a certain reference.

1. Introduction

At present, the widely used power transmission mechanism on the corn planter is still the traditional chain drive mechanism, the performance of the corn seeder transmission mechanism is one of the key technologies of seeding quality. Chain drive mechanism has been used in many times because of its reliable working performance, simple structure, low manufacturing cost and low cost, which is the most widely used driving mechanism in the field of corn seeder [1].The chain drive mechanism is the core mechanism of the sowing machine drive. The design quality of the chain drive directly affects the transmission performance, seeding efficiency and seeding quality of the planter [2]. Corn planter condition is very complex, the chain transmission mechanism is the main reason to influence the planting quality, in order to study the work of transmission mechanism of corn planter and its actual dynamic characteristics. Using SolidWorks and ADAMS as tools, this paper focuses on transmission mechanism of a corn planter.

IOP Conf. Series: Materials Science and Engineering **281** (2017) 012052 doi:10.1088/1757-899X/281/1/012052

2. Dynamic model establishment

2.1 The establishment of the whole model

In the SolidWorks 3D software, each part and part model of the seeder is bound together through certain connection relation, so that the whole machine assembly can complete the specific seeding function. Through the assembly of interference, collision checks and dynamic gap detection, to ensure that any part of the space does not appear overlap with each other. After interference check, found that there is no interference in the assembly phenomenon, to meet the assembly requirements. Corn planter machine assembly diagram, as shown in Figure 1. The main parameters are: 1750mm × 1400mm × 950mm, ground wheel diameter of 600mm.

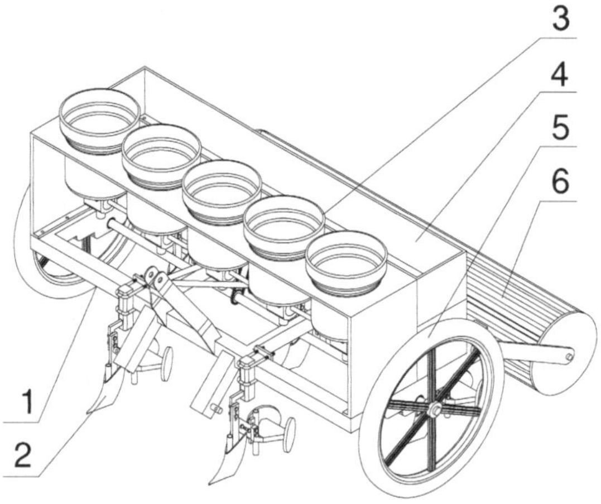

1. Frame 2.Furrow opener 3.Sowing box 4.Fertilization box 5.Wheel 6.Pressure wheel

Figure 1 The whole machine assembly drawing of the planter

2.2 The establishment of chain drive model

The roller chain drive mechanism consists of active sprocket, driven sprocket and roller chain, the roller chain is the inner and outer chain plate, pin and sleeve roller composition. This article uses the chain link 05B chain link, the main parameters shown in Table 1. Sprocket with chain number 05B sprocket, the main parameters shown in Table 1 [3]. Design chain drive with SolidWorks software, according to the design parameters designed by the chain drive three-dimensional model diagram, as shown in Figure 2 [4].

Table 1 Link and Sprocket parameters

Link parameters		Sprocket parameters	
Parameter name	Numerical value	Parameter name	Numerical value
Chain number	05B	Chain number	05B
Pitch	8.00mm	Number of teeth	20
Roller diameter	5.00mm	Pitch circle diameter	51.140mm
Shaft pin diameter	2.31mm	Root ciecle diameter	46.140mm
Chain plate thickness	0.80mm	Addendum ciecle diameter	54.820mm
Inner chain width	3.00mm		
Inner plate height	7.10mm		

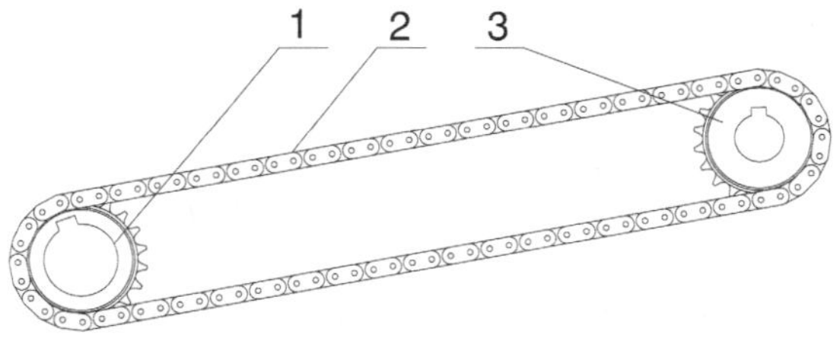

1. Active wheel 2. Chain 3. Driven wheel

Figure 2 Chain drive assembly diagram

3. Dynamic Analysis of Transmission Mechanism

3.1 Force analysis of chain drive

As shown in Figure 2, the chain drive in the work, there are tight side tension and loose pull [5]. If you do not count the dynamic load in the drive, the tight side tension and loose pull force were:

$$F_1 = F_e + F_c + F_f \tag{1}$$

$$F_2 = F_c + F_f \tag{2}$$

Where F_e—Effective circular force(N);

F_c—Centrifugal force caused by the tension(N);

F_f—Overhanging tension(N).

Effective circular force:

$$F_e = \frac{M_2}{r_1} = \frac{M_1}{r_2} \tag{3}$$

Where M_1、M_2—Active wheel,driven wheel required torque(N·m);

r_1、r_2—Active wheel,driven wheel pitch circle radius(m).

The pulling force caused by the centrifugal force is:

$$F_c = qv^2 \tag{4}$$

Where q—The quality of the chain unit length(kg/m);

v—The speed of the chain(m/s).

Assuming the lower part of the hanging chain from the middle of the chain and the chain of the starting point of the moment to take the moment, by the balance of conditions：

$$F_f f = \frac{qga}{2} \times \frac{a}{4} \tag{5}$$

Where q—Length of chain unit length(kg/m);

g—Gravitational acceleration(m/s^2);

a—Chain drive center distance(m);

f—Sagging(m).

IOP Conf. Series: Materials Science and Engineering **281** (2017) 012052 doi:10.1088/1757-899X/281/1/012052

3.2 The force calculation of chain drive

In order to ensure the quality of planting, we must use the operating speed in the operation, usually the speed of the corn planter should not be greater than 7km/h [6]. In this paper, the speed of the corn planter is 5km/h. It can be seen from the design of the whole machine. The diameter of the ground wheel is D = 600mm. The ground wheel is fixedly connected with the chain drive wheel, and the chain speed is 0.12m /s.

The required torque for the normal operation of the planter is 5N·m and the effective circumference of the chain is:

$$F_e = \frac{M_2}{r_1} \approx 195.54N \tag{6}$$

Chain material is commonly used 45Mn, chain linear density is 4.25kg/m, then the centrifugal force caused by the pull:

$$F_c = qv^2 \approx 0.06N \tag{7}$$

When the chain sag is 10mm,the tension of the chain is:

$$F_f \approx \frac{qga^2}{8f} = 40.82N \tag{8}$$

So the tight side of the chain pull fore:

$$F_1 = F_e + F_c + F_f = 236.42N \tag{9}$$

The number of sprockets is 20, the transmission ratio of the chain drive is 1, its wrap angle, so there are 10 teeth within the angle. As the sprockets and chains are rigid, the average load of each sprocket is:

$$F = \frac{2\pi F_1}{\alpha z} \approx 47.28N \tag{10}$$

4. Dynamics Simulation and Result Analysis

4.1 Import models and add constraints

In the SolidWorks software, save the chain drive model as a Parasolid (*. x_t) file, import the file in the ADAMS virtual prototyping system, and perform kinetic analysis of the chain drive. Imported simulation model, the need to set the quality of the model, centroid, inertial tensor, the model to add constraints [7]. The active wheel, the driven wheel and the earth are the rotation sub-constraints, and the link between the link and the link is a rotation pair. The sprocket is engaged with the chain to drive the whole system. Because the chain and the sprocket are engaged with the high side of the line, and the meshing of the sprocket and the chain is complicated, it is difficult to define the motion pair between them. Without affecting the simulation results under the premise of the need to define a chain between the chain and the contact, contact type for the physical contact [8 ~ 9].

After defining the constraint relationship, you also need to determine the contact parameters. Contact parameters: metal and metal stiffness between 1.0E + 008N / m; contact index Force Exponent take 1.5; Damping refers to the maximum damping, take the stiffness of 0.1%. Penetration Depth takes the system default value of 1.0E + 004. Static Coefficient and Dynamic Coefficient of the steel and steel are taken at 0.15 [10-11].

IOP Conf. Series: Materials Science and Engineering **281** (2017) 012052 doi:10.1088/1757-899X/281/1/012052

4.2 Other settings

Solver settings, integrator (General) generally choose the default GSTIFF. When choosing Formulation, select I3 or SI2 according to the different requirements of the simulation. I3 for the default value, simulation robust, high efficiency, and SI2 simulation more stable, effective control divergence, not easy to error, but the speed is 10 times slower than I3. In order to improve the simulation results, the simulation is more stable, this paper chooses SI2. Corrector selects Modified, other settings default [12]. Add the driver to make the simulation, the left sprocket to add the rotation drive. In order to make the simulation results closer to the actual situation, we need to set up the Step function for the drive sprocket, and set the Step function as STEP (time, 0,0,0.1,270d) + STEP (time, 0.9,0,1, -270d).

4.3 Simulation analysis

In order to study its kinetic properties, kinetic simulation is required. Set the simulation time to 1s, Steps set to 200, click to start the simulation. After completion of the simulation, into the post-processing module (PostProcessor), through the measurement and other functions, you can get each component speed, angular velocity, angular acceleration, force and other curves.

As can be seen from Figure 3 and Figure 4, the speed of the link between 0.1 and 0.9 s is slightly fluctuating at 0.12 m/s, approximately uniformly moving and satisfying the chain speed in the above. Indicating that the speed of the chain changes with the actual situation is basically consistent.

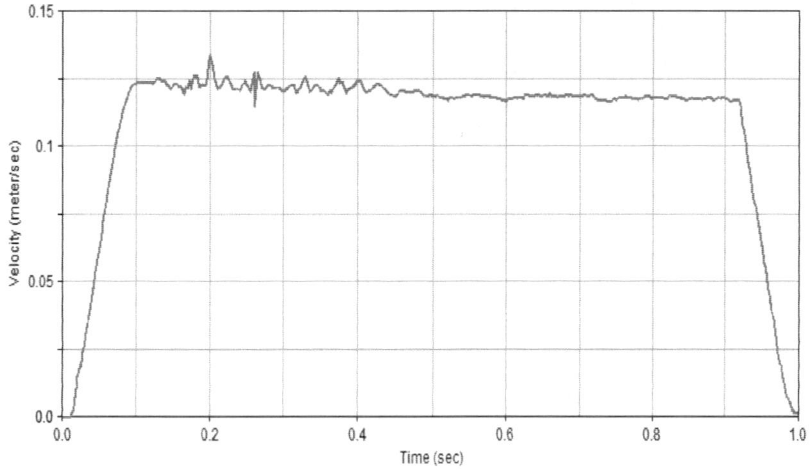

Figure 3 The speed change of the chain tight inner link 34.

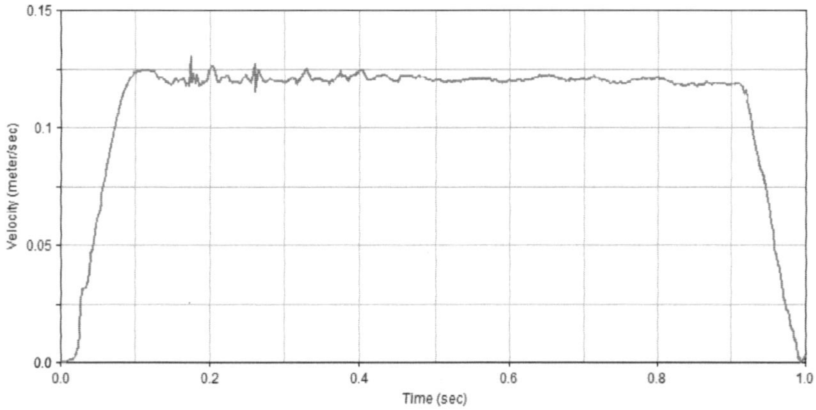

Figure 4 The speed change of the chain loose inner link 100.

As can be seen from Figure 5, the angular velocity of the driven wheel is accelerated from 0s to 0.1s, and decays to 0.9s. There is a significant change in angular velocity between 0.1 and 0.9 s, indicating a change in acceleration and additional dynamic load.

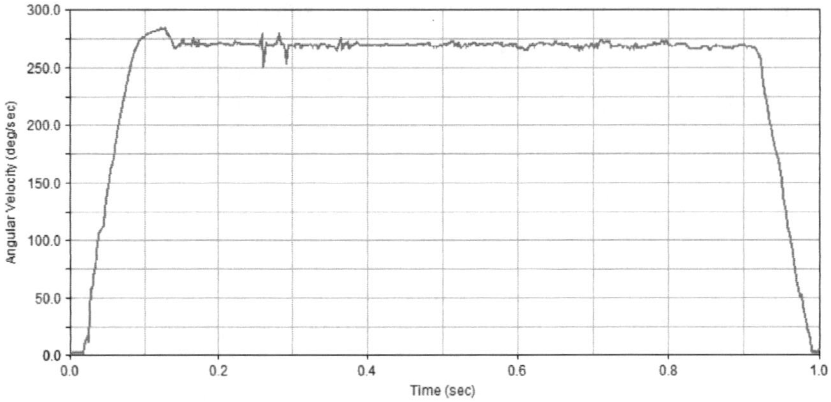

Figure 5 The angular velocity of the driven whee.

It can be seen from Figure 6, the angular acceleration of the driven wheel in the 0~0.5s when there is a significant change, the emergence of a series of vibration peaks, showing a significant impact vibration phenomenon, indicating the contact between the driven wheel and its complex and complex , The chain has a very serious damage, and to speed up the sprocket and chain wear. Therefore, the chain drive mechanism should be relatively smooth start, the maximum speed to maintain a reasonable, in order to reduce the chain and sprocket wear and damage.

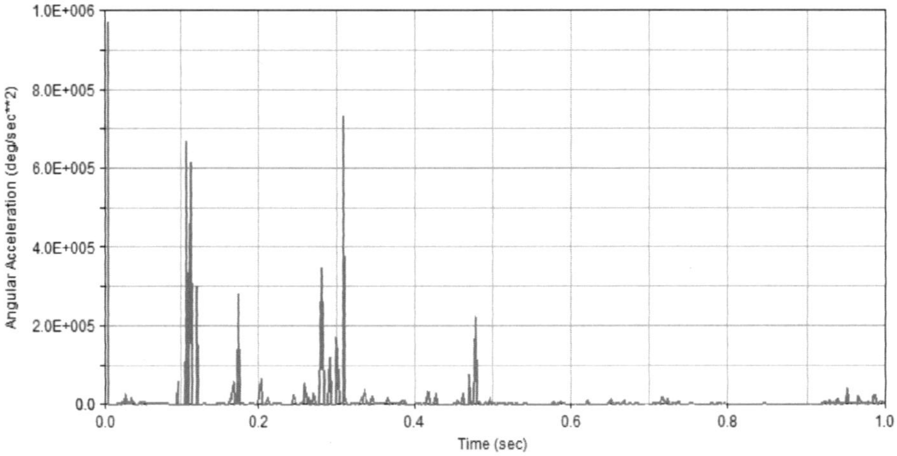

Figure 6 The angular acceleration of the driven wheel.

It can be seen from Figure 7, the contact force between the driven wheel and the inner link 85 occurs between 0.15 and 0.5 s, and the other time is not in contact with the sprocket, the curve is approximately 1/4 sine wave, the maximum is 47.28N, Which changes greatly in the 0.3s, indicating the existence of the drive wheel and the inner section of the wear and impact, resulting in the drive wheel produced a significant vibration. Therefore, in the corn planter work, should be as much as possible to maintain a reasonable uniform movement, to avoid greater acceleration, improve the quality of corn seeding sowing.

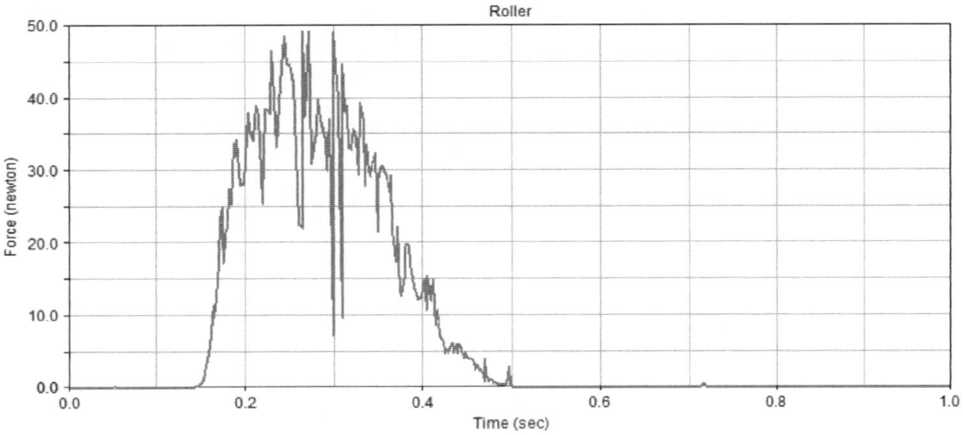

Figure 7 The contact force between the driven wheel and the inner link 85.

5. Conclusions

In this paper, the model of the corn planter and the model of the transmission mechanism are established by the three-dimensional modeling software SolidWorks. The dynamic characteristics of the transmission mechanism are studied by ADAMS. Conclusions are as follows:

1) By the analysis of dynamics of chain drive, the wheel angular velocity from 0s to 0.1s began to accelerate, accelerate to maximum, 0.9s started to slow down, until the 1s down to zero; in the 0.1~0.9s between the angular velocity of obvious change, indicating that the change has occurred in the acceleration and additional dynamic load; chain dynamics simulation of the situation and the actual situation is basically consistent.
2) By the analysis of dynamics of chain drive, the corn planter in the working process, the driven wheel angular acceleration of the chain transmission mechanism show significant changes in 0~0.5s, a series of vibration peak, have obvious impact vibration phenomenon; from the contact force of the driving wheel and the inner link of 85 occurred in the 0.15~0.5s, the maximum value is 47.28N, the complex contact forces resulting from abrasion and impact between the driving wheel and the driven wheel inner link, has obvious vibration phenomenon.
3) In the corn planter work, the amount should maintain a reasonable uniform movement to avoid a greater acceleration and improve the quality of corn seeding sowing. The kinetic analysis of the transmission mechanism of the corn seeder has shortens the development cycle improves the working efficiency, and has some reference significance.

Acknowledgments

This work is financially supported by the Southwest Forestry University Scientific Research Fund (111709), the Southwest Forestry University Science and Technology Innovation Fund Project (K16016) and Yunnan Province eight key industry brand professional construction fund project (51600631).The authors would like to thank foundations for supporting.

References

[1] Gao Y X, Zheng D C. Analysis of existing problems in precision seeding power transmission system [J]. *Shanxi agricultural science*, 2016, 44 (6): 843-847.

[2] Jiao H Z, Yang Zhaojian, Wang Shuping, et al. Structure optimization design of sprocket of scraper conveyor [J]. Journal of Mechanical Strength, 2013 (4): 419-423.

[3] *Mechanical Design Handbook Editorial Board. Mechanical Design Handbook: New Edition* [M]. China Machine Press, 2007.

[4] Yao L Z, Zhang S M. Dynamics simulation analysis of AFC sprocket transmission system and optimization [J]. *Meikuang Jixie mine machinery coal,* 2016, 37 (6): 6-8.

IOP Conf. Series: Materials Science and Engineering **281** (2017) 012052 doi:10.1088/1757-899X/281/1/012052

[5] Lu Y H, Zeng Z W, Ma G X, et al. Finite element analysis of banana harvester with multi degree of freedom and self locking [J]. *Jounrnal of of Agricultural Mechanization Research,* 2016 (3) .
[6] Hou Y Z, Dai Z J.Influence of Sowing Speed on Sowing Quality [J]. *Modernizing Agriculture,* 2012 (6): 57-58.
[7] Xie M, Li C, Mao J, et al. Contact dynamics of chain transmission system of scraper conveyer [J].*Journal of Mechanical Design,* 2017 (5): 58-64.
[8] Pu M H, Zhu X H, Zhang D L. The improvement of chain drive efficiency about the Multi-contact system Simulation based on ADAMS [J]. *Machine Design & Research,* 2016 (3): 60-64.
[9] Mao J, Dong X R, Bai Y J, et al. Reseach on gearing characteristics of the scraper conveyor sprocket and chain hoop [J]. *Journal of Mechanical Strength,* 2017 (2): 341-346.
[10] Liu F, Rui Y N, Ji S S. Simulation of step mechanism based on Pro/E and ADAMS. *Journal of Mechanical Transmission[J]* . 2013 (3): 40-44. [J].
[11] Pan S Q, Cao Z F, Zhao W N. Simulation analysis of chain drive system based on virtual prototype [J]. *Journal of Chinese Agricultural Mechanization,* 2016, 37 (5): 41-44.
[12] Ni J J, Li W B. Simulation analysis of chain transmission mechanism based on ADAMS and Pro-E [J]. *Journal of Chinese Agricultural Mechanization,* 2015, 36 (1): 242-245.

IOP Conf. Series: Materials Science and Engineering **281** (2017) 012053 doi:10.1088/1757-899X/281/1/012053

Research on the Optimization Method of Arm Movement in the Assembly Workshop Based on Ergonomics

X M Hu, H W Qu*, H J Xu, L Yang[1] and C C Yu

The Key Laboratory of Intelligent Manufacturing and Robotics, School of Mechatronic Engineering and Automation, Shanghai University, Mailbox 232, No. 149 yanchang Road, Shanghai, China, 200072.

Email: quhewei10280@163.com.

Abstract: In order to improve the work efficiency and comfortability, Ergonomics is used to research the work of the operator in the assembly workshop. An optimization algorithm of arm movement in the assembly workshop is proposed. In the algorithm, a mathematical model of arm movement is established based on multi rigid body movement model and D-H method. The solution of inverse kinematics equation on arm movement is solved through kinematics theory. The evaluation functions of each joint movement and the whole arm movement are given based on the comfortability of human body joint. The solution method of the optimal arm movement posture based on the evaluation functions is described. The software CATIA is used to verify that the optimal arm movement posture is valid in an example and the experimental result show the effectiveness of the algorithm.

1. Introduction

Workshop assembly usually needs the direct participation of people [1-2]. Due to the randomness and subjectivity of the labor itself, it has become a major problem to improve the efficiency of labor. Ergonomics, as a multidisciplinary subject, considers how to improve work efficiency, human health, safety and comfort, and other factors simultaneously so as to achieve multi-objective optimization [3-5]. In view of the advantage of Ergonomics, it is used to study the labor of operator in workshop assembly and to put forward reasonable improvement measures and suggestions [6-7].

The existing ergonomics evaluation methods collect a large number of human physiological parameters, and use the evaluation criteria, such as fatigue degree and energy consumption degree, to evaluate human movement in order to make corresponding improvement measures [8-10]. However, these evaluation methods do not study the movement of human body, which makes it difficult to improve the movement of human body by the evaluation results.

Therefore, an optimization algorithm of human arm movement in the assembly workshop based on ergonomics is proposed. The mathematical model of human arm movement is established, and the evaluation function of movement posture is defined. The optimal movement posture can be obtained from the evaluation results. Experiments and results show the effectiveness of the algorithm.

2. The optimization algorithm of upper limb movement

2.1. The upper limb kinematics model and its solution

The degrees of freedom of the arm and the coordinate systems are established based on D-H method, shown in Figure 1:

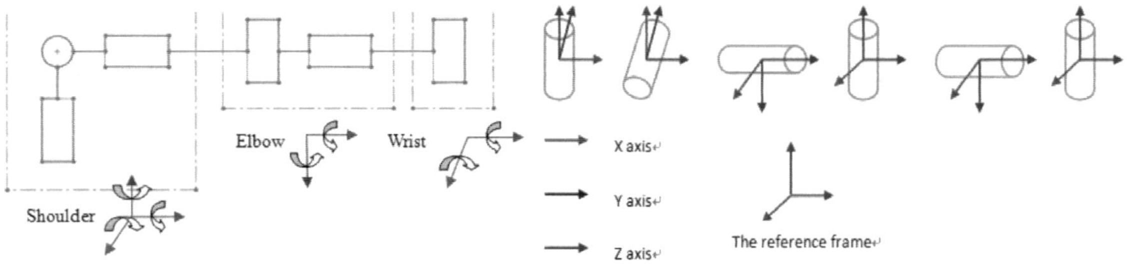

Figure 1. The degrees of freedom of the arm and the coordinate systems based on D-H method

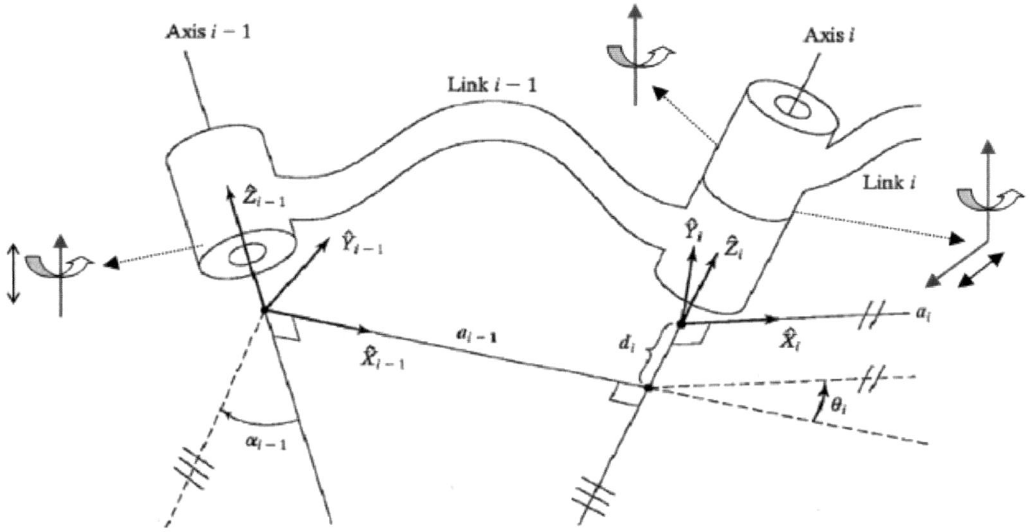

Figure 2. The D-H method of the space connecting rod

D-H method is used to establish the coordinate system of each link, shown in Figure 2. The relative position between the coordinate system $i-1$ and i can use the following four corresponding link parameters to indicate:

(1) Define the distance from the axis Z_{i-1} to the axis Z_i as the connecting rod length a_i, the direction along the axis X_i points in the positive;

(2)Define the angle from the axis Z_{i-1} to the axis Z_i as the coupler-angle α_i $(\alpha_i \in (-\pi, \pi])$, the direction of rotation in the positive of axis X_i points in the positive;

(3) Define the distance from the axis X_{i-1} to the axis X_i as the joint distance d_i, the direction along the axis Z_{i-1} points in the positive;

(4)Define the angle from the axis X_{i-1} to the axis X_i as the joint angle θ_i, the direction of rotation in the positive of axis Z_{i-1} points in the positive;

According to the Figure 1 and Figure 2, the D-H parameters of the arm can be set in Table 1:

Table 1. D-H parameters of the arm

connecting rod i	Joint degree of freedom	Coupler-angle α_{i-1}	Connecting rod length a_{i-1} (mm)	Joint distance d_i (mm)	Joint angle θ_i
1	Shoulder pitch	0°	0	0	θ_1
2	Shoulder swing	-90°	0	0	θ_2
3	Shoulder transfer	-90°	0	d_3	θ_3
4	Elbow pitch	90°	0	0	θ_4
5	Elbow transfer	-90°	0	d_5	θ_5
6	Wrist pitch	90°	0	0	θ_6

The connecting rod transformation formula is defined as follows:

$$^{i-1}T_i = \begin{bmatrix} c\theta_i & -s\theta_i & 0 & a_{i-1} \\ s\theta_i c\alpha_{i-1} & c\theta_i c\alpha_{i-1} & -s\alpha_{i-1} & -d_i s\alpha_{i-1} \\ s\theta_i s\alpha_{i-1} & c\theta_i s\alpha_{i-1} & c\alpha_{i-1} & d_i c\alpha_{i-1} \\ 0 & 0 & 0 & 1 \end{bmatrix} \tag{1}$$

There is a note that c represents $\cos$ and s represents $\sin$ in each formula in this paper.

According to the D-H parameters table and the connecting rod transformation formula, the coordinate transformation matrix of each coordinate system can be defined as follows:

$$^{0}T_1 = \begin{bmatrix} c\theta_1 & -s\theta_1 & 0 & 0 \\ s\theta_1 & c\theta_1 & 0 & 0 \\ 0 & 0 & 1 & 0 \\ 0 & 0 & 0 & 1 \end{bmatrix} \quad ^{1}T_2 = \begin{bmatrix} c\theta_2 & -s\theta_2 & 0 & 0 \\ 0 & 0 & 1 & 0 \\ -s\theta_2 & c\theta_2 & 0 & 0 \\ 0 & 0 & 0 & 1 \end{bmatrix}$$

$$^{2}T_3 = \begin{bmatrix} c\theta_3 & -s\theta_3 & 0 & 0 \\ 0 & 0 & 1 & d_3 \\ -s\theta_3 & -c\theta_3 & 0 & 0 \\ 0 & 0 & 0 & 1 \end{bmatrix} \quad ^{3}T_4 = \begin{bmatrix} c\theta_4 & -s\theta_4 & 0 & 0 \\ 0 & 0 & -1 & 0 \\ s\theta_4 & c\theta_4 & 0 & 0 \\ 0 & 0 & 0 & 1 \end{bmatrix}$$

$$^{4}T_5 = \begin{bmatrix} c\theta_5 & -s\theta_5 & 0 & 0 \\ 0 & 0 & 1 & d_5 \\ -s\theta_5 & -c\theta_5 & 0 & 0 \\ 0 & 0 & 0 & 1 \end{bmatrix} \quad ^{5}T_6 = \begin{bmatrix} c\theta_6 & -s\theta_6 & 0 & 0 \\ 0 & 0 & -1 & 0 \\ s\theta_6 & c\theta_6 & 0 & 0 \\ 0 & 0 & 0 & 1 \end{bmatrix} \tag{2}$$

The equation of the inverse kinematics solution is as follows:

$$^{0}T_6 = \begin{bmatrix} n_x & o_x & a_x & p_x \\ n_y & o_y & a_y & p_y \\ n_z & o_z & a_z & P_z \\ 0 & 0 & 0 & 1 \end{bmatrix} = {}^{0}T_1(\theta_1){}^{1}T_2(\theta_2){}^{2}T_3(\theta_3){}^{3}T_4(\theta_4){}^{4}T_5(\theta_5){}^{5}T_6(\theta_6) \tag{3}$$

IOP Conf. Series: Materials Science and Engineering **281** (2017) 012053 doi:10.1088/1757-899X/281/1/012053

By using the inverse matrix converter of both sides, reconstruct the equation (2) and equation (3), formula (4) can be derived

$$\begin{bmatrix} c_4c_5c_6+s_4s_6 & s_5s_6 & s_4c_5c_6+c_4s_6 & -s_6d_5 \\ -c_4c_5s_6+s_4c_6 & -s_5s_6 & -s_4c_5s_6+c_4s_6 & -c_6d_5 \\ s_5c_4 & -c_5 & s_4s_5 & 0 \\ 0 & 0 & 0 & 1 \end{bmatrix} = \begin{bmatrix} n_x' & o_x' & a_x' & p_x' \\ n_y' & o_y' & a_y' & p_y' \\ n_z' & o_z' & a_z' & P_z' \\ 0 & 0 & 0 & 1 \end{bmatrix} \begin{bmatrix} c_1c_2c_3+s_1s_3 & -c_1c_2c_3+s_1s_3 & -c_1s_2 & -c_1s_2d_3 \\ s_1c_2c_3-c_1s_3 & -s_1c_2c_3-c_1s_3 & -s_1s_2 & -s_1s_2d_3 \\ -s_2c_3 & s_2c_3 & -c_2 & -c_2d_3 \\ 0 & 0 & 0 & 1 \end{bmatrix} \tag{4}$$

According to formula (4), formula (5) can be conducted:

$$\begin{cases} -s_6 = -c_1s_2n_x'' - s_1s_2o_x'' - c_2a_x'' + p_x'' \\ -c_6 = -c_1s_2n_y'' - s_1s_2o_y'' - c_2a_y'' + p_y'' \\ 0 = -c_1s_2n_z'' - s_1s_2o_z'' - c_2a_z'' + p_z'' \end{cases} \tag{5}$$

In the formula (5): $n_x'' = n_x'd_3/d_5, o_x'' = o_x'd_3/d_5, a_x'' = a_x'd_3/d_5, p_x'' = p_x'/d_5$, others just analogy.

By formula (5), all the θ angles are calculated so as to solve the inverse kinematics problem of arm.

2.2. The optimization of arm movement

2.2.1. The simplified equivalent transformation of arm movement

In order to simplify the movement analysis, the arm movement in three-dimensional space is equivalent to a two-dimensional space movement in a particular plane, and the specific method is as follows:

Three Degrees Of Freedom (DOF) of the shoulder joint can be regarded as the rotational movement around three mutually perpendicular axes in a coordinate system. It can be equivalent to a rotational movement around a shaft that is through the origin of the coordinate system.

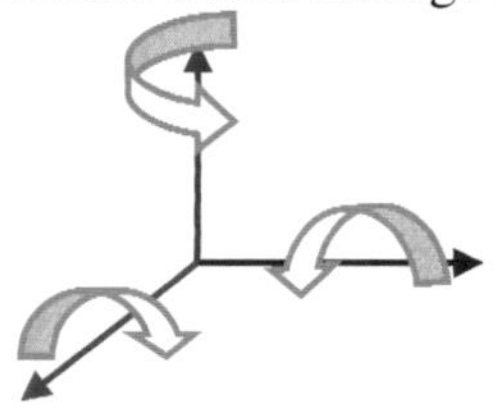

Figure 3. The reference coordinate system {A}

The rotary movement of three degrees of freedom of the shoulder joint uses the reference coordinate system {A}, as shown in Figure 3.

Let $\vec{k} = \begin{bmatrix} k_x & k_y & k_z \end{bmatrix}^T$ be a unit vector through the origin. If the rotation angle around $\vec{k}$ is θ, the rotation transformation matrix is $R(k,\theta)$. After the rotation transformation, the reference coordinate system {A} is transferred into coordinate system {B}. Then there are equations of the rotation transformation matrix as follows:

$${}^A_BR = R(k,\theta) = \begin{bmatrix} k_xk_x vers\theta + c\theta & k_yk_x vers\theta - k_zs\theta & k_zk_x vers\theta + k_ys\theta & 0 \\ k_xk_y vers\theta + k_zs\theta & k_yk_y vers\theta + c\theta & k_zk_y vers\theta - k_xs\theta & 0 \\ k_xk_z vers\theta - k_ys\theta & k_yk_z vers\theta + k_xs\theta & k_zk_z vers\theta + c\theta & 0 \\ 0 & 0 & 0 & 1 \end{bmatrix} \tag{6}$$

In the equation (6): $s\theta = \sin\theta, c\theta = \cos\theta, vers\theta = 1 - \cos\theta$;

$R(k,\theta)$ can also be obtained directly by rotation transformation, shown in equation (7):

$$\begin{aligned} {}_B^A R &= R(Z,\theta_3)R(Y,\theta_2)R(X,\theta_1) \\ &= \begin{bmatrix} \cos\theta_3 & -\sin\theta_3 & 0 & 0 \\ \sin\theta_3 & \cos\theta_3 & 0 & 0 \\ 0 & 0 & 1 & 0 \\ 0 & 0 & 0 & 1 \end{bmatrix} \begin{bmatrix} \cos\theta_2 & 0 & \sin\theta_2 & 0 \\ 0 & 1 & 0 & 0 \\ -\sin\theta_2 & 0 & \cos\theta_2 & 0 \\ 0 & 0 & 0 & 1 \end{bmatrix} \begin{bmatrix} 1 & 0 & 0 & 0 \\ 0 & \cos\theta_1 & -\sin\theta_1 & 0 \\ 0 & \sin\theta_1 & \cos\theta_1 & 0 \\ 0 & 0 & 0 & 1 \end{bmatrix} \end{aligned} \tag{7}$$

k_x, k_y, k_z and θ can be obtained by simultaneous equation (6) and (7).

2.2.2. Evaluation of arm movement based on improved Fitts's law

Fitts [11] proposed Index of Difficulty I_d and Index of Performance I_p based on one dimensional translational movement in the law, its definition is:

$$\begin{cases} I_d^{\theta} = -\log_2(W/2A) \\ I_p^{\theta} = \dfrac{1}{t_m}\log_2(W/2A) \end{cases} \tag{8}$$

In the formula(8): A is the range of movement, W is the width of object held by arm, t_m is the movement time.

Kondraske [12] further extended the Fitts's law to the angle movement, and proposed Index of Difficulty and Index of Performance of the angle movement, its definition is:

$$\begin{cases} I_d^{\theta} = -\log_2(\theta_w/2\theta_A) \\ I_p^{\theta} = \dfrac{1}{t_m}\log_2(\theta_w/2\theta_A) \end{cases} \tag{9}$$

In the formula (9): θ_A is the range of angle movement, θ_W is the object width of angle movement, t_m is the movement time.

The Fitts law based on one dimensional movement has great limitations. Therefore, an improved Fitts law is proposed to analyze and evaluate the two-dimensional movement.

The maximum range and the comfort range of movement about each joint DOF are shown in Table 2. The data in Table 2 are the statistical results obtained by sampling the large range of human movement.

Table 2. The range of movement for each degree of freedom

Joint degree of freedom	The maximum range of motion	The comfort range of motion
Shoulder pitch	0~180°	0~30°
Shoulder swing	-45°~135°	0~90°
Shoulder transfer	-135°~90°	-100°~0
Elbow pitch	0~142°	0~100°
Elbow transfer	-90°~90°	-45°~45°
Wrist pitch	-70°~90°	-25°~45°

When the arm is moving, a series of inverse solutions of θ_i are calculated. In order to evaluate the quality of solutions and select the most suitable set of values of θ, an evaluation criteria is proposed.

The simplified equivalent transformation of arm movement is established based on the parameter in Table2. The projection of the arm in the equivalent plane is shown in Figure4.

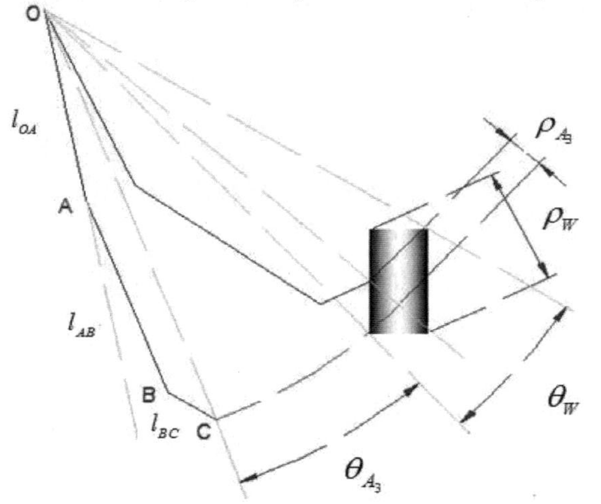

Figure 4. The projection of the arm in the equivalent plane

In Figure 4, O is the center point of the shoulder joint, A is the center point of the elbow joint, and B is the center point of the wrist joint, l_{OA} is the length of forearm, l_{AB} is the length of small arm, l_{BC} is the grasping length from the wrist joint center to the palm center; ρ_W is the maximum length of the object, θ_W is the maximum angle of the object in the equivalent plane, ρ_{A_3} is the motion amplitude of wrist, θ_{A_3} is the motion angle of wrist. In the equivalent plane, the evaluation function of each joint is obtained as follows:

The evaluation function of shoulder joint is defined as :

$$\begin{cases} I_{d_1}{}^{\theta} = -\log_2(\theta_w / 2\theta_{A_1}) \\ I_{p_1}{}^{\theta} = \dfrac{1}{t_m}\log_2(\theta_w / 2\theta_{A_1}) \end{cases} \tag{10}$$

The evaluation function of elbow joint is defined as:

$$\begin{cases} I_{d_2}{}^{\theta} = \log_2(\theta_w / 2\theta_{A_2}) \\ I_{p_2}{}^{\theta} = -\dfrac{1}{t_m}\log_2(\theta_w / 2\theta_{A_2}) \\ I_{d_2}{}^{\rho} = \log_2(\rho_w / 2\rho_{A_2}) \\ I_{p_2}{}^{\rho} = -\dfrac{1}{t_m}\log_2(\rho_w / 2\rho_{A_2}) \end{cases} \tag{11}$$

The evaluation function of wrist joints is defined as :

$$\begin{cases} I_{d_3}{}^{\theta} = -\log_2(\theta_w / 2\theta_{A_3}) \\ I_{p_3}{}^{\theta} = \dfrac{1}{t_m}\log_2(\theta_w / 2\theta_{A_3}) \\ I_{d_3}{}^{\rho} = -\log_2(\rho_w / 2\rho_{A_3}) \\ I_{p_3}{}^{\rho} = \dfrac{1}{t_m}\log_2(\rho_w / 2\rho_{A_3}) \end{cases} \tag{12}$$

in the equations (10) (11) (12), $I_{d_i}{}^{\rho}, I_{d_i}{}^{\theta}$ is the Index of Difficulty of each join, $I_{p_i}{}^{\rho}, I_{p_i}{}^{\theta}$ is the Index of Performance of each joint, ρ_{A_i}, θ_{A_i} is the range and angle of movement for each joint.

According to the evaluation function of each joint, the evaluation function of the whole arm movement can be defined as follows:

$$\begin{cases} I_d{}^{\rho} = \dfrac{I_{d_2}{}^{\rho} + I_{d_3}{}^{\rho}}{2} \\ I_p{}^{\rho} = \dfrac{I_{p_2}{}^{\rho} + I_{p_3}{}^{\rho}}{2} \\ I_d{}^{\theta} = \dfrac{I_{d_1}{}^{\theta} + I_{d_2}{}^{\theta} + I_{d_3}{}^{\theta}}{3} \\ I_p{}^{\theta} = \dfrac{I_{p_1}{}^{\theta} + I_{p_2}{}^{\theta} + I_{p_3}{}^{\theta}}{3} \end{cases} \tag{13}$$

According to the definition, the smaller $I_d{}^{\rho}, I_d{}^{\theta}$ is and the larger $I_p{}^{\rho}, I_p{}^{\theta}$ is, the higher the comfort degree of movement is. This criteria is used to select the optimal arm movement.

The minimum values of $I_d{}^{\rho}$ and $I_d{}^{\theta}$ are defined as $I_d{}^{\rho}{}_{\min}$ and $I_d{}^{\theta}{}_{\min}$; the maximum values of $I_p{}^{\rho}$ and $I_p{}^{\theta}$ are defined as $I_p{}^{\rho}{}_{\max}$ and $I_p{}^{\theta}{}_{\max}$.

The optimization goal λ of arm movement is defined as:

$$\lambda = \frac{\left(\dfrac{I_d{}^{\theta}{}_{\min}}{I_d{}^{\theta}} + \dfrac{I_d{}^{\rho}{}_{\min}}{I_d{}^{\rho}} + \dfrac{I_p{}^{\theta}}{I_p{}^{\theta}{}_{\max}} + \dfrac{I_p{}^{\rho}}{I_p{}^{\rho}{}_{\max}} \right)}{4} \tag{14}$$

According to the definition, the closer to 1 the value of λ is, the closer to the optimal solution the arm movement is.

3. Experiments and results

In order to verify the effectiveness of the arm movement optimization method, the operations in a specific station in assembly workshop are regarded as objects. Using the evaluation function of arm movement, the optimal operations in the specific station can be obtained respectively, and the results are verified in the simulation platform CATIA.

In the example, the simple contact movement in the sagittal plane is a selected as the experimental object, as shown in Figure 5. The object is moved from point A to point C.

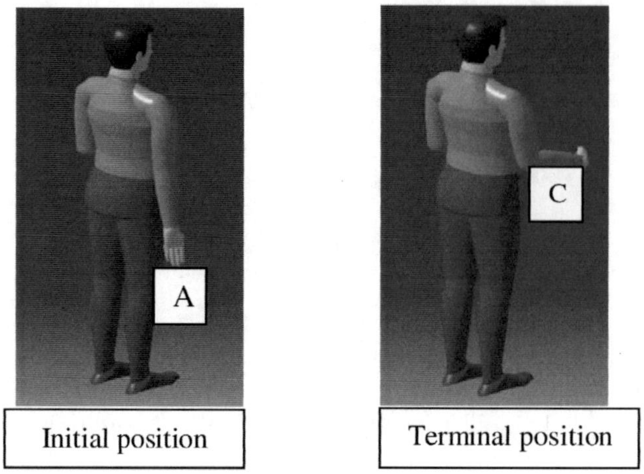

Figure 5. The contact action task

Three joint movement angles are focused on in this movement: shoulder swing angle θ_2, elbow pitch angle θ_4 and wrist pitch angle θ_6. When the movement is finished, the value of ρ_w is 12 , the value of θ_w is 15° and the value of t_m is 3.

(1) The joint angles of the six postures to achieve the task are shown in Table 3.

Table 3. The various postures of joint angle position within the range of body comfort

Posture	Shoulder swing θ_2	Elbow pitch θ_4	Wrist pitch θ_6
1	6°	78°	15°
2	9°	70°	36°
3	11°	64°	45°
4	13°	60°	62°
5	15°	56°	67°
6	17°	51°	71°

(2) The corresponding parameters of each posture projected in the equivalent two-dimensional plane are listed in Table 4:

Table 4. The parameters of each posture in the equivalent two-dimensional plane

Posture	movement angle of shoulder joint θ_{A_1}	motion range of elbow joint ρ_{A_2} (mm)	movement angle of elbow joint θ_{A_2}	motion range of wrist joint ρ_{A_3} (mm)	movement angle of wrist joint θ_{A_3}
1	6°	12.2	43°	15.5	52°
2	9°	10.2	41°	15.5	52°
3	11°	8.5	40°	15.5	52°
4	13°	7.5	39°	15.5	52°
5	15°	6.5	38.5°	15.5	52°
6	17°	5.5	38°	15.5	52°

(3)The parameters in Table 4 are put into the evaluation (10), (11), (12), (13) and (14), the evaluation results of the whole arm movement can be obtained, as shown in Table 5

IOP Conf. Series: Materials Science and Engineering **281** (2017) 012053 doi:10.1088/1757-899X/281/1/012053

Table 5. The evaluation results of the whole arm movement

Posture	I_d^θ	I_p^θ	I_d^ρ	I_p^ρ	λ
1	-0.01592	-0.00531	-1.51192	0.50400	1.00000
2	0.20197	-0.06732	-1.38277	0.46093	3.68490
3	0.31035	-0.10345	-1.25125	0.41709	5.38075
4	0.39674	-0.13225	-1.11157	0.37053	6.75029
5	0.47168	-0.15722	-1.05774	0.35258	7.93431
6	0.53248	-0.17749	-0.95650	0.31989	8.91025

In Table 5, the value of λ in posture 1 is 1. Therefore, the optimal solution is posture 1 according to the definition of evaluation function.

Take the first posture as an example, and the assessment result in CATIA is shown in Figure 6.

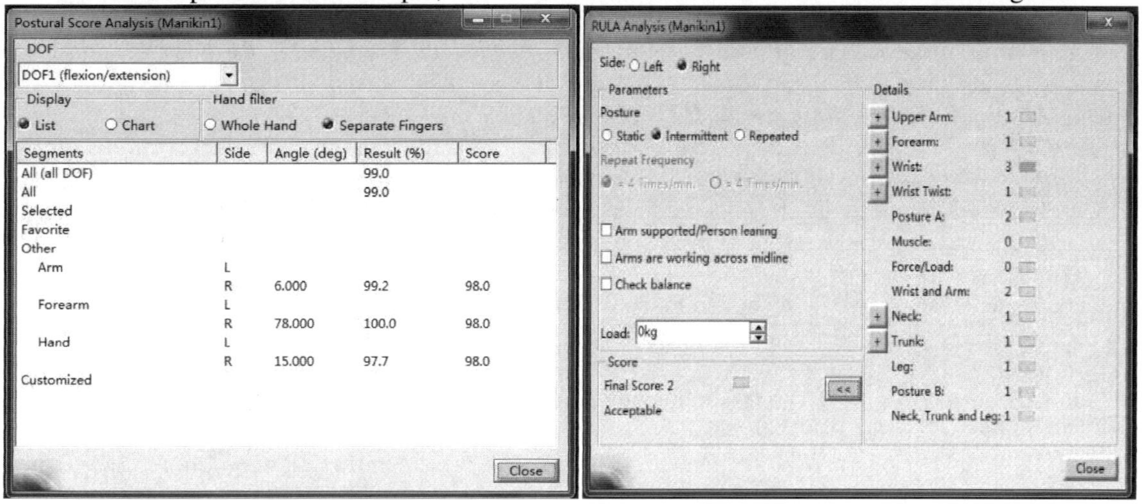

Figure 6. The assessment result of the first posture

The assessment result of six postures in CATIA is shown in Table 6.

Table 6. The assessment result of six postures in CATIA

Posture	Score of upper arm posture evaluation	Score of forearm posture evaluation	Score of wrist posture evaluation	Score of the whole posture evaluation
1	99.3	100	97.7	99.0
2	99.2	100	96.6	98.6
3	99.0	100	91.5	96.8
4	96.9	100	88.7	95.2
5	94.8	100	86.8	93.9
6	92.7	100	59.8	84.2

In Table 6, the assessment result of the first posture is the best because the first posture has the highest score, which is in accordance with the result obtained by the optimization method of arm movement.

4. Conclusions

An optimization algorithm of arm movement based on ergonomics is proposed to improve the work efficiency in the assembly workshop. The mathematical model of arm movement is established, and the evaluation function of arm movement is defined. The optimal movement posture is solved based on the evaluation function, and it is also verified by CATIA. In the future, combined with the mathematical model of human body movement, the optimization algorithm of human body movement based on ergonomics should be studied. The movement of operators can be optimized in the assembly workshop based on the algorithm to improve the work efficiency.

Acknowledgements:
This project is supported by Shanghai Grand Science and Technology Program of China (Grant No. 16111105900), Shanghai Municipal Commission of Economy and Informatization (NO.160646) and Shanghai University School Level Program.

References

[1] Zuo Y H. 2014 The status and development trends of modern manufacturing system's research. *Modern Manufacturing Engineering*. **4**, p0138-05.
[2] Pittaway L., Robertson R, Munir K, Denyer D, Neely D. 2004 Networking and innovation: a systematic review of the evidence. *International Journal of Management Reviews*.**56**, p137-168.
[3] Hu B, Ma L, Zhang W, et al. 2011 Predicting real-world ergonomic measurements by simulation in a virtual environment. *International Journal of Industrial Ergonomics*. **41**, p6471.
[4] Zhou Q X, Wang C H. 2009 Review on Body Fatigue Assessment of Human Operation.*Space Medicine & Medical Engineering*.**22**, p226-230.
[5] Arunkumar Pr, Anil M. 2003 Worker mobility and training in advanced manufacturing. *International Journal of Industrial Ergonomics*.**32**, p363-388.
[6] Cerchione R, Esposito E. 2016 A systematic review of supply chain knowledge management research: State of the art and research opportunities. *International Journal of Production Economics*. **182**, p276-292.
[7] Centobelli P, Cerchione R., Esposito E. 2017 Environmental sustainability in the service industry of transportation and logistics service providers: Systematic literature review and research directions. *Transportation Research Part D: Transport and Environment*. **53**, p454-470.
[8] Guo X B, Guo H L. 2009 The research of modern operation fatigue. *Journal of Safety Science and Technology*.**5**, p189-192.
[9] Li L R, Shen J, Hou B W. 2009 On the Atheletic Fatigue from the View of Chinese Medicine. *Sports Forum*.**3**, p72-84.
[10] Shang Y G, Du L Z. 2014 Based on the Human Factors Analysis and Optimization of the Working Environment. *Logistics Engineering and* Management, **36**, p116-118.
[11] Fitts P M. 1954 The information capacity of the human motor system in controlling the amplitude of movement. *J. Exp. Psychol*. **47**, p381-391.
[12] Kondraske G V. 1994 Angular motion Fitts' Law for human performance modeling and prediction. *Proceeding of the 17th international conference of the IEEE-EMBS*. **16**, p307-308

IOP Conf. Series: Materials Science and Engineering **281** (2017) 012054 doi:10.1088/1757-899X/281/1/012054

Analysis of the Identification Principle of Yaw Error of Five-axis Machine Tool Rotary Table in the Virtue Error Sensitive Direction Based on the Machining Test

Y Zhang[1*] and L Zhang[2]

[1] Zhejiang University of Science and Technology, Hangzhou, 310027, China
[2] Yangzhou Polytechnic College, Yangzhou, 225127, China

E-mail: yazhang1982@126.com

Abstract. The identification principle of the yaw error of five-axis machine tool rotary table in the virtue error sensitive direction by DBB is analysed. According to the measurement principle of DBB, the virtue error sensitive direction was adopted, and a machining test to identify the yaw error of the rotating table is designed and discussed. The function relationship of the yaw error and the machining error in the corresponding sensitive direction was deduced, and the analysis shows that the yaw error can be separated from the machining test.

1. Introduction

Error identification method has been used to identify the kinematic errors in single part of five-axis machine tool. Sensor and information technology has been applied in the area of error identification [1-3]. The existing identification methods of the kinematic error of five-axis machine tool almost adopted the special instruments, such as telescoping double ball bar[4-8], R-test[9,10], and so on. They have the disadvantages of computational complexity, time-consuming of operation, high cost, low universality, and so on.

Double ball bar (DBB) is a one-dimensional displacement measurement sensor, as can be seen from Figure 1. The error measurement method by DBB is shown in Figure 2.There is a high-precision steel ball in each end of DBB. One end of the ball is fixed; the other end of the ball is connected to a high-precision displacement sensor. One end of the ball is adsorbed on the spindle, while the other end of the ball is adsorbed on the table. When the rotary table of the machine tool did the circular interpolation movement relative to the spindle, the displacement change between two balls in the bar direction is transported to the computer. The characteristic of DBB measurement is that the ball bar always measuring the error movement in a certain direction, which can be called the sensitive direction of the ball bar, as can be seen in Figure 3. By designing some measurement patterns which are sensitive to the rotary errors of the machine tool table, DBB was used in the measurement of the tilt errors of five-axis machine tool table A-axis[5].

According to the measuring principle of DBB, this paper proposed the concept of virtue sensitive direction vector, and adopted the sensitive direction vector to design a machining test to identify the yaw error of the rotary table.

IOP Conf. Series: Materials Science and Engineering **281** (2017) 012054 doi:10.1088/1757-899X/281/1/012054

Figure.1 Renishaw QC20 ballbar

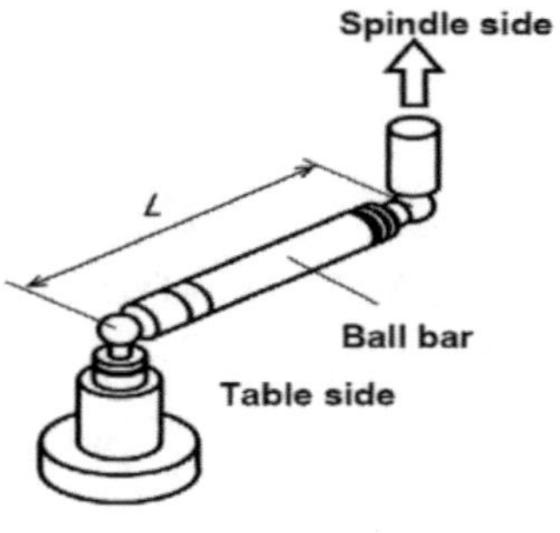

Figure.2 Error measurement method by double ball bar

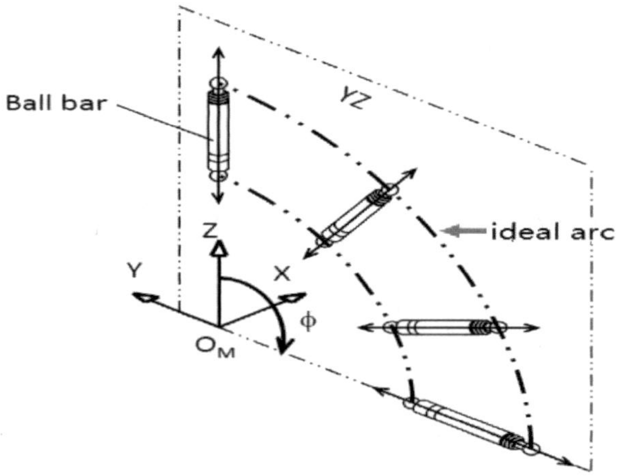

Figure.3 Error sensitive direction of DBB

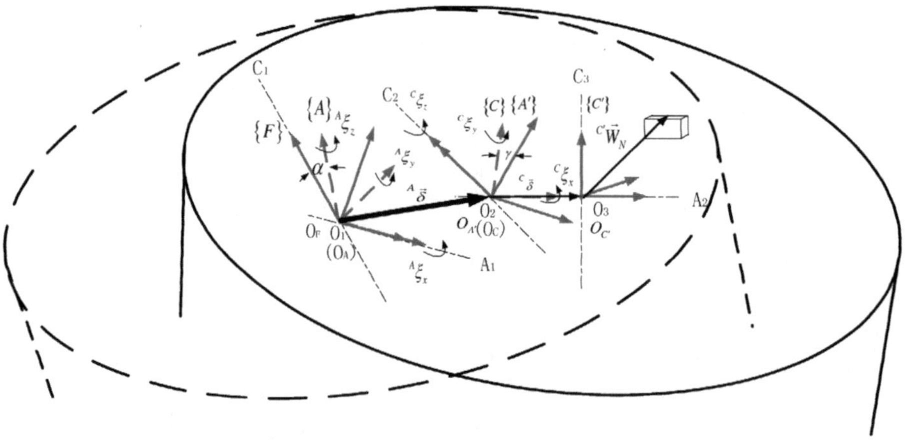

Figure.4 The tilt model of the rotary table

2. Analysis of the error sensitivity of the machining test

The tilt model of the rotary table is shown in Figure 4. The kinematic errors of C-axis can be separated from similar machining tests when A-axis remains stationary [10]. Therefore, in the kinematic chain, only the translation from the coordinate{C} to {A} should be considered. In Figure 4, the translation vector is coarsely represented. The yaw error is one of the rotary errors in the rotating table.

(1) Location of the workpiece

The initial position of the workpiece is in +Y. The distance of the workpiece from the original point in the direction of +Y is L. Suppose that the value of Z-axis of the machining plane is 0, and the workpiece position can be described as[11]

$$ {}^{C'}\vec{W}_N = \begin{bmatrix} 0 \\ L \\ 0 \end{bmatrix} \quad (1) $$

To increase the error sensitivity, the workpiece should be fit on the table as far as possible from the centre point of the table. So the value of L should be as big as possible.

(2) Machining test design and analysis

When α =90 degree, $\gamma = 0$ degree, cut the workpiece on the outer side along +X direction. When α =90 degree, $\gamma = 180$ degree, cut the workpiece on the outer side along +X direction, too. The machining difference between these two positions has the biggest error sensitivity to the kinematic errors in the virtue sensitivity direction.

According to the principle robotic kinematic, the positon vector of the workpiece in the coordinate system {F} can be obtained:

$$ {}^{F}\vec{W} = {}^{F}_{A}R[{}^{A}\vec{\delta}(\alpha) + {}^{A}_{A'}R\,{}^{A'}_{C}R[{}^{C}\vec{\delta}(\gamma) + {}^{C}_{C'}R\,{}^{C'}\vec{W}_N]] \quad (2) $$

Where

$$ {}^{F}_{A}R = \begin{bmatrix} C\alpha & -S\alpha & 0 \\ S\alpha & C\alpha & 0 \\ 0 & 0 & 1 \end{bmatrix} \quad (3) $$

$$ {}^{A}\vec{\delta}(\alpha) = \begin{bmatrix} {}^{A}\delta_x(\alpha) \\ {}^{A}\delta_y(\alpha) \\ {}^{A}\delta_z(\alpha) \end{bmatrix} \quad (4) $$

$$ {}^{A}_{A'}R = \begin{bmatrix} 1 & -{}^{A}\xi_z(\alpha) & {}^{A}\xi_y(\alpha) \\ {}^{A}\xi_z(\alpha) & 1 & -{}^{A}\xi_x(\alpha) \\ -{}^{A}\xi_y(\alpha) & {}^{A}\xi_x(\alpha) & 1 \end{bmatrix} \quad (5) $$

$$ {}^{A'}_{C}R = \begin{bmatrix} 1 & 0 & 0 \\ 0 & C\gamma & -S\gamma \\ 0 & S\gamma & C\gamma \end{bmatrix} \quad (6) $$

IOP Conf. Series: Materials Science and Engineering **281** (2017) 012054 doi:10.1088/1757-899X/281/1/012054

$$ {}_{C'}^{C}R=\begin{bmatrix} 1 & 0 & {}^{C}\xi_y(\gamma) \\ 0 & 1 & -{}^{C}\xi_x(\gamma) \\ -{}^{C}\xi_y(\gamma) & {}^{C}\xi_x(\gamma) & 1 \end{bmatrix} \tag{7} $$

$$ {}^{C}\vec{\delta}(\gamma)=\begin{bmatrix} {}^{C}\vec{\delta}_x(\gamma) \\ {}^{C}\vec{\delta}_y(\gamma) \\ 0 \end{bmatrix} \tag{8} $$

Through matrix calculation, we can obtain:

$$ {}^{F}\vec{W}=\begin{bmatrix} {}^{A}\vec{\delta}_x(\alpha)+\cos\gamma*{}^{C}\delta_x(\gamma)-\sin(\gamma)*[{}^{C}\delta_y(\gamma)+\mathrm{L}]- \\ {}^{A}\xi_z(\alpha)*\{\sin\gamma*{}^{C}\delta_x(\gamma)+\cos\gamma*[{}^{C}\delta_y(\gamma)+\mathrm{L}]\}+{}^{A}\xi_y(\alpha)*{}^{C}\xi_x(\gamma)*L; \\ \cos\alpha*\begin{pmatrix} {}^{A}\delta_y(\alpha)+{}^{A}\xi_z(\alpha)*\left(\cos\gamma*{}^{C}\vec{\delta}_x(\gamma)-\sin\gamma*\left({}^{C}\vec{\delta}_y(\gamma)+\mathrm{L}\right)\right)+ \\ \sin\gamma*{}^{C}\delta_x(\gamma)+\cos\gamma*\left({}^{C}\vec{\delta}_y(\gamma)+\mathrm{L}\right)-{}^{A}\xi_x(\alpha)*{}^{C}\xi_x(\gamma)*\mathrm{L} \end{pmatrix}- \\ \sin\alpha*\begin{pmatrix} {}^{A}\delta_z(\alpha)-{}^{A}\xi_y(\alpha)*\left(\cos\gamma*{}^{C}\delta_x(\gamma)-\sin\gamma*\left({}^{C}\vec{\delta}_y(\gamma)+\mathrm{L}\right)\right)+ \\ {}^{A}\xi_x(\alpha)*\left(\sin\gamma*{}^{C}\delta_x(\gamma)+\cos\gamma*\left({}^{C}\vec{\delta}_y(\gamma)+\mathrm{L}\right)\right)+{}^{C}\xi_x(\gamma)*\mathrm{L} \end{pmatrix}; \\ \sin\alpha*\begin{pmatrix} {}^{A}\delta_y(\alpha)+{}^{A}\xi_z(\alpha)*\left(\cos\gamma*{}^{C}\vec{\delta}_x(\gamma)-\sin\gamma*\left({}^{C}\vec{\delta}_y(\gamma)+\mathrm{L}\right)\right)+ \\ \sin\gamma*{}^{C}\delta_x(\gamma)+\cos\gamma*\left({}^{C}\vec{\delta}_y(\gamma)+\mathrm{L}\right)-{}^{A}\xi_x(\alpha)*{}^{C}\xi_x(\gamma)*\mathrm{L} \end{pmatrix}+ \\ \cos\alpha*\begin{pmatrix} {}^{A}\delta_z(\alpha)-{}^{A}\xi_y(\alpha)*\left(\cos\gamma*{}^{C}\delta_x(\gamma)-\sin\gamma*\left({}^{C}\vec{\delta}_y(\gamma)+\mathrm{L}\right)\right)+ \\ {}^{A}\xi_x(\alpha)*\left(\sin\gamma*{}^{C}\delta_x(\gamma)+\cos\gamma*\left({}^{C}\vec{\delta}_y(\gamma)+\mathrm{L}\right)\right)+{}^{C}\xi_x(\gamma)*\mathrm{L} \end{pmatrix} \end{bmatrix} \tag{9} $$

Suppose that $\vec{r}=[0 \quad -1 \quad 0]$ is the virtue sensitive direction expressed as a unit vector according to the measurement principle of DBB. Adopt the error sensitive vector in formula (9), we can obtain:

$$Z=\begin{bmatrix} {}^{A}\vec{\delta}_x(\alpha)+\cos\gamma*{}^{C}\delta_x(\gamma)-\sin(\gamma)*[{}^{C}\delta_y(\gamma)+L]- \\ {}^{A}\xi_z(\alpha)*\{\sin\gamma*{}^{C}\delta_x(\gamma)+\cos\gamma*[{}^{C}\delta_y(\gamma)+L]\}+{}^{A}\xi_y(\alpha)*{}^{C}\xi_x(\gamma)*L; \\ \cos\alpha*\begin{pmatrix} {}^{A}\delta_y(\alpha)+{}^{A}\xi_z(\alpha)*\left(\cos\gamma*{}^{C}\vec{\delta}_x(\gamma)-\sin\gamma*\left({}^{C}\vec{\delta}_y(\gamma)+L\right)\right)+ \\ \sin\gamma*{}^{C}\delta_x(\gamma)+\cos\gamma*\left({}^{C}\vec{\delta}_y(\gamma)+L\right)-{}^{A}\xi_x(\alpha)*{}^{C}\xi_x(\gamma)*L \end{pmatrix}- \\ \sin\alpha*\begin{pmatrix} {}^{A}\delta_z(\alpha)-{}^{A}\xi_y(\alpha)*\left(\cos\gamma*{}^{C}\delta_x(\gamma)-\sin\gamma*\left({}^{C}\vec{\delta}_y(\gamma)+L\right)\right)+ \\ {}^{A}\xi_x(\alpha)*\left(\sin\gamma*{}^{C}\delta_x(\gamma)+\cos\gamma*\left({}^{C}\vec{\delta}_y(\gamma)+L\right)\right)+{}^{C}\xi_x(\gamma)*L \end{pmatrix}; \\ \sin\alpha*\begin{pmatrix} {}^{A}\delta_y(\alpha)+{}^{A}\xi_z(\alpha)*\left(\cos\gamma*{}^{C}\vec{\delta}_x(\gamma)-\sin\gamma*\left({}^{C}\vec{\delta}_y(\gamma)+L\right)\right)+ \\ \sin\gamma*{}^{C}\delta_x(\gamma)+\cos\gamma*\left({}^{C}\vec{\delta}_y(\gamma)+L\right)-{}^{A}\xi_x(\alpha)*{}^{C}\xi_x(\gamma)*L \end{pmatrix}+ \\ \cos\alpha*\begin{pmatrix} {}^{A}\delta_z(\alpha)-{}^{A}\xi_y(\alpha)*\left(\cos\gamma*{}^{C}\delta_x(\gamma)-\sin\gamma*\left({}^{C}\vec{\delta}_y(\gamma)+L\right)\right)+ \\ {}^{A}\xi_x(\alpha)*\left(\sin\gamma*{}^{C}\delta_x(\gamma)+\cos\gamma*\left({}^{C}\vec{\delta}_y(\gamma)+L\right)\right)+{}^{C}\xi_x(\gamma)*L \end{pmatrix} \end{bmatrix}\cdot\begin{bmatrix}0 & -1 & 0\end{bmatrix} \quad (10)$$

Suppose that:

When α =90 degree, $\gamma=180$ degree, the center position of the workpiece is Y_2.
When α =90 degree, $\gamma=0$ degree, the center position of the workpiece is Y_1.
h= Y_2- Y_1, we can obtain:

$$h=-2\cdot{}^{A}\xi_x({}^{C}\delta_y+L)+2\cdot{}^{A}\xi_y\cdot{}^{C}\delta_x \quad (11)$$

Where, the value of ${}^{A}\xi_y\cdot{}^{C}\delta_x$ is tiny and can be ignored. The value of ${}^{C}\delta_y$ can be identified by our previous work [12]. So formula (11) only include one kinematic error ${}^{A}\xi_x(\alpha)$ of A-axis.

The physical meaning of the machining test is analyzed as follows:
Formula (11) is the distance between the centre point of the workpiece when the rotary angle is 0 degree and the centre point of the workpiece when the rotary angle is 180 degree. Therefore, in theory, ${}^{A}\xi_x(\alpha)$ can be identified from this machining test. There may be several possible machining tests which are sensitive to one kinematic error in theory. To improve the efficiency and precision of the error identification, further analysis and experiment should be carried out to compare and choose a better machining test pattern.

3. Conclusions

To identify the kinematic error of the yaw error of five-axis machine tool rotary table, a machining test pattern was designed to identify the yaw error of the rotating table in the virtue error sensitive direction. The virtue error sensitive direction was adopted to separate the error, and the separation principle of the yaw error is analysed. The function relationship of the yaw error and the machining error in the corresponding sensitive direction was deduced, and the yaw error can be separated from the machining test in theory. The physical meaning of the identification result was analyzed. It provides a theoretical evidence to design the machining tests for the rotary error identification for five-axis machine tool.

IOP Conf. Series: Materials Science and Engineering **281** (2017) 012054 doi:10.1088/1757-899X/281/1/012054

Acknowledgment
This work was financially supported by Startup Foundation of Zhejiang University of Science and Technology (F701102H02).

References

[1] Nitulescu T, Talu S, *Applications of descriptive geometry and computer aided design in engineering graphics*, Cluj-Napoca, Romania, Risoprint Publishing house, 2001. ISBN 973-656-102-X.
[2] Tsutsumi M. and Saito A, Identification and compensation of systematic deviations particular to 5-axis machining centers. *International Journal of Machine Tools and Manufacture,* 2003. 43(8): p. 771-780.
[3] Mahbubur R M, Heikkala J, Lappalainen K, and Karjalainen J A, Positioning accuracy improvement in five-axis milling by post processing. *Int. J. of Machine Tools and Manufacture*, 1997.37(2): p. 223-236.
[4] Abbaszaheh-Mir Y, Mayer J R R, Clotier G, and Fortin C, Theory and simulation for the identification of the link geometric errors for a five-axis machine tool using a telescoping magnetic ball bar. *Int. J. of Production Research*, 2002. 40(18): p. 4781-4797.
[5] Zargarbashi S.and Mayer J, Assessment of machine tool trunnion axis motion error, using magnetic double ball bar. *Int. J. of Machine Tools and Manufacture*, 2006. 46(14): p.1823-1834.
[6] Lei W, Sung M, Liu W, and Chuang Y, Double ball bar test for the rotary axes of five-axis CNC machine tools. *Int. J. of Machine Tools and Manufacture,* 2007.47(2): p. 273-285.
[7] Dassanayake K, Cui C and Tsutsumi M, Accuracy evaluation method for multi-tasking turning centre. *Int. J. Mechatronics and Manufacturing Systems,* 2011.4(3/4): p. 285-303.
[8] Yukitoshi I, Ball Bar Measurement on Machine Tools with Rotary Axes. *Int. J. of Automation Technology*, 2012.6(2): p. 180-187.
[9] Bringmann B and Knapp W, Model-based 'Chase-the-Ball' calibration of a 5-axis machining center. *CIRP Annals – Manufacturing Technology,* 2006. 55(1): p.531–534.
[10] Weikert S, R-test, a new device for accuracy measurements on five axis machine tools. *Cirp Annals-Manufacturing Technology*, 2004. 53(1): p. 429-432.
[11] Xiong Y L (1993) *Robotics*. China Machine Press, Beijing
[12] Zhang Y, Fu J Z, Chen Z C. Machining tests to identify kinematic errors of machine tool table rotation axis based on sensitive directions[J]. *The International Journal of Advanced Manufacturing Technology,* 2013, 67(1–4): 495–500.

Optimal the Process Variables of Internal Grinding for Shrink Fit Tool Holder

Y C Lin[1,*], G H Yang[2], H M Chow[1] and L T Tsai[1]

[1]Department of Mechanical Engineering, Nan Kai University of Technology, Nantou County 54243, Taiwan
[2]Department of Industrial Management, Nan Kai University of Technology, Nantou County 54243, Taiwan

[*]Email: ycline@nkut.edu.tw

Abstract. The goal of this study was to investigate the optimal variables settings in internal grinding for shrink fit tool holder. The effects of the main grinding variables on finishing responses were comprehensively investigated based on Taguchi method. The main process variables such as abrasive grain size (AGS), grinding wheel speed (GWS), feed rate for rough (FRR), depth for rough (DR), feed rate for finishing (FRF), depth for finishing (DF), dressing speed (DS), and workpiece rotational speed (WRS) were chosen to determine their effects on grinding performances related to roundness and surface roughness in the internal grinding of the shrink fit tool holder. The experimental response values were transferred to signal-to-noise (S/N) ratios, and then the significant process variables associated with the grinding performance were examined by analysis of variance (ANOVA). In addition, the optimal combination levels of the grinding variables were also obtained from the response graphs of S/N ratios to improve the surface quality of internal grinding. The experimental results show that the significant grinding variable affecting the roundness was FRF. In addition, the significant variables with regard to surface roughness were AGS, GWS, FRR, DR, DF, and DS. The optimal grinding variable settings were achieved. Consequently, the finishing characteristics of internal grinding could be enhanced to meet the requirements of modern manufacturing applications.

1. Introduction

The shrink fit tool holder utilizes the expansion and contraction properties of metal to provide extremely powerful tool holding. The tool holder will be expanded due to heating operation, and then the cutting tool fixed within the tool holder will be fastened and clamped as the cooling procedure. The cutting tool released from the holder should be reiterated the heating and cooling operations. Thus, the tool holder materials should have the properties such as splendid hardness, strength, wear resistance, and thermal property to withstand the elevated temperature and cutting force during the cutting applications. The depth of grinding and the feed rate in grinding process for tool holder finishing should be tightly adjusted to obtain the desirable finishing effects. Thus, the grinding process is a time-consuming, and the operation cost is inevitably increasing. The small holes grinding would be faced more challenges due to the lack of the rigidity with slender wheel stem. The shape and dimensional accuracy would encounter severe challenges and would receive the failure results occasionally with the small hole grinding process.

The grinding is the most important finishing process for precision components to attain the high surface quality and well dimensional accuracy. The investigation for understanding the machining mechanisms and characteristics in grinding process is an interesting issue for precision engineering manufacture [1-6]. Zhong et al.[7] conducted a study on grinding and pointed out that the grinding process could ensure the components with high dimensional and shape accuracy, so the grinding process was widely used in surface finishing. Qian [8] explored the machining effects on grinding wheel with electrolyte dressing and truing, and then discussed the machining performance by using the grinding wheel for ceramics and metals. The experimental results indicated that the grinding process would obtain the mirror-like surface integrities. Opoz and Chen [9] analyzed the scratch types using cubic boron nitride (CBN) abrasive on workpiece surface. They also discussed the removed mechanism of grinding process and tried to ameliorate the grinding effects and surface quality in grinding process. Gu et al.[10] conducted a scratching test on BK7 optical glass to analysis the chip formation and the material removed mechanisms during the grinding process. Liu et al.[11] examined the grinding effects of process variables on machining characteristics and operation cost by using CBN and Al_2O_3 grinding wheel for nickel-based super alloy. Fathalla et al.[12] studied the finishing effects of various abrasives and cooling methods with different grinding speed for AISI D2 tool steel. The residual stress, surface quality, and machining efficiency were comprehensively discussed in their experimental works. Several researchers also used optimization approach for grinding process to investigate the effects of process variables on grinding characteristics [13-16]. The optimal grinding variables could be obtained from their optimization approach. In general, the grinding process was a complicated machining process and the amount of controlled variables was enormous. The actions were very complex during the grinding process. In this study, the significant variables and optimal combination levels of grinding variables would be attained economically and timely to improve the machining efficiency and surface quality. The Taguchi method was a powerful approach used in the design of experiment and in determining the effect of each machining variable. Moreover, Taguchi method used a specially designed regime called orthogonal array, which nowadays is a predominant approach in the design of experiment. It has been proved that the experiment conducted using Taguchi method is more efficient and less costly to investigate the effects of the entire machining variables. The variable design via Taguchi method can also optimize the machining characteristics through setting process variables and reducing the sensitivity of the system performance to sources of variation. Therefore, high quality of machining characteristics can be achieved without increasing the operating cost [17,18]. In this investigation, an L18 orthogonal array based on the Taguchi experimental design method was employed in determining the effects of essential input variables of the internal grinding for shrink fit tool holder on the roundness and the surface roughness. Moreover, the optimal machining input variables for grinding SKD 61 tool steels could be achieved. A sophisticated grinding process with high efficiency and high quality of surface integrity was attained to fit the modern industrial requirements.

2. Experimental method

The grinding characteristics of internal face regarding shrink fit tool holder were investigated in this experimental study. The experimental works were implemented according to an L18 orthogonal array based on Taguchi method. The main process variables such as AGS, GWS, FRR, DR, FRF, DF, DS, and WRS were chosen to determine their effects on grinding characteristics relating to the roundness and the surface roughness in the internal surface grinding of the shrink fit tool holder. Moreover, the experimental response values were transferred to signal-to-noise (S/N) ratios, and then the significant input variables were determined by ANOVA. In addition, the optimal combination levels of the grinding input variables were also achieved from the response graph of S/N ratios.

2.1. Experimental materials

The shrink fit tool holder used the expansion and contraction properties of metal to provide extremely powerful tool holding. The SKD 61 tool steel had excellent toughness, wear resistance, and thermal property for frequently heating and cooling. Therefore, the SKD 61 steel was widely employed in the tool manufacture industry. In this experimental study, the workpiece adopted SKD 61 tool steel with hardness HRC 58-62 after a specific heat treatment. The initial hole diameter was ϕ5.7 mm with depth 31 mm, and the desired hole diameter after grinding process was ϕ6 mm. Moreover, the internal grinding wheels were made of CBN abrasives with diameter ϕ5.7 mm and thickness 6 mm.

2.2. Experimental equipment and procedures

The machining characteristics such as the roundness and the surface roughness were chosen to evaluate the effects of input variables associated with the internal grinding for shrink fit tool holder. The experimental works was conducted by a high speed internal grinding machine (Tsugami UCG). The response values of roundness were measured by a roundness tester (Tokyo Semitsu Co., RONDCOM 44 DX/SD) at specific hole depths (2 mm, 10 mm, and 18 mm), and each depth implemented five roundness values. In addition, the surface roughness was measured by a precision profile meter (DIAVITE DH-5) to evaluate the surface quality of the machined surface. The measuring length was 4.8 mm and the cut-off was set at 0.25 mm. The value of surface roughness (Rz) was obtained by taking the average of five measurements that were stochastically acquired from the different positions of the internal grinding surface for SKD 61 at each machining condition.

2.3. Taguchi method

A systematic and statistical approach was effective and efficient for optimizing the machining variables of a complicated process. Therefore, a design of experiment method was adopted to optimize the variable settings in this work. The experimental design was developed according to an L18 orthogonal array based on the Taguchi method while using the Taguchi orthogonal array would markedly reduce the experimental numbers. The L18 orthogonal array has eight columns and 18 rows, so it has 17 degrees of freedom to manipulate one variable with two levels and seven variables with three levels. Thus, eight machining variables can be appointed to the columns and the rows designated by 18 experiments with various combination levels of the machining variables. The main process variables such as AGS, GWS, FRR, DR, FRF, DF, DS, and WRS were chosen to determine their effects on grinding performance relating to the roundness and the surface roughness in the internal surface grinding of the shrink fit tool holder. In this investigation, only six machining variables were considered, so two columns were empty. The orthogonality was preserved even if two columns of the array remained empty.

The observed values of the roundness and the surface roughness in the internal surface grinding of the shrink fit tool holder were explored. The levels of each input variable were set in accordance with the L18 orthogonal array. The S/N ratios are calculated from the observed values to determine the significant input variables, and the optimal combination levels of input variables are obtained. The response values and the levels of input variables assigned in L18 orthogonal array were listed in the Table 1.

Table 1 Experimental response values and the levels of input variables

Response Values	Input variables	Levels		
● Roundness (μm) ● Surface roughness (Rz/μm)	A. abrasive grain size, AGS (#)	60#		120#
	B. grinding wheel speed, GWS (m/min)	900	1350	1500
	C. feed rate for rough, FRR (mm/min)	500	1000	1500
	D. depth for rough, DR (mm)	0.0015	0.0025	0.0035
	E. feed rate for finishing, FRF (mm/min)	300	600	900
	F. depth for finishing, DF (mm)	0.0015	0.0025	0.0035
	G. dressing speed, DS (mm/min)	200	500	800
	H. wokpiece rotational speed,WRS (rpm)	377	739	1001

Based on the Taguchi method, the S/N ratio calculation was decided as "the higher the better, HB" and "the lower the better, LB" as given in the following equations:

$$\text{HB}: \eta = -10\log\left[\frac{1}{n}\sum_{i=1}^{n} yi^{-2}\right] \quad (1)$$

$$\text{LB}: \eta = -10\log\left[\frac{1}{n}\sum_{i=1}^{n} yi^{2}\right] \quad (2)$$

where η denotes the S/N ratio calculated from the observed values (unit: dB), *yi* represents the experimentally observed value of the i th experiment, and n is the repeated number of each experiment. In this work, the experimentally observed values of the roundness and the surface roughness are "the lower the better" (LB). Therefore, the optimal observed roundness and surface roughness values were the minimum value. Notably, each experiment in the L18 array is conducted three times in this investigation.

3. Result and discussion

3.1. Orthogonal array and S/N ratios

Table 2 presents the S/N ratios calculated from the experimentally response values and the levels of the input variables according to an L18 orthogonal array based on Taguchi method. The optimal combination levels of the input variables with regard to the internal grinding for shrink fit tool holder with a small value roundness and surface roughness were determined. Moreover, the significant input variables associated with roundness and surface roughness were also analyzed by the S/N ratios.

Table 2 L18 orthogonal array, control parameters and S/N ratios.

	Input variables								Experimental results	
	AGS	GWS	FRR	DR	FRF	DF	DS	WRS	S/N ratios(dB)	
No.	(A)	(B)	(C)	(D)	(E)	(F)	(G)	(H)	Roundness	Surface roughness
1	1	1	1	1	1	1	1	1	4.8673	-5.4832
2	1	1	2	2	2	2	2	2	-2.7787	-7.3843
3	1	1	3	3	3	3	3	3	6.4296	-1.8684
4	1	2	1	1	2	2	3	3	0.0000	-6.4444
5	1	2	2	2	3	3	1	1	-1.0153	-6.5267
6	1	2	3	3	1	1	2	2	4.9775	-2.2789
7	1	3	1	2	1	3	2	3	7.2302	-12.1491
8	1	3	2	3	2	1	3	1	2.6861	-4.4022
9	1	3	3	1	3	2	1	2	2.7813	-2.2789
10	2	1	1	3	3	2	2	1	-1.0462	-2.7976
11	2	1	2	1	1	3	3	2	7.0523	-0.8279
12	2	1	3	2	2	1	1	3	3.8493	0.6772
13	2	2	1	2	3	1	3	2	3.3882	-1.5836
14	2	2	2	3	1	2	1	3	5.8827	0.7242
15	2	2	3	1	2	3	2	1	0.6490	-4.5062
16	2	3	1	3	2	3	1	2	-1.1488	-4.4022
17	2	3	2	1	3	1	2	3	4.4631	-3.1672
18	2	3	3	2	1	2	3	1	5.6416	-1.4376

Table 3 ANOVA and F test of roundness.

Variables (A)	Degree (f_A)	Square sum (S_A)	Variance (V_A)	F_{A0}	$F_{0.05,n1,n2}$
AGS	1	7.9190	7.9190	8.3761	18.51
GWS	2	2.4687	1.2344	1.3056	19.00
FRR	2	31.4320	15.7160	16.6232	19.00
DR	2	13.3498	6.6749	7.0602	19.00
FRF	2	48.4443	24.2222	25.6204*	19.00
DF	2	20.6209	10.3104	10.9056	19.00
DS	2	33.0161	16.5081	17.4610	19.00
WRS	2	5.4009	2.7005	2.8563	19.00
Error	2	1.8909	0.9454		
Total	17	164.5427			

*, Significant variable

3.2. Roundness

Table 3 indicates the ANOVA and F test of roundness obtained from the internal surface grinding of the shrink fit tool holder. The analysis results indicate that the significant input variable affecting the roundness for SKD 61 steel was FRF. Figure 1 shows the S/N ratios response graph of the roundness. The response graph indicates that the optimal combination levels of the input variables for roundness were: AGS 120#, GWS 900 m/min, FRR 1500 mm/min, DR 0.0015 mm, FRF 300 mm/min, DF 0.0015 mm, DS 800 rpm, and WRS 1001 rpm.

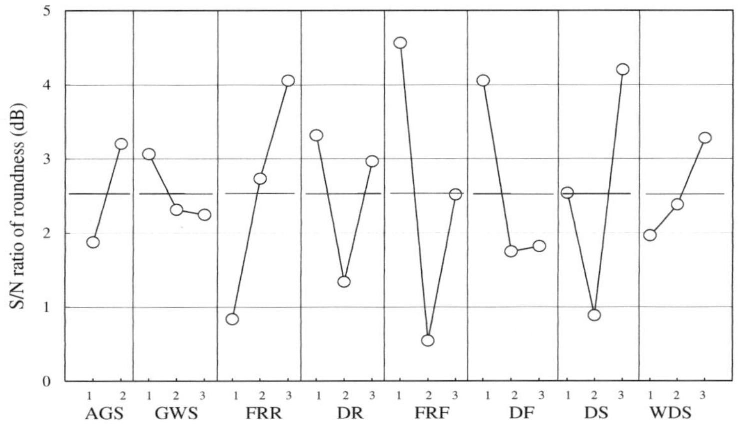

Figure 1 S/N ratios response graph of roundness.

Table 4 ANOVA and F test of surface roughness.

Variables (A)	Degree (f_A)	Square sum (S_A)	Variance (V_A)	F_{A0}	$F_{0.05,n1,n2}$
AGS	1	55.1079	55.1079	265.0283*	18.51
GWS	2	9.1016	4.5508	21.8859*	19.00
FRR	2	37.3907	18.6954	89.9109*	19.00
DR	2	15.0267	7.5133	36.1336*	19.00
FRF	2	5.7455	2.8728	13.8159	19.00
DF	2	17.9055	8.9527	43.0560*	19.00
DS	2	26.2463	13.1232	63.1127*	19.00
WRS	2	3.4192	1.7096	8.2220	19.00
Error	2	0.4159	0.2079		
Total	17	170.3593			

*, Significant variable

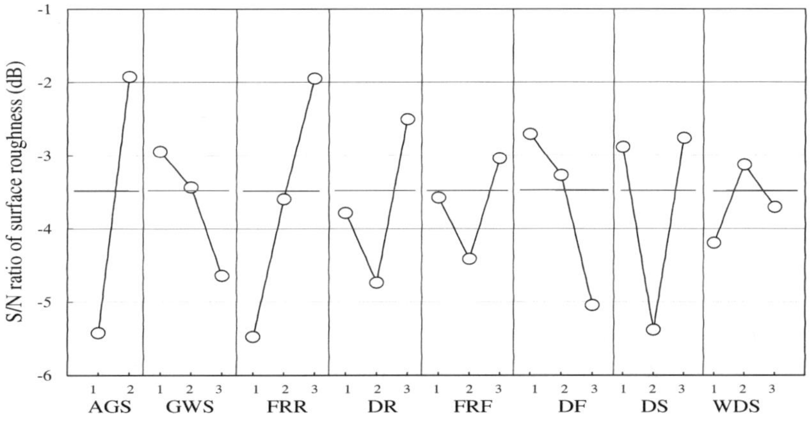

Figure 2 S/N ratios response graph of surface roughness.

3.3. Surface roughness

Table 4 lists the ANOVA and F test of surface roughness obtained from the internal surface grinding of the shrink fit tool holder. The analysis results reveal that the significant input variables affecting the surface roughness for SKD 61 steel were AGS, GWS, FRR, DR, DF, DS. Figure 2 illustrates the S/N ratios response plot of surface roughness. The response graph indicates that the optimal combination levels of the input variables for surface roughness were: AGS 120#, GWS 900 m/min, FRR 1500 mm/min, DR 0.0035 mm, FRF 900 mm/min, DF 0.0015 mm, DS 800 rpm, and WRS 739 rpm. As the grinding conditions set at the optimal combination levels, the internal surface grinding of the shrink fit tool holder could obtain the best response value such as surface roughness.

3.4. Confirmation Experiments

The optimal combination levels of the input variables were determined and confirmed as follows. The estimated S/N ratios are calculated as,

$$\hat{\eta} = \overline{\eta}_m + \sum_{i=1}^{n_o} (\overline{\eta}_i - \overline{\eta}_m) \tag{3}$$

$\hat{\eta}$: Estimated S/N ratio for optimal combination levels of input variables.
$\overline{\eta}_m$: Total mean S/N ratio.
n_0 : The number of significant variables.
$\overline{\eta}_i$: Mean S/N ratio at the optimal level.

Table 5 illustrates the results of confirmation experiments. As the analysis results indicate below, the roundness increased from 2.75 dB to 7.51 dB, and the surface roughness advanced from -3.64 dB to 1.51 dB. In addition, the S/N ratios correlated with roundness and surface roughness for the optimal combination levels of machining variables are 4.76 dB and 5.15 dB larger than those obtained at the initial experimental conditions; the initial conditions were set at AGS (120#), GWS (1350 m/min), FRR (1000 m/min), DR (0.0025 mm), FRF (600 mm/min), DF (0.0025 mm), DS (500 mm/min) and WRS (739 rmp), ($A_2B_2C_2D_2E_2F_2G_2$). The experimental results confirm that the input variables of the internal grinding for shrink fit tool holder would be optimized for roundness and surface roughness, so the observed values would thus be significantly improved.

Table 5 Results of the confirmation experiments

		Initial levels of input variables	Optimal combination levels of input variables	
			Prediction	Experiment
Roundness	Level	$A_2B_2C_2D_2E_2$ $F_2G_2H_2$	$A_2B_1C_3D_1E_1$ $F_1G_3H_3$	$A_2B_1C_3D_1E_1$ $F_1G_3H_3$
	Response values (μm)	0.728	-	0.421
	S/N ratio (dB)	2.75	4.03	7.51
Surface roughness	Level	$A_2B_2C_2D_2E_2$ $F_2G_2H_2$	$A_2B_1C_3D_3E_3$ $F_1G_3H_2$	$A_2B_1C_3D_3E_3$ $F_1G_3H_2$
	Response values (Rz/ μm)	1.52	-	1.32
	S/N ratio (dB)	-3.64	0.98	1.51

4. Conclusions

The effects of essential variables associated with the internal grinding for shrink fit tool holder on roundness and surface roughness were explored, and the optimal input variables were determined based on Taguchi method for SKD 61 steel. According to the experimental results and imperative analysis, the following conclusions have been drawn.

(1) The significant variable affecting roundness is FRF associated with the internal grinding for shrink fit tool holder. The optimal combination levels of input variable for roundness are AGS 120#, GWS 900 m/min, FRR 1500 mm/min, DR 0.0015 mm, FRF 300 mm/min, DF 0.0015 mm, DS 800 rpm, and WRS 1001 rpm.

(2) The significant variables affecting surface roughness are AGS, GWS, FRR, DR, DF, and DS associated with the internal grinding for shrink fit tool holder. Moreover, the optimal combination levels of input variable for surface roughness obtained from the S/N ratio response graph are AGS 120#, GWS 900 m/min, FRR 1500 mm/min, DR 0.0035 mm, FRF 900 mm/min, DF 0.0015 mm, DS 800 rpm, and WRS 739 rpm.

(3) The S/N ratios of roundness and surface roughness are improved as 4.76 dB and 5.15 dB based on the optimal approach of Taguchi method. The roughness ameliorated form 0.728 μm to 0.421 μm, and the surface roughness meliorated from 1.52 μm to 1.32 μm. The roundness and surface roughness in internal grinding for shrink fit tool holder were obviously improved at the optimal setting levels of input variables obtained from Taguchi experimental analysis.

Acknowledgments
The authors would like to thank the Action Tools Corp, Taiwan, for technical supporting the experimental works.

References

[1] Brinksmeler E and Giwerzew A 2003 *CIRP Annuals -Manuf Technol.* **52(1)** 253.
[2] Durgumahanti U S P, Singh V and Rao PV 2010 *Int. J. Mach. Tool. Manuf.* **50** 231.
[3] Ghosh S, Chattopadhyay AB and Paul S 2008 *Int. J. Mach. Tool. Manuf.* **48** 1242.
[4] Jackson MJ, Khangar A, Chen X, Robinson G M, Venkatesh V C and Dahotre N B 2007 *J. Mater. Process. Technol.* **185** 17.
[5] Krajnik P, Drazumeric R and Badger J 2013 *CIRP Annuals –Manuf. Technol.* **62(1)** 347.
[6] Wang X Y, Wu Y B, Wang J, Xu W J and Kato M 2005 *J Mater Process Technol*, **164-165** 1128.
[7] Zhong Z W and Venkatesh V C 2009 *Int. J. Adv. Manuf. Technol.* **41** 468.
[8] Qian J, Li W and Ohmori H 2000 *J. Materia.l Process. Technol.* **105** 80.
[9] Opoz T T and Chen X 2012 *Int. J. Mach. Tool. Manuf.* **63** 32.
[10] Gu W, Yao Z and Liang X 2011 *Wear* **270** 241.
[11] Liu Q, Chen X and Gindy N 2007 *Int. J. Adv. Manuf. Technol.* **33** 940.
[12] Fathallah B B, Fredj N B, Sidhom H, Braham C and Ichida Y 2009 *Int. J. Mach. Tool. Manuf.* **49** 261.
[13] Gopal A V and Rao P V 2003 *Int. J. Adv. Manuf. Technol.* **22** 475.
[14] Krishna A G and Rao K M 2009 *Int. J. Adv. Manuf. Technol.* **26** 475.
[15] Lee T S, Ting T O, Lin Y J and Htay T 2007 *Int. J. Adv. Manuf. Technol.* **33** 468.
[16] Saravanan R, Asokan P and Sachidanandam M 2002 *Int. J. Mach. Tool. Manuf.* **42** 1327.
[17] Lin Y C, Chow H M, Tsui H P and Chen Y F 2013 *Adv. Mater. Res.* **675** 365.
[18] Wang D A, Lin Y C, Chow H M, Fan S F and Wang A C 2012 *Adv. Maters. Res.* **459** 170.

IOP Conf. Series: Materials Science and Engineering **281** (2017) 012056 doi:10.1088/1757-899X/281/1/012056

Working String Axial Force in L-Shaped Oil Well

L S Wang*, B K Gao and L Gao

Department of Petroleum Engineering, China University of Petroleum-Beijing, Beijing, 102249, China

Email: pewangls@163.com.

Abstract. The bottom section of a L-shaped well warps upward, so it is a different type against normal horizontal well and should be designed carefully. With regard to the peculiar geometry of borehole, in this paper, the working string axial force is calculated and the feasibility of string running in and out the hole is discussed. String buckling and bottom sticking are considered. This papers suggested that the tip-tilted section makes string tripping in more difficult, while makes tripping out easier; during tripping out, the ratio between the increase of hook load and sticking force is a certain value; string buckling begins at the kick of point firstly and at the target secondly during tripping in; the friction caused by buckling in the L-shaped well is more obvious than in normal horizontal well; the smaller the string size is, the shorter the largest extending length will be.

1. Introduction

For the exploration of unconventional reservoirs, some special well types have been developed, such as directional wells, horizontal wells, large extended wells and etc. In China, horizontal well is usually the well which deviation angle is not less than 86° and extended to certain length in the reservoir [1]. Currently, although the deviation angle of most horizontal wells is 90°, the deviation angle of certain well is greater than 90°, which is defined as L-shaped well in this paper.

Tubular string, such as drill stem, tubing and casing, plays an important role in the process of well drilling and completion. Working string confronts various difficulties because borehole is narrow and down-hole condition is complicated. In the area of tubular string mechanics, many scholars have made a large number of researches: In 1950, Lubinski researched the buckling behavior of strings in the vertical borehole and calculated the critical load of buckling ignoring the gravity[2]; In 1984, Johan Sick analyzed the mechanics of strings in the whole wellbore and calculated the axial force by the method of iteration from bottom to the wellhead[3]. Sheppard revised Johan Sick's method by taking drilling fluid viscosity and string inner and outer pressure differential into consideration[4]; Without ignoring the effect of gravity, Yu-che Chen, in 1990, calculated the critical load of buckling in horizontal well[5]; Gao Deli and Gao Baokui calculated the critical load of helical buckling in deviated well section and contact force between pipe and wellbore after buckling[6]; Han Zhiyong studied the method to calculate the real axial force in vertical, horizontal, and deviated wellbore[7]. In 2008, S.Menand established mechanic model of minor-sized string and studied the influence on critical load when string rotated [8].

In horizontal well, focuses of string mechanics investigation are mainly on strength check, friction calculating, tripping feasibility and extending limits. L-shaped well faces more mechanic difficulties compared with conventional horizontal wells since its bottom section warps upward against the horizontal plane. In the process of tripping in, working strings is easier to get buckling because of the increase of friction and compression; while in the process of tripping out, it's easier to extract the strings. The transfer coefficient of hook load and bottom hole sticking load is different between L-shaped well and normal horizontal well when sticking occur. However, the previous researches didn't focus on L-shaped well. Therefore, it's necessary to calculate the friction and axial force of the working strings in L-shaped well, and analyze their tripping capacity in order to direct the practical construction.

2. Analyses of String Mechanics

2.1. Description of wellbore and assumptions

A typical L-shaped well trajectory is shown in Figure1 a), where AB is vertical section, BC is curved section and CD is tip-tilted deviated section.

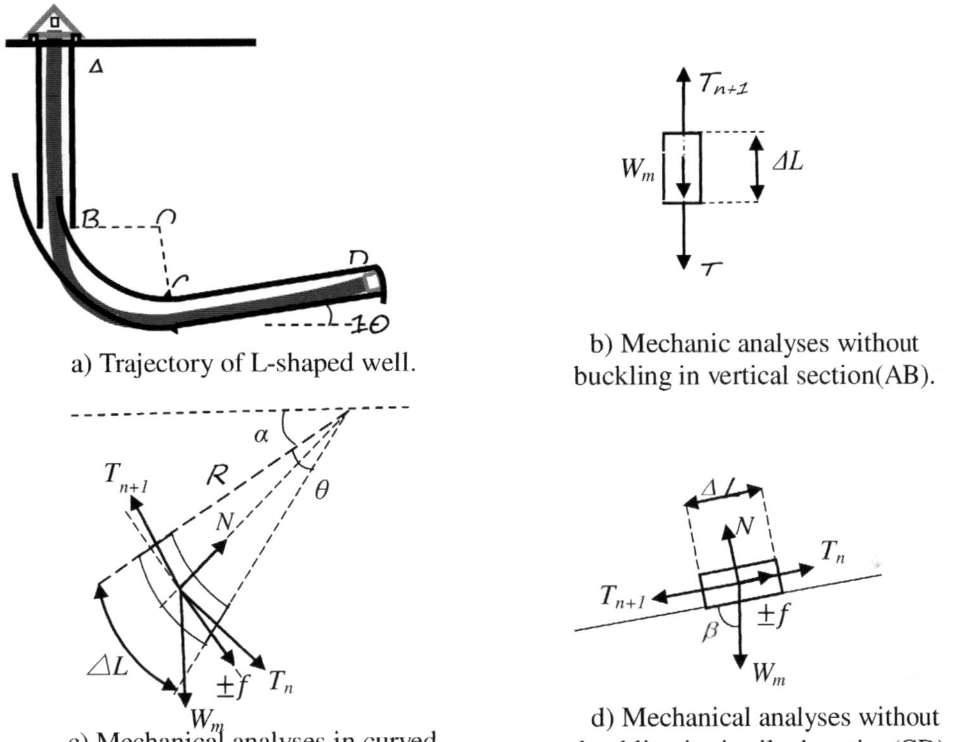

a) Trajectory of L-shaped well.

b) Mechanic analyses without buckling in vertical section(AB).

c) Mechanical analyses in curved section(BC).

d) Mechanical analyses without buckling in tip-tilted section(CD).

Figure 1. Trajectory of conceptual L-shaped well and Mechanic analyses

In this paper the following assumptions are used: the influence of pipe couplings and steps of are ignored; the friction coefficients in casing section and bare section are constant respectively; axial tension is positive while compression is negative; and no pipe buckling occurs in curved borehole section.

2.2. String mechanic analyses in vertical section

Taking a micro-section from string in the vertical section, then the mechanic analyses without buckling is shown in Figure 1 b).

In Figure 1 b), T_{n+1} is axial load in upper section, kN; T_n is axial load in lower section,kN; W_m is unit weight of string in drilling fluid,kN/m; $W_m=W_a\times(1-\rho_m/\rho_s)$; W_a is unit weight of string in air,kN/m; ρ_m is fluid density, g/cm^3; ρ_s is steel density, g/cm^3; ΔL is the length of micro-section, m.

According to the mechanic equivalent

$$T_{n+1}=T_n+W_m\times\Delta L \tag{1}$$

Buckling is likely to happen when the string axial force is compressive and greater than the critical force. String buckling, which will increase the contact force and friction, should be considered for calculating axial force distribution.

The critical force of helical buckling in the vertical section is[5]:

$$F_{hel}=5.5\sqrt[3]{EIW_m^{\ 2}} \tag{2}$$

Where F_{hel} is the critical force in vertical section, kN; EI is bending rigidity, kPa·m^4.

The contact force between string and borehole wall caused by buckling is

$$N=rF^2/(4EI)+W_m\sin\beta \tag{3}$$

Where N is the contact force, kN/m; r is the clearance between string and bore wall, m; F is the axial force, kN; β is deviation angle, rad.

The friction caused by buckling is

$$f=\mu\times N \tag{4}$$

Where f is the friction caused by buckling, kN; μ is the friction coefficient.

2.3. *String mechanical analyses in curved section*

In curved section, string may partly contact with upper wall and the rest contact with the lower wall. Without loss of generality, this paper takes a micro-section n, with angle θ, which contacts with the lower wall, to analyze the mechanical condition, as shown in Figure 1 c)[9].

In Figure 1 c), α is the angle between upper section and horizontal level, rad; θ is the angle of micro-section, rad; R is radius of the curve section, m; N is unit contact force between micro-section and bore wall, kN/m.

From normal force equilibrium, we have

$$W_m\times R\times\theta\times\sin(\alpha+\frac{\theta}{2})=N\times R\times\theta+(T_{n+1}+T_n)\times\sin(\frac{\theta}{2}) \tag{5}$$

The contact force N can be derived from equation (5)

$$N=W_m\times\sin(\alpha)-\frac{T(\alpha)}{R} \tag{6}$$

When $N\geq 0$, string contacts with the lower wall; when $N<0$, string contacts with upper wall.
From tangential force equilibrium, we have

$$T_{n+1}\times\cos(\frac{\theta}{2})=T_n\times\cos(\frac{\theta}{2})+W_m\times R\times\theta\times\cos(\alpha+\frac{\theta}{2})\pm|f| \tag{7}$$

In the last term of equation (7), positive sign means tripping out, while negative sign means tripping in.

2.4. *String mechanical analyses in deviated section*

Take a micro string section from tip-tilted deviated section, as shown in Figure 1 d), β is the deviation angle.

From normal force equilibrium, we have

$$N = W_{\mathrm{m}} \times \square L \times \sin\beta \tag{8}$$

From axial force equilibrium, we have

$$T_{n+1} + W_{\mathrm{m}} \times \square L \times \cos\beta = T_n \pm f \tag{9}$$

So the axial force, T_{n+1}, can be derived from equation (8) and (9)

$$T_{n+1} = W_{\mathrm{m}} \times \square L \times (\pm \mu \times \sin\beta - \cos\beta) + T_n \tag{10}$$

In the second term of equation (10), positive sign means tripping out, while negative sign means tripping in.

If the axial compressive force is greater than the critical force, the friction caused by buckling should be taken into consideration.

The critical force of helical buckling in deviated section is[5]

$$F_{\mathrm{hel}} = 2\sqrt{\frac{2EIW_{\mathrm{m}}\sin\beta}{r}} \tag{11}$$

Where F_{hel} is the critical force in deviated section, kN.

The equations of contact force and friction are same as equation (2) and (3).

3. Case Study

Take a conceptual L-shaped well for example, as shown in Figure1 a). Main parameters of wellbore and working string are listed in Table 1.

Table 1. Main parameters of wellbore and working string

Parameter	Value
Length of vertical section AB (m)	2000
Radius of curvature of curved section BC (m)	150
Length of section CD (m)	2000
Inclinational angle of section CD (°)	100
Hole diameter of section CD (mm)	152.4
Friction coefficient in section CD	0.25
Casing diameter of section AC (mm)	177.8
Wall thickness of casing section AC (mm)	8.05
Friction coefficient in section AC	0.15
Density of working fluid (g/cm^3)	1.05
External diameter of tubing (mm)	114.3
Internal diameter of tubing (mm)	88.9
Unit weight of tubing (kg/m)	32.14
Density of tubing material (g/cm^3)	7.85

3.1. Axial force sensitivity to deviation

The string axial force distributions while tripping in and tripping out are drawn in Figure 2 a) and Figure 2 b) respectively. In order to investigate the influence of inclination of deviated section, Figure 2 gives the axial force curves corresponding to deviation angle of 80°, 90°, 100° and 105°.

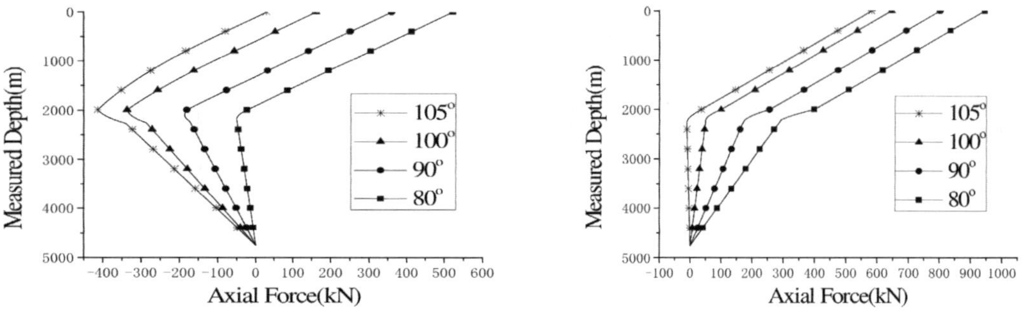

a) Tripping in. b) Tripping out.

Figure 2. Axial force under different inclination

Figure 2 a) illustrates that, in the process of tripping in, the hook load decreases with the increase of inclination. Hook load approaches zero when inclination increases to about 105°, and it's hard to tripping in in this condition. This means that the tip-tilted section diminishes the capacity of tripping in.

Figure 2 b) shows that, in the process of tripping out, hook load decreases when inclination increases, which means it's easier to tripping out.

3.2. Axial force sensitivity to sticking force

Axial force will be changed if string is stuck at bottom during tripping in or out. The axial forces correspond to different sticking force is shown in Figure 3.

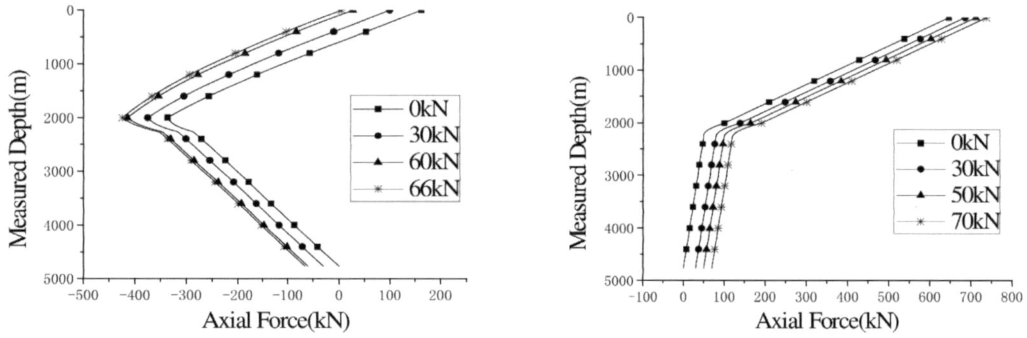

a) Tripping in. b) Tripping out.

Figure 3. Axial force under different sticking force

In Figure 3 a), hook load is diminishing with the increase of sticking force. Additionally, compression is pretty large during tripping in. Thus the larger sticking force is, the severer buckling will be. It should be noticed that ratio of hook load and sticking force is not a constant value because of buckling.

From Figure 3 b), it is obvious that, in the process of tripping out, hook load increases with the increase of sticking force, and the ratio between hook load change and sticking force is constant. The reason why the ratio is constant is that buckling doesn't happen.

3.3. Axial force sensitive to string size

Now, change the string dimension, as listed in Table 2.

IOP Conf. Series: Materials Science and Engineering **281** (2017) 012056 doi:10.1088/1757-899X/281/1/012056

Table 2. String parameters

External diameter (mm)	Internal diameter (mm)	Unit weight in air (Nm^{-1})
114.3	88.9	315.293
88.9	80.98	82.89
73.03	65.11	67.42

According to the consequence of axial force calculating, the largest extending length of 114.3mm string is 5340m. However, if drilling operators use 88.9mm string or 73.03mm string, the largest extending length will diminish because the severe buckling happens in vertical and deviated section. The consequence of calculating shows that 88.9mm string's largest extending length is 4410m, and 73.03mm string is 3883m. The axial force distributions of tripping in and out of different string size are shown in Figure 4.

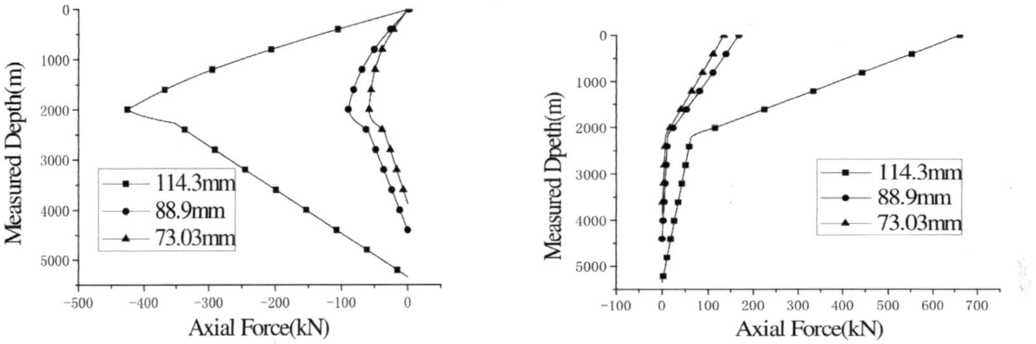

a) Tripping in. b) Tripping out.

Figure 4. Axial force with different string sizes

4. Conclusions and Suggestions

According to the analysis above, certain conclusions can be drawn:

(1) The design of L-shaped well should take string tripping feasibility into consideration. String is harder to tripping in in L-shaped well compared with conventional horizontal well, but easier to tripping out.

(2)If string is stuck in L well during the process of tripping out, the ratio between changes of hook load and sticking force is constant. But the ratio is not constant during tripping in because of severe buckling.

(3)The string size used in L-shaped well also should be chosen carefully, since the smaller the diameter is, the shorter the largest extending length will be.

(4)In L-shaped well, tripping out doesn't lead to buckling. However, during tripping in, buckling is more likely to happen compared to conventional horizontal well both at vertical and deviated section.

Acknowledgments

This work is supported by "National Basic Research Program of China" (973 Program, 2015CB251205)

References

[1] Xiao L W 2013 The research of drilling technic's current condition and developing tendency in horizontal well *J. China Petroleum and Chemical Standard and Quality* **10** 7-9

[2] Lubinski A 1950 A Study of the buckling of rotary drilling strings *J. API Drill And Prod Prae* **35** 178-214

IOP Conf. Series: Materials Science and Engineering **281** (2017) 012056 doi:10.1088/1757-899X/281/1/012056

[3] Johansick C A and Friesen D B 1984 Torque and drag in directional wells-prediction and measurement J. Journal of Petroleum Technology **36** 987-992

[4] Sheppard M C and Wick C 1987 Designing well paths to reduce drag and torque *J.SPE Drilling Engineering* **2** 344-350

[5] Chen Y C and Lin Y H, Cheatham J B 1990 Tubing and casing buckling in horizontal wells *J. Journal of Petroleum Technology* **42** 140-141

[6] Gao D L and Gao B K 2000 Analyses of strings' bucking and friction in horizontal well *J. Journal of China University of Petroleum* **24** 2-5

[7] Han Z Y 2011 *String mechanics in liquid pressure* (Beijing: Petroleum industry press) pp 35-168

[8] Menand S.2008 How Drilltring Rotation affects critical buckling load *Conf. IADC/SPE Drilling conference* (Orlando: Florida, USA/ Society of Petroleum Engineers) p 135

[9] Xie B 1995 The mechanic analyses and practical calculation of running tubings in horizontal well *J. Xinjiang Petroleum Science & Technology* **4** 107-112

IOP Conf. Series: Materials Science and Engineering **281** (2017) 012057 doi:10.1088/1757-899X/281/1/012057

The B Life Assessment of Auto Chassis Components and Application Based on Weibull Distribution

X F Zhang*, **Z P Zhang, W Liu and X C Wei**
China Automotive Technology & Research Center, No.68 Xianfeng East Road, Dongli District, Tianjin, China

Email:zhangxinfeng@catarc.ac.cn

Abstract. The reliability life of auto chassis components is the important part of auto safety, and B life is theuniversal evaluation index. The essay established a method of fatigue reliability B life based on weibull theory. The method have been used in the actual test and got the B life, which give the reference to the reliability life of auto chassis components.

1. Introduction
Recently,the automotive industry grownrapidly,The requirements of the market for vehicle comfort and safety are getting higher and higher.Car chassis system as an important part of the overall structure of the car,Its safety,especially reliability aspects is the important evaluation basis of vehicle safety performance[1].

At present,fatigue life verification for chassis components,Most of theauto companies andparts suppliers only do experimental verification for a few parts,Test through the design life on it. But in the actual use of the process,the force situation of car chassis parts is complicated,environmental conditions is bad,A single design life or laboratory test life is not sufficient to fully demonstrate its reliability life.The cost of vehicle reliability testingis too high,which is maximum short point.Part of the automotive industry standards regulate thatthe fatigue life of the car chassis is to meet the life expectancy.Bx is that the sample of x% reaches the specified failure or overhaul indicator within the specified time zone or mileage[2]. The evaluation of the B-life is based on the reliability of the Weibull distribution, and this method is widely used in the automotive industry.But the calculation of B value life is complex,which is the difficulty of hindering its normal application.

In this paper, the steering knuckle in the car as the object of study,relying on the measured data ofreliability laboratory hydraulic servo system,based on Weibull theory, to estimate the Weibull distribution parameter in terms of the fatigue life,and to establish a reasonable B value life calculation method,so as to provide reference for the reliability life prediction and test judgment of automobile chassis parts.

2. Introduction ofweibulltheory
Weibull distribution is one of the random variable distributions,using the probability value to derive its distribution parameters,which isone of the most widely used distributions in failure data analysis and be widelyused in various life test data processing[3].

Weibull distribution can be defined mathematicallyasfollows:

$$F(t) = 1 - e^{-((t-t_0)/\eta)^\beta} \tag{1}$$

$F(t)$—Faultdensity distribution function; t—Time of failure; t_0—Distribution starting point or origin; η—Characteristic life or scale parameters; β—Slope or shape parameters; e—Natural constants.

$F(t)$ defines the cumulative probability of a set of components that will fail at time t, and 1-$F(t)$ is the probability of no failure,expressed by reliability $R(t)$.

$$R(t) = e^{-((t-t_0)/\eta)^\beta} \tag{2}$$

The Weibullian analysis studiestherelationship between the life time of a component and its reliability by plotting thelife data of a single failure mode on a Weibullprobabilitygraph.

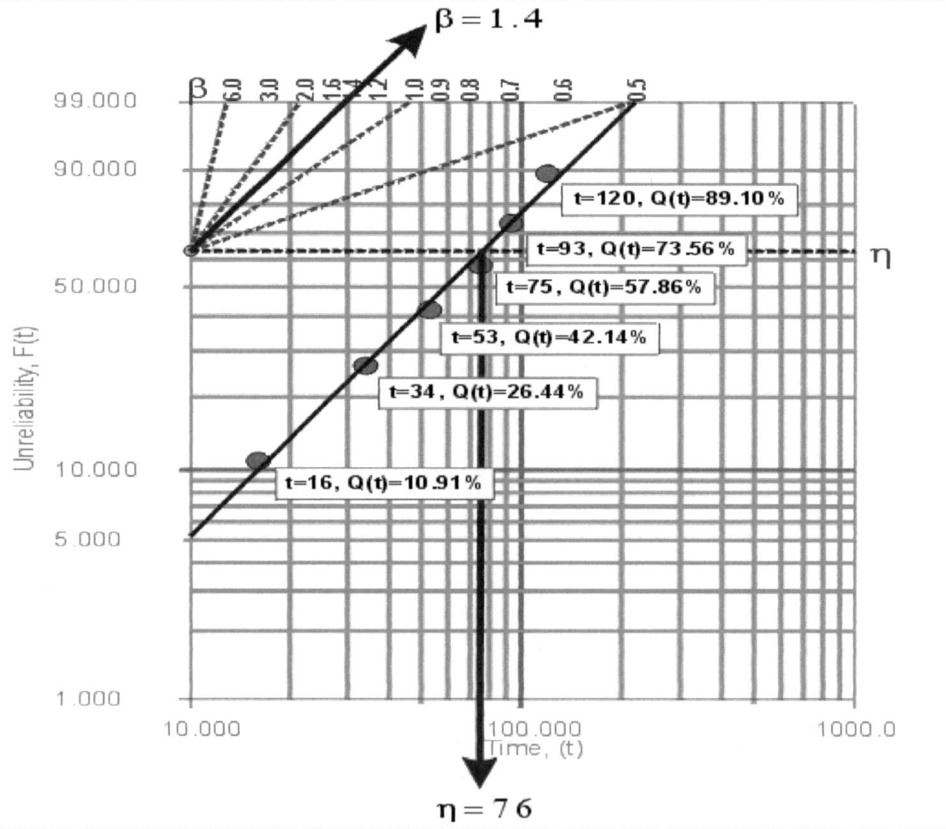

Figure 1. Weibull probability figure.

That is the Weibull probability figure[3] in Figure 1, where the abscissa is the time t, and the ordinate is the corresponding cumulative probability. The slope of the straight lineis β, its value is different, representing differentfailure categories, and corresponding to different failure reasons.

3. Reliability B value life algorithm

The B life is obtained by the Weibull cumulativeprobability density distribution function[4]. Weibull fault density distribution function expression is formula (1).

then,the reliability function is

$$R(t) = 1 - F(t) = e^{-((t-t_0)/\eta)^\beta} \tag{3}$$

t_0 isthe distribution of origin, when $t_0 = 0$,get

$$1 - F(t) = e^{-(t/\eta)^\beta} \tag{4}$$

take the logarithm , get

$$\ln \ln((1 - F(t))^{-1}) = \beta \ln t - \beta \ln \eta \tag{5}$$

The above equation is represented by a straight line equation,

$$Y = BX + A \tag{6}$$

$$Y = \ln \ln((1 - F(t)^{-1}) \tag{7}$$

$$X = \ln(t) \tag{8}$$

$$B = \beta \tag{9}$$

$$A = -\beta \ln \eta \tag{10}$$

For the linear equation, the least squares method is used to obtain the estimated value $\hat{B}$, $\hat{A}$

$$\hat{B} = \sum_{i=1}^{n} X_i Y_i - \frac{(\sum_{i=1}^{n} X_i)(\sum_{i=1}^{n} Y_i)}{n} (\sum_{i=1}^{n} X_i^2 - \frac{(\sum_{i=1}^{n} X_i)^2}{n})^{-1} \tag{11}$$

$$\hat{A} = \frac{1}{N} \sum_{i=1}^{n} Y_i - \hat{B} \frac{1}{N} \sum_{i=1}^{n} X_i \tag{12}$$

To bring it into the above formula and to find the value β and η, the expression of the Weibull fault density distribution function is obtained.For B life(take B_x for example),make

$$x\% = F(t) = 1 - e^{-(t/\eta)^\beta} \tag{13}$$

Find it,that is life of B_x , which means that the time or lifetime is t when the sampleof x% reaches the specified failure criteria.

4. The test of life assessment and analysis

The car steering knuckle is an important part of the chassis system. An automobile company is required to carry out fatigue and durability tests for the knuckle components to verify its fatigue and reliability performance. The number of samples is 10, In the chassis reliability test system using hydraulic servo for single load fatigue life test, The loading conditions (Figure 2) and test photo (Figure 3) are as follows.

The test bench is hydraulic servo system. It was made by IST. The biggest load is 6.3t and the biggest frequency is 50Hz at special load.

IOP Conf. Series: Materials Science and Engineering **281** (2017) 012057 doi:10.1088/1757-899X/281/1/012057

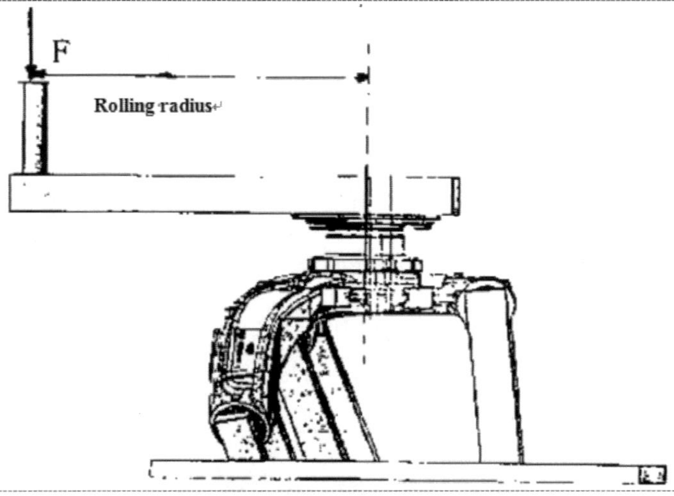

Figure 2. Loading conditions

Figure 3. Test photo

Test conditions: According to the standard requirements (to meet the torque requirements) fixed samples, test load±3900N,load frequency 5Hz。After 1 million timestest, check the knuckle for cracks, cracks and other failure forms. The test load increased to 1.1 times to continue the test 700,000 times, the steering knuckle does not appear cracks, cracks and other failure forms. And then the test load increased to 1.6 times to continue the test 500,000 times, the steering section of local cracks, cracks and other failure forms. During the test, any sample appeared in the crack and the length was greater than 10mm, or the sample appeared to break, then, the sample failed and the test stopped. According to the test requirements, 10 samples were tested in succession. The sample number and the corresponding test life were shown in Table 1.

Table 1. Test sample life

Sample No	Sample life / million times
1	200. 0
2	125.7
3	159.4
4	168.9
5	165.0
6	152.4
7	202.1
8	152.6
9	108.8
10	100.0

The sample size of the test sample is small, and the fatigue life of the chassis parts is evaluated for B_{10} life. According toformula (5) —formula (12), The Weibull probability density function $F(t)$ is obtained, and then put $x = 10$ into the formula (13), you can find the B_{10} life of the knuckle sample. First of all, the test data in accordance with the life cycle from small to large rearrangement, and serial number. According to the literature [3, 5], the median rank value is obtained by using the median rank of Benard's formuls.

$$P = \frac{i - 0.3}{N + 0.4} \tag{14}$$

Where N is the total number of samples and i is the adjustment rank (the number after the test life is adjusted).

Using the formula (1) - formula (10), the data are shown in Table 2.

Table 2. Date sheet

Serial number	The original serial number	Test life	Adjust the rank value	Median rank value	Xi	Yi
1	10	100.0	1	6.73%	4.6052	-2.6638
2	9	108.8	2	16.35%	4.6895	-1.7233
3	2	125.7	3	25.96%	4.8339	-1.2020
4	6	152.4	4	35.58%	5.0265	-0.8217
5	8	152.6	5	45.19%	5.0278	-0.5086
6	3	159.4	6	54.81%	5.0714	-0.2304
7	5	165.0	7	64.42%	5.1059	0.0329
8	4	168.9	8	74.04%	5.1293	0.2990
9	1	200.0	9	83.65%	5.2983	0.5940
10	7	202.1	10	93.27%	5.3088	0.9927

Then, using the least squares method, the data is taken into the formula (11), the formula (12),get the minimum estimate $\hat{B} = \hat{\beta} = 4.64$, $\hat{\eta} = e^{-\hat{A}/\hat{\beta}}$ =167.73,the Weibull probability density function is obtained,

$$F(t) = 1 - e^{-(t/167.73)^{4.64}} \tag{15}$$

In the function,Slope or shape parameters $\beta = 4.64$,It can be seen from the literature [3], [6], the steering section failure is a rapid loss, the reasons of may cause the loss: the aging parts; material properties have changed or defect; manufacturing process problems.

Thus, the relationship between the available life and the Weibull distribution probability is shown in Figure 4.

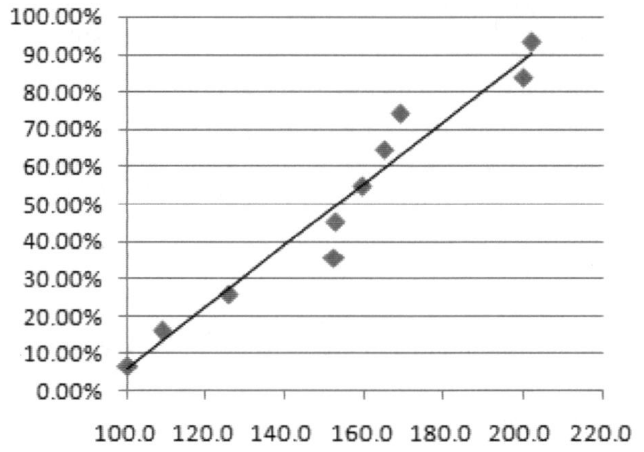

Figure 4. Weibull probability line

Among them, the abscissa is the experimental life, the ordinate is the corresponding Weibull distribution cumulative probability, the distribution data correlation coefficient r = 0.98.

put $x = 10$, $\beta = 4.64$, $\eta = 167.73$ into formula (13), get the relable life span $B_{10} = 103.29$ million times.

From the distribution data correlation coefficient r and B_{10} we can see that Weibull can be reasonably applied to the reliability of data processing, more in line with the actual situation, and can be applied to thelife assessment of automotive chassis fatigue reliability B value.

5. Conclusions

Based on the Weibull distribution theory, this paper deduces the reliability B value life evaluation algorithm of the automobile chassis. Weibull distribution is widely used for its powerful failure analysis processing and forecasting capabilities. The application of Weibull distribution and B value life is summarized as follows.

1) The B-life assessment can be used to evaluate the reliability of the chassis parts, providing a great deal of usefulness for product design life and verification of life assessment.
2) Weibull distribution is suitable for engineering life assessment and analysis, its β value can be a good analysis of fault data and the reasons and can be widely applied to engineering practice.
3) The B life assessment method is very necessary for the chassis parts, this method could give the components accurate quality status.
4) Weibull distribution is powerful, parameter estimation methods are more, the accuracy is also different, we can seek high precision and easy to implement the method to improve the calculation accuracy.

References

[1] Zhu W Y. *Mechanical Reliability Design* [M]. Shanghai: Shanghai Jiaotong University Press, 1992: 61 - 68.

[2] Zhou F G, Zhao L F.Evaluation of the reliability of a rear brake foot pad based on B10 life [J]. *Journal of Hefei University of Technology (Natural Science Edition)*, 2014, 37 (11): 1290 - 1293.

[3] Abernethy R B. *Weibull Analysis Manual* [M]. Beijing: Beijing University of Aeronautics and Astronautics, 1992: 12-19, 129-131.

[4] Wang L L, Zhang Z P, Liu W. Study on mechanical reliability of automobile parts [J]. *Research and Development,* 2014: 49-51.

[5] Pascovici D S, Konstantinos G. Kyprianidis. Weibull Distributions Applied to Cost and Risk Analysis for Aero Engines [J] .*ASME TURBO EXPO 2008 Proceedings*, GT2008-51060.
[6] Ling D, Huang H Z, Zhang X L, and Jiang G L. Parameters Estimation for Mixed Weibull DistributionUsing L-M Algorithm [J]. *Journal of University of Electronic Science and Technology of China.* 2008(04).

IOP Conf. Series: Materials Science and Engineering **281** (2017) 012058 doi:10.1088/1757-899X/281/1/012058

Comprehensive Deployment Method for Technical Characteristics Base on Multi-failure Modes Correlation Analysis

W Zheng[1*], J M Gao[1], R X Wang[1,2], K Chen[1] and Y Jiang[1]

[1] State Key Laboratory for Manufacturing Systems Engineering, Xi'an Jiaotong University, Xi'an 710049, China
[2] Collaborative Innovation Centre of High-End Manufacturing Equipment, Xi'an Jiaotong University, Xi'an 710049, China

Email:zw_liaoya@163.com

Abstract. This paper put forward a new method of technical characteristics deployment based on Reliability Function Deployment (RFD) by analysing the advantages and shortages of related research works on mechanical reliability design. The matrix decomposition structure of RFD was used to describe the correlative relation between failure mechanisms, soft failures and hard failures. By considering the correlation of multiple failure modes, the reliability loss of one failure mode to the whole part was defined, and a calculation and analysis model for reliability loss was presented. According to the reliability loss, the reliability index value of the whole part was allocated to each failure mode. On the basis of the deployment of reliability index value, the inverse reliability method was employed to acquire the values of technology characteristics. The feasibility and validity of proposed method were illustrated by a development case of machining centre's transmission system.

1. Introduction

The cause of the failure for mechanical parts simultaneously comes from the failure mechanisms [1-3], the soft failures and the hard failures, just like wear degradation, corrosion, fatigue, fracture, overload and so on. Failure modes can be divided into soft failures and hard failures. There are complicated correlative relations between failure mechanisms and failure progresses, interlocking failures from one part may show a network spreading phenomenon just like the spreading of failures in electromechanical system [4]. For a long time, due to lacking of effective tools for descripting and analysing the complicated relations above, there is a disconnect between reliability design and performance design, that makes deployment values of design parameters cannot guarantee the reliability of products.

Reliability design is closely related to uncertain factors, the design method rely on probability was widely used in mechanical reliability design after the general method first/second-order reliability was put forward [5, 6]. The stress-intensity interfere model based on probability[7] is a reliability design method which is general used at present time, researchers have made a lot of studies on the method for calculating the reliability of parts and systems. However, in practical application, the analysing tools like FTA, FMEA/FMECA[8-10] and so on which are used in the stress-intensity interfere model have their limitations such as the expression is not rigorous, lack of consistency and they only consider

IOP Conf. Series: Materials Science and Engineering **281** (2017) 012058 doi:10.1088/1757-899X/281/1/012058

single failure mode, especially not considering the correlations between failure modes[11]. Meanwhile the method above is only for the main failure mode for probability design, which ignores the correlation between multiple failure modes. That makes it unreasonable.

In the nineties of the 20th century, the concept of reliability function deployment (RFD) emerged, it makes up the lack of QFD technology[12] in product planning with considering the product failures and their impact[13]. The article [14] makes qualitative and quantitative description, analysis for the fault propagation path and trend between components using the domino matrix in RFD, and improves the traditional reliability deployment method based on it. It can be found the advantage of RFD is that it converts the reliability analysis method from filling in the forms to analysing the matrix structure correlation, controlling product reliability from top to bottom using the method of decomposing matrixes. RFD's matrix structure can indicate different processing level for the system complexity, the description for the relation between design factors is clearer and more concise. Meanwhile, RFD process also can use the numeric operations and normalization method existing in QFD for numerical operation. So using RFD matrix structure to describe correlations of failure mechanisms and failure modes has the advantages of tight expression and convenient calculation, and it can effectively evaluate and calculate the interrelated effect of multiple failure modes.

In summary, this paper, which is aiming at overcoming the limitation of traditional method, puts forward a comprehensive deployment method for technical characteristics considering the interrelated and coupled relations between failure modes. First, build a reliability deployment space for failure modes which is made up with houses of reliability. Second, comprehensive considering the interrelation between failure modes, the total reliability index value is decomposed and allocated to failure modes of the part. Next, establish a connection equation or limit state equation, build the technical characteristics deployment space, solve the design parameters of technical characteristics using the inverse reliability method [15, 16]. Last, considering interrelated effect of design parameters of parts, the deployment design parameters are imported into the space for matching design parameters and weakening conflicts, output the final deployment value.

2. Build the house of reliability and synthetically allocating the technical characteristics

In order to considering the correlative effects of different failure modes, the house of reliability will be established. And through complex processes, the reliability index of one part or component will be allocating to the design parameters finally.

2.1. 2.1 Build the house of reliability for failure modes

The relation of failure mechanisms, hard failures and soft failures is shown in figure 1. It can be found that there are complex relationships between failure modes and failure mechanisms. Meanwhile it can be concluded that the reliability losses caused by different failure modes are different.

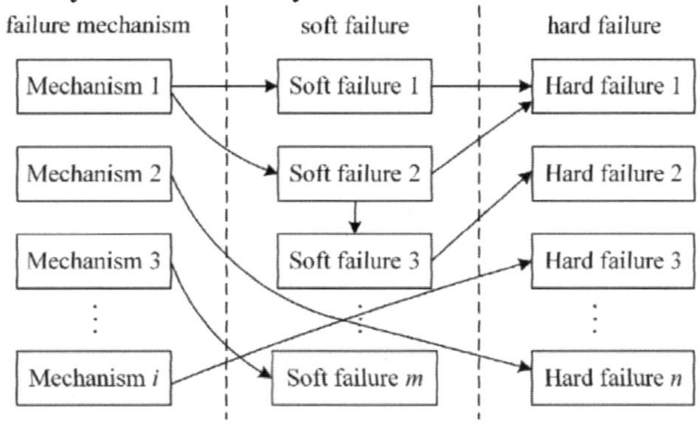

Figure 1. The relation between failure mechanism, hard failure and soft failure

This paper only establishes the failure mode reliability deployment house according to some main failure mechanisms. The reliability deployment house consists of three matrices:

(a) Relationship matrix between failure mechanism and failure mode of parts, $\mathbf{U}\{u_{ij}\}$.

The element u_{ij} expresses the relationship between failure mechanisms and soft and hard failure modes. In this paper, the assessment value of u_{ij} is divided into weak, medium and strong three grades, respectively take the integer 1, 3 and 9.

(b) Failure mode autocorrelation matrix, $\mathbf{P}\{p_{ij}\}$.

The element p_{ij} represents the correlation coefficient between failure mode p_i and p_j on the same component.

(c) Reliability loss weight matrix of failure mechanisms, $\mathbf{W}\{w_i\}$.

The element w_i represents the magnitude of the loss of reliability caused by failure mechanism *i*, it can be obtained by the evaluator according to the relevant reliability statistics. The weight is expressed by integers 1 to 9, according to the reliability loss from small to large.

2.2 *Comprehensive deployment process for technical characteristics*

According to the basic principles of RFD, and considering the interaction of the design value of technical characteristics. The whole process can be divided into the following three stages:

(a) First phase, on the basis of product reliability allocation and FMECA analysis, the failure mode reliability allocation of each component is established to construct the failure mode reliability deployment space. And the reliability loss of each failure mode is calculated with the matrix structure of the failure mode reliability deployment house of each component. According to the size of the reliability loss, the reliability index values are assigned to the failure modes of the component, and output the main failure mechanisms corresponding to each failure mode.

(b) Second phase, input the results of the first phase, the comprehensive deployment space of the technical characteristics of each part was established. In this space, the connection equation or the limit state equation is established for the main failure modes of each component according to its main failure mechanisms, and the design parameter value can be found.

(c) Third phase, input the result of the second phase, and the matching and conflict resolution space for design parameters is established. In this space, according to certain rules, the deployment design parameters of each part are matched and resolved conflicts, and get the final design parameter deployment value.

The above configuration process is integrated with the RFD theory, inverse reliability method based on probabilistic design and the conflict resolution technique, and the reliability design process of mechanical parts is integrated into staged data processing and calculation process. By using the matrix decomposition structure of RFD method, the reliability index of part is further allocated to various failure modes, finally, the inverse reliability method is used to establish the mapping relationship between reliability index and mechanical component design parameter, which makes the reliability design and performance design of mechanical parts more closely integrated, so the process are systematic, hierarchical and practical.

3. Mathematical model for allocating technical characteristics based on multi-failure modes correlation

After the total reliability of product is allocated to components, if the distribution of stress and intensity is known, the reliability index β corresponding to the reliability value of components can be obtained by looking up tables. β is used as the input of the failure mode reliability deployment space, the relationship matrix between failure mechanisms and fail modes will be established, the reliability loss of each failure mode will be assessed with considering of the impact of the correlation between failure modes. Then according to the size of the loss of reliability, the total reliability index β of the component will be allocated to failure modes.

If a failure mode i is independent, the causing reliability loss is FL_i, there is

$$FL_i = w_j u_{ij} \tag{1}$$

w_j is reliability loss weight of failure mechanism j corresponding with the failure mode i; u_{ij} is the correlation coefficient between failure mode i and failure mechanism j.
Suppose soft failure mode i directly leads to k association failures $k=(1,2,\cdots,m)$ corresponding to t failure mechanisms $t=(1,2,\cdots,m)$, considering the single layer chain effect, the calculating equation for reliability loss of failure mode i can be obtained.

$$FL_i = w_i \cdot u_{ij} + \sum_{\substack{k=1 \\ t=1}}^{m} p_{ik} \cdot w_t \cdot u_{kt} \tag{2}$$

Where, $p_{ik}=1, \forall i=k$; p_{ik} is the probability that soft failure mode i raises the associated failure mode k; w_t is the reliability loss weight of failure mechanism t corresponding to failure mode k; u_{kt} is the correlation coefficient between failure mode k and failure mechanism t.
Considering the matrix structure of the house of reliability, and considering the propagation effects of multilayer failure modes, equation (2) can be expressed in the form of matrices

$$\boldsymbol{FL} = \left[\boldsymbol{W}^T \boldsymbol{U}\right]\left[\boldsymbol{P} + \left(\boldsymbol{P}^2 + \cdots + \boldsymbol{P}^f\right)\left(1 - \boldsymbol{I}_{mn}\right)\right] \tag{3}$$

$p_{ik}=1, \forall i=k$; **FL** is $1\times m$ reliability loss vector; **P** is $1\times m$ autocorrelation matrix; **W** is $n\times 1$ weight column vector; **W** is $n\times m$ relation matrix. f is the layer of the failure chain.

It can be seen from equation (3) that the longer the failure chain is, the greater the accumulated reliability loss is, for a failure mode which is the initiator of the failure chain. The reliability index allocation coefficient can be acquired by normalizing the obtained reliability loss of failure modes as shown in equation (4). The greater the reliability loss of failure mode is, the larger the reliability index allocation coefficient is.

$$\alpha_i = 1 + \frac{FL_i - FL_{i0}}{\sum_{i=1}^{s} FL_i} \tag{4}$$

Where FL_{i0} represents the loss of reliability when the correlation of failure modes is not be considered, s is the number of failure modes that constitute the propagation chain, the deployment value β_i^* of reliability index for each failure mode can be calculate by equation (5)

$$\beta_i^* = \alpha_i \cdot \beta \tag{5}$$

β is the reliability index of the whole part. According to the information of HOR, the main failure mechanisms corresponding to β_i^* can be obtained, the β_i^* and the corresponding main failure mechanisms are input into the technical characteristics deployment space, and then deploy the technical characteristic values using the inverse reliability method.

4. Case study

In this paper, a spline shaft in the driving system of one vertical machining centre is selected as an example. Through the analysis of the failure data of the machining centre collected from August 2008 to December 2009 and the FMECA, the corresponding house of reliability of the failure modes and house for allocating technical characteristics of the components can be established.

Known conditions of the spline shaft: the transmitting torque: $T=9500N\square m$, σ_T =400N·m. The bending moment of dangerous section is 5710N·m, σ_M =250N·m. The shaft material is 45 steel quenched and tempered, material strength is μ_S =600MPa, σ_S =60MPa. The shaft is subjected to stable cyclic strain, the reliability of the spline shaft during 1000h operation is R= 0.993688. It is assumed that the stress and strength distribution of the spline shaft are normal distribution, and the reliability index value of spline shaft can be obtained by looking up the table: β =2.49.

4.1 Build the house of reliability for failure modes of spline drive shaft

The 8 failure modes of the spline shaft are obtained by FMECA, and the house of reliability of the spline shaft failure modes is established as shown in figure 2.

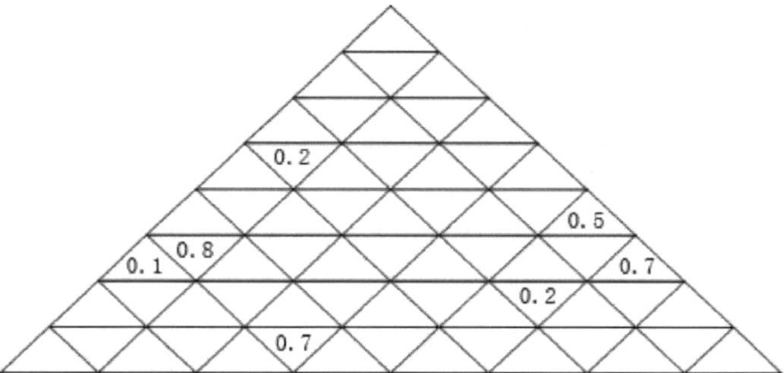

● Strong (9) ○ Medium (3) ▽ Weak (1)		Reliability loss weight	Surface pitting	Surface corrosion	Spline shaft crack	Fatigue crack	Plastic deformation	Spline wear	Torsional stiffness	Large transmission error
Failure Mechanism	Overload (FM1)	0.28			○		○			
	Fatigue failure (FM2)	0.25				●				
	Pitting (FM3)	0.15	●							
	Wear (FM4)	0.15						●		
	Corrosion (FM5)	0.05		●						
	Poor stiffness (FM6)	0.12					▽		○	●
	Loss of reliability without considering failure correlation		1.35	0.45	0.84	2.25	0.96	1.35	0.36	1.08
	Loss of reliability considering failure correlation		1.43	0.45	0.84	4.00	1.5	2.40	0.66	1.08
	Reliability index allocation coefficient considering failure correlation		1.01	1	1	1.14	1.04	1.09	1.02	1
	Allocation reliability index value of failure mode		2.51	2.49	2.49	2.84	2.60	2.72	2.55	2.49
	Main failure mechanism corresponding to the failure mode		FM3	FM5	FM1	FM2	FM1	FM4	FM6	FM6

Figure 2. The house of reliability for spline shaft failure modes

HoR assigns the reliability index value β =2.49 to the failure modes through matrix operation, and finally output the deployment value β_j^* of failure modes reliability index and the corresponding main failure mechanisms.

The relationship between the failure mechanisms of the spline shaft and the soft and hard failure modes can be clearly expressed in the $n \times m$ matrix, which is shown in figure 2. In the two-dimensional plane of autocorrelation matrix, there are following propagation paths:

(a) Fatigue crack → Spline shaft crack
(b) Fatigue crack → Surface pitting → Spline shaft crack
(c) Spline wear → Surface pitting → Spline shaft crack
(d) Spline wear → Large transmission error
(e) Torsional stiffness → Plastic deformation → Large transmission error

In the above failure paths, the maximum order of cascading failure modes is f = 2. Using the equation (4) to calculate the reliability loss FL of each failure mode under the consideration of correlations, it can be found that the three failure modes: fatigue crack, torsional stiffness, and spline wear have a greater increase for loss of reliability.

4.2 Technical characteristics integrated deployment for spline shaft

According to the output of reliability deployment house of the spline shaft, it can be found that the fatigue crack and spline wear make greater loss to reliability of spline shaft, and the spline shaft diameter d and the life for wear of spline shaft t are the corresponding design parameters. Therefore, this paper established the connection equation for failure modes fatigue crack and wear of the spline transmission shaft, solved the deployment value of corresponding design parameters using the inverse reliability method.

4.2.1 Technical characteristics deployment of reliability for fatigue crack

In this case, the area which is easy to form fatigue crack is the spline. Due to the specific parameters of the spline shaft (such as the number of teeth, diameter, minor diameter, etc.) is not determined, the average diameter of the spline part d is chosen as the deployment parameter.

The limit state equation can be expressed as

$$G(X) = S - s = S - \frac{32}{\pi d_2^3}(M^2 + \alpha T^2)^{\frac{1}{2}} \tag{6}$$

Where the α represents the coefficient of the type of the cyclic stress. According to 3σ law, the strength limit of the material is $\delta = 600 \pm 180MPa$, $S = \delta_{\min} = 420MPa$. The shaft is rotated in one direction and thus subjected to pulsating cyclic stresses, α=0.6.

Known conditions: $\delta_T = 400N \cdot m$, $\sigma_m = 250N \cdot m$, $\beta^* = 2.84$. The mean value of M and T is the initial value for iteration, $T_0^* = 9500N \cdot m$, $M_0^* = 5710N \cdot m$, take the shaft diameter $d^* = 200mm$ as initial value for iteration, repeat the iteration until the calculated value of the shaft diameter meets the condition of convergence, finally get the deployment value of shaft diameter is 199.7204mm.

Table 1 is a comparison of the calculated results with traditional probabilistic design method and the probabilistic design method considering multiple failure modes. It can be found that due to considering the extra reliability loss caused by the correlation of failure modes, deployment value of shaft diameter increase from 187.8311mm to 199.7204mm, increased by 6.2%.

Table 1. Comparison of calculation results(unit: shaft diameter-mm)

		Allocation Method	
		Traditional Probabilistic Design	Probabilistic Design for Multiple Failure Modes
Original Reliability	R	0.993688	0.993688
Original Reliability Index	β	2.49	2.49
Allocation Coefficient of Reliability Index	α_i	1	1.14
Allocation Value of Reliability Index	β^*	2.49	2.84
Shaft Diameter	d	187.8311	199.7204

4.2.2 Reliability technical characteristics deployment for wear failure

Through the same method, the wear life of the spline shaft was calculated. The wear life declines from $331h$ with traditional probabilistic design method to $318h$, reduced by 3.8%.

5. Conclusion

This paper proposed a comprehensive deployment method for reliability technical characteristics based on the correlation analysis of multiple failure modes. This method used the matrix structure to describe the transmission relationship among failure mechanisms. And a mathematical model calculating the reliability loss of each failure mode was established with considering the correlation between them. The reliability losses were normalized to obtain reliability index coefficient of each failure mode, established the technical characteristics deployment space and finally assigned reliability index to failure modes. The inverse reliability method was introduced to the calculation model for technical characteristics to do the comprehensive deployment for the technical characteristic. And the method was applied to the reliability design of the spline shaft in the transmission system of a machining centre, the example shows that the method can effectively adjust and optimize the results of the conventional probabilistic design. The following research will focus on the matching of design parameters and conflict resolution, as well as the corresponding algorithm implementation.

Funding

Supported by the First UHVAC GIL Special Project of China (Grant No. SGJSJX00YJJS1602361), National Nature Science Foundation of China (Grant No.: 51375375)

Reference

[1] Cui W Y, et al., *Failure Spreading Mechanism of Weighted Complex Networks Based on a Tunable Edge Weight Model.* 2016 First Ieee International Conference on Computer Communication and the Internet (Iccci 2016), 2016: p. 275-278.

[2] Chen Y, et al., *Failure* mechanism *dependence and reliability evaluation of non-repairable system.* Reliability Engineering & System Safety, 2015. 138: p. 273-283.

[3] Li Y Y, et al., Modeling *and Analysis of Failure Mechanism Dependence Based on Petri Net*, in 2016 Prognostics and System Health Management Conference, M.J. Zuo, et al., Editors. 2016, Ieee: New York.

[4] Wang R X, et al., Complex network theory-based condition recognition of electromechanical system in process industry. Science China-Technological Sciences, 2016. 59(4): p. 604-617.

[5] Kang B J, J.H. Kim, and Y. Kim, *Engineering criticality analysis on an offshore structure using the first- and second-order reliability method.* International Journal of Naval Architecture and Ocean Engineering, 2016. 8(6): p. 577-588.

[6] Lu Z H, Hu D Z and Zhao Y G, *Second-Order Fourth-Moment Method for Structural Reliability.* Journal of Engineering Mechanics, 2017. 143(4): p. 5.

IOP Conf. Series: Materials Science and Engineering **281** (2017) 012058 doi:10.1088/1757-899X/281/1/012058

[7] Chien Y H, A number-dependent replacement policy for a system with continuous preventive maintenance and random lead times. Applied Mathematical Modelling, 2009. 33(3): p. 1708-1718.
[8] Che J G, et al., *Equipment Systems Reliability Analysis Based on FTA*. 2012 International Conference on Quality, Reliability, Risk, Maintenance, and Safety Engineering (Icqr2mse), 2012: p. 293-296.
[9] Catelani M, et al., *Toward a New Definition of FMECA Approach*. 2015 Ieee International Instrumentation and Measurement Technology Conference (I2mtc), 2015: p. 981-986.
[10] Sharma R K and Sharma P, *Qualitative and quantitative approaches to analyse reliability of a mechatronic system: a case*. Journal of Industrial Engineering International, 2015. 11(2): p. 253-268.
[11] Ashrafi M, Davoudpour H and Khodakarami V, *Risk assessment of wind turbines: Transition from pure mechanistic paradigm to modern complexity paradigm*. Renewable & Sustainable Energy Reviews, 2015. 51: p. 347-355.
[12] Dewi D R S., Rahaju D E S and Ieee, *An Integrated QFD and Kano's Model to Determine the Optimal Target Specifications*. 2016 International Conference on Industrial Engineering, Management Science and Applications (Icimsa), 2016: p. 5.
[13] Kato T and Matsuoka Y, Quality Function Deployment Using Improved Interpretive Structural Modeling, in Human Interface and the Management of Information: Information and Knowledge Design and Evaluation, Pt I, S. Yamamoto, Editor. 2014, Springer-Verlag Berlin: Berlin. p. 352-363.
[14] Du L, Gao J M. and Chen K., *Reliability deployment based on fault correlation analysis*. Computer Integrated Manufacturing Systems, 2011. 17(9): p. 1973-1980.
[15] Wang L., Peng J and Shao Y, Multidisciplinary Collaborative Reliability Analysis of the Gear Reducer based on Inverse Reliability Strategy. Journal of Mechanical Transmission, 2015. 39(7): p. 16-19.
[16] Yi P and Zhu Z, *Step length adjustment iterative algorithm for inverse reliability analysis*. Structural and Multidisciplinary Optimization, 2016. 54(4): p. 999-1009.

Tribological Behavior of Coating Cr Layer on 40Cr after Surface Electron Beam Pretreatment

J J Hu[a*], J Wang[b], P Jiang, H B Xu, H Li and T F Hou

Chongqing University of Technology, Banan, Chongqing400054, China

Email:[a] 5671538@qq.com,[b]w1181748426@163.com

Abstract. In this study,the friction and wear behavior of PVD coatings which were treated by 5 different processes,based on gear material-40Cr. Analyzing the effects of treating the gear material with electron beam in combination with magnetron sputtering on it,for dry friction and wear properties.The result showed that the electron beam pretreated substrate was useful to improve the tribological performance of coating material.Furthermore, the surface roughness of coating ,the bonding force between substrate and coating as well as the load are the main factors affecting the tribological performance of this coating. Most importantly, the contribution of plowing effect on friction coefficient should be considered when the surface roughness is high.

1. Introduction

Gear transmission as the largest proportion of transmission form in the mechanical products has complex stress situation and a diversity of failure forms. Statistics showed that the gear transmission failure caused by the damage of the gear surface has been up to 74%[1,2]. This suggests that improving the gear surface performance is an effective way to improve the performance and service life of gear transmission.

Electron beam material surface modification can make the material surface cool after rapid melting by it , so it can change material's structure, improve its surface hardness and so on[3,4]. In this paper, the friction coefficient, the three-dimensional topography of worn surface, SEM and local area EDS analysis of worn surface will be used for exploring the effects of electron beam pretreatment on the tribological performance of PVD coating and discussing the wear mechanisms[5-7].

2. Material and experimental

The common gear material—40Cr(C 0.39 %,Si 0.23%,Mn 0.70%,Cr 0.80%,Ni≤0.30%） was used in this experiment. There are five samples, after them have been modulated, sample 1 sputtered 2 μm Cr coating,and others treated by electron beam irradiation at the same voltage(27kV)while different times(sample 2,3,4 and 5 are 5times,15times,25times,35times respectively)then sputtered 2 μm Cr coating on their surface.The reciprocating dry sliding tests were carried out on a ball-on-flat tribometer (MFT-R4000), which has been described in detail in the literature[8] The friction pair is a Al_2O_3 ceramic ball (hardness: 1900 HV) with 4mm in diameter.

The Carle Zeiss vertical metallographic microscope Axio Imager (A10), super depth filed microscopy (Kean VR-3000) and scanning electron microscope (S-3700N) was used to observe the morphology of the wear tracks. Ambios profiler was used to measure their cross-sectional contour and wear volume. The energy dispersive spectrometer (EDS) in combination with the SEM morphology were also used to analyze the composition of worn surfaces and wear debris.

3. The experimental results

3.1. The Coefficient of Friction

The friction behavior of PVD coatings on 40Cr with various normal loads are illustrated in Figure 1. In general, the friction coefficient increased during the early stage of the test and then rapidly climbed to a relatively stable stage value and fluctuated around this stage at different degrees. After experiencing certain cycles, it began to decline obviously but eventually approached to a stable value. This suggests that the wear of the coating got into a dynamic equilibrium. The pictures reveal that the friction coefficient in relatively stable stage experiences more and more serious fluctuation with the increasing load except for the sample 5, moreover, this trend in sample 1 is most obvious. The mean stable friction coefficient of all samples increased with increasing load except for the sample 3,and it approximately increased with the increasing pre-irradiated times ,in general(as shown in Figure 2). This reveals that the load is not the only factor affecting the coating's friction coefficient, and the friction coefficient of the samples which were pretreated by electron beam were not constantly increasing with the increasing load.

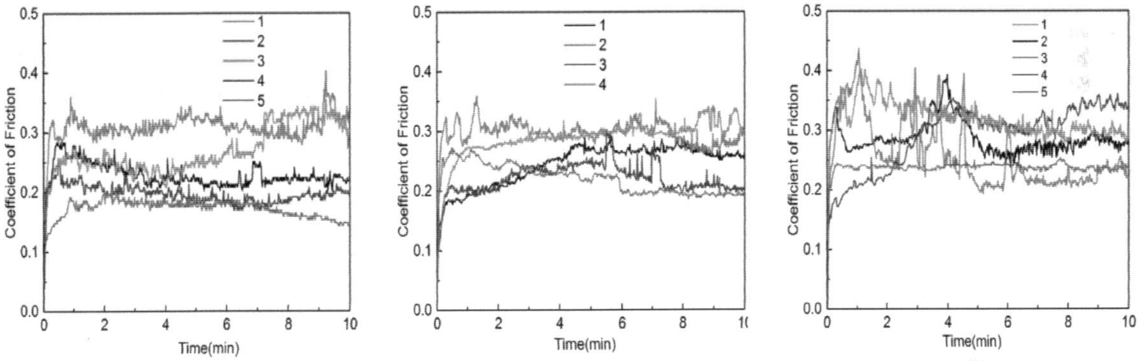

(a) load 5N (b) load10N (c) load 15N

Figure 1 Effect of applied load on the friction coefficient

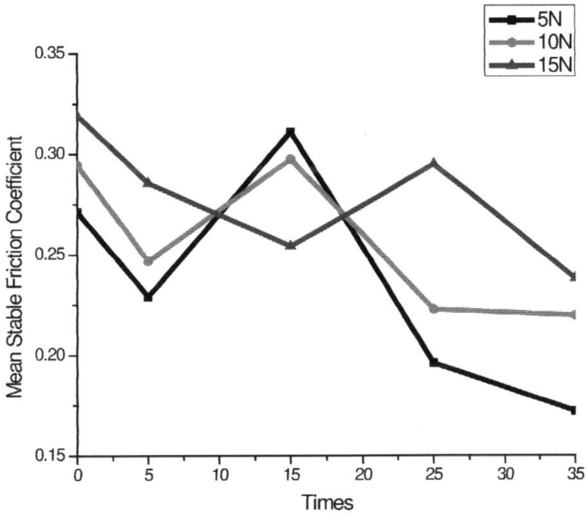

Figure 2 Mean stable friction coefficient at different load

IOP Conf. Series: Materials Science and Engineering **281** (2017) 012059 doi:10.1088/1757-899X/281/1/012059

3.2. Three Dimensional Morphology

The three-dimensional morphology of the wear tracks (as shown in Figure 3) indicates that all samples wear tracks morphologies had different shape and fluctuation degree, and although the width and depth of them changed a lot, that of sample 5 remained basically unchanged. The results suggest that the wear of samples gradually extended to the substrate from coating with the increase in the load except for the sample 5. Moreover, the morphology of wear tracks of samples 1, 3 and 4 are experienced a greater change and less stable wear than other 2 samples at various normal loads.

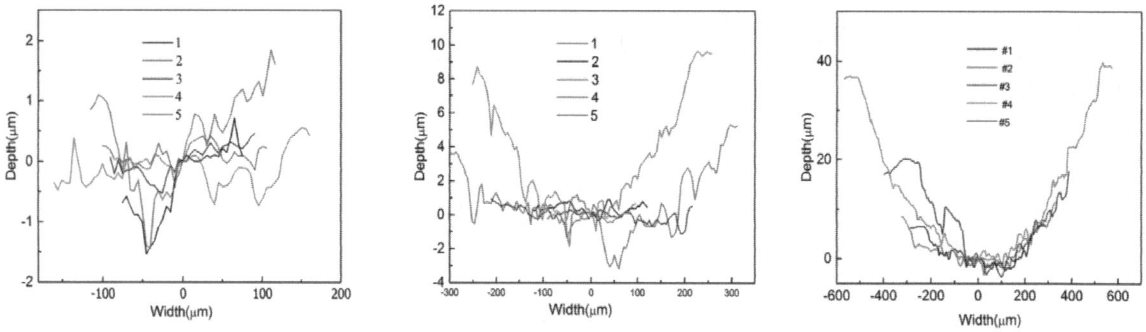

(a) load of 5N (b) load of 10N (c) load of 15N

Figure 3 The three-dimensional morphology of worn samples

3.3 The Fracture Morphology of Worn Surfaces

As we can see from the SEM micrograph of fracture morphology (as shown in Figure 4), five main types of wear morphologies were observed on the worn surfaces of all samples at various normal load. They were sheet spalling and delamination, continuous ploughing grooving, sheet pit, crack, sheet and filamentous shallow traces. At the different load situations,it's easily found that sample 5 had the highest degree of wear, and it always kept the simple ploughing lines morphology ,while others not only were worn severely but also had a very complex morphology.

4. Analyses and discussion

4.1. The Wear Mechanism

In order to more clearly clarify the wear mechanism of the five sample groups more clearly, the energy spectrum was used to analyze the oxygen content value in the representative worn areas of each sample at different load situations, and the results are shown in Table1. It indicates that wear tracks of of each sample have oxide under all load situations.The dominate wear mechanism of coating in which the substrate hadn't pretreated was oxidation wear and fatigue wear.

The binding force of coating improved after the substrate had been treated by electron beam, the coating was too hard to chipped off under the repeated stress. With the increase of binding force, the spalling speed of coating was reduced,mainly abrasive wear was observed.

Therefore, it can be concluded that the main wear form for samples which were modified by electron beam pretreatment and PVD coating composite process is the abrasive wear. However,the bonding force between the coating and the substrate is the main factor which decides the wear mechanism of coating at a higher load. When the binding force is lower, oxidation wear is more likely to be observed.

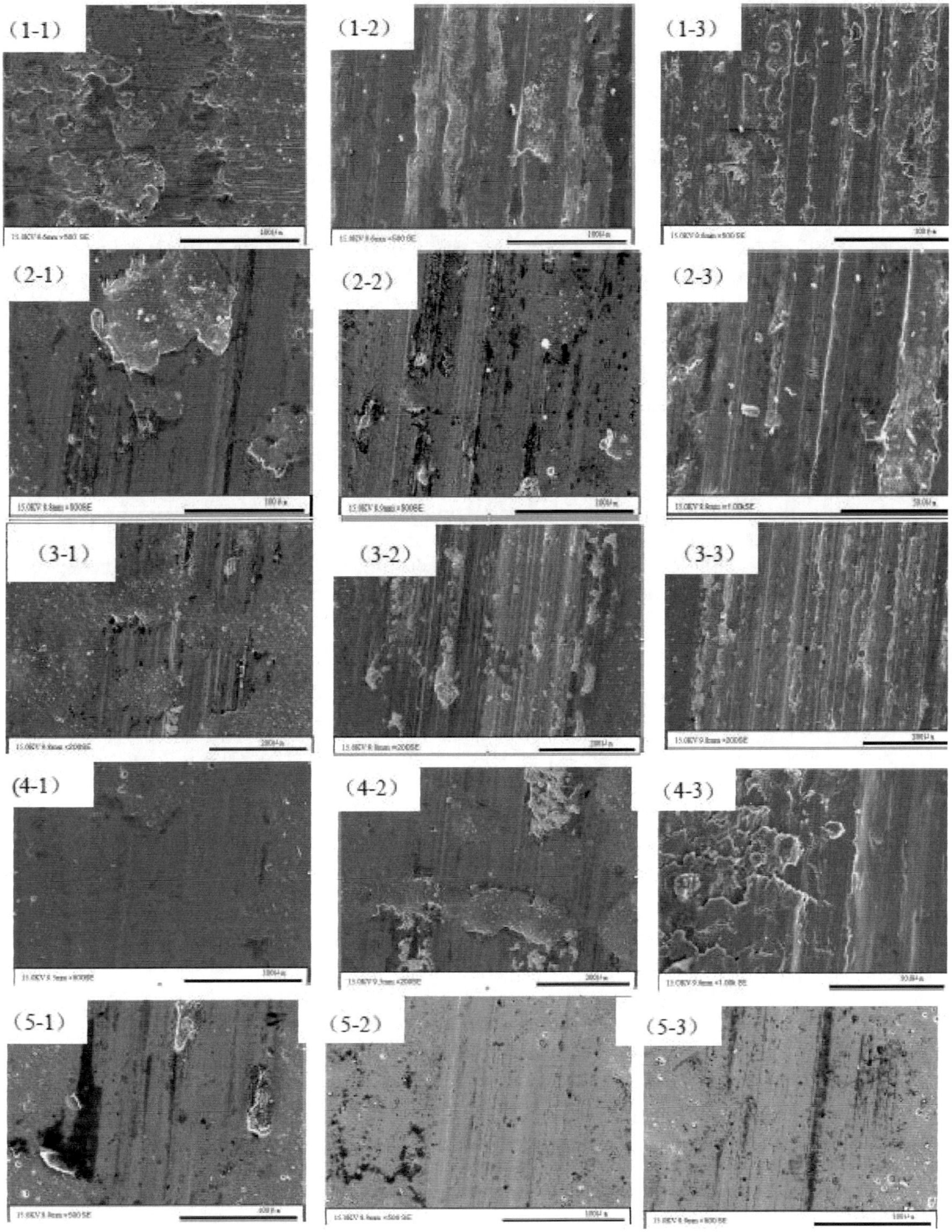

Figure 4 Representative SEM fractographs of worn surface for different samples (a)sample 1,(b)sample 2,(c) sample 3,(d)sample 4,(e)sample 5 under 5N,10N and 15N loads respectively

Table 1 Each sample's energy spectrum analysis of the oxygen content value on its representative areas at various loads (Wt %)

Load / Sample	5N	10N	15N
1	16.02	17.72	3.54
2	4.57	6.60	14.52
3	6.75	5.51	3.30
4	4.07	4.74	3.80
5	4.46	3.92	0.78

And with the gradually increase of binding force the fatigue wear and abrasive wear are also more susceptible to observed. However, the samples for which substrate were not pretreated by electron beam, fatigue wear and oxidation wear are observed at a lower load due to their poor bonding force.

4.2. Effects of Electron Beam Pretreatment on the Wear Resistance of the Coating

According to the research carried by Kenneth Holmberg [9], the coating-to-substrate hardness relationship, the thickness of the coating, the surface roughness and the contact debris are the four main factors affecting the macro wear mechanism of the coating.

In this study, however, the electron beam was used to pretreat substrate so these four factors mentioned above are also the main factors that affected the tribological properties of the coating. The hardnessincreased with the increasing irradiation time but the roughness increases first and then decreased with it on 27kV(as shown in Figure 5 and Figure 6). In addition, the bonding force between the substrateand the coating can also be improved by electron beam pretreatment. For example, the bonding force of 40Cr +Cr coating without electron beam pretreatment was about 12N,but it was improved afterthe electron beam pretreatement. Among them, the bonding force of samples 2 and 3 increased with 33.3% ,probably reaching 16N,and the bonding force of 4 and 5 increased with 41.7%, probably reaching 17N[10].

For the coating material itself, the bonding force between substrate and coating and the surface roughness determined by electron beam parameters have significant influence on friction coefficient of coating. So, in theory, the dry friction coefficient should decrease with the decrease of surface roughness under the elastic or elastic-plastic contact conditions [11].

So we can conclude that the bonding properties of coating and substrate are the main factors affecting the friction coefficient of the coating material when the binding force between coating and the substrate is poor, and under the same load, the bonding properties are poorer, the friction coefficient is higher. Furthermore, the roughness plays a key role when the roughness is too high. The friction coefficient increases with the increase of roughness in this situation. However, the friction coefficient of the same sample decreases with the increase of normal load. This is mainly because the high roughness surface is more likely to generate three body friction at a higher load.

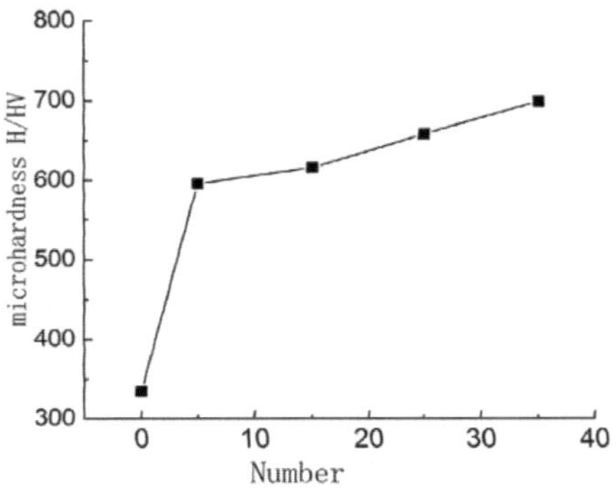

Figure 5. The hardness of pretreated substrates change with the irradiation times

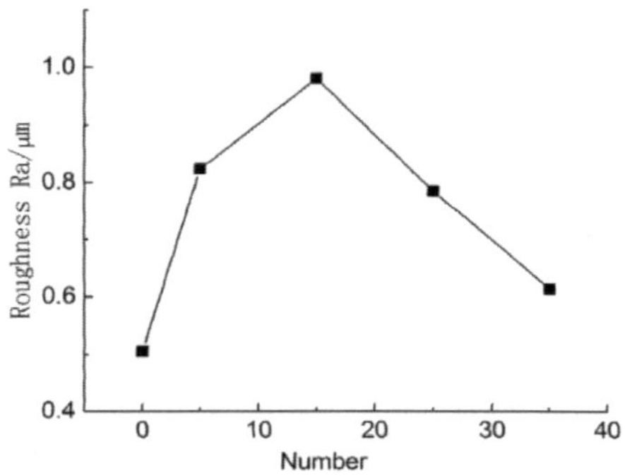

Figure 6. Roughness of samples change with irradiation times

5. Conclusions

Ball-on-flat reciprocating dry sliding wear tests with flats gear coating which was pretreated by electron beam and then sputtered Cr coating with PVD method was investigated against Al_2O_3 ceramic ball counterface at the normal load of 5N, 10N and 15 N. The main results can be summarized as following:

1) Generally speaking, the tribological properties of coating material could be improved by electron beam pretreated 40Cr substrate before it was sputtered to obtain Cr coating. And the sample 5 had the best tribological performance in this study.
2) At a lower load, the main wear mechanism was the abrasive wear. With the increase of load, oxidation wear was observed when the bonding force was poor while the fatigue and abrasive wear were observed when the bonding force was strong.
3) Besides the load, the bonding force between coating and substrate, and surface roughness are the main factors affecting the friction coefficient of the coating material. At the same load, bonding force played a dominant role and made the friction coefficient become larger when the bonding force was poor, however, the roughness played a dominant role and made the friction coefficient became larger when the roughness was too high.
4) The ploughing effect must be considered when the surface roughness is high and it can seriously affect the friction coefficient.

Acknowledgments

This study was supported by the National Natural Science Foundation of China (51575073, 51501158 and 51275548), International Cooperation Special Project in Science and Technology of China (2015DFR70480), and Scientific and Technological Research Program of Chongqing (cstc2014gjhz70003 and cstc2017jcyjBX0031)

References

[1] Liu P D. Study on Failure Analysis and Failure Prevention of Gear. *Gansu Science and Technology* 2013; 02:28-31.
[2] Li R Z, Zhao Q H, *The Technical Information Publishing Division of Japan Society of Mechanical Engineers. Strength Design of Gear Data[M].* Translate. Beijing: Machinery Industry Press 1984.
[3] Xu B S and Liu S C. Surface Engineering[M]. Beijing:Machinery Industry Press 2000.
[4] Li H. *Study on Process of Titanium Nitride Deposition onto Gears and Effects on Load-Carrying Ability of Gears[D].* Chongqing University.2007.
[5] Dong C, Wu A, Hao S, Zou J, Liu Z, Zhong P, Zhang A, Xu T, Chen J, Xu J, Liu Q and Zhou Z. Surface treatment by high current pulsed electron beam[J]. *Surface and Coatings Technology* 2003,163-164:620-624.
[6] GaoY K. Use of low-energy, high-current electron beams for surface treatment of materials [J]. *Applied Surface Science* 2011;257(17):7455-7460.
[7] Guan Q F, Yang P L, Zou H and Zou G T Nanocrystalline and amorphous surface structure of 0.45%C steel produced by high current pulsed electron beam[J]. *Materials Science* 2006;41(2):479-483.
[8] Zhou F, Chen K M, Wang M L, Xu X J, Meng H, Yu M and Dai Z D. Friction and wear properties of CrN coatings sliding against Si3N4 balls in water and air[J]. *Wear* 2008;265(7-8)1029-1037.
[9] Holmberg K., Ronkainen H and Matthews A. Tribology of thin coatings[J]. *Ceramics International* 2000;26(7):787-795.
[10] Zhou Y. The study of technology and properties based on 40 Cr surface coating preparation by electron beam[D]. *Chongqing University of Technology* 2014.
[11] Liu Z L. Principles of Metal Tribology(The third edition)[M]. *Higher Education Press* 2009.

IOP Conf. Series: Materials Science and Engineering **281** (2017) 012060 doi:10.1088/1757-899X/281/1/012060

Bio-Tribology Properties of Bionic Carp Scale Morphology on Ti6A14V Surface

W Wang[1*], X Y Wei[1], K Meng[1], L H Zhong[2], Y Wang[1*] and X H Yu[3*]

[1] College of Mechanical and Manufacturing Engineering, Southwest Forestry University, No. 300 Bailong temple, Panlong District, Kunming, Yunnan, China [2] College of Large Data and Information Engineering, Southwest Forestry University, No. 300 Bailong temple, Panlong District, Kunming, Yunnan, China

[3] Faculty of Materials Science and Engineering, Kunming University of Science and Technology, No. 253, Xuefu Road, Wuhua District, Kunming, Yunnan, China

*Email: 1075642893@qq.com

Abstract. In order to improve the bio-tribology properties of Ti6A14V surface, the bionic carp scale appearance pattern on Ti6A14V surface was prepared by laser surface texturing technology. The ball-disc reciprocating linear tribological experiment under different lubricants with dry friction was carried out by MRTR multifunction friction and wear testing machine using ZrO2/Ti6A14V as friction pair. The wear scar morphology of the sample surface was observed by SEM. The results show that for dry friction, the friction factor of the bionic carp scale morphology Ti6A14V reduces by 0.23 than those without bionic carp scale morphology, a decline of 45%. Under different lubrication conditions, the friction factors of samples with the bionic carp scale are increased in varying degrees with the increase of size of bionic texturing. The friction factor with same specimen under different lubrication conditions according to the ascending order are 0.5g/dl of sodium hyaluronate +0.5g/dl-γglobulin and 0.5g/dl mixed aqueous solution of sodium hyaluronate solution and artificial saliva. The wear volume also showed a similar variation.

1. Introduction

Ti6Al4V alloy has many advantages, such as high specific strength, good corrosion resistance, and high temperature resistance and so on. It is widely used in the manufacture of high temperature parts such as Aeroengine Blades [1, 2]. At the same time, titanium alloy has many disadvantages, such as high friction coefficient and poor wear resistance, which seriously affects the safety and reliability of the structure [3]. Therefore, researchers are now focusing on improving the surface of the material to reduce friction factors and improve tribological properties.

Zhang et al [4] found that laser surface treatment can effectively improve the hardness, wear resistance and corrosion resistance of the material. Lian et al [5] found that the hydrophobic Ti6Al4V surface conformed to the Wenzel state after laser processing and low surface energy modification. Under the condition of loading, the friction coefficient decreases with the increase of surface hydrophobicity. Sorin-Cristian et al [6] found that surface texture can significantly reduce the friction coefficient and can effectively act on the actual mechanical components. Tang et al [7] developed a biomimetic robotic fish tail fin that gave the robotic fish a greater thrust and was more stable with

IOP Conf. Series: Materials Science and Engineering **281** (2017) 012060 doi:10.1088/1757-899X/281/1/012060

amplitude and frequency. Li et al [8] established a model of bionic jet surface by imitating the gummy shark skin, in order to achieve drag reduction and energy-saving effect. These studies show that the bionic technology has great potential at present.

To sum up, it is not difficult to find that the surface structure of many biological tissues can be applied to the material surface to optimize the performance of drag reduction and friction reduction. The non-smooth structure of the surface of animal body is common, and the carp scale is beneficial to buffer its external force, and it has a protective effect on the body. At present, studies on the tribological properties of carp scale have not been reported [9-11]. Therefore, this paper aims at using a new technique to fabricate the shape of carp like scales, so as to reduce the friction effect.

2. Experimental materials and methods

2.1. Selection and simplification of scale shape of bionic morphology

Figure 1 (a) is the local tissue topography of carp, It can be observed that the carp scales are arranged in an orderly arrangement of irregular ellipsoids.At present, the studies on the morphology of carp scale mainly focus on the ellipsoids structure and arrangement [12].Therefore, this paper takes the scales of elliptical carp as the object of study, and the irregular ellipsoids structure of carp scale is simplified into regular ellipsoids structure,As shown in Figure 1 (b).

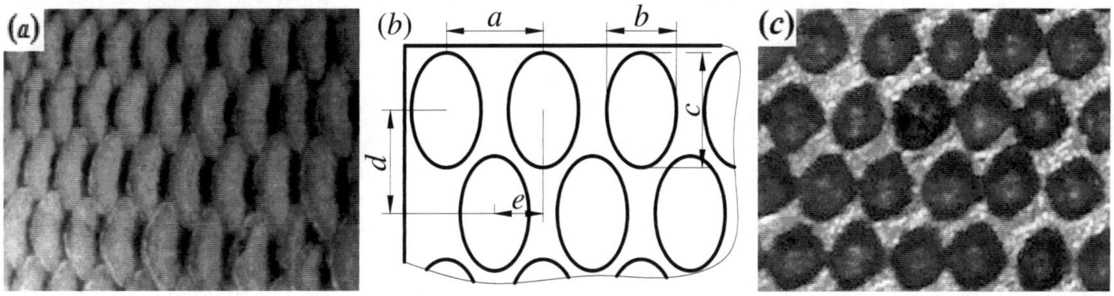

Figure1. Design of the bionic pattern (a) Local scales of carp (b) Bionic design sketch (c) The bionic morphology after polishing

2.2. specimen preparation

Ti6A14V alloy was processed into the sample with the size of 10×10×5 mm, which then underwent accurate grinding by 600#, 1200#, and 2000# sandpaper, respectively. Next, diamond paste was used to polish it to mirror surface. The sample surface was under bionic surfacing treatment by HGL-LSY50F laser process equipment, whose laser wavelength was 1064 nm, operating voltage AC220 V, output power 75 W, and output current 12.5 A. a total of 4 groups of samples, respectively, numbered A, B, C, D, detailed parameters in Table 1. Four groups of samples of bionic carp scales was polished slightly so that laser process residues were removed. Afterwards, the samples underwent ultrasonic cleaning for 30 min in acetone solution, followed by hot-wind drying before it was placed in a dryer at the ready. Figure 1 (c) shows the micro-morphology of the bionic carp scales sample.

Table 1.The sample number and design size /mm

Samples	*a*	*b*	*c*	*d*	*e*
A	0.07	0.05	0.08	0.072	0.035
B	0.14	0.1	0.16	0.144	0.07
C	0.21	0.15	0.24	0.219	0.105
D	Non-textured sample				

2.3. Lubrication conditions

Artificial saliva (produced in accordance with the ISO1021 Standard), 0.5 g/L sodium hyalurate, and 0.5 g/L sodium hyalurate +γ-globulin were used separately as the lubrication for the bio-tribological test.There are four test conditions for dry friction and three kinds of lubricating fluid, which are numbered No.1, No.2, No.3 and No.4 respectively.

2.4. Bio-tribological tests

Bio-tribological tests at room temperature, the ball-plate reciprocating bio-tribological test was undertaken on MRTR multi function tribometer, and the friction pair chosen ZrO2 ball with 72~74 HRC and the diameter of 5 mm. The external load was 10 N, the one-way distance 5 mm, the reciprocating velocity 100/min, and the test period 10min. Under the same test parameters, four groups of samples were subjected to friction and wear tests under dry friction and three different lubrication conditions.

3. Results and discussion

3.1. Friction Properties of Ti6Al4V Alloy with Morphology of Bionic carp scales

Figure 2 is the changing curves of friction coefficients for A, B, C and D samples with sliding time under No.1, No.2, No.3 and No.4 lubrication , respectively. Table 2 is the steady-state friction coefficients that are calculated from the changing curves in Figure 2. The friction coefficient of D sample under dry friction fluctuates between 0.47~0.59, The steady-state friction coefficient is 0.51; the friction factor of B sample under dry friction is between 0.23~0.32, and the steady friction coefficient is 0.28.Under the condition of dry friction, the fluctuation range of friction coefficient of B sample is smaller than that of D sample, and the steady-state friction coefficient drops by 45%.The pits of the bionic carp scales have the effect of collecting debris under dry friction[13,14], and the debris also acts as a solid lubricating particle, thereby significantly reducing the fluctuation range of the friction factor[15,16].

Under the condition of No.1, No.2 and No.3 lubrication, the steady - state friction factors and fluctuation range of biomimetic carp scales increased with the increase of bionic structure size. Among them, under No.1 lubrication, the fluctuation ranges of friction factors of A, B and C samples are 0.26~0.41, 0.30~0.45 and 0.26~0.48, respectively. The steady-state friction factors are 0.35, 0.38 and 0.39 respectively; Under No.2 lubrication, the fluctuation ranges of friction factors of A, B and C samples are 0.26~0.32, 0.25~0.36 and 0.27~0.41, respectively. The steady-state friction factors are 0.29, 0.30 and 0.33 respectively; Under No.3 lubrication, the fluctuation ranges of friction factors of A, B and C samples are 0.31~0.41, 0.32~0.45 and 0.37~0.45 respectively, and the steady friction factors are 0.37, 0.39 and 0.43 respectively; The size of the bionic structure is small, the whole area of the bionic texture is large, the cumulative effect of the lubricating medium in the hole area and the pumping effect will be more obvious, which leads to the strong dynamic lubrication effect, and the dynamic pressure lubrication will reduce the bionic Structural Friction Factors and Fluctuation Range[17,18].

The same sample under No.1, No.2 and No.3 lubrication, the steady state friction factors are related to the lubricant, As can be seen from the data in Table 2, the friction steady-state factors were the lowest under the No.2 lubrication, and the friction steady-state factors were the highest under No.3 lubrication. The reason for the minimum friction factor in No.2 lubricating fluid is that the bionic scales have a certain streamline shape and the frictional resistance of water and scale is rapidly converted into water with the action of γ-globulin water mucus Frictional resistance of mucus. A water film is formed between the water and the mucus, reducing the fluid resistance to the scale, thereby reducing the friction factor of the sample [19-21].

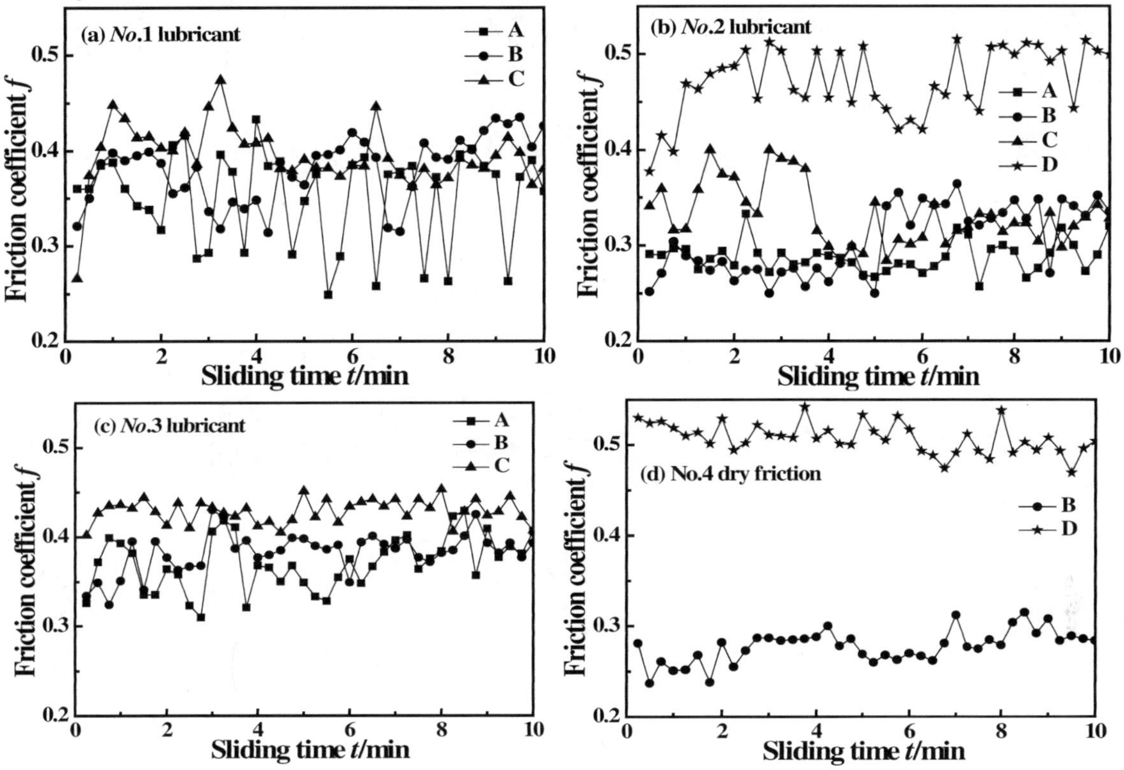

Figure 2 The friction factor of sample changing with time under different lubrication conditions

Table 2. The steady state factor of scale sample of bionic carp under different lubrication conditions

Samples	No.1	No.2	No.3	No.4
A	0.35	0.29	0.37	×
B	0.38	0.30	0.39	0.28
C	0.39	0.33	0.43	×
D	×	0.47	×	0.51

3.2. Analysis of the scratch morphology

Figure 3 is the scratch morphology of A,B,C,D samples under No. 1, No. 2, No. 3 lubrication , respectively. As shown in Figure 3 (g) and (L), the surfaces of the non-textured specimen are obvious scratches. In most areas there are furrow scratches. As shown in Figure 3 From 3 (a) to (c) and (h) to (j), the bionic morphology of the sample surface is covered by the debris, which proves that No.1 and No.3 are only part of the anti-friction effect. The flow of lubricant leads to the sliding of debris with sliding friction, which affects the deposition of debris. As can be seen from Figure3 (d) to (f), specimen surface scratch the smallest degree, the bionic cavities to maintain the most complete, in the process of friction and scratch, No. 3 in the γ-globulin on the surface of the biomimetic sample has a very good anti-friction effect. In addition, compared with figure 3 (d), (e) and (f), with the increase of the surface area and size of the specimen, the surface cavities are gradually increased the degree of damage. This indicates that under the same lubrication conditions, the smaller the surface and size of the biomimetic sample, the better the scratch resistance in the lubricating fluid.

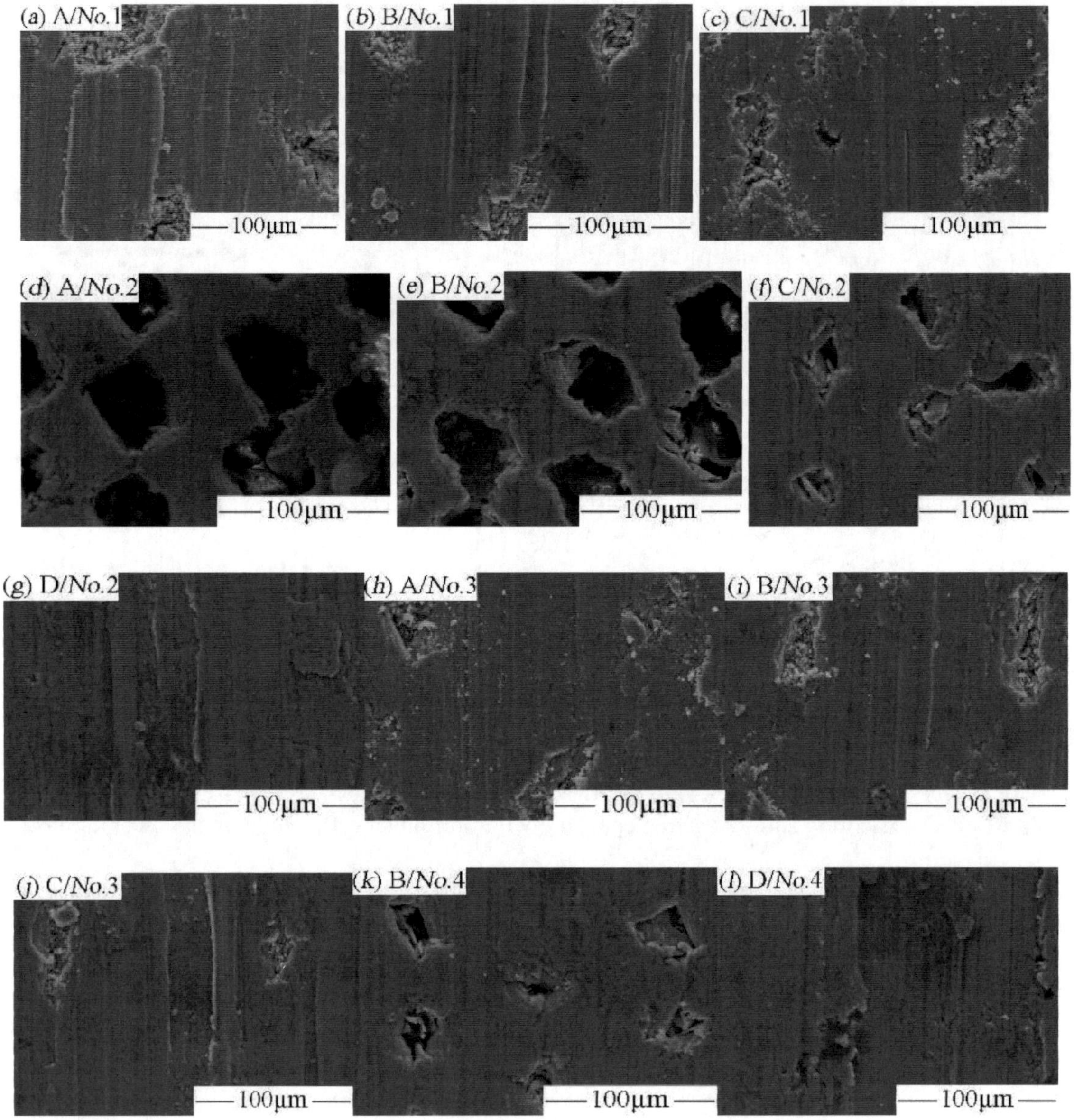

Figure 3 Scarlet morphology of Ti6Al4V bionic carp surface under different lubrication

As can be seen from figure 3 (k) and (l), under the dry friction condition, the bionic morphology of the sample B is better, and the surface of the sample D is seriously damaged. The surface showing a large number of furrow scratch.

4. Conclusions

1) Under dry friction, the steady-state friction factor of D sample is as high as 0.51, and the steady-state friction factor of B sample is only 0.28. The friction reduction properties of Ti6Al4V surface can be improved significantly by simulating the morphology of carp scale.

2) Under the lubrication of No.1, No.2 and No.3, the friction factor of the sample with the shape of carp scale increased with the increase of the size of bionic structure. The size of bionic topography significantly affects its friction factors.
3) Under the lubrication of No.1, No.2 and No.3 in the same sample, the friction factor under No.2 lubrication is the smallest, followed by No.1 lubrication, the maximum under No.3 lubrication, and the wear volume also showed a similar variation.

Acknowledgements
This project have been supported by the Chinese National nature science foundation(Grant .51301144) and Yunnan eight key industrial brand professional construction project foundation(Grant .51600631)

References
[1] Zhang T Z, Zhang C, Li J, Zhang H C and Lu J 2017 *Acta. Optica. Sin.* **37** 0214001-1-10
[2] Chassaing G, Faure L, Philippon S, Coulibaly M, Tidu A, Chevrier P and Meriaux J 2014 *Wear* .**320** 25-33
[3] Sun S Y and Lv W J 2016 *Rare Metal. Mat. Eng.* **5** 1138-41
[4] Wang Z F, Wang Y, Guo J, Zhang P and Wang L 2015 *Rare Metal. Mat. Eng.* **44** 247-54
[5] Lian F, Ren H, Guan S and Zhang H 2016 *Rare Metal. Mat. Eng.* **45** 2182-88
[6] Vlădescu S C, Olver A V, Pegg I G and Reddyhoff T 2016 *Wear.* **358-359** 51-61
[7] Tang L, Deng J, Zhang J and Shao X M 2016 *J. Hydrodyn.* **31** 56-62
[8] Li F, Zhao G and Liu W X 2015 *J. Harbin .U.Eng.* **36** 222-27
[9] Gao J Q, Chen D H, Sun J Y and Tong J 2006 *J. Agri.Mech.* **11** 147-50
[10] Etsion I. 2004 *Tribol. Lett.* **17** 733-37
[11] Gualtieri E, Borghi A, Calabri L, Pugno N and Valeri S 2009 *Tribol. Int.* **42** 699-05
[12] Summers A P. 2004 *Nature.* **429** 31-33
[13] Wan Y and Xiong D S 2006 *tribology.* **26** 603-07
[14] Wang B, Chang Q Y and Qi Y 2013 *Lubrication. Oil.* **12** 11-14.
[15] Pettersson L and Stalfan J 2003 *Tribol. Int.* **36** 857-64
[16] Wang X, Kato K, Adachi K and Aizawa K 2003 *Tribol. Int.* **36** 189-97
[17] Pettersson U and Jacobson S 2004 *Tribol. Lett.* **17** 553-59
[18] Cheng X P, Kang L P, Zhang Y L, X-K.Meng, Peng Xudong, D. Wang and Y. Fu 2015 *Tribology.* **35** *658-64*
[19] Li Z, Qiu N and Yang G 2009 *J. Membrane. Sci.* **326** 533-38.
[20] Zhao W J, Wang L P and Xue Q J 2011 *Tribology.* **31** 622-31
[21] Niţulescu T and Ţălu Ş. 2001 *Pub. House.Riso.* **973** 656-02

IOP Conf. Series: Materials Science and Engineering **281** (2017) 012061 doi:10.1088/1757-899X/281/1/012061

Iris Location Algorithm Based on the CANNY Operator and Gradient Hough Transform

L H Zhong[1*], K Meng[2], Y Wang[2], Z Q Dai[1] and S Li[1]

[1]College of Big Data and Intelligent Engineering, Southwest Forestry University, Kunming, 650224, China
[2]College of Mechanical and Manufacturing Engineering, Southwest Forestry University, Kunming, 650224, China

* E-mail:zhongsimple@sina.com

Abstract. In the iris recognition system, the accuracy of the localization of the inner and outer edges of the iris directly affects the performance of the recognition system， so iris localization has important research meaning. Our iris data contain eyelid, eyelashes, light spot and other noise, even the gray transformation of the images is not obvious, so the general methods of iris location are unable to realize the iris location. The method of the iris location based on Canny operator and gradient Hough transform is proposed. Firstly, the images are pre-processed; then, calculating the gradient information of images, the inner and outer edges of iris are coarse positioned using Canny operator; finally, according to the gradient Hough transform to realize precise localization of the inner and outer edge of iris. The experimental results show that our algorithm can achieve the localization of the inner and outer edges of the iris well, and the algorithm has strong anti-interference ability, can greatly reduce the location time and has higher accuracy and stability.

1. Introduction

The human eye is made of three parts: the sclera, iris and pupil. The iris which is located between the sclera and the pupil contains the most abundant texture information, occupying 65% of the total texture information of the human eye. The iris is determined by genetic genes, so the characteristics of the iris, such as unique, stability and irreversible, are the material basis for the iris to be used for identification. Iris recognition is the most convenient and accurate method in the all biological recognition technology recently, and become the most promising biometric authentication technology in the twenty-first Century[1]. In the future, iris recognition will be used in the field of security, national defense and electronic commerce, etc. Iris recognition technology consists of iris image acquisition, location, feature extraction and matching. Iris localization is the key step of the iris recognition, and the accurate localization will directly affect the final recognition result. So the research of iris location is of great significance[2].

In recent years, iris localization algorithms include gray threshold segmentation method[3-4], active contour method[5], Integro-Differential Operators method[6-7]and Hough transform method[8], etc. Gray threshold segmentation is to compare the gray values of each pixel with the threshold, and to classify the pixels in the image according to whether or not it exceeds the threshold value. Threshold segmentation is a really nice way to separate the pupil, but this method can only be used to the iris image with uniform illumination[9-10]. Active contour method, that is snake model, is an effective method for image segmentation and edge detection, but it has large computational complexity and poor anti-interference. The Integro-Differential Operators method uses the blind search method based on the annular geometric characteristics of the iris, so the detection time is long. Dr. Wildes, from the MIT Artificial Intelligence Lab, proposed the Hough transform algorithm, which gets the binary iris image, and then uses the Hough transform to locate the iris edge. The method needs to search the center and the radius of the circle according to the change of the three parameters, and the computation and storage are exponentially related to the parameter space, which makes the search become complex and takes a long time, so it doesn't work with a lot of images[11]. The iris localization algorithm proposed by Wang Yunhong and Tan Tieniu of the Institute of automation of Chinese Academy of Sciences uses the binary image to separate the pupil, and the Canny operator is used to extract the edge of the image. Then the outer boundary is fitted by the least square algorithm. However, due to the presence of low gray areas beyond the pupil, the method still fails to locate the pupil accurately. In the study of iris localization algorithms, researchers have to keep improving. With the continuous development, a variety of iris localization algorithms have been developed[12-13], but there are always some defects and deficiencies, and a universal detection algorithm has not been found yet[14].

The data in this paper were obtained by the pupil detection instrument which was developed by Kunming Animal Research Institute of Chinese Academy of Sciences. The obtained images contain noise such as eyelids and eyelashes, and the whole image is not obvious gray change, so it is very difficult to locate the inner and outer edges of iris. In this paper, a Canny operator and gradient Hough transform algorithm were proposed for locating the inner and outer edges of iris. Firstly, Gauss low-pass filtering was used to smooth the image. Then, the hysteresis threshold Canny operator was used to roughly locate the inner and outer edges of iris based on the gradient information of the image. Finally, the gradient Hough transform was used to realize the precise localization of the inner and outer edges of iris according to the rough positioning image.

2. The rough positioning for the inner and outer edges of iris

2.1. Data acquisition

In highly non-invasive systems, the position and size of the iris in the image will change. At the same time, the change in the size of the iris's inner border (the pupil) will cause deformation of the iris texture. The data in this paper take into account the actual situation, the subjects were stimulated by visual stimuli in the acquisition process, so the pupil in the collected data would scale, and that is, the iris of the same subject was different in size and location. The data were collected using a pupil detection instrument, which was developed by Kunming Animal Research Institute of Chinese Academy of Sciences. The subjects were all 18~23 years old, non-neurological, and healthy university students. The involved subjects are 300, 150 of them are male, and the other half are female. The used dataset has150000 images and the environment condition during acquisition is bright.The collected iris image is shown in Figure 1.

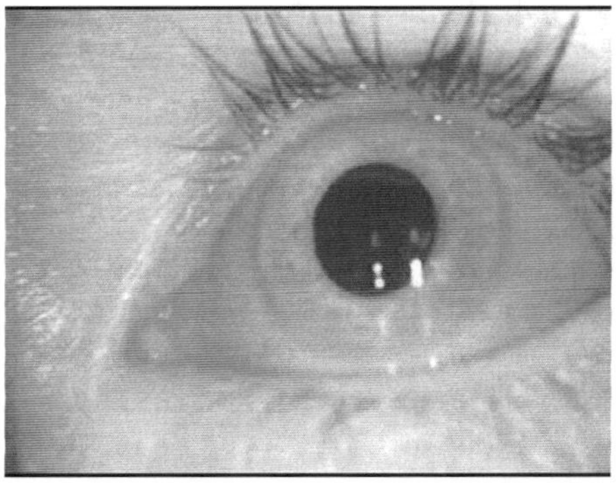

Figure 1. The original image obtained

As is shown in Figure 1, the noise in the image is very high. Noises include eyelids, eyelashes, border lines and spots in the pupils. The whole image is concentrated in the low gray area; the iris is

IOP Conf. Series: Materials Science and Engineering **281** (2017) 012061 doi:10.1088/1757-899X/281/1/012061

almost indistinguishable from the muscle gray around it, so it is very difficult to locate the inner and outer edges of iris.

2.2. Preprocessing of iris images
Firstly, in order to enhance the iris part of the iris image, remove the interference of black lines from the upper and lower boundaries of the original image, so crop the image. The cropped image is shown in Figure 2. Then, because the gray level of the whole image is relatively concentrated, leading to the outer edge of iris is not obvious, and further preprocessing is necessary. The Gauss function was used to smooth the image, which can eliminate the local noise and protect the contour information of the object. The image after Gauss smoothing filtering is shown in Figure 3.

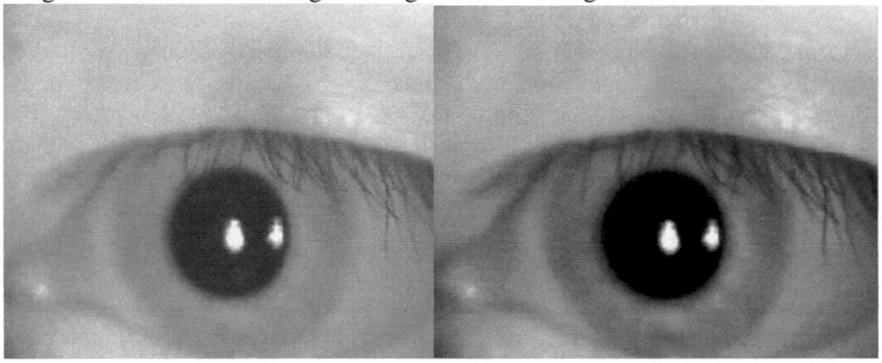

Figure 2. Pre-processing 1-cropped image **Figure 3.** Pre-processing 2-image after Gauss filtering

The comparison between Figure 1 and Figure 3 shows that the interference of black lines at the boundary has been removed and that the iris portion of the image has been significantly enhanced at the same time.

2.3. The rough positioning for iris edges of Canny operator
Canny operator edge detection method has many advantages, such as no loss of crucial edges, without false edges, the deviation between the actual edge and the detected position at edges is minimum, the edge accuracy can be set by parameter to adapt to different classes of images, strong noise immunity and keep the weak edge. Canny operator is very suitable for edge detection of noisy images, so this paper uses Canny operator to roughly locate the inner and outer edges of iris.

2.3.1. The basic process of the Canny operator
Firstly, the annular two-dimensional Gauss function was used to smooth the image. Secondly, the gradient image and the angle image of each pixel in the image were calculated. Thirdly, the gradient amplitude image was suppressed with the non-maximum value according to the degree of the four directions of the gradient angle and the gradient value of each pixel point. Finally, hysteresis threshold processing and connection analysis were used to detect and connect edges.

2.3.2. Realization of the rough positioning for the inner and outer edges of iris
The results of edge location by the Canny operator are shown in Figure 4. As can be seen from figure 4, Canny operator is used to locate iris edge, and the location of inner edge is better than that of outer edge. But there's still a lot of noise.

3. The accurate positioning for the inner and outer edges of iris- gradient hough transform
Iris boundaries can be defined as circular model. Hough transform circle detection algorithm is affected by noise and boundary discontinuities very small, and the positioning accuracy is higher. So Hough transform circle detection is a good algorithm to locate the iris boundary. The method needs to search the center and the radius of the circle according to the change of the three parameters, and the computation and storage are exponentially related to the parameter space, which results in complex

search. The time consumption and memory requirement of the Hough transform are dramatically increased, which is almost impossible to achieve in practice.

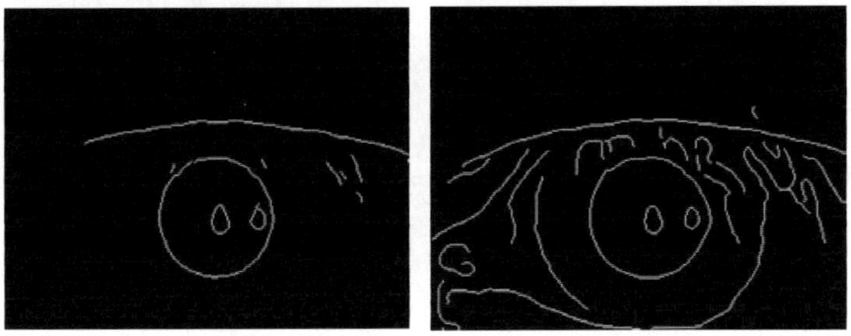

(a) Rough positioning of iris inner edge of Canny operator

(b) Rough positioning of iris outer edge of Canny operator

Figure 4. The results of edge location by the Canny operator

The gradient Hough transform can reduce the computational complexity and solve the problem of invalid accumulation when the edge information is detected by using the information of the edge gradient direction and the continuity of the curve. By introducing the gradient direction information, the accumulated matrix of the circular contour measurement can be reduced from three dimensions to two dimensions, and then the Hough transform algorithm can be improved. The result of the accurate positioning for the inner and outer edges of iris using the gradient Hough transform algorithm are shown in Figure 5.

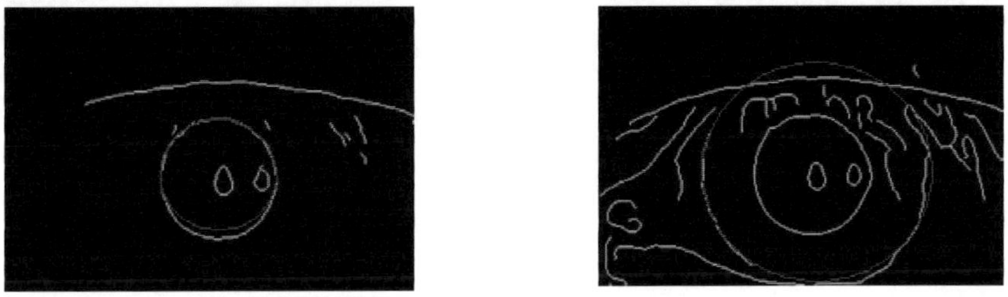

(a) the results of gradient Hough transform algorithm for accurate positioning of iris inner edge

(b) the results of gradient Hough transform algorithm for accurate positioning of iris outer edge

Figure 5. The accurate positioning of iris edges with gradient Hough transform

As can be seen from the figure, the gradient Hough transform can well locate the inner and outer edges of iris. There is a slight error in the results of Figure 5a because we assume that the inner and outer edges of iris are circular.

As can be seen from Figure 6, the algorithm used in this paper can well locate the inner and outer edges of iris.

4. Experimental results

When we collected data, we gave the subjects visual stimulation, so the pupil of each iris image was collected had a scaling process. The algorithm is applied to female subjects, male subjects and data which has large noise in eye movements. The result is shown in Figure 7-9.

IOP Conf. Series: Materials Science and Engineering **281** (2017) 012061 doi:10.1088/1757-899X/281/1/012061

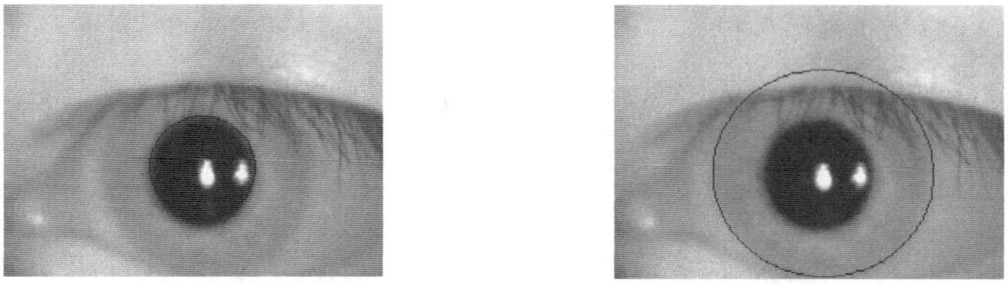

(a) localization result of original image inner edge

(b) localization result of original image outer edge

Figure 6. The localization results for iris edges of the algorithm

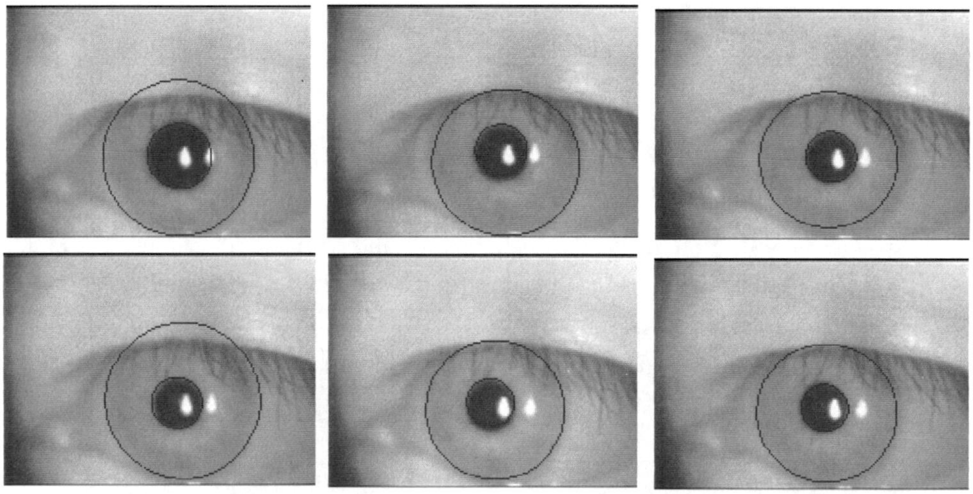

Figure 7. The localization results of iris edges in female subjects

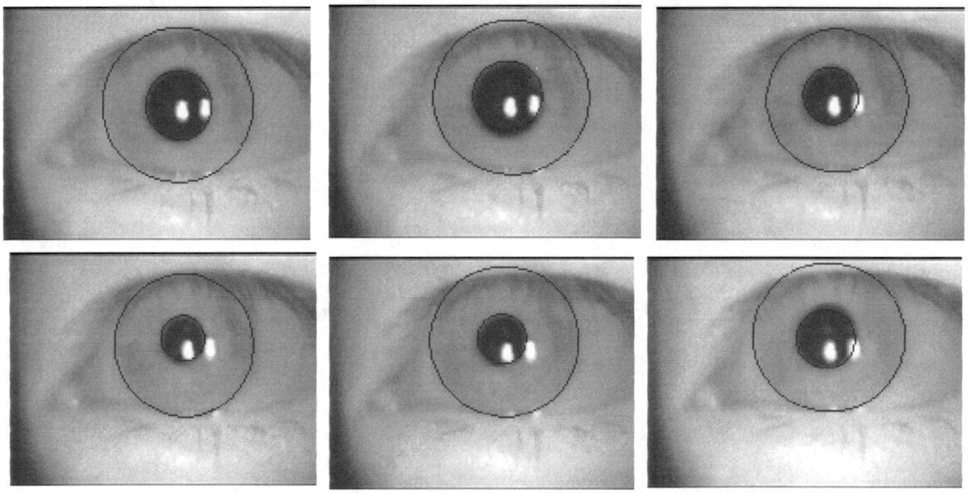

Figure 8. The localization results of iris edges in male subjects

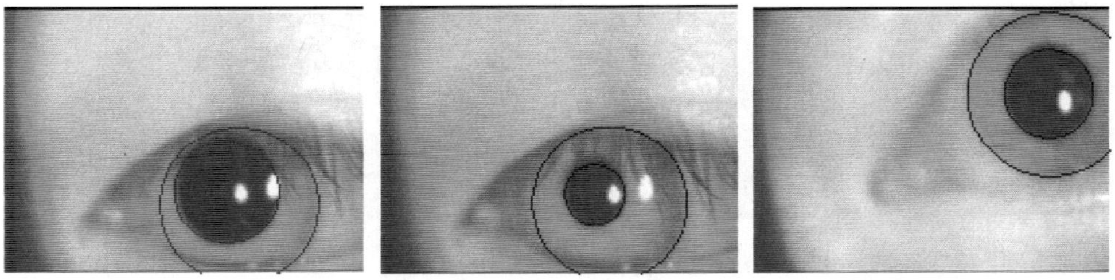

Figure 9. The localization results of iris edges in large noise iris images

In Figure 7 and 8, subjects with different sexes can locate the inner and outer edges of iris in different pupil sizes. In Figure 9, the noises of three images are particularly large. Due to the impact of factors such as eye movement during data acquisition, iris outer edge is not successfully obtained. Moreover, in the case of eyelash interference, the algorithm can still locate the inner and outer edges of iris very well. In summary, the algorithm in this paper can realize the localization of iris both inner and outer edges, the percent of the iris that are correctly localized is 98%, and the anti-interference ability of this algorithm is very strong, which shows the superiority and accuracy of this localization algorithm.

5. Conclusions

For data containing eyelashes, spots and eyelids and other noise, especially the whole image of the gray concentration of the situation, on the basis of the original random Hough transform, a simpler and more accurate algorithm based on Canny operator and gradient Hough transform was proposed. First, the image was cropped and Gauss filtered. Then, through the transformation of image gradient, the Canny operator with hysteresis threshold was adopted to realize the rough positioning for the inner and outer edges of iris. Finally, the gradient Hough transform algorithm is used to locate the image accurately. In this paper, the algorithm overcomes the disadvantages of the traditional random Hough transform, such as the large amount of calculation and selection of many invalid points, avoid blind search, and effectively shorten the positioning time, more accurate positioning of the inner and outer edges of iris. Meanwhile, this algorithm can also achieve excellent inner and outer edge localization for noisy large iris images.

Acknowledgments

The authors would like to acknowledge the financial support provided by the research centre of Kunming forestry information engineering technology and the construction fund of eight key industrial brand professionals in Yunnan Province (Grant No. 51600613).

References

[1] Wang L J, Xun Z Y, Sun Q C 2014 *The Human Iris Image Information Processing and Pecognition Technology* (Beijing: China Water Power Press)
[2] Zhao X P 2011 *Criminal Technique* 06L44
[3] Chen S, Zhu X Q, Li Z M and Pan T H 2011*Computer Engineering and Applications* 47(15)L197
[4] Shi C Q and Jin L X 2010 *Computer Scienc* 37(9)L264
[5] Zhang X D, Dong X P and Wang Q 2011 *Journal of Northeastern University(Natural Science)*, 32(11)L1550
[6] Li X 2014 *Research on the Iris Location and Recognition Algorithm* (Gang Zhou: South China University of Technolog)
[7] Li X G, Sun Z N and Tan T N 2014 *Journal of Image and Graphics*,06L813
[8] Sun W W, Zhou M X and Bai C J 2016 *Computer& Digital Engineerin* 44(1)L113
[9] Zhao J 2013 *Computer Technology and Development* 23(4)L51
[10] Zhang Z F 2009 *Research and Implementation of Iris Recognition Algorithm* (Lan Zhou: Lan Zhou University)
[11] Tan T N, He Z F and Sun Z N 2010 *Image and Vision Computing* 28(2)L223

IOP Conf. Series: Materials Science and Engineering **281** (2017) 012061 doi:10.1088/1757-899X/281/1/012061

[12] Sankowski W, Grabowski K, and Zubert M 2009 *Image and Vision Computing* 26(2)L231
[13] Chen Y, Adjouadi M and Hart C G 2009 *Image and Vision Computing* 26(2)L261
[14] Li P H, Liu X M and Xiao L J 2010 *Image and Vision Computing* 28(2)L246

IOP Conf. Series: Materials Science and Engineering **281** (2017) 012062 doi:10.1088/1757-899X/281/1/012062

Statistical Analysis and Prediction on Tensile Strength of 316L-SS Joints at High Temperature Based on Weibull Distribution

Z L An[1],T Chen*[1], D L Cheng[2], T H Chen[3] and Z Y Wang[1]

[1] School of Railway Transportation, Shanghai Institute of Technology, Shanghai, China
[2] School of Mechanical Engineering, Shanghai Institute of Technology, Shanghai, China
[3] Shanghai Shentong Metro group limited company

* Corresponding email: chentao931@163.com

Abstract. In this work, the prediction on average tensile strength of 316L stainless steel is statistically analyzed by Weibull distribution method. Direct diffusion bonding of 316L-SS was performed at high temperature of 550°C and 8 tension tests were carried out. The results obtained vary between 87.8MPa and 160.8MPa. The probability distribution of material failure is obtained by using the Weibull distribution.

1. Introduction

Diffusion bonding technology is a solid state bonding process, involving grinding and polishing the metal surface, and applying the appropriate amount of nominal pressure. It usually applied in micro heat exchangers and micro reactors [1]. With the development of bonding technology, diffusion bonding has become a preferred fabricating method due to its high processing precision and convenient manufacture [2,3]. As an austenitic stainless steel, 316L stainless steel (316L-SS) possesses more unique merits, such as excellent mechanical behavior and oxidation resistance at high temperature, therefore, it was selected as the base metal in the diffusion bonding technology. The research on room temperature performance of diffusion bonding joint between 316L-SS and heterogeneous material is relatively thorough. However, there are few studies about diffusion bonding connection of homogeneous material of 316L-SS, particularly about the design standard of high temperature performance [4-6]. Moreover, the predicted average tensile strength value of the direct diffusion bonding joints of 316L-SS has not been considered in the studies mentioned.

Weibull distribution is a theoretical basis of reliability analysis and life testing evaluation， and it is widely used in reliability engineering, the cumulative distribution of the form of failure. It can be easily inferred from the probability value of its distribution parameters, and it is also widely used in a variety of fatigue life test data processing [7-10].

In the present work, the tensile strength of the direct diffusion bonding joints was tested at a high temperature of 550°C. The validity of the Weibull distribution was confirmed and the reliability of the 316L-SS in terms of its tensile strength was analysed by experimental data, and the average tensile strength value of 316L-SS diffusion bonding joint is predicted.

IOP Conf. Series: Materials Science and Engineering **281** (2017) 012062 doi:10.1088/1757-899X/281/1/012062

2. Experiments and theoretical model

The chemical composition of 316L-SS is shown in Table 1.

Table 1. Chemical composition of 316L-SS

element	wt %
C	0.01
Fe	67.10
Ni	12.43
Si	0.41
Mo	2.16
S	0.01
Cr	17.84
P	0.04

The length of 50mm, 70mm diameter cylindrical blank made of 316L stainless steel rods. The bonded surface was polished to a roughness of 0.8μm and then cleaned in acetone prior to diffusion bonding. The tensile tests at 550°C are performed on the Shimadzu material tester with a temperature control in high temperature furnace according to national standards GB/T 228-2002 (entitled as Metallic Materials-Tensile Testing at Ambient Temperature) and GB/T 4338-2002 (entitled as Metallic Materials-Tensile Testing at Elevated Temperature). [11-13]. The result of tensile test was shown in Table 2.

Table 2. Tensile strength of 316L-SS diffusion bonding specimens at 550℃

No.	1	2	3	4	5	6	7	8
Tensile strength [MPa]	87.8	102.4	122	123.9	140	149.1	159.7	160.8

Weibull distribution is being used to moderate extreme values in fracture strength and failure times. Two form of this distribution are two-parameter and three-parameter Weibull distribution. The distribution function and probability density function (PDF) in this case can be written respectively as follows [14]:

$$F(x)=1-\exp\left[-\left(\frac{x-\gamma}{\eta}\right)^{\beta}\right],\quad \gamma\geq 0,\quad \eta\geq 0,\quad \beta\geq 0 \tag{1}$$

$$f(x)=\frac{\beta}{\eta}\left(\frac{x-\gamma}{\eta}\right)^{\beta-1}\exp\left[-\left(\frac{x-\gamma}{\eta}\right)^{\beta}\right],\quad \gamma\geq 0,\quad \eta\geq 0,\quad \beta\geq 0 \tag{2}$$

Whereγ is the position parameter, when γ=0, the model degrade into two-parameter Weibull distribution; η is the scale parameter and β is the shape parameter. Three-parameter Weibull distribution applies to more data. In this study, the two-parameter Weibull distribution will be considered. The distribution function in this case can be written as:

$$F(x) = 1 - \exp\left[-\left(\frac{x}{\eta}\right)^{\beta}\right], \quad \eta \geq 0, \quad \beta \geq 0 \tag{3}$$

In the context of this study, F (x), represents the probability that the tensile strength is equal to or less than x. Using the equality F (x) + R (x) = 1, the reliability R (x), that is, the probability that the tensile strength is at least t, is defined as:

$$R(x) = \exp\left[-\left(\frac{x}{\eta}\right)^{\beta}\right], \quad \eta \geq 0, \quad \beta \geq 0 \tag{4}$$

If the natural logarithm of both sides of the Eq.(1) is taken, the following Eq.(4) can be written.

$$\ln\left\{\ln\left[\frac{1}{1-F(x)}\right]\right\} = \beta \ln(t-\gamma) - \beta \ln \eta \tag{5}$$

When the Eq.(5) is rearranged as linear equation, Y=ln{ln[1/1-F(x)]},X=ln(x), a=β and b = −βlnηis written. Therefore, a linear regression model in the form of Eq.(6) is obtained as:

$$Y = aX + b \tag{6}$$

$$\eta = \exp(-b/\beta) \tag{7}$$

A good estimator of F(x) is Bernard' s Median Rank formula:

$$F(x) = \frac{i-0.3}{n+0.4} \tag{8}$$

where i is failure serial number and n is total test number of samples.

3. Results and discussion

The result of tensile test was shown in Table 2 and the linear regression model with the regression line in Figure 1 is obtained, then we can compute β and η. The slope of the line is 4.68, which is the value of the shape parameter β, according to Eq.(7), the scale parameter η value is computed as η=148.94. β< 1:0 indicates that the material has a decreasing failure rate. Similarly β= 0 indicates constant failure rate and β> 1:0 indicates an increasing failure rate. Therefore, β=4.68 indicates that the 316L-SS tends to crush with higher tension. The scale parameter η measures the spread in the distribution of data. As a theoretical property R (x)=0.368, therefore, when x=η, R (x)=0.368,that is 36.8% of the tested specimens have a tensile strength of at least 148.94 MPa.

IOP Conf. Series: Materials Science and Engineering **281** (2017) 012062 doi:10.1088/1757-899X/281/1/012062

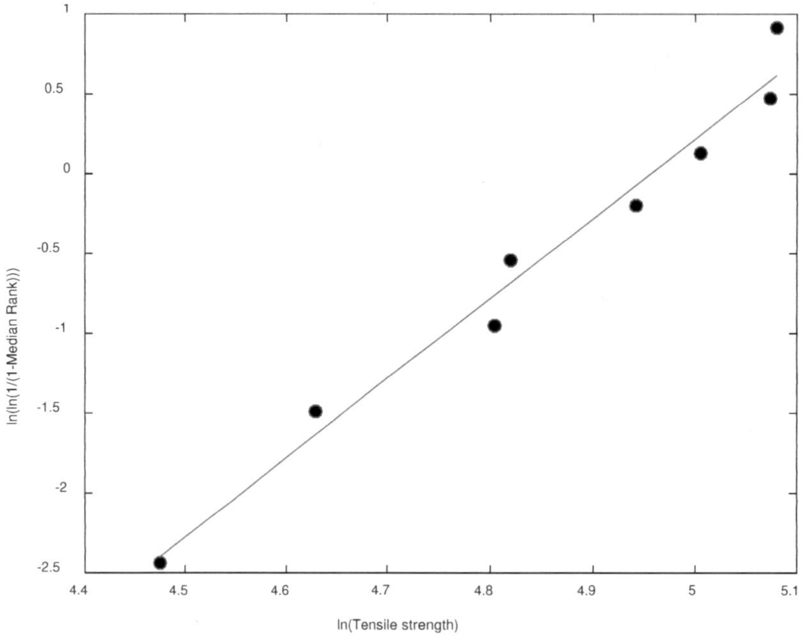

Figure 1. Regressionline

The plot of R (x) is shown in Figure 2. The reliability curve in Figure 2 shows that tensile strength value less than or equal to 50 MPa will provide high to be reliability. For a more certain assessment, consider 0.90 reliability levels. When this value are put as R (x) in Eq.(4) and the equation is solved for x, the tensile strength value 92.08 is obtained. In another word, this material will crush with 0.90 probability for a tension of 92.08 MPa.

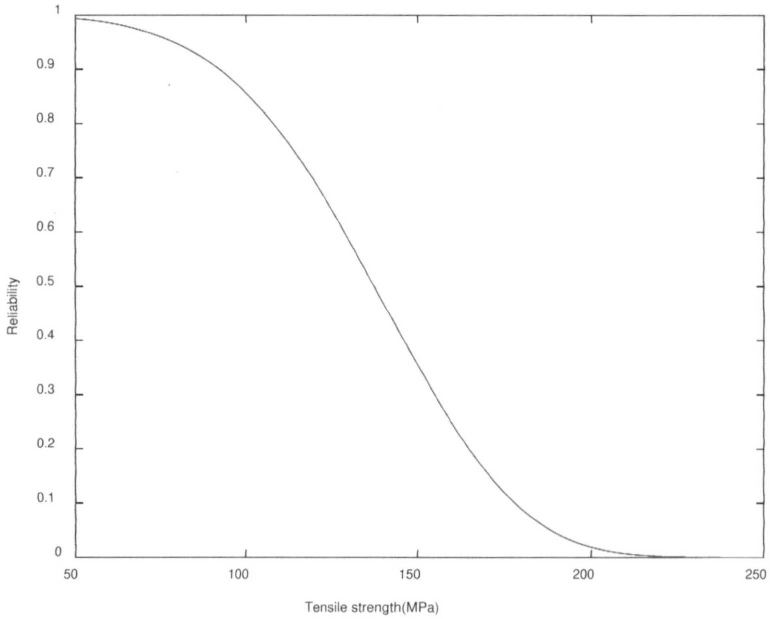

Figure 2. Weibull reliability distribution

The cumulative distribution probability for the tensile strength data of 316L stainless steel diffusion welding joint is given in the following formula:

$$\int_0^{+\infty} R_{(t)} dt = \gamma + \eta * \Gamma(\frac{1}{\beta} + 1) \tag{9}$$

Where Γ is the gamma function. And the reliability curve at high temperature is provided in Figure 3.

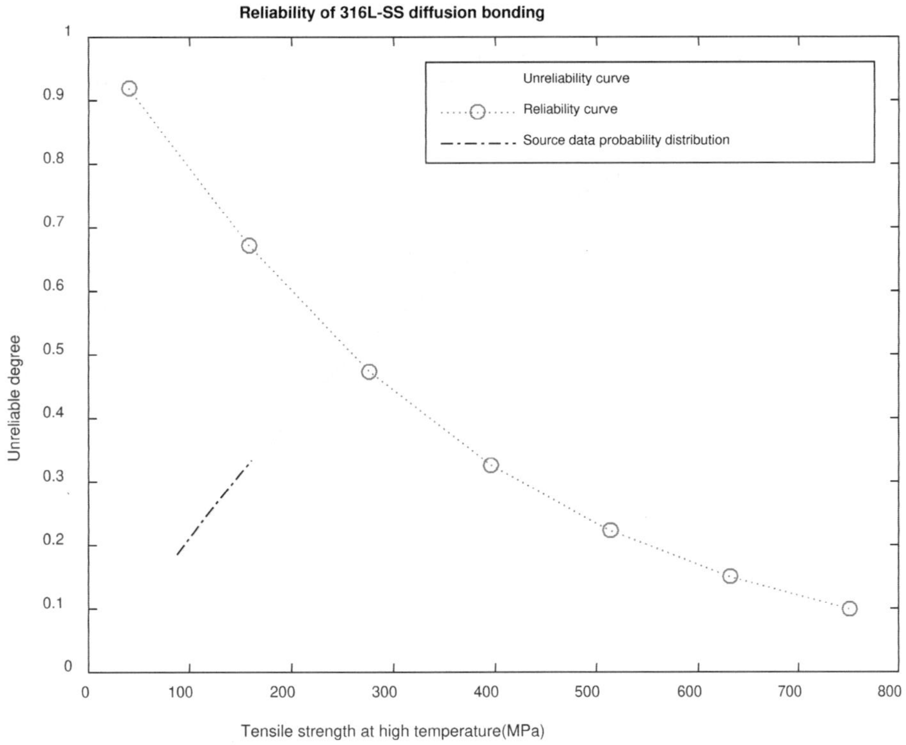

Figure 3. Reliability curveat high temperature

Average tensile strength can be predicted by the formulation (9). The predicted average tensile strength value of 316L-SS diffusion bonding joint is 156 MPa

4. Conclusions

316L-SS is generally used in diffusion bonding, the tensile strength variation in the tension has been modeled using Weibull distribution. The Weibull distribution provides a way to predict the reasonable tension when the 316L-SS was working. It also allows researchers to describe the tensile strength of a material in terms of a reliability function. The length of 50mm and diameter of 70mm 316L-SS round bar test sample follows the Weibull distribution withscale parameter of 148.94 and shape parameter of 4.68 at high temperature; among which the average bonding life at normal temperature is tensile strength of 156MPa.

References

[1] An Z L, Luan W L, Xuan F Z, et al. High Temperature Performance of 316L-SS Joint Produced by Diffusion Bonding[J]. *Key Engineering Materials*, 2005, 297-300:2795-2799.

[2] Nishi H, Araki T, Eto M. Diffusion bonding of alumina dispersion-strengthened copper to 316 stainless steel with interlayer metals[J]. *Fusion Engineering & Design*, 1998, s 39 - 40(98):505-511.

[3] Nishi H, Kikuchi K. Influence of brazing conditions on the strength of brazed joints of alumina dispersion-strengthened copper to 316 stainless steel[J]. J*ournal of Nuclear Materials,* 1998, 258-263(98):281-288.

[4] Kliauga A M, Travessa D, Ferrante M. Al2O3/Ti interlayer/AISI304 diffusion bonded joint: Microstructural Characterization of the two interfaces[J].*Materials Characterization,*2001, 46(1): 65-74.

[5] Yang De. *Experiment Design and Analysis[M].*Beijing: China Agriculture Press, 2002. 171-171.

[6] Huang ChunYue, Zhou DeJian, Wu ZhaoHua. Study on the Relationships between Solder Joint Process Parameters and Reliability of Plastic Ball Grid Array Component Based on the Orthogonal Experiment Design[J]. *Electronic Journal,* 2005, 33(5): 788-792.

[7] Hallinan A J. A review of the Weibull distribution[J]. *Journal of Quality Technology,* 1993, 25(2):85-93.

[8] Sharif M N, Islam M N. The Weibull distribution as a general model for forecasting technological change[J]. *Technological Forecasting & Social Change, 1*980, 18(3):247－256.

[9] Stout J C, Grove J, Kitchingman A, et al. Using the Weibull distribution to improve the description of riverine wood loads[J]. *Earth Surface Processes & Landforms,* 2016.

[10] Revol B P, Thomassey M, Ruch F, et al. Influence of the sample number for the prediction of the tensile strength of high tenacity viscose fibres using a two parameters Weibull distribution[J]. *Cellulose,* 2016, 23(4):2701-2713.

[11] Xuan FuZhen, Zhang Bo, Li ShuXin. Numerical simulation of the resistance method for the evaluation of micro hole defects in diffusion welded joints[J]. *Welding Journal,* 2007, 28(4): 9-12.

[12] Han WenBo, Zhang KaiFeng, Wang GuoFeng. Study on superplastic forming and diffusion bonding process of Ti-6Al-4V alloy with multi plate structure[J]. *Journal of Aeronautical Materials*, 2005, 25(6): 29-32.

[13] Zhu Hanliang, Zhao Bing, Li Zhiqiang, et al. Superplasticity and superplastic diffusion bonding of a fine-grained TiAl alloy[J]. *Materials Transactions,* 2005, 46(10): 2150-2155.

[14] Dirikolu M H, Aktaş A, Birgören B. Statistical analysis of fracture strength of composite materials using Weibull distribution[J]. *Turkish Journal of Engineering & Environmental Sciences,* 2002, 26(1):45-48.

IOP Conf. Series: Materials Science and Engineering **281** (2017) 012063 doi:10.1088/1757-899X/281/1/012063

Development of the Noise-Resistant and Sound Focusing Accessory of Ultrasonic Leak Detector for Spacecraft on Orbit

W Sun*, R X Yan, L C Sun and R P Shao

Beijing Institute of Spacecraft Environment Engineering, 104 Youyi Road, Haidian District, Beijing, China

*corresponding e-mail address: sunwei84@163.com (W Sun)

Abstract: Ultrasonic signal produced by the gas leak is so week that it is difficult to detect, and easily interfered. So developing the noise-resistant and sound focusing accessory for the ultrasonic leak detector is very important for improving ultrasonic leak detector sensitivity and noise-resistant capability. Based on the theory analysis of the leak ultrasonic signal reverberation and anacampsis, the 5A06 aluminium alloy and nylon were selected as the material of noise-resistant and sound focusing accessory by calculation and compare. Then the circular cone trumpet structure was design as the accessory main structure, and the nylon expansion port, nylon shrinking port and aluminium alloy expansion port structures were manufactured. The different structure characters were shown by the contrasting experiment. The results indicate that the nylon expansion circular cone trumpet structure has better sound focusing performance and it can improve the testing sound pressure amplitude 10 bigger than the detector without the accessory. And the aluminium alloy expansion circular cone trumpet structure has better noise-resistant ability than others. These conclusions are very important for the spacecraft leak detection and it can provide some references for the design of the noise-resistant and sound focusing structure.

Key Words: ultrasonic leak detection, noise-resistant, sound- focusing

1. Introduction

Sound environment in spacecraft cabin is utterly complex, and the inboard noise is very important space environment factor[1-3], and it seriously affects the ultrasonic leak testing sensitivity. So it is necessary to develop a noise-resistant and sound focusing accessory loading on the ultrasonic leak detector for improving the detector sensitivity and reducing noise environment influence. At the same time ,NASA have developed an ultrasonic leak detector for spacecraft on-orbit leak testing with a ABS plastics expansion port circular cone structure as the noise-resistant and Sound focusing accessory[4], and the UP9000 Ultrasonic leak detector made in USA has a similar throat structure accessory[5] and Jiang Yang developed an aluminum alloy parabola expansion port structure as the accessory[6].This article researched new noise-resistant and Sound focusing structure and testing these performances for providing some references.

IOP Conf. Series: Materials Science and Engineering **281** (2017) 012063 doi:10.1088/1757-899X/281/1/012063

2. Theory analysis of the leak ultrasonic signal reverberation and anacampsis

According on aeroacoustics theory, gas leak can lead to the ultrasonic signal production. And the leak ultrasonic propagates in air. When the sound reaches solid boundary, reverberation and anacampsis are happened as the Figure1 shown. In the Fig.1, incident wave P_i and reflected wave P_r both are the longitudinal wave, and their sound speed is c_1 ,and the air acoustic impedance is $\rho_0 c_0$, incidence angle and reflected angle are the same($\theta_i = \theta_r$). But in solid medium, there are refracted longitudinal wave P_{zz} and refracted transverse wave P_{zh} . At the same time, because the longitudinal wave speed c_{zz} is approximately the double of the c_{zh} transverse wave, longitudinal wave refracted angle θ_{zz} is more than the transverse wave refracted angle θ_{zh} . Hence, based on the wave reverberation and anacampsis rules, the Formula 1 can be gotten.

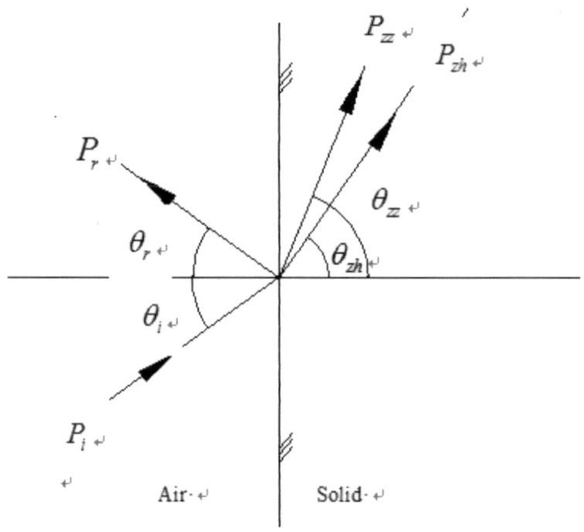

Figure 1. Leak ultrasonic signal reverberation and anacampsis between air and solid boundary

$$\theta_i = \theta_r ; \quad \frac{\sin\theta_r}{c_r} = \frac{\sin\theta_i}{c_i} = \frac{\sin\theta_{zz}}{c_{zz}} = \frac{\sin\theta_{zh}}{c_{zh}} \tag{1}$$

When the incidence angle θ_i is increasing to some value, θ_{zh} will be 90° and there is no refracted wave to achieve the all reverberation.

For reducing the noise environment influence and focusing the leak ultrasonic signal , Designing one structure can make the incident wave happen total reflection on the solid surface as the Figure 2 shown.

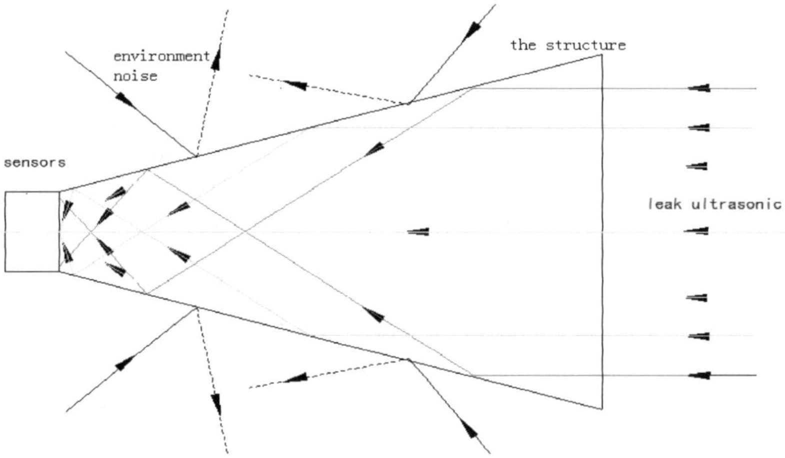

Figure 2. Total reflection on the structure solid surface

IOP Conf. Series: Materials Science and Engineering **281** (2017) 012063 doi:10.1088/1757-899X/281/1/012063

The horn structure is a variable cross-section acoustic tube, which can focus the leak ultrasonic and reflect the environment noise by the total reflection design.

Depending on Webster equation, the relationship between the sound pressure and variable cross-section can be described as the following:

$$\frac{1}{S}\frac{\partial}{\partial x}(S\frac{\partial p}{\partial x})+k^2 p=0 \tag{2}$$

Where S is as the sectional area, m^2. P denotes the sound pressure, Pa. x denotes the axial distance, m. $k=w/c$, and w is angular frequency, c is sound speed, m/s。

The pyramid type structure area is described as the following:

$$S=S_0(1+\frac{x}{x_0})^2 \tag{3}$$

Where S_0 denotes the throat sectional area, m^2. x_0 is the distance from throat to the top of pyramid. And the sound pressure can be gained by formula (2) and formula(3).

$$p=\frac{P_0}{x+x_0}e^{j(wt-kx)} \tag{4}$$

The particle velocity can be described as the following equation:

$$u=\frac{1}{\rho_0 c_0}[1+\frac{c_0/(jw)}{x+x_0}]p \tag{5}$$

And the acoustic impedance Za can be gain as the equation (6) shown:

$$Za=\frac{p}{u}=\frac{\rho_0 c_0}{1+c_0/(jwx_0)}=[\frac{\rho_0 c_0}{1+(c_0/(wx_0))^2}+\frac{c_0}{wx_0}j] \tag{6}$$

Hence, for realizing the total reflection result, the noise-resistant and Sound focusing accessory need have the high reflection and the low refraction characteristics, and its acoustic impedance need be matching with environment air acoustic impedance $\rho_0 c_0$. When $f>f_0=c_0/2\pi x_0$ the Acoustic impedance is level off to $\rho_0 c_0$.

3. The material selection

Depending on the equation (1), the Second critical incidence angle need be so little that the total reflection is happened. Hence, ultrasonic propagation velocity must be much bigger than propagation velocity in air.

In ultrasonic application, Metal often is used as the flared horn material. And then in the sound application, ABS plastic, nylon often is used as sound material. At same time ,according to process ability and mass, Aluminum alloy、ABS plastic and nylon are analyzed. The total reflection angle can be gained by comparing with ultrasonic speed in air as 343.38m/s, some data is shown as the table 1.

Table 1. Acoustic material character and total reflection critical angle

material	Density kg/m3	velocity of longitudinal wave m/s	velocity of transverse waves m/s	Attenuation coefficient Np/m	total reflection critical angle	Transmission incidence range
5A06	2640	6420	3040	230	6.48°	12.96°
ABS plastic	1050	2130	950	1.7	21.19°	42.38°
Nylon 1010	1046	2300	990	1.0	20.29°	40.58°

Through comprehensive analysis, the 5A06 (aluminum alloy) and Nylon 1010 are selected as the structure material.

4. The noise-resistant and Sound focusing structure design

The conical horn type structure is used as the main structure of the noise-resistant and sound focusing accessory by integral forming.

According to the engineering design of the conical horn maximum directivity coefficient as the following equation, the Optimum length and caliber can be gained.

$$L = \frac{D^2}{2.4\lambda} - 0.15\lambda \tag{7}$$

Where L is the optimum length, mm. D denotes the conical horn diameter, mm. and the λ denotes the ultrasonic wave length, mm.

According to the experience, when L=nλ/4, sound power is higher.at the same time, the leak ultrasonic features frequency is 40kHz, and its wave speed is 343.38m/s in air, so its wave length is 8.58mm.the formula can be described as the following equation.

$$D = \sqrt{0.6n + 0.36} \times \lambda \tag{8}$$

At present, ultrasonic sensor diameter is 70mm, n is 110.5 by according to the formula (7) for 40 kHz ultrasonic wave. Hence, the structure diameter is more than sensor's diameter, n must be larger than 111.

And then ,the acoustic impedance characteristics of the conical horn noise-resistant and sound focusing structure is related to the distance x_0 the distance from the throat to the cone and the frequency, the leak ultrasonic wave frequency should be $f >> f_0 = c_0 / 2\pi x_0$, so x_0 >>14mm.

The focusing performance is related with the cone angle. The focusing distance and the diameter D have the $F / D < D / 4\lambda$ relationship, the focusing ability is better. So the structure parameters are designed as the Figure 3 shown.

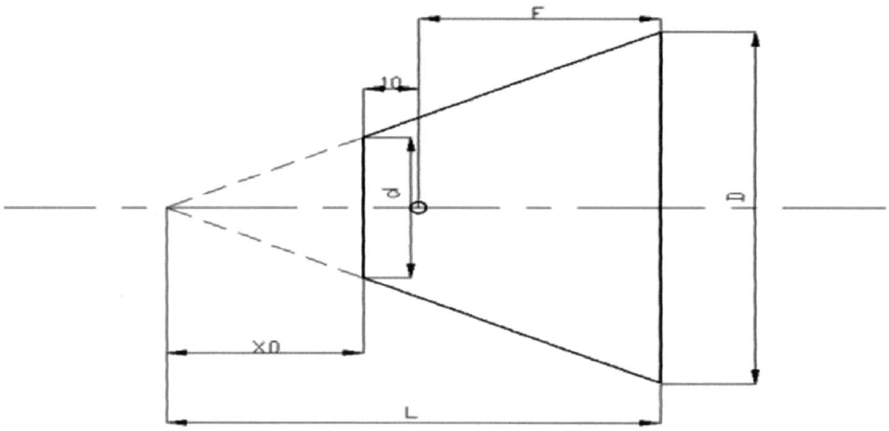

Figure 3. The structure design parameters

The structure throat diameter is7 0mm, the whole length is L-X0<200mm because the structure mass cannot be too high. Selecting the D=100mm, the λ is Calculated as 8.58mm, so L=484.9mm;And the X0=339.43mm as the following formula.

$$\frac{d}{D} = \frac{X_0}{L} \tag{9}$$

Hence, the structure length is l=L-X0=145.47mm<200mm.

And the structure can focus the sound to the sensor, F=l-10=135.47mm. and $F / D = 1.35 < D / 4\lambda = 2.91$, it can have better focusing performance. And then, for the total reflection ,when X0=339.43mm>>14mm and $f = 40kHz >> f_0 = c_0 / 2\pi x_0 = 161Hz$, the structure can have better impedance characteristic for the 40kHz leak ultrasonic wave.
Through the calculation, the design parameters can be gained as the Table 2 shown.

Table 2. The structure design parameters

Structure parameters mm	throat diameter d	Diameter D	Length l	Cone length L	Distance from throat to the cone top X_0	Focusing distance F
data	70	100	146	484.9	340.67	136

The nylon expansion port, nylon shrinking port and aluminium alloy expansion port structures are manufactured.

5. Experimental analysis

For validating the designed structure performances, some experiments are developed as the Figure 4 and Figure 5 shown. The vacuum pump draws out air from the vacuum vessel to keep vacuum environment inside of the vessel for simulating the space environment, and the external air flows into the vacuum vessel from the leak and emits ultrasonic which is diffusing through air. The leak ultrasonic is tested by the leak ultrasonic detector which consists of sensor, amplifier, AD sample and processor, the leak ultrasonic sound pressure can be analyzed. In the Figure 4, the square part entitled 'noise' is the external noise signal generator which can produce the different frequency ultrasonic signals to simulate background environment noise or other disturbing leak detection noise for validating the noise-resistant performance.

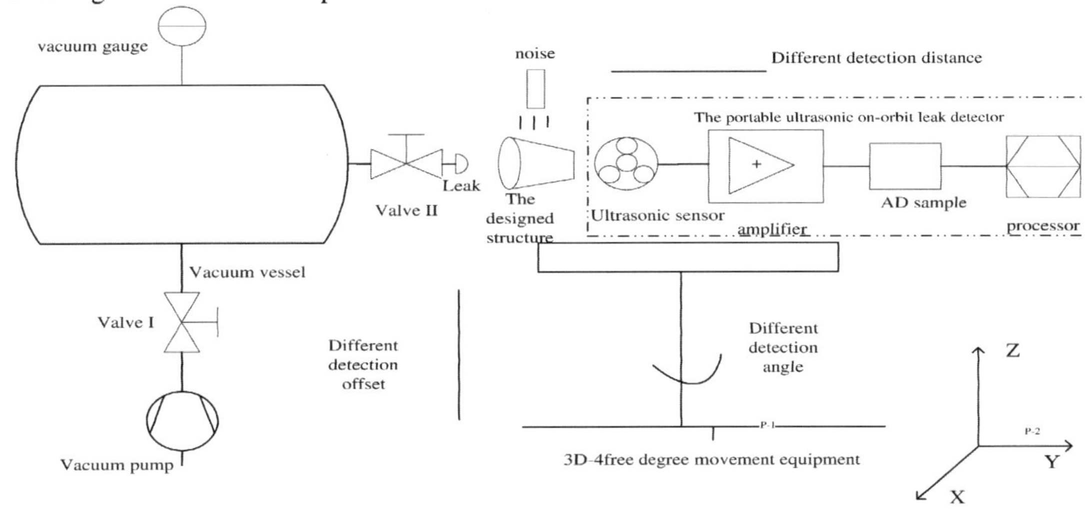

Figure 4. Experiment scheme

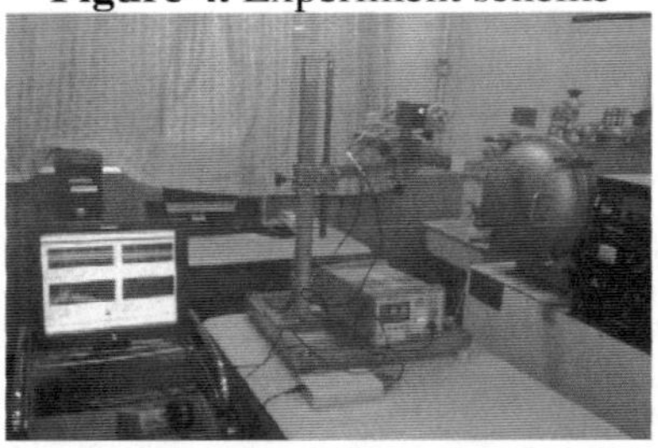

Figure 5. Experiment system photograph

With vacuum tank simulating the space environment, the air would leak from the atmosphere into the inside of the tank in which gas pressure can be 1×10^{-1}Pa by the vacuum pump. And the leaks' geometry is an cylinder structure with Ø0.3mm diameter and 3mm length.

The nylon expansion port, nylon shrinking port and aluminum alloy expansion port structures are tested by detecting different leaks. And the noise-resistant and sound focusing performances are compared between the leak detection with the noise-resistant and sound focusing structure and without.

The distance from the structure to the leak is set as 100mm, and the Ø0.3mm leak is used as the leak ultrasonic source, and air is used as the leak gas medium. Changing the different noise-resistant and sound focusing structures, the ultrasonic signal is detected thirty times. And the signal can be gained and stored, by the FFT, Chebyshev band-pass filter, the 40 kHz signal is used as leak ultrasonic main characteristic. Through analyzing sound pressure in the frequency, these differences can be gained. And the maximum value, average value and minimum value are shown as the Figure.6.

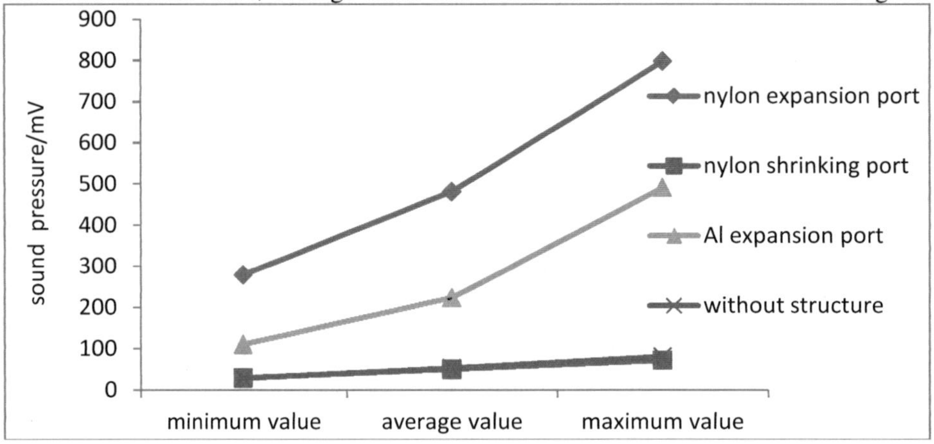

Figure 6. Focusing performance testing results

In the noise-resistant experiment, 40 kHz ultrasonic source is used as the noise interfere source is perpendicular to leak ultrasonic detecting direction. The leak ultrasonic with noise can be tested through analyzing sound pressure change in the frequency between leak with noise and without, the designed structures' noise-resistant character will be shown. The structure can reflect external noise. And the results for Ø0.3mm leaks detection are shown as the Figure 7.

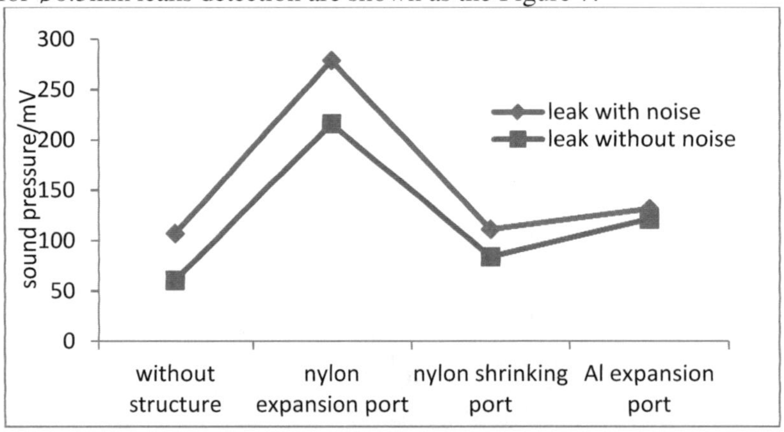

Figure 7. The noise-resistant experiment results

Through the noise-resistant and the sound focusing experiments, some results can be shown as the following:

1. Expansion port type structure has better focusing performance, and the shrinking port type structure has better noise-resistant performance.

2. Nylon structure has better sound focusing performance, but the noise-resistant is not better than others. And then, the aluminum alloy structure has better noise-resistant and focusing performances.

6. Conclusions

Because the ultrasonic leak detector is used on orbit, whose mass is more little the better. According to the study results, the nylon can be selected as the noise-resistant and sound focusing accessory, and the expansion port type and shrinking port type structures are manufactured. And then the expansion port type structure can be used when the environment noise is little. But the shrinking port type structure can be used when the environment noise is large. And the ultrasonic leak detector with the structure has higher sensitivity and lower noise influence than the without. This paper study conclusions are very important for the spacecraft leak detection and it can improve the ultrasonic leak detector sensitivity and reduce space noise environment influence for leak detection and provide some references for the design of the noise-resistant and sound focusing structure.

Acknowledgments

The authors thank the Beijing Institute of Spacecraft Environment Engineering (BISEE) for supporting this work as part of W.Y PhD work. We are grateful to Li Weidan for data analyses. Thanks also go to Qi Lei and Ren Guohua for their help in the field.

Reference

[1] You J, Hao Ping and Gu Wei 2016 Noise Prediction and Acoustic Design Optimization of Spacecraft's Equipment Cabins,*Noise and Vibration Control[in Chinese]* vol 36 pp 80-81
[2] Zhang F, Yang J,Feng Y Q and Yao F 2014 Current state of art and suggestions about noise evaluation and control for the space station *Spacecraft Environment Engineering[in Chinese]* vol 32 pp 223-224
[3] Christopher S Allen 2015 International Space Station Acoustics – A Status Report *45th International Conference on Environmental Systems* Bellevue Washington pp 1-21;
[4] NASA 1998 Low Level Leaks *Spinoff* p 96;
[5] http://www.davidson.com.au/products/ultrasonics/ue;
[6] Jiang Y 2006 Theoretics and system of using ultrasonics to detect the leakage amount [D] *ShangHai Jiao Tong University[in Chinese]* pp 44-46.

IOP Conf. Series: Materials Science and Engineering **281** (2017) 012064 doi:10.1088/1757-899X/281/1/012064

Capturing Revolute Motion and Revolute Joint Parameters with Optical Tracking

C Antonya

Department of Automotive and Transport Engineering, Transilvania University of Brasov, 29 B-dul Eroilor, 500036 Brasov, Romania

Email:antonya@unitbv.ro

Abstract. Optical tracking of users and various technical systems are becoming more and more popular. It consists of analysing sequence of recorded images using video capturing devices and image processing algorithms. The returned data contains mainly point-clouds, coordinates of markers or coordinates of point of interest. These data can be used for retrieving information related to the geometry of the objects, but also to extract parameters for the analytical model of the system useful in a variety of computer aided engineering simulations. The parameter identification of joints deals with extraction of physical parameters (mainly geometric parameters) for the purpose of constructing accurate kinematic and dynamic models. The input data are the time-series of the marker's position. The least square method was used for fitting the data into different geometrical shapes (ellipse, circle, plane) and for obtaining the position and orientation of revolute joins.

1. Introduction

The model identification process of a mechanical or biomechanical system can be a tedious process when different kinds of parameters are required to be computed. In order to develop a computer model, the required parameters are related to the physical properties of each of the component (geometry, size, mass, inertia, stiffness) and the connection with neighbouring elements which will impose relative displacement and motion constraint.

Tracking and motion capturing is the process of recording the movement (position) of objects or people [1,2,3]. These systems can be optical, magnetic, acoustic, gyroscopic or mechanical. Commercial systems with optical tracking and reflective markers works mainly in a constrained space and the tracked system or person has to wear markers. These 3D motion capture systems are fast and accurate, but are costly and not available in outdoor environment. Inertial sensors (like gyroscopes and accelerometers) are also common for motion capturing, these electromagnetic tracking systems are fast, but they behave poorly in the presence of metallic objects.

The major two components of a mechanical or biomechanical system are the links (bodies with certain geometry, of rigid or elastic nature) and joints. The joints are imposing restrictions on the relative motion of bodies of the system. The revolute joint has one degree of freedom, if an anchor point is defined on the connected link, the bodies can move so that these points are always in the same place and the relative rotation of the bodies is not restricted. Revolute motion and revolute joint are present in a large number of technical systems and biomechanical models (e.g. robot arms, human knee joint, journal–bearing joint, car suspension mechanism). The parameter identification of joints

IOP Conf. Series: Materials Science and Engineering **281** (2017) 012064 doi:10.1088/1757-899X/281/1/012064

deals with extraction of physical parameters (mainly geometric parameters) for the purpose of construction of accurate kinematic and dynamic model. Having a precise 3D kinematic (shape and joint) model of a mechanical system is very useful in a variety of computer aided engineering simulations. The measured data contains noise due to the tracking imperfections. To reason about the uncertainty of the belief multiple techniques exist, like Bayes rule [4], divide and conquer strategy, maximum-likelihood estimate or least square regression. In this paper the linear least square algorithm was chosen, because it is fast and accurate for this kind of applications.

2. Optical tracking

Optical motion tracking are used in a wide range of applications [1,2,5], which require accurate and robust measurement of the 3D position and orientation of one or more objects in a user-defined coordinate system.

Marker-based tracking is a popular method in virtual reality applications, performance evaluation for athletes, rehabilitation, training, films and video games to create realistic movements and special effects. However, it requires physical markers to be placed on the subjects. Markerless and vision-based motion capture techniques have gained an increasing interest recently and become available thanks to increasing computational power of mobile devices. Markerless tracking has some limitations in the way that data is recorded and collected, due to the settings, environment and number of cameras. Two-dimensional (2D) video cameras with image processing and analyzing software are more affordable, but they are less accurate and are light sensitive. In [1] the motion capture data combined with contact forces measurements allowed the estimation of inertial parameters of the human body. Combined with various anthropometric biometrics, the captured full-body motion is used for person identification [6], joint angle measurement [7,8], capturing and identification of a wide range of human movements [3,9] and shape feature extraction for motion pattern [10].

2.1. Optical tracking with markers and image processing

Optical tracking consist of obtaining geometric information by analysing sequence of recorded images using video capturing devices and image processing algorithms. The tracking system is providing the temporal position of the marker(s), but even if the motion capture system is calibrated and the tracking data is recorded with high frequency, the uncertainty remains, especially due to the image processing methods, deformations or other tracking deficiencies [4].

2.2. Optical tracking with reflective markers

The most efficient and fast motion capturing systems are using retro-reflective markers (usually small balls). The cameras are equipped with an infrared pass filter in front of the lens and infrared LEDs around the lens illuminate the space. The markers reflect the incoming light back to the camera and the image is internally processed by the tracking system. This system calculates the 2D markers position in image coordinates with high precision. Using multiple cameras, the 3D position of each marker can be derived. The degree of accuracy of this system can be under 0.1 mm [2]. The optical tracking has been used in measurements of 3D displacements in various fields, from body parts (head, arms, fingers) [5] to buildings and other types of civil structures [2].

3. Least square method for retrieving the revolute motion and revolute joint parameters

The input data for the least square method is the time-series of the marker's position. The markers are attached to a link with revolute join, so their path is circular and the position and orientation of the revolute joint can be found from this circle's geometrical parameters. Depending of the optical tracking method, the markers' coordinates are expressed in the camera plane (in case the position is obtained from the image of one camera) or in the user-defined 3D reference frame, if more cameras and reflective markers are used.

IOP Conf. Series: Materials Science and Engineering **281** (2017) 012064 doi:10.1088/1757-899X/281/1/012064

In case the coordinates of the markers (x_M, y_M) are obtained in the camera's reference frame, the path of circular motion of the marker is seen as an ellipse in the image plane. The revolute joint parameters can be obtained by fitting the coordinates of the markers on an ellipse.

If the markers' coordinates are in the user-defined reference frame, first these positions have to be fitted on a plane, then (with the new coordinates in the plane) on a circle.

The equations for fitting the data to an ellipse (1), plane (2) and circle (3) can be written as:

$$F(x,y) = x^2 + \omega_1 y^2 + \omega_2 xy + \omega_3 x + \omega_4 y + \omega_5 = 0, \tag{1}$$

$$F(x,y,z) = x + \omega_1 y + \omega_2 z + \omega_3 = 0, \tag{2}$$

$$F(x,y) = x^2 + \omega_1 y^2 + \omega_2 x + \omega_3 y + \omega_4 = 0. \tag{3}$$

The unknown are (4):

$$\boldsymbol{\omega} = [\omega_1 \quad \omega_2 \quad \dots \quad \dots \quad \omega_i]^T \tag{4}$$

with i=5 for the ellipse, i=3 for the plane and i=4 for the circle.

The recursive least square requires a training set of N samples [11,12]. The following two matrices can be defined for the circle and ellipse (5), and for the plane (6):

$$\mathbf{X}_N = [x_i^2], \tag{5}$$

$$\mathbf{X}_N = [x_i], \quad i = 1....N \tag{6}$$

respectively for the ellipse (7), circle (8) and plane (9):

$$\mathbf{Z}_N = \lfloor y_i^2 \quad x_i y_i \quad x_i \quad y_i \quad 1 \rfloor \quad i = 1....N \tag{7}$$

$$\mathbf{Z}_N = [y_i \quad z_i \quad 1] \quad i = 1....N \tag{8}$$

$$\mathbf{Z}_N = \lfloor y_i^2 \quad x_i \quad y_i \quad 1 \rfloor \quad i = 1....N \tag{9}$$

The unknown, which will describe the required parameters will be:

$$\boldsymbol{\omega} = \left(\mathbf{Z}_N^T \mathbf{Z}_N\right)^{-1} \mathbf{Z}_N^T \mathbf{X}_N \tag{10}$$

In case a new measurement is acquired (x_{N+1}, y_{N+1}) or (x_{N+1}, y_{N+1}, z_{N+1}), the unknowns will be updated with:

$$\boldsymbol{\omega}_{N+1} = \left(\begin{bmatrix} \mathbf{Z}_N \\ \mathbf{z}_{N+1} \end{bmatrix}^T \begin{bmatrix} \mathbf{Z}_N \\ \mathbf{z}_{N+1} \end{bmatrix} \right)^{-1} \begin{bmatrix} \mathbf{Z}_N \\ \mathbf{z}_{N+1} \end{bmatrix}^T \begin{bmatrix} \mathbf{X}_N \\ x_{N+1} \end{bmatrix}. \tag{11}$$

Using the matrix inversion rules for avoiding the computation of the inverse of big matrices, the recursive algorithm will correct the unknowns:

$$\boldsymbol{\omega}_{N+1} = \boldsymbol{\omega}_N - [\mathbf{K}_{N+1}][\mathbf{Z}_N]\left\{error : x_{N+1}^2 (or \;\; x_{N+1}) + z_{N+1}^T \boldsymbol{\omega}_N\right\}, \text{ where:} \tag{12}$$

$$\mathbf{K}_{N+1} = \mathbf{K}_N - \frac{\mathbf{K}_N \mathbf{z}_{N+1} \mathbf{z}_{N+1}^T \mathbf{K}_N}{1 + \mathbf{z}_{N+1}^T \mathbf{K}_N \mathbf{z}_{N+1}}, \quad with \;\; \mathbf{K}_N = \mathbf{Z}_N^T \mathbf{Z}_N. \tag{13}$$

The obtained vector **ω** is describing in parametric form the position of the ellipse and its major axis (which will be the diameter of the circular path of the marker), and the centre point of circle and its radius.

4. Obtaining the revolute motion of the link of a four bar mechanism

To one link of a four-bar mechanism a coloured marker is attached (figure 1). The path of this marker is circular, and the proposed algorithm's aim is to obtain the centre point and radius of this circle. In the camera plane, this circle will have elliptical shape, so its centre point and its major axis will have to be found.

Figure 1. The four-bar mechanism with the marker.

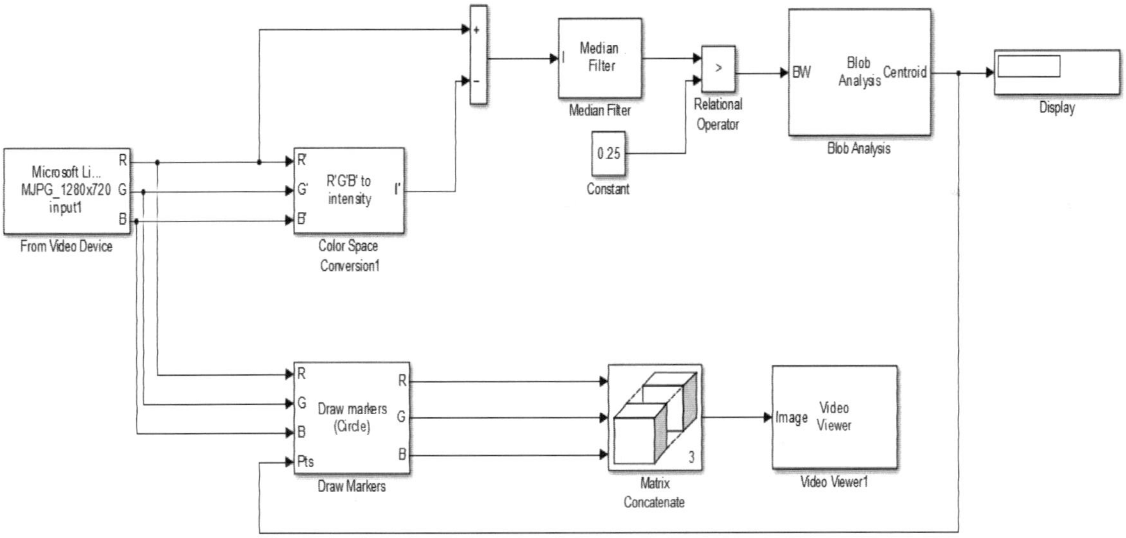

Figure 2. Image processing for the centre point of the marker.

A high-definition camera is recording the movement, and the image processing algorithm is retrieving the x and y coordinates of the centre of the round shaped coloured marker, defined in the image plane (figure 2). The algorithm is transforming the captured image in grey-scale, followed by median filtering and blob analysis of the round shape. In figure 3 are presented two ellipses identified in different positions of the camera, together with their centre and major axis, corresponding to a revolute motion of the crank of the four-bar mechanism.

5. Conclusions

Information regarding the parameters of the analytical model of a mechanical device can be obtained by measuring coordinates of markers in a sequence of recorded images using video capturing devices and image processing algorithms. Using the least square method for fitting the data into different geometrical shapes (ellipse, circle, plane), the parameters of the revolute joint can be obtained. A precise 3D kinematic (shape and joint) model of a mechanical system is very useful in a variety of computer aided engineering simulations.

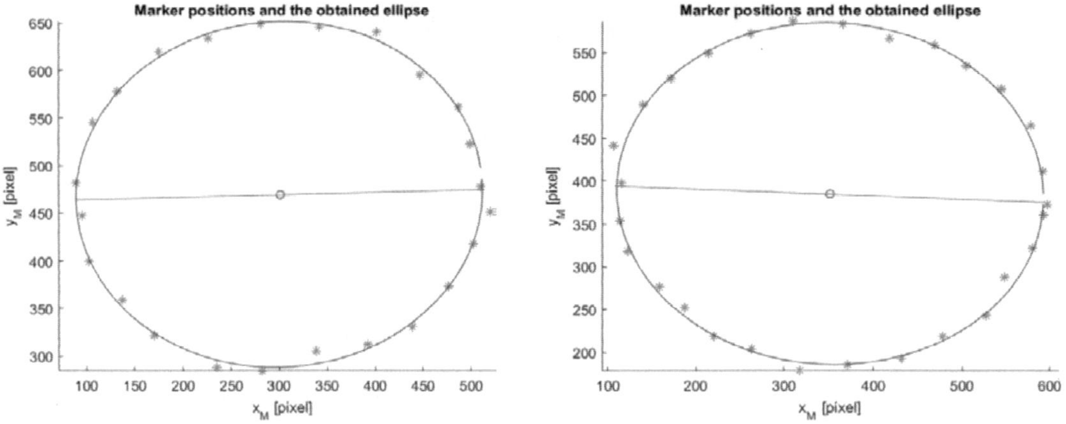

Figure 3. The obtained ellipses and their major axis.

References

[1] Venture G, Ayusawa K and Nakamura Y 2008 Motion capture based identification of the human body inertial parameters *Engineering in Medicine and Biology Society*, EMBS 2008. 30th Annual International Conference of the IEEE pp 4575-78

[2] Park S W, Park H S, Kim J H and Adeli H 2015 3D displacement measurement model for health monitoring of structures using a motion capture system. *Measurement*, **59**, 352-362.

[3] Bernard J, Wilhelm N, Krüger B, May T, Schreck T and Kohlhammer J. 2013 Motionexplorer: Exploratory search in human motion capture data based on hierarchical aggregation. *IEEE Transactions on visualization and computer graphics*, **19**(12), 2257-2266.

[4] Antonya Cs, Butnariu S and Beles H 2015 Parameter Estimation from Motion Tracking Data, *Digital Human Modeling. Applications in Health, Safety, Ergonomics and Risk Management: Ergonomics and Health*, Springer LNCS 9185 pp 113-121

[5] Lim C H, Vats E and Chan C S 2015 Fuzzy human motion analysis: A review. *Pattern Recognition* 48(5) 1773-96

[6] Munsell B C, Temlyakov A, Qu C and Wang S 2012 Person identification using full-body motion and anthropometric biometrics from kinect videos. *European Conference on Computer Vision* pp. 91-100 (Springer, Berlin, Heidelberg)

[7] Schmitz A, Ye M, Shapiro R, Yang R and Noehren B 2014 Accuracy and repeatability of jointangles measured using a single camera markerless motion capture system. *Journal of biomechanics*, **47**(2), 587-591

[8] Vincze M, Prenninger J P and Gander H 1994 A laser tracking system to measure position and orientation of robot end effectors under motion. *The International Journal of Robotics Research* **13**(4) 305-314

IOP Conf. Series: Materials Science and Engineering **281** (2017) 012064 doi:10.1088/1757-899X/281/1/012064

[9] Wei X, Zhang P and Chai J 2012 Accurate realtime full-body motion capture using a single depth camera. *ACM Transactions on Graphics* **31**(6) 188 pp 1-12
[10] Zhou H, Hu H, Liu H and Tang J 2012 Classification of upper limb motion trajectories using shape features. *IEEE Transactions on Systems, Man, and Cybernetics*, Part C (Applications and Reviews) **42**(6) 970-982
[11] Umbach D and Jones K N 2003 A few methods for fitting circles to data. *IEEE Transactions on instrumentation and measurement* **52**(6) 1881-85.
[12] Gander W, Golub G H and Strebel R. 1996 Least-squares fitting of circles and ellipses. *Bulletin of the Belgian Mathematical Society Simon Stevin* **3**(5) 63-84

Institute of Physics
Dirac House, Temple Back
Bristol BS1 6BE UK

ISSN: 1757-8981
ISBN 978-1-5108-5496-3